A MONOGRAPH

OF THE

BRITISH FOSSIL CORALS.

BY

H. MILNE EDWARDS,

DEAN OF THE FACULTY OF SCIENCES OF PARIS; PROFESSOR AT THE MUSEUM OF NATURAL HISTORY;
MEMBER OF THE INSTITUTE OF FRANCE;
FOREIGN MEMBER OF THE ROYAL SOCIETY OF LONDON, OF THE ACADEMIES OF BERLIN, STOCKHOLM, ST. PETERSBURG
COPENHAGEN, VIENNA, KONIGSBERG, MOSCOW, BRUXELLES, HAARLEM, BOSTON, PHILADELPHIA, ETC.,

AND

JULES HAIME.

LONDON:
PRINTED FOR THE PALÆONTOGRAPHICAL SOCIETY.
1850—1854.

J. E. ADLARD, PRINTER, BARTHOLOMEW CLOSE

A MONOGRAPH

OF THE

BRITISH FOSSIL CORALS.

BY

H. MILNE EDWARDS,

DEAN OF THE FACULTY OF SCIENCES OF PARIS; PROFESSOR AT THE MUSEUM OF NATURAL HISTORY;
MEMBER OF THE INSTITUT OF FRANCE;
FOREIGN MEMBER OF THE ROYAL SOCIETY OF LONDON, OF THE ACADEMIES OF BERLIN, STOCKHOLM, ST. PETERSBURG,
VIENNA, KÖNIGSBERG, MOSCOW, BOSTON, PHILADELPHIA, ETC.

AND

JULES HAIME.

FIRST PART.

INTRODUCTION; CORALS FROM THE TERTIARY AND CRETACEOUS FORMATIONS.

LONDON:

PRINTED FOR THE PALÆONTOGRAPHICAL SOCIETY.

1850.

PRINTED BY C. AND J. ADLARD, BARTHOLOMEW CLOSE.

A MONOGRAPH

OF

THE BRITISH FOSSIL CORALS.

INTRODUCTION.

§ I.[1]

NATURALISTS often designate under the general name of *Coral*, not only the stony substance of a vivid red which is found on the coast of Barbary, and has been long used for ornamental purposes, but also a vast number of other marine productions, which have a calcareous structure, and are considered as appertaining to Zoophytes, more or less analogous to the Polypi that form the *Isis nobilis* of Linnæus, or real Mediterranean Coral. The remains of the minute plant-like animals which abound in most tropical seas, and constitute in some parts of the globe extensive reefs, or even large clusters of islands, have thus been very properly called *Corals.* But the same appellation has been erroneously given to the lapidified teguments of many beings which differ most essentially from all Zoophytes, and belong some to the great Mollusca tribe, some to the family of Sponges, and others to the Vegetable kingdom. In all Natural classifications it is necessary to separate that which is fundamentally different, and to unite that which is in reality similar. Zoologists must, therefore, be more reserved in the use of this expression, and cannot, without impropriety, continue to comprehend under the same name all the natural productions which are com-

[1] In writing this Monograph in English, a language with which I am not so familiar as I could wish, I much fear that the incorrectness of the phraseology will often strike the reader. I preferred, however, not having recourse to a translator, for the meaning of an author is often misrepresented by those who lend him their pen, and I thought that in a work of this kind accuracy of description would be preferable to elegance of style. Before commencing the task I have undertaken, I must also beg leave to express publicly my grateful feelings for the kind and liberal manner in which Sir H. De la Beche, Mr. Stokes, Mr. J. S. Bowerbank, Professor John Phillips, Mr. Frederick Edwards, Mr. Searles Wood, Mr. Dixon, Mr. Pratt, Mr. Sharpe, Dr. Battersby, Mr. F. W. Fletcher, Mr. J. Gray, and the Council of the Geological Society of London, have communicated to me the palæontological treasures belonging to their respective collections.—H. MILNE EDWARDS.

monly thus blended together. For us the word *Coral*, or *Corallum*, must be synonymous with *Polypidom*,[1] and signify the hard or ossified parts of the body of a Polyp.

In treating of the "Fossil Corals of Great Britain," we must, therefore, exclude from our investigation the various organic remains which bear a certain resemblance to Polypidoms, but which do not in reality belong to beings of the same structure, and we must circumscribe our researches within the boundaries of the group of Zoophytes, which, in a Natural arrangement of the Animal Kingdom, is represented by the CLASS OF POLYPI.[2]

These Zoophytes are closely allied to Medusæ, and in the actual state of science there is some uncertainty respecting the natural limits which separate these two groups; but the mode of organization common to both is so characteristic, that the most superficial anatomical investigation will always enable the zoologist to distinguish a Polyp or an Acaleph from the Bryozoa and the Spongidæ, which, till lately, have been erroneously considered as belonging to the class of Corals. Polypi have a radiate structure; a protractile mouth, surrounded by non-ciliate tentacula; a large and well-organized digestive cavity; but have no anus. In Spongidæ no appearance of tentacula or of a stomach is ever met with; and in Bryozoa an intestinal canal, much resembling that of ordinary Mollusca, is always provided with two distinct openings, a mouth and an anus, the first of which is encircled by ciliated tentacula. The structure of the digestive organs is, therefore, characteristic in all these animals, and in most instances the radiate form of the tegumentary system will alone suffice to render the diagnosis of Polypi an easy task. But when the Polypidom is reduced to its most simple condition, it sometimes bears great resemblance to the calcareous or horny covering of certain Bryozoa, or to the reticulate skeleton of some of the Spongidæ; and the Polypidom being the only part of these animals which is found in the fossil state, it is sometimes hard for the palæontologist to decide whether the organic remains that assume this form are in reality Corals, or whether they do not belong to one of the other above-mentioned Zoological divisions.

Polypidoms may present two very distinct forms. Some, belonging to aggregate Polypi, are developed on the basal surface of these Zoophytes, and constitute a sort of stem in the

---

[1] In translating the French expression *Polypier* by the word *Polypidom*, which has of late been adopted by some of the most eminent English zoophytologists, we deem it necessary to guard the reader against the erroneous ideas which the etymology of that name might lead to. Till of late the nature of Corals was in general misunderstood; they were supposed to be produced by a plastic exudation moulded round the body of the Polyp, and serving as a dwelling for these singular beings, but not forming a part of their organism. Such is far from being the case; the corallum is a part of the animal, in the same way as the coating of the armadillo or the shell of the lobster belong to the structure of these beings. The words "Polypidom," *Polypier*, &c., might therefore be objected to, if their meaning was not generally known, and had not become independent of their etymology.

[2] The class of Polypi, reduced to its natural limits, corresponds to the *Anthozoa* of M. Ehrenberg, and to the sub-class of *Radiated Zoophytes* of Mr. Johnston. In the excellent work recently published by Mr. Dana, the same group is designated by the name of *Zoophytes*, which is usually employed in a much wider acceptation, and had long ago been given by Cuvier to the great division of radiate animals, comprising Echinoderma and Acalephæ, as well as Polypi, etc.

centre of the ramified mass produced by the multiplication of these plant-like animals. The dendroid red Coral of the Mediterranean Sea and the horny skeleton of Gorgonia are thus inclosed in the axis of cylindrical branches, formed by the thick coriaceous tegumentary tissue belonging to the whole community of aggregate Polypi, and studded, as it were, by the radiate protractile heads of the many individual Zoophytes thus united. Other Corals, appertaining either to simple or to compound Polypi, are, on the contrary, produced by the ossification of this tegumentary tissue itself, and instead of forming a sort of stem, constitute a sheath, or an assemblage of calcareous tubes, each of which belong to an individual Zoophyte, correspond to the lower part of its digestive cavity, and serve as a kind of cell or lodge into which the anterior portion of the animal's body recedes when in a contracted state.

The basal or stalk-like Corals are in general well characterised by their dendroid form, compact tissue, and concentric layers. At first sight they may bear a slight resemblance to certain Bryozoa that have attained a very advanced age;[1] but even then the remains of some non-obliterated cells will always enable an attentive observer to recognise the latter, and the absence of all trace of any such cavities can easily be ascertained, by grinding down or fracturing the stem of the above-mentioned Zoophytes. In some few instances these basal Polypidoms are more like the reticulated skeleton of certain foliaceous Spongidæ; but the concentric lamellæ of their stem contrasting with the fibrous structure of the tissue of the Sponge, will still render them recognisable.

Dermal Corals are in general characterised by features of a more striking aspect, and it is only when these Polypidoms are reduced to their most simple and degraded form, that they can be mistaken for the tegumentary skeleton of some of the lowest Bryozoa, or the reticulate, stony tissue of some highly-organized Spongidæ. In all well-developed Corals of this kind, the central cavity or visceral chamber is more or less completely divided by a certain number of vertical plates, which project from its walls towards its axis, and produce that radiate structure which is so remarkable in the Astrean tribe. In most Bryozoa the mouth, or cephalic aperture of the tegumentary cell, is provided with a horny operculum,[2] but no such organ ever exists in a true Coral; and, on the other hand, the radiate septa which we have just alluded to as being conspicuous in most Polypidoms, never exist in the cells of Bryozoa. The absence of an operculum, or of vertical septa, will not, however, enable the observer to decide whether the coral-like organic remains submitted to his investigation belong to the one or to the other of the two great zoological divisions, for it is a well-known fact that, in many of the inferior forms among recent Bryozoa, the tegumentary skeleton is reduced to a simple non-operculated tubular sheath, and that in certain Polypi (the Tubipora for example), no longitudinal septa are to be found; and the Polypidom is equally reduced to a calcareous tube, tapering and closed at its base, open and more or less enlarged at its upper end.

[1] The *Millepora truncata* of Ellis and Solander, for example.
[2] See "Recherches sur les Eschares," Annales des Sciences Naturelles, 2ᵐᵉ série, t. vi, pl. i.

In cases of this kind the distinction between the Polypi and the Bryozoa is always rendered easy by the most superficial examination of the soft parts of the animal; but it is sometimes a matter of great difficulty for the palæontologist, who is necessarily deprived of all such resources, and can only be guided by the peculiarities observable in the ossified tissues.

In general, the distinction between Corals and Spongidæ is also very easy, for the lamellar structure, so prevalent among the former, is never met with in the latter; but in some Polypidoms (certain Milleporidæ for example), the vertical plates disappear, and the mural tissue becomes extremely porous, irregular, and abundant, so as to resemble much the reticulated mass formed by the stony skeleton of some Spongidæ, where the oscula and aquiferous canals are on the contrary more regular than usual. In cases of this kind it may be necessary to seek for distinctive characters in the internal structure of the Zoophyte; and, independently of the benefit to be obtained by the microscopical investigation of the tissue itself, it will sometimes be found useful to examine the form of the tubular cavities which pervade the mass, and correspond either to the visceral chambers of the Polypi, or to the great aquiferous ducts of the Spongidæ; for in the first instance they are always simple, whereas in the latter they are more or less ramified.

§ II.

The external forms of Corals vary considerably, but are in general more dependent on the mode of aggregation of the different individuals produced by a common parent than on the mode of organization peculiar to the animals to which these tegumentary skeletons belong. Characters derived from these forms can therefore be but of little avail for the natural arrangement of Polypi; and the classification of these Zoophytes, like that of the higher animals, must be founded on the principal modifications observable in their structure. It would lead us too far from the special object of this Monograph, if we were to enter on the investigation of the anatomical facts which alone can furnish satisfactory elements for such a classification; but in order to facilitate the study of the Corals about to be described, it may be useful for us to revert to a few of the leading points in the structure of Polypi, and to define some of the expressions which we shall often have to employ.[1]

The SCLERENCHYMA, or hardened tissue of Polypi, by which Corals are formed, is always a portion of the tegumentary system of these Zoophytes, but, as we have already stated, it may be produced in two very different ways. In some cases it is the result of a sort of ossification of the chorion or principal tunic of the Polypi; in others it grows on

---

[1] For more ample details on this subject we must refer to our " Memoir on the Structure and Development of Corals," published in the Annales des Sciences Naturelles, 3ᵐᵉ série, t. ix.

certain parts of the surface of that membrane in a manner somewhat similar to that in which calciferous epidermis covers the skin of Crustacea and Mollusca. This *epidermic scleren-chyma* constitutes the tissue which Mr. Dana has designated by the name of "foot-secretion," and is the only anatomical element employed by nature in the formation of the common red Coral, and the horny tubes of Sertulariæ; but in most Polypidoms it is of secondary importance, and the structure is essentially made up with the *dermic sclerenchyma*, or ossified chorion. The calcification of this tegumentary tissue always commences in the centre of the inferior part of the Polyp, and, spreading gradually, rises as the animal grows, so as to inclose the lower part of the gastric cavity, and to constitute a sort of cup or cell, which is sometimes broad and shallow, sometimes long and tubular.

In general the fundamental part of these Corals corresponds to the parietes of the great gastric or visceral cavity of the Polyp, and forms what may be called the *walls* of the Polypidom. The basal disc, the spreading cup, or the columnar sheath so produced, very seldom remains in this simple condition, and in general soon gives rise to a certain number of laminate processes, which converge towards the axis of the body, and divide the central cavity into so many radiating *loculi*. These vertical laminæ, to which we shall exclusively apply the name of *septa*, cover the upper surface of the wall when this spreads out in the form of a disc (as in Fungiæ); but in general they are more or less completely inclosed in the cup-shaped or tubular cell produced by the growth of this wall around the visceral cavity, which pervades the body of the Polyp from top to bottom. In some Corals the septa remain free all along their inner edge; in other species they adhere to a sort of central style or plate, which rises from the bottom of the same cavity, and which M. Ehrenberg has proposed calling the *columella*. The loculi, or interseptal spaces, are then completely separated; and in many Polypidoms, where there is no true columella, the same result is produced by a greater development of the septa, which become united by means of irregular trabiculæ branching off from their inner edge, and forming a *spurious columella*, the structure of which is usually loose and spongy.

Other lamellar or styliform processes, quite distinct from the septa and the columella, are in some Corals interposed between these organs, and form around the central style a sort of circular palisade, somewhat like the staminæ which in most flowers surround the pistil. These additional elements of the Polypidom have been designated by the name of *pali*, and form sometimes one, sometimes two or three, circular rows or *coronets*.

In most Corals other lamellar or spiniform processes extend from the walls outward, and constitute the parts which we propose calling the *costæ* of the Polypidom. In general they correspond exactly to the septa; and in many cases they seem to be mere prolongations of these organs through the sort of sheath formed by the walls. Sometimes, indeed, the walls themselves are no longer composed of a distinct, independent, calcified lamina, and are made up by a slight thickening and cementing of the septa along the line corresponding to the boundaries of the gastric cavity and the inner margin of the costæ.

The cavity thus circumscribed by the walls of the corallum, and subdivided by the septa, the pali, and the columella, is always closed at its bottom and open at its upper extremity, where it usually presents the appearance of a sort of radiated cup, and constitutes the *calice*. In some species, this central cavity, or *visceral chamber*, remains completely pervious from one extremity of the corallum to the other; and the membranous appendices containing the reproductive organs, and situated in the loculi, extend to its basis, without encountering any obstacle; but in other species a certain number of transverse *trabiculæ* or *synapticulæ* extend from one septum to another at various heights, and fill up, more or less completely, the inferior part of the loculi. In other cases, horizontal or oblique laminæ occupy the same position, and subdivide the loculi into a series of small, irregular cells; and sometimes these partitions are developed to such an extent that no direct communication is preserved between the lower and the upper parts of the visceral chamber, so that the calice, instead of resembling a deep tubular cup, is reduced to the form of a shallow basin. In general, these transversal laminæ, to which the name of *dissepiments* has been given, grow from the sides of the septa in an irregular manner, and do not unite so as to constitute complete horizontal tabulæ, extending from wall to wall; but in some Corals, where the septal apparatus is even rudimentary, the bottom of the visceral chamber is incessantly raised by the formation of new floors or *tabulæ*, which extend horizontally through the centre of the Polypidom, and constitute, under the calices, a vertical series of secondary chambers.

Intercostal dissepiments are frequently met with on the outside of the walls of the corallum and in compound Polypidoms, where the costæ are highly developed, a thick cellular mass is thus formed, and often assumes the appearance of a *cœnenchyma*, or common tissue. In other instances, the calcified derm continues to extend exteriorly without constituting distinct costæ, and forms a dense or a reticulate tissue, which, in certain aggregate Corals, is nowhere referable to any individual Polyp, and produces a sort of intermediate mass or true cœnenchyma.

It is also to be remarked, that the exterior surface of most Corals is covered by a layer of epithelic sclerenchyma, which is sometimes thick and spongy, but in general thin and dense, and then constitutes a species of coating, which may be called the *epitheca*.

These different constitutive parts of the Polypidom furnish the principal characters employed in the classification of Corals; but the mode of multiplication of the Polypi must also be attended to in the methodical arrangement of these Zoophytes. In some species, the young are only produced by the ova, and each corallum is formed by the skeleton of a single individual; but in most, reproduction also takes place by fissiparity or by gemmation, and in those cases the young usually remain adherent to the body of their parent, and thus produce *compound Polypidoms*. The manner in which the different individual Polypidoms, or *corallites* thus united, are grouped together, varies very much, and furnishes also useful zoological characters. It is equally necessary not to neglect studying the changes which take place in the structure of Polypidoms by the progress of age. Corals, when young, are in general much less complicated than in the adult state, and the manner in which the multiplication of their constituent parts is effected is often a subject of great interest for classifiers as well as for physiologists.

The natural affinities of recent Corals can, in general, be easily recognised by means of facts obtained from these different sources; but the study of fossil Polypidoms presents greater difficulties, and the palæontologist must also direct his attention to the modifications which may have taken place after the death of the Zoophyte, and have been produced by the slow, but long-continued action of solvent or lapidescent fluids. Changes of this kind sometimes efface the most important features of these organic remains, for it often happens that the different parts of a corallum are not modified with an equal degree of facility, and the complete destruction of certain organs in specimens, where other parts are well preserved, may give rise to most delusive appearances. Even generic divisions have thus been established by some palæontologists, on accidental changes due to fossilization alone, and it is indeed often very difficult to avoid errors of this kind in the distinction of species, when the observer is not able to compare a sufficient number of specimens.

§ III.

This Monograph being intended principally for the use of Geologists, we have thought it advisable not to follow the Zoological classification of Corals in describing the species belonging to the Fossil Fauna of Great Britain, but to distribute them in reference to the different Formations in which they are found. We must, however, not lose sight of the Natural arrangement of these Zoophytes, and before entering on the specific history of the organic remains which we have to study, it is necessary that we should make known to the reader the system of classification which we have adopted for Polypi in general. The following Synopsis will suffice for that purpose, and will serve as a sort of framework illustrative of the divers Zoological divisions to which we shall often have to revert as we proceed in the descriptive part of our work.

CLASSIFICATION OF POLYPI.

SUB-KINGDOM ZOOPHYTA; SECTION RADIATA.

CLASS

POLYPI.

Animals of the sub-kingdom of ZOOPHYTA, and of the section of RADIATA,[1] organized for a sedentary mode of life, having no locomotive organs, and being provided with a circle of retractile tentaculæ around the mouth, and a central gastric cavity, not communicating with an anus, and containing the reproductive organs when these exist; in general fissiparous, or multiplying by buds as well as by ovules.

The systems adopted by Cuvier, Lamarck, Lamouroux, and their contemporaries, for the subdivision of the class of Polypi, were founded on external characters of very little value, and were quite artificial. In a Memoir, published about twenty years ago,[2] a first attempt was made to establish this classification on anatomical facts, and the Zoophytes presenting the above-mentioned structure were distributed in two groups, characterised by the presence or the absence of internal ovaria, and a membranaceous tube leading from the mouth to the great gastric cavity. Subsequent observations have confirmed these views, and Mr. Dana, whose recent work[3] is one of the most valuable contributions which America has yet made to Natural History, divides in a similar manner the class of Polypi into two secondary groups. We shall continue adopting this classification here; but the name of *Actinoidea*, which Mr. Dana applies to the first of the two sub-classes thus established, having been previously employed by other zoologists in a much narrower acceptation, we have thought it advisable not to make use of it here, and we propose substituting for it that of *Corallaria*. The second group comprises the *Sertularian Polypi* (Milne Edw.), and may be designated by the name of *Hydraria*.

---

[1] The sub-kingdom of Zoophytes may be divided into two natural groups: the one comprising all the true Radiate animals (Echinoderma, Acalephæ, and Polypi); the other containing the spheroidal or amorphous Zoophytes (such as Spongidæ and certain Infusoria). The first may retain the name of *Radiata*; the second has been designated by that of *Sarcodaria* (Milne Edwards, Cours élémentaire de Zoologie).

[2] Recherches sur les Animaux sans Vertèbres, faites aux îles Chausay, par MM. Audouin et Milne Edwards (Annales des Sciences Naturelles, première série, t. xv, p. 18, Septembre, 1828).

[3] United States Exploring Expedition; Zoophytes. Philadelphia, 1846.

Sub-class 1.
CORALLARIA.

Actinoidea, Dana. Op. cit., p. 16, 1846.

Polypi possessing distinct internal reproductive organs, and having the gastric or visceral cavity surrounded by vertical, radiating, membranaceous lamellæ.

In this division of the class of Polypi, the Corallum is in general calcareous, and may be either tubular, cyathoid, discoidal, or basal; but never assumes the form of cylindrical, tubular, horny sprigs, bearing simple bell-shaped cells, for the reception of the contracted tentacula, as we usually find in the sub-class of Hydraria.

Corallaria present three principal structural modifications, and must therefore be subdivided into three corresponding groups or orders: Zoantharia, Alcyonaria, and Podactinaria.

Order 1.
ZOANTHARIA.

Zoanthaires (Zoantha), Blainville. Manuel d'Actinologie, p. 308, 1834.

Zoanthaires (Zoantharia), Milne Edwards. Elém. de Zoologie, p. 1045, 1835; Annot. de Lamarck, Anim. sans Vertèb., tom. ii, p. 106, 1836.

Zoophyta helianthoidea, Johnston; in Mag. of Zool. and Bot., vol. i, p. 448, 1837; Hist. of British Zoophytes, p. 207, 1838.

Zoantharia, J. E. Gray. Synop. Brit. Mus., 1842.

Actinaria, Dana. United States Exploring Expedition, Zoophytes, p. 112, 1846.

Anthozoa helianthoidea, Johnston. Hist. of Brit. Zooph., 2d ed., vol. i, p. 181, 1847.

Polypi with conical, tubular, simple or arborescent, but not bipinnate, tentacula, and with numerous perigastric membranaceous laminæ, containing the reproductive organs.

Zoantharia are in general coralligenous, and almost all the known fossil Polypidoms belong to this natural group of Zoophytes.

These Corals are very seldom essentially composed of epidermic tissues, nor do they scarcely ever constitute basal stems, as is usually the case in Alcyonaria. They are almost always formed of calcified dermic sclerenchyma, and inclose, more or less completely, the inferior portion of the great visceral or gastric cavity of the Polyp. Each individual has in general the form of a deep cup or a tubular sheath, the cavity of which is subdivided into a circle of loculi, by vertical septa affecting a radiate disposition. No trace of any such septa is ever met with in Corals belonging to other animals of the same class, and although these parts are sometimes rudimentary in Zoantharia, the starlike appearance of the calice pro-

2

duced by their existence must be considered as one of the most striking features of this zoological division. The septa are developed successively, as the Polyp grows, and in general six of these vertical laminæ constitute the primary or fundamental cyclum. Shortly afterwards a second circle, equally composed of six septa, appears, and the twelve loculi situated between these secondary septa and the primary ones are next subdivided by a third row or cyclum of twelve younger septa. The number of the septa often augments still more, and is sometimes carried very high; but in general the primary septa continue to be more developed than the others, and thus divide the whole of the radiate structure into six distinct groups or systems. In some instances, however, the secondary, or even the tertiary, septa grow so rapidly, that they soon exactly resemble those of the first cyclum, and in such cases the number of the systems is apparently much greater.[1] Sometimes the number of the primary septa is, on the contrary, reduced to four, or perhaps even to two, but never reaches eight, as would be the case if the Polypi of this order had ever eight tentacula and eight perigastric lamellæ, a structure which is always met with in the order of Alcyonaria. It is also to be noted, that the septa vary considerably in their structure, and thus furnish most important characters, not only for the distinction of species and genera, but even for the formation of higher zoological divisions in this order of Polypi.

Zoantharia may be divided into two principal groups, characterised by the structure of the parietes of their body. One of these sections comprises the species in which the dermal tissue remains soft and flexible; the other contains those the teguments of which assume an osseous structure and constitute a calcareous Polypidom.

The SCLERENCHYMATOUS ZOANTHARIA are the only Zoophytes of this order which we shall have to mention in the sequel of this work; it would, therefore, be superfluous for us to treat of the classification of Malacodermous Zoantharia; but it is necessary that we should give a detailed account of the methodical arrangement of the first of these groups. Little is known concerning the anatomical modifications of the soft parts in the different representatives of this zoological form; but the structure of the Polypidom offers great variety, and furnishes, to an attentive observer, data which appear sufficient for the natural classification of Sclerenchymatous Zoantharia. The principal characters which we have made use of for that purpose, are derived from the dense or porous structure of the sclerenchyma; the predominance of the septal apparatus, the mural tissue or the tabular system in the formation of the corallum; the existence or the absence of dissepiments uniting the septa and subdividing the loculi, and the mode of development of the Polypi. Five principal divisions may be thus established in this section, and may be designated by the following appellations: *Zoantharia aporosa*, *Zoantharia perforata*, *Zoantharia tabulata*, *Zoantharia rugosa*, and *Zoantharia cauliculata*.

---

[1] The laws by which the development of the septal apparatus appears to be regulated, have been laid down in our memoir on the Structure of Corals, published in the Annales des Sciences Naturelles, 3ᵐᵉ série, tom. ix, 1848.

Sub-order 1.
ZOANTHARIA APOROSA.

Corallum composed essentially of lamellar dermic sclerenchyma, with the septal apparatus highly developed, completely lamellar, and primitively composed of six elements; no tabulæ.

The foliaceous or lamellar structure of the calcified tissue, which furnishes one of the principal characters of these Corals, is always recognisable in the exterior part of the septa; these organs are never composed of irregular trabiculæ, as is the case in Porites, or even perforated, excepting near their inner margin. The walls are also very seldom porous, and usually constitute an uninterrupted theca, so as to admit of no communication between the visceral chamber and the exterior, except by the calice. The septa form the most important part of the Polypidom; they augment more or less in number as the Polyp rises, but in general remain unequally developed, and are disposed in groups corresponding to the six primitive radii, or to a multiple of that number, but never present a quaternary arrangement, as is often the case in Cyathophyllidæ. The visceral chamber remains open from top to bottom, or is only subdivided by synapticulæ, or by irregular dissepiments, which extend from one septum to another without joining together, so as to form a series of distinct tabulæ or discoid floors; a mode of structure which is on the contrary prevalent, and very remarkable in most of the Corals belonging to our third and fourth sections.

The *Zoantharia aporosa* are the most lamelliferous and stelliform of all the Corallaria; they are very numerous, and belong to four principal families: the *Turbinolidæ*, the *Oculinidæ*, the *Astreidæ*, and the *Fungidæ*; but some few of them cannot find a proper place in any of these natural divisions, and appear to constitute a certain number of satellite or transitional minor groups, which partake of some of the characters of two or more of the above-mentioned principal forms, without possessing any structural peculiarity of sufficient importance to make us consider them as the representatives of a special type; these groups are therefore not of the same zoological value as the preceding, and in order to point out their aberrant nature, we shall designate them by names indicative at once of their principal affinities and their dependent character: *Pseudastreidæ* and *Pseudoturbinolidæ* for example.

Family I.
TURBINOLIDÆ.

Milne Edwards and Jules Haime, Recherches sur les Polypiers; Annales des Sciences Naturelles, 3ᵐᵉ série, tom. ix, p. 211, 1848.

Corallum in general simple, never fissiparous, and multiplying by lateral gemmation in compound species. Interseptal loculi extending from the top to the bottom of the visceral chamber, and containing neither dissepiments, as in the Astreidæ, nor synapticulæ, as in the Fungidæ. *Walls* thin, lamellar, and imperforated. *Septa* highly developed, simple, compact, in general regularly granulated on each side, and never denticulated or lobulated at their apex. *Costæ* in general well marked and straight. No cœnenchyma in the compound Polypidoms.

First Tribe—CYATHININÆ.

Milne Edwards and J. Haime, loc. cit., p. 289, 1848.

Calicule presenting one or more rows of pali, placed between the columella and the septa.

§ 1. A single coronet of pali.

1. *Genus* CYATHINA.

Caryophyllia, Stokes. Zool. Journ., vol. iii, p. 486, 1828.
Cyathina, Ehrenberg. Corall. des Rothen Meeres, p. 76, 1834 ; Milne Edwards and J. Haime, op. cit., p. 285.

Corallum simple, never gemmiparous, subturbinate and adherent. *Calice* circular or nearly so, with a broad but not very deep central fossula. *Columella* fasciculate, composed of a certain number (3 to 20) of vertical, narrow, and twisted lamellar processes, and terminated by a convex, crispate surface. *Pali* broad, entire, free in a considerable part of their length, and equally developed. *Septa* straight, broad, exsert, and forming six systems, which are in general unequally developed, and become in appearance much more numerous. *Costæ* straight, slightly prominent near the calice, more or less obsolete lower down, delicately granulated, and never armed with tubercles, crests, or spines.

Typical species, *Cyathina cyathus,* Ehrenb., loc. cit. ; Milne Edwards and J. Haime, Ann. des Sc. Nat., 3ᵐᵉ série, tom. ix, tab. iv, fig. 1.

2. *Genus* CŒNOCYATHUS.

Milne Edwards and J. Haime, Ann. des Sc. Nat., 3ᵐᵉ série, tom. ix, p. 297, 1848.

Corallum composite and adherent ; the *corallites* sub-cylindrical, rather tall, segregate (united near their basis, but free in the greatest part of their length), and not grouped in rows. *Calice* circular ; fossula not very deep. *Columella* composed of a few twisted, lamellar, vertical processes. *Pali* entire, equidistant from the centre, and similar in size. *Septa* rather broad, not projecting much above the walls, and forming four cycla, the last of which is incomplete in one of the six systems. *Costæ* distinct near the calice only, straight, flat, broad, and delicately granulated.

These Corals have great affinity to Cyathina, from which they differ principally by their gemmiparous mode of multiplication, and the permanent union of the young to the parent.

Typ. sp., *Cœnocyathus cylindricus,* Milne Edw. and J. Haime, loc. cit., tab. ix, fig. 8.

3. *Genus* ACANTHOCYATHUS.

Milne Edw. and J. Haime, loc. cit., p. 292, 1848.

Corallum simple, free, subturbinate, slightly compressed, and subpedicellate. *Calice* more or less oval. *Columella* and *pali* as in Cyathina. *Septa* broad, exsert, and forming five cycla; systems unequally developed, so as to form sixteen groups. *Costæ* partly armed with crests or spines.

Typ. sp., *Acanthocyathus Grayi*, Milne Edw. and J. Haime, loc. cit., tab. ix, fig. 2.

4. *Genus* BATHYCYATHUS.

Milne Edw. and J. Haime, loc. cit., p. 294, 1848.

Corallum simple, adherent by a broad basis, tall, subturbinate, and slightly compressed. *Calice* subelliptical, with a broad and very deep fossula. *Columella* small and crispate. *Pali* narrow, feeble, entire, and closely united to the septa. *Septa* exsert, thin, closely set, and forming apparently twelve equally developed systems; five cycla, the last of which is more developed than the penultimate one, the septa of which are closely approximated towards the wall, or even cemented to those of the primary, secondary, and ternary cycla. *Costæ* very narrow, straight, unarmed, delicately granulated, and distinct down to the basis of the corallum.

Typ. sp., *Bathycyathus chilensis*, Milne Edw. and J. Haime, loc. cit., tab. ix, fig. 5.

5. *Genus* BRACHYCYATHUS.

Milne Edw. and J. Haime, loc. cit., p. 295, 1848.

Corallum simple, extremely short, widening very rapidly, and becoming free in the adult state. *Calice* circular, and very slightly excavated. *Columella* very thick, fasciculate, and terminated by circular papillæ. *Pali* very broad, entire. *Septa* exsert, narrow, and forming four cycla; the systems equally developed, and apparently twelve in number. *Costæ* unarmed.

Typ. sp., *Brachycyathus Orbignyanus*, Milne Edw. and J. Haime, loc. cit., tab. ix, fig. 6.

6. *Genus* DISCOCYATHUS.

Milne Edw. and J. Haime, loc. cit., p. 296, 1848.

Corallum simple, free, and discoidal. *Calice* circular and slightly convex. *Columella* formed by a single vertical lamina; its apex smooth and undivided. *Pali* free and corresponding to the septa of the antepenultimate cyclum. *Septa* very exsert, broad, and striated laterally near their apex. *Wall* horizontal, and covered with an *epitheca* presenting some concentric striæ.

Typ. sp., *Discocyathus Eudesii*, Milne Edw. and J. Haime, loc. cit., tab. ix, fig. 7.

7. *Genus* CYCLOCYATHUS.

Corallum simple, discoidal, and having the same characters as the preceding genus, except that the *columella* is fasciculate and papillous.

Typ. sp., *Cyclocyathus Fittonii*, nob.

§ 2. Pali of divers orders, forming two or more coronets.

8. *Genus* TROCHOCYATHUS.

Milne Edw. and J. Haime, loc. cit., p. 300, 1848.

Corallum simple, pediculate or sub-pediculate, but free in the adult state. *Calice* with a broad but not very deep fossula. *Columella* well developed, and composed of prismatic or twisted processes disposed fascicularly or in a single row. *Pali* well developed, entire, free on both edges, and differing in breadth according to the coronet to which they belong. *Septa* very exsert, broad, thick near the wall, striated laterally, and forming from four to six cycla. *Costæ* often armed.

Typ. sp., *Trochocyathus mitratus*, nob. (*T. mitratus* et *T. plicatus*, Milne Edw. and J. Haime, loc. cit., p. 303); *Turbinolia mitrata*, Goldfuss, op. cit., pl. xv, fig. 5; *Turbinolia plicata*, Michelotti, Specim. Zooph. dil., tab. ii, fig. 9.

9. *Genus* LEPTOCYATHUS.

Corallum presenting most of the characters of the preceding genus, from which it differs by its subdiscoid form, and its not showing any trace of adherence.

Typ. sp., *Leptocyathus elegans*, nob.

10. *Genus* THECOCYATHUS.

Milne Edw. and J. Haime, loc. cit., p. 317, 1848.

Corallum simple, very short, and adherent, at least when young. *Calice* circular, with the fossula shallow. *Columella* very large, fasciculate, formed by a great number of prismatic processes, and terminated by a flat papillous apex. *Pali* thick, narrow, short, and entire, those corresponding to the penultimate cyclum of septa the most developed. *Septa* not exsert, thick, closely set, and almost equally developed; systems equally developed. *Wall* covered by a complete *epitheca*, slightly striated transversely, and constituting around the calice a small projecting ring.

Typ. sp., *Thecocyathus tintinnabulum*, Milne Edw. and J. Haime, loc. cit., p. 317; *Cyathophyllum tintinnabulum*, Goldfuss, Petref. Germ., tab. xvi, fig. 6.

11. *Genus* PARACYATHUS.

Milne Edw. and J. Haime, loc. cit., p. 318, 1848.

Corallum simple, subturbinate, and having a broad adherent basis. *Calice* with a large but not very deep fossula. *Columella* very broad, terminated by a papillous surface, and formed by processes that appear to arise from the lower part of the inner edge of the septa. *Pali* in general lobulated at their apex, narrow, tall, and appearing also to proceed from the inferior part of the margin of the septa, their size diminishing as they approach nearer to the columella. *Septa* nearly equal, very slightly exsert, and closely set, their lateral surface strongly granulated, and presenting sometimes traces of imperfect dissepiments; four or five cycla; systems equally developed. *Costæ* nearly equal, straight, closely set, projecting very little, and delicately granulated.

Typ. sp., *Paracyathus procumbens*, Milne Edw. and J. Haime, loc. cit., tab. x, fig. 6.

12. *Genus* HETEROCYATHUS.

Milne Edw. and J. Haime, loc. cit., p. 323, 1848.

Corallum simple, sub-cylindrical, extremely short, and adherent by a basis at least as broad as the calice, but appearing free, because in the adult state it imbeds in its tissue the small shell to which it is fixed. *Calice* circular, or nearly so, with a broad, deep fossula; *Columella* small, and composed of very slender vertical styli. *Pali* broad, thin, and denticulate. *Septa* very exsert, broad, thick, and covered with conical granulations arranged in radiate series; four or five cycla, the last of which is more developed than the penultimate one, and composed of septa that diverge from the older septa as they advance towards the centre of the visceral chamber. *Costæ* straight, thick, closely set, and strongly granulated.

Typ. sp., *Heterocyathus æquicostatus*, Milne Edw. and J. Haime, loc. cit., tab. x, fig. 8.

13. *Genus* DELTOCYATHUS.

Milne Edw. and J. Haime, loc. cit., p. 325, 1848.

Corallum short, conical, free, and presenting no trace indicating its having been adherent when young. *Calice* circular, and almost flat. *Columella* multipartite. *Pali* highly developed, and very unequal, those of the penultimate circle the largest, and turned towards those of the antepenultimate row, so as to form with them a series of deltæ. *Septa* slightly exsert. *Costæ* straight, unequal, distinct down to the basis of the corallum, and strongly granulated, so as to assume a moniliform appearance.

Typ. sp., *Deltocyathus italicus*, Milne Edw. and J. Haime, op. cit., tab. x, fig. 11; *Stephanophyllia italica*, Michelin, Icon. Zooph., tab. viii, fig. 3.

14. *Genus* TROPIDOCYATHUS.

Milne Edw. and J. Haime, loc. cit., p. 326, 1848.

Corallum simple, free, presenting no trace of former adherence, compressed, and having at its basis a large, thick, transverse, vertical crest, or two projecting lobes, resembling wings, or the fins of sepia. *Calice* elliptic and arched, its small axis being much higher than its long axis; fossula not very deep. *Columella* oblong and multipartite. *Pali* entire; those corresponding to the penultimate cyclum of septa taller and broader than the others. *Septa* exsert; the six systems equally developed. *Costæ* well marked, especially at the upper part of the wall, and covered with small granulations.

Typ. sp., *Tropidocyathus Lessonii*, Milne Edw. and J. Haime, loc. cit.; *Flabellum Lessonii*, Michelin, in Guerin's Mag. de Zool., 1843, tab. vi.

15. *Genus* PLACOCYATHUS.

Milne Edw. and J. Haime, loc. cit., p. 327, 1848.

Corallum simple, pedicellate, and slightly compressed. *Columella* lamellar, with its apical margin straight. *Pali* thin, resembling lobes of the septa, and corresponding only to the septa of the penultimate and antipenultimate cycla; those facing the latter more developed than the others (a disposition which forms an exception to the common rule). *Septa* numerous, thin, broad, and slightly exsert. *Costæ* distinct from the top to the bottom of the walls, but projecting very little, nearly equal, and appearing to bifurcate towards the upper part of the corallum.

Typ. sp., *Placocyathus apertus*, Milne Edw. and J. Haime, loc. cit., tab. x, fig. 10.

Second Tribe—TURBINOLINÆ.

Milne Edw. and J. Haime, loc. cit., p. 235, 1848.

Corallum destitute of pali; the septa extending to the columella, or meeting in the centre of the visceral chamber.

§ 1. Wall naked, or having only an incomplete epitheca.

16. *Genus* TURBINOLIA.

Turbinolia (*in parte*), Lamarck, An. sans Vert., vol. ii, p. 359, 1816; *Turbinolia* (*in parte*), Ehrenberg, op. cit., p. 53, 1834; Dana, op. cit., p. 374; *Turbinolia*, Milne Edw. and J. Haime, loc. cit., p. 235, 1848.

Corallum simple, conical, straight, and presenting no trace of adherence. *Calice* circular. *Columella* styliform. *Septa* exsert, those of the last cyclum bent toward the neighbouring ones and united to them. *Costæ* lamellar, straight, entire, and very projecting; the intercostal grooves presenting a double series of small dimples, resembling pores.

Typ. sp., *Turbinolia sulcata*, Lamarck, Hist. Anim. sans Vert., vol. ii, p. 231; Cuvier and Brongniart, Géographie Minéral. des Envir. de Paris, tab. ii, fig. 3.

17. *Genus* SPHENOTROCHUS.

Milne Edw. and J. Haime, loc. cit., p. 240, 1848.

Corallum simple, presenting no trace of adherence, straight, and cuneiform. *Calice* elliptical. *Columella* lamellar, and occupying the great axis of the calice; its upper margin flexuous and bilobate. *Septa* broad, slightly exsert, and forming three cycla; apparently twelve systems in the adult. *Costæ* broad, not very prominent, in general crispate, or represented by series of papillous tubercles.

Typ. sp., *Sphenotrochus crispus*, Milne Edw. and J. Haime, loc. cit., p. 241; *Turbinolia crispa*, Lamarck, op. cit., vol. ii, p. 231; Milne Edwards, Atlas du Règne Animal de Cuvier, Zooph., pl. lxxxii, fig. 4.

18. *Genus* PLATYTROCHUS.

Milne Edw. and J. Haime, loc. cit., p. 246, 1848.

Corallum simple, straight, cuneiform, and presenting no trace of adherence. *Calice* elliptical. *Columella* fasciculate, and terminated by papillæ. *Septa* exsert, very broad, nearly equal, and very strongly granulated; three cycla; systems equally developed. *Costæ* of two sorts, those that occupy the middle of each side of the corallum enlarging as they ascend; the lateral ones larger and much broader at their bases than near the calice, so as to render the lateral edges of the corallum almost parallel.

Typ. sp., *Platytrochus Stokesii*, Milne Edw. and J. Haime, loc. cit. tab. vii, fig. 7 ; *Turbinolia Stokesii*, Lea, Contrib. to Geol., tab. vi, fig. 207.

19. *Genus* CERATOTROCHUS.

Milne Edw. and J. Haime, loc. cit., p. 248, 1848.

Corallum simple, subpedicellate, free in the adult state, and recurved towards its basis. *Calice* circular, or nearly so. *Columella* fasciculate, and highly developed. *Septa* straight, broad, and exsert. *Costæ* partly armed with spines, crests, or small lobular processes.

Typ. sp., *Ceratotrochus multiserialis*, Milne Edw. and J. Haime, loc. cit., tab. vii, fig. 5 ; *Turbinolia multiserialis*, Michelotti, Spec. Zool. tab. ii, fig. 7.

20. *Genus* DISCOTROCHUS.

Milne Edw. and J. Haime, loc. cit., p. 251, 1848.

Corallum simple, discoidal, and presenting no trace of adherence. *Calice* circular, and almost flat. *Columella* fasciculate, and terminated by papillæ equal in size. *Septa* straight, very broad, and projecting but little laterally. *Wall* horizontal. *Costæ* straight and simple.

Typ. sp., *Discotrochus Orbignyanus*, Milne Edw. and J. Haime, loc. cit., tab. vii, fig. 6.

21. *Genus* DESMOPHYLLUM.

Ehrenberg, op. cit., p. 75, 1834.

Corallum simple, and adherent by a broad basis. *Calice* with a very deep fossula. No *columella* (a character which distinguishes this group from all the preceding Turbinolinæ). *Septa* broad, very exsert, free almost all along their inner edge, and grouped in fasciculæ; those of the last cyclum taller than those of the penultimate cyclum, and cemented exteriorly to the older septa. *Costæ* distinct near the calice, but obsolete on the lower part of the wall, where there are only a few granulations.

Typ. sp., *Desmophyllum crista-galli*, Milne Edw. and J. Haime, loc. cit., tab. vii, fig. 10.

3

§ 2. Wall completely covered by a pellicular epitheca.

22. *Genus* FLABELLUM.

Lesson, Illustr. de Zoologie, 1831 ; *Phyllodes,* Philippi, Neues Jahrbuch für Miner. Geol. 1841.

Corallum simple, compressed, and in general free in the adult state. *Calice* usually elliptic, very strongly arched in the direction of its long axis ; fossula narrow, and very deep. *Columella* spurious, and formed by marginal trabiculæ of the septa ; very little developed, or even quite rudimentary. *Septa* in general very numerous, appertaining in reality to six primitive systems, but forming in appearance a much greater number of systems ; not projecting above the margin of the wall, and presenting laterally regular rows of well-developed granulations. *Walls* completely covered with a thin, slightly-striated epitheca, and in general armed laterally with long spiniform processes, corresponding with the direction of the long axis of the calice. No radiciform appendices.

Typ. sp., *Flabellum pavoninum,* Lesson, op. cit., pl. xiv.

23. *Genus* PLACOTROCHUS.

Milne Edw. and J. Haime, loc. cit., p. 282, 1848.

Corallum resembling much those of the preceding genus, but having a lamellar *columella.*

Typ. sp., *Placotrochus lævis,* Milne Edw. and J. Haime, loc. cit., tab. viii, fig. 15.

24. *Genus* BLASTOTROCHUS.

Milne Edw. and J. Haime, p. 282, 1848.

Corallum resembling those of the genus Flabellum, but gemmiparous ; the young produced by buds placed along the lateral edges of the corallum, and becoming free by the progress of their development.

Typ. sp., *Blastotrochus nutrix,* Milne Edw. and J. Haime, loc. cit., tab. viii, fig. 14.

25. *Genus* RHIZOTROCHUS.

Milne Edw. and J. Haime, loc. cit., p. 281, 1848.

Corallum simple, trochoid, and adherent by means of cylindrical radiciform appendices, which proceed from the wall, at different heights, and descend to embrace the extraneous body on which the Zoophyte lives. *Calice* almost oval, with a very narrow and very deep fossula. No *columella. Septa* extending to the middle of the visceral chamber, where they unite without presenting any trabiculæ.

Typ. sp., *Rhizotrochus typus,* Milne Edw. and J. Haime, loc. cit., tab. 8, fig. 16.

Aberrant Group.

PSEUDOTURBINOLIDÆ.

Corallum simple, with the loculi open and devoid of synapticulæ or dissepiments, as in Turbinolidæ, but having the septa represented by groups of three vertical laminæ, not adhering together, excepting near their external margin, where they are united by a common costa; a mode of structure, which is quite anormal in the whole order of Zoantharia.

Genus DASMIA.

Milne Edw. and J. Haime, op. cit., p. 328, 1848.

Corallum subturbinate, and appearing not to be free. *Septa* strongly granulated. *Costæ* thick, equal, not numerous, and separated by deep grooves.

Typ. sp., *Dasmia Sowerbyi*, Milne Edw. and J. Haime, loc. cit., tab. vii, fig. 8.

Family II.

OCULINIDÆ.

Corallum composite, produced by gemmation, and presenting in general an abundant, compact cœnenchyma or common tissue, the surface of which is smooth, delicately striate near the calices, or slightly granular, but never echinulate. *Walls* of the corallites complete (that is to say, presenting no perforations), not distinct from the cœnenchyma, and increasing by their internal surface, so as to invade progressively the inferior part of the visceral cavity, and to fill it up more or less completely in old age. Loculi imperfectly divided by a few dissepiments; no synapticulæ. *Septa* entire, or having their upper edge slightly divided.

§ 1. Septa of various sizes, forming distinct cycla.

1. *Genus* OCULINA.

(*Pars*) Lamarck, Hist. des An. sans Vert., t. ii, p. 283, 1816; Milne Edw. and J. Haime, Comptes rend. de l'Ac. des Sc., t. xxix, p. 68, 1849.

Corallum in general arborescent; gemmation irregular or affecting a spiral disposition; cœnenchyma highly developed; its surface smooth, excepting near the calices, where it presents slight radiating striæ. *Corallites* with the calice very deep; a *columella* well developed, papillose at its apex, and becoming compact towards its basis. *Pali* corresponding to all the septa, excepting those of the last cyclum. *Septa* almost entire, slightly exsert, and very unequally developed.

Typ. sp., *Oculina virginea*, Lamarck, An. sans Vert., p. 289; *Madrepora virginea*, Ellis and Sol., tab. xxxvi.

2. *Genus* TRYMHELIA.

Milne Edw. and J. Haime, Comptes rend. de l'Académie des Sciences, t. xxix, p. 68, 1849.

Corallum arborescent, differing from Oculina by the non-existence of a *columella*, and the great development of the *pali*, which are cemented together, so as to form a vertical tube.

Typ. sp., *Trymhelia eburnea*, Milne Edw. and J. Haime, loc. cit., p. 68.

3. *Genus* CYATHELIA.

Milne Edw. and J. Haime, loc. cit., p. 68, 1849.

Corallum arborescent; gemmation terminal and regularly opposite. *Corallites* free to a considerable distance from the calice, which are grouped in a way similar to that of flowers constituting a dichotomous cyme. *Columella* large and papillose. *Pali* well developed. *Septa* entire, exsert, and strongly granulated.

Typ. sp., *Cyathelia axillaris*, nob.; *Madrepora axillaris*, Ellis and Solander, tab. xiii, fig. 5.

4. *Genus* ASTRHELIA.

Milne Edw. and J. Haime, loc. cit., p. 68, 1849.

Corallum in general arborescent, and resembling Oculina by its form and its mode of gemmation, but differing from the three preceding genera by the non-existence of *pali*. *Calice* with a deep central fossula. *Columella* septal; edges of the *septa* denticulated.

Typ. sp., *Astrhelia palmata*, nob.; *Madrepora palmata*, Goldfuss, tab. xxv, fig, 6.

5. *Genus* SYNHELIA.

Milne Edw. and J. Haime, loc. cit., p. 68, 1849.

Corallum arborescent, with thick branches; gemmation irregular. *Calices* very shallow, their border scarcely projecting above the surface of the cœnenchyma, and united by common striæ. *Columella* compact, styliform, and terminated by a small tubercle. *Septa* scarcely exsert.

Typ. sp., *Synhelia gibbosa*, nob.; *Lithodendron gibbosum*, Goldfuss, op. cit., tab. xxxvii, fig. 9.

6. *Genus* ACRHELIA.

Milne Edw. and J. Haime, op. cit., p. 69, 1849.

Corallum arborescent, or forming a ramified cluster; gemmation pretty regularly spiral. Surface of the cœnenchyma smooth, excepting in the immediate vicinity of the calices, where slight traces of radiating costæ are perceptible. *Septa* extremely exsert, lanceolate, and entire; the principal ones uniting towards the lower part of their inner edge, without there being either a columella or pali in the centre of the visceral chamber.

Typ. sp., *Acrhelia Sebæ*, Milne Edw. and J. Haime, loc. cit., p. 69; Seba, Thes., vol. iii, tab. cxvii, fig. 5.

7. *Genus* LOPHELIA.

Milne Edw. and J. Haime, loc. cit., p. 69, 1849.

Corallum arborescent, segregate, with coalescent branches; no true cœnenchyma, but walls very thick; gemmation irregularly alternate and subterminal. *Calices* with a reverted lamellar border. *Septa* entire, exsert, and uniting at the bottom of the visceral chamber as in the preceding genus. No *columella* nor *pali*.

Typ. sp., *Lophelia prolifera*, nob.; Ellis and Sol., tab. xxxii, fig. 2; *Oculina prolifera*, Lamarck, An. sans Vert., vol. ii, p. 286.

8. *Genus* AMPHELIA.

Milne Edw. and J. Haime, loc. cit., p. 69, 1849.

Corallum arborescent, with coalescent branches, and well-developed cœnenchyma in aged parts; gemmation subterminal, regularly alternate. *Calice* deep. *Columella* rudimentary. *Septa* slightly exsert, entire, and small. No distinct *costæ*; the surface of the corallum smooth or very delicately striated.

Typ. sp., *Amphelia oculata*, nob.; *Madrepora oculata*, Esper, tab. xii.

9. *Genus* DIPLHELIA.

Corallum resembling Amphelia, but having a large *columella* and denticulated *septa*.

Typ. sp., *Diplhelia raristella*, nob. *Oculina raristella*, Defrance, Dict. des Sc. Nat., vol. xxxv, p. 356.

10. *Genus* ENALLHELIA.

D'Orbigny MSS.; Milne Edw. and J. Haime, loc. cit., p. 69, 1849.

Differs from Amphelia by the shallowness of the calices, a greater development of the septa, and the existence of long costal striæ.

Typ. sp., *Enallhelia compressa*, D'Orbigny; *Lithodendron compressum*, Goldfuss, op. cit., tab. xxxvii, fig. 11.

§ 2. Septa equally developed, and forming apparently a single cyclum.

11. *Genus* AXHELIA.

Milne Edw. and J. Haime, Compt. rend., t. xxix, p. 69, 1849.

Corallum arborescent, with coalescent branches, and a well-developed cœnenchyma, the surface of which is entirely covered with sub-granulose striæ. *Calices* very shallow. *Columella* compact, very thick, and terminated by a rounded tubercle. No *pali*. *Septa* exsert, entire.

Typ. sp., *Axhelia myriaster*, nob.; *Oculina myriaster*, Valenciennes MSS., Catal. of the Museum of Nat. Hist. of Paris.

12. *Genus* CRYPTHELIA.

Milne Edw. and J. Haime, loc. cit., p. 69, 1849.

Corallum arborescent, flabellate, and unifacial, all the corallites opening on one of the surfaces of the flabellum; surface of the branches quite smooth. *Calices* very prominent, pediculate, explanate, and folded in two. No *columella* nor *pali*.

Typ. sp., *Crypthelia pudica*, Milne Edw. and J. Haime, loc. cit., p. 69.

13. *Genus* ENDHELIA.

Milne Edw. and J. Haime, loc. cit., p. 69, 1849.

Corallum of the same general form as in the preceding genus, but with the corallites alternate on the branches, which are thick and coalescent. *Calices* immersed; their border not projecting, but armed with a tongue-shaped process. No *columella* nor *pali*.

Typ. sp., *Endhelia Japonica*, Milne Edw. and J. Haime, loc. cit., p. 69 (Mus. of Leyden).

14. *Genus* STYLASTER.

Gray, Zool. Miscel., p. 36, 1831; *Allopora*, Ehrenb., Cor. Roth. Meeres, p. 147, 1834; Dana, op. cit., p. 693, 1846; Milne Edw. and J. Haime, loc. cit., p. 69, 1849.

Corallum arborescent and subflabellate; cœnenchyma highly developed, smooth, and presenting certain excrescences or tubercles, the nature of which is problematic. *Calices* rare and not projecting much. Neither *columella* nor *pali*.

Typ. sp., *Stylaster rosea*, Gray, loc. cit.; *Oculina rosea*, Lamarck, op. cit., t. xi, p. 287; Esper., tab. xxxvi.

Transitional Group.
PSEUDOCULINIDÆ.

Corallum composite, with a highly developed, spongy, or cellulose, echinulate, dermic, cœnenchyma. Costal apparatus rudimentary. Walls imperforate, and never invading the visceral cavity. Septal apparatus well developed; dissepiments few in number.

This small group participates of the characters belonging to the *Oculinidæ* and the *Astreidæ*, but differs essentially from both. It does not, however, present any important structural peculiarity, and does not appear to be derived from a special zoological type.

1. *Genus* MADRACIS.

Milne Edw. and J. Haime, Comptes rend. de l'Acad. des Sc., t. xxix, p. 70, 1849.

Corallum arborescent; cœnenchyma almost compact, and highly echinulated. *Calices* unarmed. *Columella* styliform. *Septa* exsert and equally developed.

Typ. sp., *Madracis asperula*, Milne Edw. and J. Haime, loc. cit., p. 70; *Dentipora asperula*, Gray, MSS. British Museum.

2. *Genus* STYLOPHORA.

Schweigger, Beobacht. auf Natur., t. v, 1819; *Sideropora* and *Stylopora*, Blainville, Manuel d'Actinologie, p. 348, 1830; *Sideropora*, Milne Edw. and J. Haime, loc. cit., p. 70, 1849.

Corallum arborescent; cœnenchyma sub-compact, with a granulated surface. *Calices* armed with a labial process near the upper part of their margin. *Columella* styliform.

Typ. sp., *Stylophora pistillaris*, Schweigger, loc. cit.; *Madrepora pistillaris*, Esper., tab. lx.

3. *Genus* DENDRACIS.

Milne Edw. and J. Haime, Comp. rend., t. xxix, p. 70, 1849.

Corallum arborescent; cœnenchyma almost compact, with its surface granulated. *Calices* sub-mammiform. No *columella*. *Septa* not exsert, or only very slightly so; nearly equal.

Typ. sp., *Dendracis Gervillii*, nob.; *Madrepora Gervillii*, Defrance, Dict. des Sc. Nat., vol. xxviii, p. 8; Michelin, Icon., Zooph., pl. xlix, fig. 8.

4. *Genus* ARÆACIS.

Milne Edw. and J. Haime, loc. cit., p. 70, 1849.

Corallum massive; cœnenchyma spongy, with its surface echinulate. *Calices* with a thin projecting margin. No *columella*. *Septa* unequally developed, entire.

Typ. sp. *Aræacis sphæroidalis*, nob.; *Astrea sphæroidalis*, Michelin, pl. xliv, fig. 9.

Family III.

ASTREIDÆ.

Dana, Exploring Expedition, Zooph., p. 194, 1846.

Corallum composite or simple, circumscribed by imperforated walls, and often increasing by fissiparity. Corallites becoming tall by the progress of their growth; each individual or series of individuals well defined, and separated from the others by perfect walls. Cœnenchyma not existing, or being formed either by the development of the costæ and their dissepiments, or by the epithecal tissue alone, and not forming a compact mass as in the Oculinidæ. The visceral chamber never obliterated inferiorly by the growth of the walls, but subdivided and more or less completely closed up by the interseptal dissepiments, which are in general very abundant; never any synapticulæ like those of the Fungidæ.

First Tribe—EUSMILINÆ.

Septa completely developed and entire (that is to say, with their apical margin neither lobate nor denticulate). *Costæ* always unarmed. *Columella* often compact, or even styliform.

Section I.—EUSMILINÆ PROPRIÆ.

Corallum simple or composite, and in that case formed by distinct corallites, affecting an arborescent disposition, fasciculate, or presenting a linear arrangement; free laterally, at least in a great part of their length, and never having their calices blended together. Reproduction usually fissiparous in the compound species.

1. *Genus* CYLICOSMILIA.

Milne Edw. and J. Haime, Ann. des Sc. Nat., 3ᵐᵉ série, t. x, p. 232, 1848.

Corallum simple, adherent. *Columella* well developed, and of a spongy structure. *Septa* thin, slightly exsert, covered laterally with small granulations, and closely set. *Dissepiments* very abundant. *Wall* thin, with a rudimentary epitheca. *Costæ* simple, not ramified, and distinct down to the basis of the corallum.

Typ. sp., *Cylicosmilia altavillinsis*, Milne Edw. and J. Haime, loc. cit., p. 233 ; *Caryophyllia altavillinsis*, Defrance ; Michelin, Icon. Zooph., tab. lxxiv, fig. 2.

2. *Genus* PLACOSMILIA.

Milne Edw. and J. Haime, loc. cit., p. 233, 1848.

Corallum simple, compressed, free, and subpediculate. *Calice* more or less elliptical. *Columella* lamellar. *Septa* numerous, closely set, slightly exsert, and not much granulated ; systems apparently very numerous. *Dissepiments* abundant. *Wall* naked, or with a rudimentary epitheca. *Costæ* simple, not ramified, and distinct from the basis of the corallum.

Typ. sp., *Placosmilia cymbula*, Milne Edw. and J. Haime, loc. cit., p. 234 ; *Turbinolia cymbula*, Michelin, Icon., pl. lxvii, fig. 1.

3. *Genus* TROCHOSMILIA.

Milne Edw. and J. Haime, loc. cit., p. 236, 1848.

Corallum simple, subpedicellate or adherent. *Calice* nearly horizontal. No *columella*. *Septa* meeting in the centre of the visceral chamber, numerous, and closely set ; systems apparently very numerous. *Dissepiments* abundant. *Wall* naked, or with a rudimentary epitheca. *Costæ* simple, granulated, delicate, usually distinct from the basis, and never ramified.

Typ. sp., *Trochosmilia Faujasii*, Milne Edw. and J. Haime, loc. cit., tab. v, fig. 6.

4. *Genus* PARASMILIA.

Milne Edw. and J. Haime, loc. cit., p. 243, 1848.

Corallum simple, adherent or pedicellate, tall, subturbinate, and presenting in general indications of an intermittent growth. *Calice* nearly circular; fossula not very deep. *Columella* spongy. *Septa* exsert, very granular laterally, and arched at their apex. *Dissepiments* not abundant, and existing only in the inferior part of the loculi. *Wall* naked, or with a rudimentary epitheca. *Costæ* straight, simple, not ramified, somewhat granulated, and in general projecting more near the calice than in the lower part of the coral.

Typ. sp., *Parasmilia centralis*, Milne Edw. and J. Haime, loc. cit. ; *Madrepora centralis*, Mantell, Geol. of Sussex, tab. xvi, figs. 2, 4.

5. *Genus* CŒLOSMILIA.

Differs from Parasmilia by not having any rudiments of a columella.

Typ. sp., *Cœlosmilia poculum ; Parasmilia poculum,* Milne Edw. and J. Haime, loc. cit., tab. v, fig. 5.

6. *Genus* LOPHOSMILIA.

Milne Edw. and J. Haime, loc. cit., p. 246, 1848.

Corallum simple, subturbinate, adherent. *Calice* almost circular. *Columella* lamellar, small. *Septa* very exsert, unequal; their apical margin highly arched, and their sides granular; the six systems equally developed. *Wall* naked. *Costæ* simple, and but slightly marked ; growth not intermittent.

Typ. sp., *Lophosmilia rotundifolia,* Milne Edw. and J. Haime, loc. cit., tab. v, fig. 3.

7. *Genus* DIPLOCTENIUM.

Goldfuss, Petref. Germ., p. 50, 1826-30.

Corallum simple, extremely compressed, flabelliform, free, but retaining a thick peduncle. *Calice* representing a very long ellipse, arched so strongly that the extremities of its long axis descend much below the level of its small axis ; fossula very narrow, very long, and shallow. No *columella. Septa* extremely numerous, nearly equal, thin, very closely set, and slightly exsert. *Dissepiments* simple and numerous. *Walls* naked. *Costæ* extremely numerous, narrow, crowded, nearly equal, distinct from the basis, and dichotomosing, or even dividing into three branches as they rise.

Typ. sp., *Diploctenium lunatum,* Michelin, Icon. Zooph., tab. lxv, fig. 8 ; *Madrepora lunata,* Bruguière, Journ. d'Hist. Nat., vol. i, tab. xxiv, figs. 5, 6.

8. *Genus* MONTLIVALTIA.

Lamouroux, Exposit. Méthod. des Genres de Polypiers, p. 78 ; Milne Edw. and J. Haime, loc. cit., p. 250.

Corallum simple, adherent, or sub-pedicellate. No *columella. Septa* exsert, in general numerous and crowded, very broad, and forming apparently twelve or more cycla. *Wall* covered by a highly-developed membraniform epitheca ; growth not intermittent.

Typ. sp., *Montlivaltia caryophyllata,* Lamouroux, op. cit., tab. lxxix, figs. 8, 9, 10 ; Michelin, op. cit., tab. liv, fig. 2.

9. *Genus* PEPLOSMILIA.

Corallum resembling Montlivaltia, but having a large, lamelliform columella.

Typ. sp., *Peplosmilia Austenii,* nob.

4

10. *Genus* AXOSMILIA.

Milne Edw. and J. Haime, loc. cit., p. 261, 1848.

Corallum simple, free in the adult state, tall, turbinate. *Calice* circular; fossula large and deep. *Columella* styliform, large, and slightly compressed. *Septa* neither exsert nor crowded, delicately granulated, and all, excepting those of the youngest cyclum, cemented to the columella; loculi deep. *Walls* entirely covered by a membraniform epitheca, presenting strong transverse folds, and extending to the edge of the calice.

Typ. sp., *Axosmilia extinctorium*, Milne Edw. and J. Haime, loc. cit.; *Caryophyllia extinctorium*, Michelin, op. cit., tab. ix, fig. 3ª.

11. *Genus* EUSMILIA.

Milne Edw. and J. Haime, loc. cit., p. 262, 1848.

Corallum composite, cespitose, with dichotomous or trichotomous branches, and a stem that does not thicken much by the progress of age. *Corallites* multiplying by fissiparity, becoming rapidly segregate, and not remaining disposed in series at their calicular extremity. *Calices* rather irregular in form, but in general nearly circular; fossula deep. *Columella* of a loose lamello-spongiate texture. *Septa* exsert, broad, thin, straight, not crowded, with their apex strongly arched, and their surface almost smooth. *Dissepiments* well formed, but not very abundant. *Walls* naked or covered inferiorly by a slight pellicular epitheca. *Costæ* indistinct towards the basis of the corallites, but becoming sub-cristiform near the calice.

Typ. sp., *Eusmilia fastigiata*, Milne Edw. and J. Haime, loc. cit., tab. v, fig. 1; *Madrepora fastigiata*, Pallas, Eleuch. Zooph., p. 301.

12. *Genus* APLOSMILIA.

D'Orbigny MSS.

Corallum composite, and having the characters of Eusmilia, excepting that the *columella* is lamellar.

Typ. sp., *Aplosmilia aspera*, D'Orbigny MSS.; *Lobophyllia aspera*, Michelin, Icon., tab. xx, fig. 4; *Eusmilia* (?) *aspera*, Milne Edw. and J. Haime, loc. cit., p. 266.

13. *Genus* LEPTOSMILIA.

Milne Edw. and J. Haime, loc. cit., p. 267, 1848.

Corallum composite, cespitose, fissiparous, and presenting the same general disposition as in the preceding genus. No *columella*. *Septa* extremely thin, crowded, broad, very slightly exsert, with their apex slightly arched, and their lateral surfaces sub-glabrous. *Dissepiments* very abundant. *Walls* very thin, plain towards the basis, and costulate near the calices.

Typ. sp., *Leptosmilia ramosa*, Milne Edw. and J. Haime, loc. cit., tab. vi, fig. 1.

14. *Genus* THECOSMILIA.

Milne Edw. and J. Haime, loc. cit., p. 270, 1848.

Corallum composite, cespitose, fissiparous, and affecting the same general disposition as in the two preceding genera. No *columella*. *Septa* closely set, not remarkably thin, slightly exsert, and granulate. *Walls* covered with a strong epitheca, reaching almost to the margin of the calices.

Typ. sp., *Thecosmilia trichotoma*, Milne Edw. and J. Haime, loc. cit.; *Lithodendron trichotomum*, Goldfuss, Petref. Germ., tab. xiii, fig. 6.

15. *Genus* BARYSMILIA.

Milne Edw. and J. Haime, loc. cit., p. 273, 1848.

Corallum composite, increasing by fissiparity, and forming a very thick stem, on the apex of which the corallites become distinct, and are disposed in transverse series. *Columella* rudimentary or not existing. *Septa* closely set. *Walls* very thick, naked, and covered with delicate costal lines, which are nearly equal and granulate.

Typ. sp., *Barysmilia Cordieri*, Milne Edw. and J. Haime, loc. cit., tab. v, fig. 4.

16. *Genus* DENDROSMILIA.

Milne Edw. and J. Haime, loc. cit., p. 274, 1848.

Corallum composite, somewhat arborescent, and increasing by lateral gemmation. *Corallites* with large *septa*, and a spongious *columella*.

Typ. sp., *Dendrosmilia Duvaliana*, Milne Edw. and J. Haime, loc. cit., p. 274.

17. *Genus* STYLOSMILIA.

Milne Edw. and J. Haime, loc. cit., p. 275, 1848.

Corallum composite, fasciculate, and increasing by lateral gemmation. *Corallites* tall, with a small number of thick *septa*, and a styliform *columella*. *Walls* thick, with obsolete *costæ*.

Typ. sp., *Stylosmilia Michelinii*, Milne Edw. and J. Haime, loc. cit., p. 275, pl. vi, figs. 2, 2ª.

18. *Genus* PLACOPHYLLIA.

D'Orbigny MSS.

Corallum composite, segregate, and increasing by gemmation, which is almost basal. *Corallites* cylindrical and low. *Columella* well developed. *Septa* probably entire. *Walls* completely covered with a membraniform epitheca, presenting thick transverse folds.

Typ. sp., *Placophyllia dianthus*, D'Orbigny MSS.; *Lithodendron dianthus*, Goldfuss, Petref. Germ., tab. xiii, fig. 8.

Section II.—EUSMILINÆ CONFLUENTES.

Corallum composite, and presenting no separation between the corallites, united in rows, so as to assume a meandriform disposition; multiplication essentially fissiparous.

19. *Genus* CTENOPHYLLIA.

Dana, Zoophytes, p. 169, 1846.

Corallum pedunculate, but increasing very little by its basis, and terminated by a large oval, almost flat, calicular surface ; the different series of corallites intimately united together by means of their common walls, and without there being in general any cœnenchyma ; the gyri or calicular grooves very long, and the mural ridges thin. *Columella* lamellar, and almost uninterrupted from one end of the gyrus to the other. In general, some traces of *pali*. *Septa* rather closely set, slightly exsert, and delicately granulated. *Dissepiments* very abundant, arched, and oblique ; sometimes simple, but in general producing a vesicular mass. The common epitheca rudimentary, and covering only the inferior part of the common exterior walls, in the upper part of which are *costæ*, nearly equal, and more or less cristiform near the margin of the calicular surface.

Typ. sp., *Ctenophyllia mæandrites*, Milne Edw. and J. Haime, Ann. des Sc. Nat., 3me série, vol. x, p. 277 ; *Meandrina pectinata*, Lamarck ; *Madrep. mæandrites*, Solander and Ellis, Zooph., tab. xlviii, fig. 1.

20. *Genus* DENDROGYRA.

Ehrenberg, Corall. des Roth. Meeres, p. 100, 1834.

Corallum composite, having the form of a thick, massive, vertical column, in which the corallites are placed perpendicularly to the axis, and constitute very tortuous gyri, completely united by their walls ; mural ridges broad, flat, and compact ; grooves shallow. *Columella* highly developed, and formed by a series of very compact, enlarged processes. *Septa* very thick and closely set. *Dissepiments* large, but not crowded.

Typ. sp., *Dendrogyra cylindrus*, Ehrenb., op. cit. ; Milne Edw. and J. Haime, Ann. des Sc. Nat., 3me série, Zool., t. x, pl. vi, fig. 9.

21. *Genus* RHIPIDOGYRA.

Milne Edw. and J. Haime, loc. cit., p. 281, 1848.

Corallum composed of a single series of corallites and constituting a flabelliform or tall tortuous mass, the lateral walls of which are always free from top to bottom. *Columella* lamellar, but almost rudimentary. *Septa* exsert and crowded. *Dissepiments* abundant. *Costæ* delicate, in general subcristate near the margin of the calice. No *epitheca*, or only a rudimentary one.

Typ. sp., *Rhipidogyra flabellum*, Milne Edw. and J. Haime, loc. cit. ; *Lobophyllia flabellum*, Michelin, Icon., tab. xviii, fig. 1.

22. *Genus* PACHYGYRA.

Milne Edw. and J. Haime, loc. cit., p. 468, 1848.

Corallum adherent by a very thick peduncle ; gyri with a narrow calicular groove, and united by a very broad mass of dense cœnenchyma. *Columella* lamellar. *Septa* crowded. *Costæ* delicate and granulated ; little or no epitheca.

Typ. sp., *Pachygyra labyrinthica*, Milne Edw. and J. Haime, loc. cit. ; *Lobophyllia labyrinthica*, Michelin, Icon., pl. lxvi, fig. 3.

23. *Genus* PLEROGYRA.

Milne Edw. and J. Haime, loc. cit., p. 284, 1848.

Corallum composed of long, thick, slightly ramified gyri, united laterally by their lower part, and free only near the calicular margin. No *columella*. *Septa* exsert, and broad; interseptal loculi very broad, and almost entirely filled up with large vesicular dissepiments, constituting a cellular mass. *Walls* presenting some *costal* striæ near the calicular margin, but covered in all the other parts by a vesicular structure, which becomes highly developed between the gyri.

Typ. sp., *Plerogyra laxa*, Milne Edw. and J. Haime, loc. cit., tab. vi, fig. 8.

Section III.—EUSMILINÆ AGGREGATÆ.

Corallum composite and massive, in which the corallites are not arranged in series, and although remaining quite distinct, are united together by their walls, by a costal cœnenchyma, or by mural annular expansions.

This group corresponds to the division of the *Astreinæ aggregatæ* of the second tribe of this family, and constitutes with these the great genus *Astrea* of most authors.

24. *Genus* STYLINA.

Lamarck, Hist. des Anim. sans Vert., t. ii, p. 220, 1816; *Fascicularia*, Lamarck, Extrait du Cours, 1812.

Corallum glomerate, astreiform. *Corallites* very tall, united by means of the costal system and its dissepiments, and having the appearance of small truncate cones at their upper end. *Calices* circular, with their margin free; usually distant from each other. *Columella* styliform and projecting. *Septa* exsert, arched at their apex; in general not numerous, and forming as usual six systems. No *pali*. *Walls* thick.

Typ. sp., *Stylina echinulata*, Lamarck, loc. cit.; Milne Edw., Atlas du Règne Animal de Cuvier, Zooph., pl. lxxxv, fig. 3.

25. *Genus* STYLOCŒNIA.

Milne Edw. and J. Haime, An. des Sc. Nat., 3me série, t. x, p. 298, 1848.

Corallum having the form of a very thick sheet, convex or bent in different ways; covered inferiorly by a finely-striated epitheca; and increasing by marginal gemmation. *Corallites* united by their walls, which are thin and prismatic. *Calices* polygonal, their margins simple, and bearing at their angles small, columnar, grooved processes. *Columella* styliform, projecting. *Septa* very thin, not exsert, nor numerous, and forming six systems.

Typ. sp., *Stylocœnia emarciata*, Milne Edw. and J. Haime, loc. cit., tab. vii, fig. 2; *Astrea emarciata*, Lamarck, op. cit., t. ii, p. 266.

26. *Genus* ASTROCŒNIA.

Milne Edw. and J. Haime, loc. cit., p. 296, 1848.

Corallum very dense, and not bearing columnar processes, as in the preceding genus. *Calices* polygonal. *Columella* styliform, not projecting much. No *pali*. *Septa* thick; apparently eight or ten systems, two or four of the secondary septa being as much developed as the six primary ones. *Walls* thick and united, as in Stylocœnia.

Typ. sp., *Astrocœnia Orbignyana*, Milne Edw. and J. Haime, loc. cit., p. 297; *Astrea formosissima*, Michelin, Icon., pl. lxxii, fig. 9.

27. *Genus* STEPHANOCŒNIA.

Milne Edw. and J. Haime, loc. cit., p. 300, 1848.

Corallum glomerulate; the *corallites* united by their walls, which are thick and compact; gemmation lateral and marginal. *Calices* subpolygonal. *Columella* styliform, and not projecting much. A coronet of *pali*, corresponding to the septa of the older cycla. *Septa* scarcely exsert, granulated on their sides, and forming six systems, which are in general equally developed.

Typ. sp., *Stephanocœnia intersepta*, Milne Edw. and J. Haime, loc. cit., tab. vii, fig. 1; *Astrea intersepta*, Lamarck, Anim. sans Vert., t. ii, p. 266.

28. *Genus* PHYLLOCŒNIA.

Milne Edw. and J. Haime, loc. cit., p. 469, 1848.

Corallum glomerate, astreiform. *Corallites* united by the costæ and the exotheca, which are highly developed. *Calices* with a free margin, slightly elevated. No *columella*, or only traces of a rudimentary one. No *pali*. *Septa* very broad, exsert, and forming six systems; gemmation lateral.

Typ. sp., *Phyllocœnia irradians*, Milne Edw. and J. Haime, loc. cit.; *Astrea radiata*, Michelin, pl. xii, fig. 4.

29. *Genus* DICHOCŒNIA.

Milne Edw. and J. Haime, loc. cit., p. 305, 1848.

Corallum glomerate, astreiform. *Corallites* united by a very abundant and dense costal cœnenchyma, the upper surface of which is subgranulate. *Calices* circular or elliptical, with a projecting margin. *Columella* small. *Pali* corresponding to most of the *septa*, which are exsert and granulated. Multiplication fissiparous.

Typ. sp., *Dichocœnia porcata*, Milne Edw. and J. Haime, loc. cit.; *Astrea porcata*, Lamarck, Anim. sans Vert., t. ii, p. 260.

30. *Genus* HETEROCŒNIA.

Milne Edw. and J. Haime, loc. cit., p. 308, 1848.

Corallum resembling that of Sarcinula, but differing from all the preceding genera by the small number and the unequal development of the *septa*, which form in appearance only three systems. In general, one of the three large primary septa is more developed than the others, and remains sometimes alone in fossil species. *Calices* circular, with a projecting free margin. No *columella* nor *pali*. *Septa* exsert; cœnenchyma abundant, of a foliate structure, and having a granular surface.

Typ. sp., *Heterocœnia exiguis*, Milne Edw. and J. Haime, loc. cit., tab. ix, fig. 13 ; *Lithodendron exigue*, Michelin, Icon., Zooph. tab. lxxii, fig. 7.

Section IV.—EUSMILINÆ IMMERSÆ.

Corallum composite. *Corallites* disposed as in the preceding Section, but imbedded in an epithecal cellular tissue, and not united by costal laminæ or mural annular expansions; gemmation lateral and basal ; reproduction never fissiparous.

31. *Genus* SARCINULA.

(*In parte*) Lamarck, Hist. des Anim. sans Vert., t. ii, p. 222, 1816 ; *Anthophyllum*, Ehrenb., op. cit., p. 89, 1834.

Corallum fasciculate, and almost massive. *Corallites* tall, free towards their upper end, which projects more or less above the surface of the cellular exotheca. *Walls* strong, with costæ but little developed. No *columella*, or only a rudimentary one. *Septa* very exsert. *Dissepiments* in general simple, and not abundant.

Typ. sp., *Sarcinula organum*, Lamarck, loc. cit., p. 223 ; Milne Edw., Atlas du Règne Animal de Cuvier, Zooph., pl. lxxxv, fig. 1.

Second Tribe—ASTREINÆ.

Septa having their upper edge lobulated, dentate, or armed with spines, and often imperfect near their inner edge. *Costæ* also spinulous, dentate or crenulate, but never forming simple cristæ, as is often the case in Eusmilinæ. *Columella* in general spongy, rarely lamellar, and never styliform. Corallum in general massive.

Section I.—ASTREINÆ HIRTÆ.

Corallum simple or composite, and then formed by perfectly delineated corallites, produced by fissiparity, or by calicular gemmation.

32. *Genus* CARYOPHYLLIA.

(*In parte*) Lamarck, Hist. des Anim. sans Vert., t. ii, p. 224, 1816 ; Milne Edw. and J. Haime, Comptes rend. de l'Ac. des Sc., t. xxvii, p. 491, 1848.

Corallum simple, and adherent by a broad basis. *Calice* circular, or almost so. *Columella* well developed, spongy, and composed of twisted lamellæ, that advance one over the other. *Septa* broad, exsert, numerous, close set, and armed with spines, the size of which augments from the centre of the calice towards its margin. *Dissepiments* vesicular and abundant. *Wall* presenting *costæ*, formed by a series of spines; epitheca rudimentary.

Typ. sp., *Caryophyllia lacera*, Milne Edw. and J. Haime, An. des Sc. Nat., 3ᵐᵉ série, t. xi, p. 237 ; *Madrepora lacera*, Esper, Pflanz., tab. xxv, fig. 2.

33. *Genus* CIRCOPHYLLIA.

Milne Edw. and J. Haime, Comptes rend., t. xxvii, p. 491, 1848.

Corallum simple, subturbinate. *Columella* large and papillose. *Septa* broad, numerous, exsert, with their calicular edge divided in small obtuse lobes. *Dissepiments* abundant, vesicular, and arranged in spiral concentric lines. *Costæ* thin, nearly equal, simple, and delicately granulated.

Typ. sp., *Circophyllia truncata*, Milne Edw. and J. Haime, Ann. des Sc. Nat., 3ᵐᵉ série, t. x, tab. viii, fig. 3 ; *Anthophyllum truncatum*, Goldfuss, Petref., tab. xiii, fig. 9.

34. *Genus* THECOPHYLLIA.

Milne Edw. and J. Haime, Comptes rend., t. xxvii, p. 491, 1848.

Corallum simple, adherent, or sub-pedicellate. *Calice* circular, or nearly so. No *columella*. *Septa* very broad, in general slightly exsert, numerous, and armed with nearly equal spiniform teeth. *Wall* covered with a thick, membraniform epitheca.

Typ. sp., *Thecophyllia decipiens*, Milne Edw. and J. Haime, Ann., t. xi, p. 241 ; *Anthophyllum decipiens*, Goldfuss, Petref., tab. lxv, fig. 3.

35. *Genus* LOBOPHYLLIA.

(*Pars*) Blainville, Dict. des Sc. Nat., t. lx, 1830 ; Milne Edw. and J. Haime, Ann., t. xi, p. 244.

Corallum composite, tall, and increasing by fissiparity. *Corallites* segregate, or united in series, which are always simple, and free laterally. *Calice* with a deep fossula. *Columella* spongy. *Septa* numerous, exsert, very granular, and armed with strong marginal teeth, the most external of which are the largest ; loculi shallow. *Walls* striated longitudinally, and armed with spines ; epitheca rudimentary.

§ 1. *Lobophyllia cymosæ*. Typ. sp., *Lobophyllia angulosa*, Blainv. ; *Caryophyllia angulosa*, Lamarck.— § 2. *Lobophyllia gyrosæ*. Typ. sp., *Lobophyllia multilobata*, Milne Edw. and J. Haime, loc. cit., p. 250 ; *Fungus marinus*, Seba, Rer. Nat. Thes., vol. iii, tab. cix, No. 4.

36. *Genus* SYMPHYLLIA.

Milne Edw. and J. Haime, Comptes rend. de l'Ac. des Sc., t. xxvii, p. 491, 1848.

Corallum composite, massive, short, and increasing by fissiparity. *Corallites* having distinct calicula, but united in linear series, which are cemented together laterally. The other characters as in the preceding genus.

Typ. sp., *Symphyllia sinuosa*, Milne Edw. and J. Haime, Ann. des Sc. Nat., vol. x, tab. viii, fig. 7.

37. *Genus* MYCETOPHYLLIA.

Milne Edw. and J. Haime, Comptes rend., t. xxvii, p. 491, 1848.

Corallum massive, composed of corallites intimately united in series by their walls, which are very thin. *Exterior common walls* lobulate, spinulous, and presenting but rudiments of an epitheca. Calicular grooves, very shallow. No *columella*, or only rudiments of one. *Septa* not numerous, scarcely exsert, strongly dentate, and confluent. *Dissepiments* vesicular, large, and abundant; loculi closed almost to their top.

Typ. sp., *Mycetophyllia Lamarckiana*, Milne Edw. and J. Haime, Ann. des Sc. Nat., vol. x, tab. viii, fig. 6.

38. *Genus* EUNOMIA.

Lamouroux, Exposit. Méthod. des Polypiers, p. 83, 1824.

Corallum cespitose, fissiparous; *Corallites* segregate, tall, cylindroid. *Calices* almost circular. *Columella* rudimentary. *Septa* not very numerous. *Walls* covered with a complete membraniform epitheca, strongly striated transversely.

Typ. sp., *Eunomia radiata*, Lamouroux, op. cit., p. 83; *Lithodendron Eunomia*, Michelin, Icon., pl. xxxiv, fig. 6; *Eunomia lævis*, Milne Edw. and J. Haime, Ann., t. xi, p. 260; *Lithod. læve*, Michelin, loc. cit., pl. xix, fig. 8.

39. *Genus* CALAMOPHYLLIA.

Calamites, Guettard, Mém. sur les Sc. et les Arts, vol. ii, p. 404, 1770; *Calamophyllia*, Blainville, Dict. des Sc. Nat., t. lx, p. 312, 1830.

Corallum fasciculate, cespitose, and dichotomous. *Corallites* very long and segregate. *Calices* not very deep. *Columella* rudimentary or not existing. *Septa* thin, numerous, crowded, and armed with apical teeth, the size of which increases from the margin towards the centre of the calice. *Dissepiments* very oblique and crowded. *Walls* delicately striated, devoid of epitheca, but presenting at certain points circular foliaceous expansions.

Typ. sp., *Calamophyllia striata*, Blainville, Dict. des Sc. Nat., pl. cccxii; *Calamite strié*, Guettard, Mém. sur les Sc., t. iii, pl. xxxiv.

40. *Genus* DASYPHYLLIA.

Milne Edw. and J. Haime, Comptes rend., t. xxvii, p. 492, 1848.

Corallum fasciculate, cespitose, and dichotomous. *Corallites* very long and segregate. *Columella* spongy. *Septa* thin, slightly exsert, and armed with apical teeth, the size of which is much greater near the columella than towards the margin of the calice.

Typ. sp., *Dasyphyllia echinulata*, Milne Edw. and J. Haime, Ann. des Sc. Nat., 3me série, t. x, pl. viii, fig. 5.

41. *Genus* COLPOPHYLLIA.

Milne Edw. and J. Haime, Comptes rend. de l'Acad. des Sc., t. xxvii, p. 492, 1848.

Corallum sub-glomerate, remarkably light and fragile, composed of series of corallites cemented together laterally, without their respective walls ceasing to be distinct on the calicular surface, where they are parallel, very thin, and constitute a double ridge on each side of the calicular trench. *Calices* individualized by the direction of their septa. *Columella* rudimentary, or not existing. *Septa* extremely thin, broad, and slightly exsert; their apical edge armed with small delicate teeth, and emarginate near the middle. *Dissepiments* very abundant, and closing up the loculi almost to the margin of the calice, and forming a vesicular mass. *Common exterior walls* of the corallum or plate presenting small, lamellar, nearly equal, denticulate costæ; *epitheca* rudimentary.

Typ. sp., *Colpophyllia gyrosa*, Milne Edw. and J. Haime, Ann., t. xi, p. 266; *Madrepora gyrosa*, Ellis and Solander, op. cit.; tab. li, fig. 2; *Manicina gyrosa*, Ehrenberg, op. cit.; *Mussa gyrosa*, Dana, op. cit., p. 186.

42. *Genus* OULOPHYLLIA.

Milne Edw. and J. Haime, Comptes rend., t. xxvii, p. 492, 1848.

Corallum composed of a series of corallites, intimately united by their lateral walls, which constitute simple ridges between the trenches formed by the aggregate calices. *Columella* spongy, and in general not highly developed. *Septa* thin, slightly exsert, closely set, and armed with numerous long, sharp, apical teeth, the size of which augments towards the centre of the corallite. *Common exterior walls* sometimes covered with a thin epitheca; multiplication fissiparous.

Typ. sp., *Oulophyllia Stokesiana*, Milne Edw. and J. Haime, Ann. des Sc. Nat., vol. x, tab. viii, fig. 10.

43. *Genus* LATOMEANDRA.

D'Orbigny MSS.; Milne Edw. and J. Haime, Ann. des Sc. Nat., 3me série, t. xi, p. 270, 1849.

Corallum having most of the characters of the preceding genus, but increasing by calicular gemmation. The gyri in general short, the marginal ones distinct, and not forming a common rim. No epitheca.

Typ. sp., *Latomeandra plicata*, Milne Edw. and J. Haime, loc. cit., p. 271; *Lithodendron plicatum*, Goldfuss, Petref. Germ., tab. xiii, fig. 5.

44. *Genus* TRIDACOPHYLLIA.

Blainville, Dict. des Sc. Nat., vol. lx, p. 327, 1830.

Corallum short, and composed of corallites, arranged in series intimately united by their lateral walls, which, instead of forming a simple ridge as in the preceding genera, constitute very tall, foliaceous expansions, variously twisted, and terminated by a sub-crenulate margin ; the calicular trenches broad, very deep, and winding. *Columella* quite rudimentary, but the calicular centres very distinct. *Septa* projecting very little, thin, nearly equal, and serrate. *Dissepiments* abundant, very oblique, convex, and forming long vesicules. *Plate* or exterior surface of the common wall of the corallum covered with lamellar costæ, which extend from the basis of the mass, project slightly, and are irregularly denticulate.

Typ. sp., *Tridacophyllia lactuca*, Blainville, loc. cit. ; *Concha fungiformis*, Seba, Thes., v. iii, tab. cxxxix, No. 10; *Pavonia lactuca*, Lamarck, An. sans Vert., vol. ii, p. 239; *Manicina lactuca*, Ehrenberg, op. cit., p. 103.

45. *Genus* TRACHYPHYLLIA.

Milne Edw. and J. Haime, Comptes rend., t. xxvii, p. 492, 1848.

Corallum short, increasing by fissiparity, and composed of very flexuous series of corallites, free laterally. Common *walls* strongly echinulate. *Epitheca* rudimentary. *Columella* well developed, but of a very loose, spongy texture. *Septa* numerous, crowded, exsert, and strongly granulated laterally.

Typ. sp., *Trachyphyllia amarantum*, Milne Edw. and J. Haime, Ann. des Sc. Nat., 3ᵐᵉ sér., vol. xi, p. 275 ; *Amarantum saxum*, &c. Rumph. Amb. Hort. vi, tab. lxxxvii, fig. 1.

46. *Genus* ASPIDISCUS.

König, Icon. Foss. Sect., p. 1, 1825 ; *Cyclophyllia*, Milne Edw. and J. Haime, Comptes rend. de l'Acad. des Sc., vol. xxvii, p. 492, 1848.

Corallum discoidal, with its inferior surface flat, and its upper surface convex. *Corallites* arranged in radiating series, separated by thick and simple, crest-like mural ridges, excepting towards the margin of the calicular surface, where the young individuals spread out so as to form a broad, continuous, lamello-striate border. *Columella* rudimentary, but the caliculus well individualized. *Septa* very thin and crowded, but not numerous in each corallite. common *plate* covered with a thick epitheca, presenting concentric striæ or folds.

Typ. sp., *Aspidiscus cristatus*, Milne Edw. and J. Haime, Ann. des Sc. Nat., 3ᵐᵉ série, vol. xi, p. 277 ; *Aspidiscus Shawi*, König, Icon. Foss. Sect., pl. i, fig. 6 ; *Cyclolites cristata*, Lamarck, Anim. sans Vert., t. ii, p. 234.

47. *Genus* SCAPOPHYLLIA.

Milne Edw. and J. Haime, Comptes rend. de l'Acad. des Sc., t. xxvii, p. 492, 1848.

Corallum columnar, erect, very dense, and formed of corallites arranged in series, completely united laterally. *Columella* tubercular, somewhat compact. *Septa* very thick, neither closely set nor numerous ; with their sides very echinulate, and the apex denticulate. *Dissepiments* simple and distant.

Typ. sp., *Scapophyllia cylindrica*, Milne Edw. and J. Haime, Ann. des Sc. Nat., 3ᵐᵉ série, vol. x, tab. viii, fig. 8.

Section II.—ASTREINÆ CONFLUENTES.

Corallum massive, increasing by fissiparity, and formed by a series of corallites, the individuality of which is not distinct. The calices, thus united in a common trench, have their septa arranged in a parallel manner in two lines; and the columella, when existing, is continuous in the whole length of the series.

These meandriform Corals much resemble the confluent Eusmilinæ, and in fossils where the apical teeth of the septa may be worn away, it is often difficult to distinguish them. It may therefore be useful to mention that, in the confluent Astreinæ, the gyri are always completely united laterally, and never more or less segregate, which is sometimes the case with the confluent Eusmilinæ; that the columella, which is generally spongy in the latter, never presents that loose structure in this section; and when it is lamellar, the septa are united to it by an undivided margin in the confluent Eusmilinæ, and by a series of trabiculæ or processes in the confluent Astreinæ; lastly, that the sides of the septa are more or less granulated in all these Astreinæ, and are on the contrary almost glabrous in the meandroid Eusmilinæ.

48. *Genus* MEANDRINA.

(*Pars*) Lamarck, Hist. des Anim. sans Vert., t. ii, p. 244, 1816; Milne Edw. and J. Haime, Comptes rend., t. xxvii, p. 493.

Corallum glomerate, adherent by a very broad basis, and having a very dense structure. *Gyri* intimately united by their lateral walls, which constitute simple, compact ridges, with a cristate apex. Calicular trenches very long. *Columella* much developed, spongy and essential (that is to say, not arising from the septa, and distinct from the bottom of the visceral chamber). *Septa* crowded, enlarging near the columella, and not presenting any appearance of a paliform lobe. *Plate* or exterior common walls of the corallum covered with a complete delicate epitheca.

Typ. sp., *Meandrina filograna*, Lamarck, loc. cit., vol. ii, p. 248; Michelin, Icon., pl. xi, fig. 7.

49. *Genus* MANICINA.

(*In parte*) Ehrenberg, Corall. des Roth. Meeres, p. 101, 1834; Dana, op. cit,, p. 188, 1846; Milne Edw. and J. Haime, Ann. des Sc. Nat., vol. xi, p. 285, 1849.

Corallum free or sub-pedicellate, in the adult state; sub-turbinate when young, but becoming convex, and massive. *Gyri* very long, and united by their walls, so as to form simple ridges, as in the preceding genus; the apex of the ridge cristate or sulcate. Calicular trench broad and deep. *Columella* spongy, and even more developed than in Meandrina. *Septa* thin, crowded, strongly granulated, and armed with delicate, equal teeth; a well-characterised paliform lobe arising from the edge of the principal septa near the columella. *Plate* or exterior common wall covered with thin and very delicately-serrated costæ; its inferior part having an incomplete epitheca.

Typ. sp., *Manicina areolata*, Ehrenberg, loc. cit.; *Madrepora areolata*, Ellis and Solander, op. cit., tab. xlvii, fig. 5.

50. *Genus* DIPLORIA.

Milne Edw. and J. Haime, Comptes rend., t. xxvii, p. 493, 1848.

Corallum glomerate, adherent by a broad basis, and of a dense structure. *Gyri* long, very sinuous, and united by highly-developed costæ, and not by the walls themselves; ridges complex, presenting on each side a mural crest, and in the middle a broad concave groove or ambulacrum, formed by the costæ and their dissepiments. *Columella* spongy, essential, and well developed. *Septa* strong, exsert, and armed with closely-set teeth, the largest of which are near the walls.

Typ. sp., *Diploria cerebriformis*, Milne Edw. and J. Haime, Ann., t. xi, p. 289; *Meandrina cerebriformis*, Lamarck, op. cit., vol. ii, p. 246; Seba Thes., vol. iii, tab. cxii, No. 6.

51. *Genus* LEPTORIA.

Milne Edw. and J. Haime, Comptes rend., t. xxvii, p. 493, 1848.

Corallum glomerate, of a light spongy structure. *Gyri* very long, and limited by their walls, which are thin or cellulose, and form simple intercalicular ridges. *Columella* lamellar; its upper edge projecting slightly, and regularly lobated. *Septa* united to the columella by means of marginal trabiculæ; their upper edge slightly exsert, and armed with very small irregular teeth. *Plate* covered with a thin but complete common epitheca.

Typ. sp., *Leptoria tenuis*, Milne Edw. and Haime, Ann. des Sc. Nat., 3me sér., vol. x, tab. viii, fig. 11; *Meandrina cerebriformis*, Quoy and Gaimard, Voy. de l'Astrol., Zooph., pl. xviii, figs. 2, 3; *Meandrina tenuis*, Dana, op. cit., p. 262; Milne Edw., Atlas du Règne An. de Cuvier, Zooph., pl. lxxxiv ter, fig. 2.

52. *Genus* CŒLORIA.

Milne Edw. and J. Haime, Comptes rend., t. xxvii, p. 493, 1848.

Corallum resembling much the true Meandrina, but differing from the four preceding genera by its rudimentary *columella*, which is not essential, but septal, and formed by trabiculæ, springing from the margin of the septa. *Gyri* long, and united by their *walls*, the tissue of which is cellular; ridges simple and continuous. *Septa* delicate, and having neither a paliform lobe nor a lateral expansion near the columella.

Typ. sp., *Cœloria labyrinthica*, Milne Edw. and J. Haime, Ann. des Sc. Nat., 3me série, t. xi, p. 194; *Madrepora labyrinthica*, Ellis and Solander, op. cit., tab. xlvi, figs. 3, 4.

53. *Genus* ASTRORIA.

Milne Edw. and J. Haime, Comptes rend., t. xxvii, p. 493, 1848.

Corallum having the same structure as in the preceding genus, but formed of very short gyri, the corallites tending to individualization. This form is intermediate between the ordinary confluent Astreidæ (or Meandrinæ) and the agglomerated Astreidæ, such as true Astrea.

Typ. sp., *Astroria dædalea*, Milne Edw. and J. Haime, Ann. des Sc. Nat., t. xi, p. 297; *Madrepora dædalea*, Ellis and Solander, op. cit., tab. xlvi, figs. 1, 2.

54. *Genus* HYDNOPHORA.

Fischer de Waldheim, Descrip. du Mus. Démidoff, vol. iii, p. 295, 1810; *Monticularia*, Lamarck, Hist. des Anim. sans Vert., t. ii, p. 248, 1816.

Corallum formed of irregular series of corallites, united by their walls, which are thick, compact, and constitute ridges, divided longitudinally, so as to represent rows of conical prominences, or monticulæ. The calicular trenches are transversal as well as longitudinal, and there is no columella. *Septa* nearly equal, and rising to the apex of the conical mural monticulæ. General form sometimes massive and sub-globose or gibbous; sometimes sub-explanate.

Typ. sp., *Hydnophora Demidovii*, Fischer, Oryct. du Gouv. de Moscou, pl. xxxii.

Section III.—ASTREINÆ DENDROIDÆ.

Corallum always increasing by lateral gemmation. The corallites segregate, and having an arborescent or fasciculate arrangement. *Septa* regularly and delicately serrated; those of the principal cycla always bearing pali.

55. *Genus* CLADOCORA.

(*In parte*) Ehrenberg, Corall. des Roth. Meeres, p. 85, 1834; *Caryophyllia*, Dana, Zoophytes, p. 378, 1846.

Corallum arborescent, forming branched clumps. *Corallites* cylindrical, very long, and completely free laterally. *Calices* circular, or almost so. *Columella* papillose. *Pali* well developed, and corresponding to all the septa, except those of the last cyclum. *Septa* slightly exsert, nearly equal, granulated, and having their apex arched and delicately serrated. *Walls* compact, with simple, granulated, or echinulated costæ, and an incomplete epitheca, which often expands into circular, horizontal leaves, extending to the neighbouring corallites.

Typ. sp., *Cladocora cespitosa*, Milne Edw. and J. Haime; *Madrepora flexuosa*, Solander and Ellis, tab. xxxi, figs. 5, 6.

56. *Genus* PLEUROCORA.

Milne Edw. and J. Haime, Comptes rend. de l'Ac. des Sc., t. xxvii, p. 494, 1848.

Corallum sub-dendroid. *Corallites* cylindrical, very short; united by their basal part, and free towards their upper end. *Columella, pali,* and *septa* much as in the preceding genus. *Walls* compact, extremely thick, and never presenting any traces of an epitheca. *Costæ* distinct from one end of the corallites to the other, and vermiculate.

This genus approximates in some degree to Dendrophyllia and to Oculina.

Typ. sp., *Pleurocora explanata*, Milne Edw. and J. Haime, Ann. des Sc. Nat., 3ᵐᵉ série, vol. x, tab. vii, fig. 10.

Section IV.—ASTREINÆ AGGREGATÆ.

Corallum composite, massive, increasing by gemmation or by fissiparity, and in that case not presenting a linear mode of arrangement of the corallites, which are always completely united laterally, but remain well defined, and never lose their individuality, as in the confluent Astreina.

57. *Genus* ASTREA.

(*In parte*) Lamarck, Syst. des Anim. sans Vert., p. 371, 1801; Milne Edw. and J. Haime, Comptes rend., t. xxvii, p. 494, 1848.

Corallum massive, in general convex or sub-globose. Gemmation extra-calicular. *Corallites* tall. *Calicules* having a free, exsert, obtuse, circular margin; fossula not very deep. *Columella* spongy, and not projecting at the bottom of the calicule. No *pali*. *Septa* complete, exsert, broad, and strongly dentated or lobated; the largest of their apical teeth near the columella; their sides strongly granulated. *Costæ* highly developed, and composed of lamellæ; in general perforated, and united by numerous dissepiments.

Typ. sp., *Astrea cavernosa*, Milne Edw. and J. Haime, loc. cit., vol. x, tab. ix, fig. 1; *Madrepora cavernosa*, Esper, Pflanz. Suppl. Mad., tab. xxxvii; *Astrea argus*, Lamarck, Hist. des Anim. sans Vert., t. xi, p. 259.

58. *Genus* CYPHASTREA.

Milne Edw. and J. Haime, Comptes rend. de l'Acad. des Sc., vol. xxvii, p. 494, 1848.

Corallum massive, convex, and globose. Gemmation extra-calicular. *Corallites* united by a compact septal cœnenchyma, the surface of which is strongly granulated, or even echinulated. *Calicular* rims as in the preceding genus. *Columella* papillose, and well developed. *Septa* lamellar near the wall, but cribriform towards the columella, where they are formed by a series of oblique processes, representing a sort of lattice; their calicular teeth rather larger towards the calice than near the walls.

Typ. sp., *Cyphastrea microphtholma*, Milne Edw. and J. Haime, Ann. des Sc. Nat., 3me série, vol. x, tab. ix, fig. 5; *Astrea microphthalma*, Lamarck, op. cit.

59. *Genus* OULASTREA.

Milne Edw. and J. Haime, Comptes rend., t. xxvii, p. 495, 1848.

Corallum massive and incrustating. Gemmation extra-calicular. *Corallites* low. *Calices* circular, with a free margin. *Columella* papillose, and appearing to be formed by the inner apical teeth of the septa. No *pali*. *Septa* with a crispate, denticulated, apical margin, and echinulate sides. *Costæ* also echinulate and crispate.

Typ. sp., *Oulastrea crispata*, Milne Edw. and J. Haime, Ann. des Sc. Nat., 3me série, vol. x, tab. ix, fig. 4; *Astrea crispata*, Lamarck, Hist. des Anim. sans Vert., vol. ii, p. 265.

60. *Genus* PLESIASTREA.

Milne Edw. and J. Haime, Comptes rend., t. xxvii, p. 494, 1848.

Corallum globose; its under surface having the form of a naked costulated plate. Gemmation extra-calicular. *Calices* with a free margin, and a fossula rather shallow. *Columella* spongy. *Pali* well developed, and corresponding to all the septa except those of the last cyclum. *Septa* exsert, formed by a well-developed lamina, and having a delicately-serrated apex. *Costæ* and their dissepiments in general well developed.

Typ. sp., *Plesiastrea Urvillii*, Milne Edw. and J. Haime, Ann. des Sc. Nat., vol. x, tab. ix, fig. 2 ; *Astrea galaxea*, Quoy and Gaim., Voyage de l'Astrolabe, Zooph., pl. xvii, figs. 10-14.

61. *Genus* LEPTASTREA.

Milne Edw. and J. Haime, Comptes rend., t. xxvii, p. 494, 1848.

Corallum very dense and incrusting, and increasing by fissiparity, as well as by extra-calicular gemmation. Costal cœnenchyma quite compact. *Calices* in general much crowded together, but preserving their margins distinct. *Columella* papillose. *Septa* thin, closely set, exsert, delicately granulated, and having their apical margin almost entire near the walls, but delicately denticulated towards the columella. *Dissepiments* not very abundant. *Costæ* rather indistinct.

Typ. sp., *Leptastrea Roissyana*, Milne Edw. and J. Haime, Ann. Sc. Nat., 3me série, vol. x, tab. ix, fig. 6.

62. *Genus* SOLENASTREA.

Milne Edw. and J. Haime, Comptes rend., t. xxvii, p. 494, 1848.

Corallum forming in general a convex mass, of a light and cellular structure. Gemmation extra-calicular. *Corallites* long, slender, and united by an exothecal structure, and not by the *costæ*, which do not meet, and are often rudimentary. *Calices* circular, with an exsert margin. *Columella* spongy, and in general small. *Septa* very thin; their margin denticulated. *Dissepiments* simple, numerous, and closely set.

Typ. sp., *Solenastrea Turonensis*, nob. ; *Astrea Turonensis*, Michelin, Icon., pl. lxxv, figs. 1, 2.

63. *Genus* PHYMASTREA.

Milne Edw. and J. Haime, Comptes rend., t. xxvii, p. 494, 1848.

Corallum forming a convex or a horizontal mass. *Corallites* prismatical; surrounded from top to bottom by a thin epitheca; very nearly approximated to each other, but not united by their walls, and cemented together by means of a certain number of large wart-like processes, so as to leave an empty space between them. Gemmation extracalicular. *Calices* sub-polygonal, with a free margin. *Columella* spongy, well developed. *Septa* large, slightly exsert, and strongly dentated. *Walls* thick; no trace of costæ.

Typ. sp., *Phymastrea Valenciennesii*, Milne Edw. and J. Haime, Ann. Sc. Nat., 3me série, vol. x, tab. ix, fig. 3.

64. *Genus* ASTROIDES.

Quoy and Gaimard, Ann. des Sc. Nat., 1ʳᵉ série, vol. x, p. 187, 1827 ; *Astroitis*, Dana, Zooph., p. 405, 1846.

Corallum incrusting, and formed of corallites very unequally approximated ; some almost entirely free, others crowded so as to become polygonal, but always separated by a more or less developed epitheca. Gemmation extra-calicular. *Calices* deep. *Columella* spongy, large, and projecting very much at the bottom of the fossula, a character which does not exist in any of the preceding Astreinæ. *Septa* not much developed, very thin, not exsert, irregularly and delicately denticulated. *Dissepiments* very abundant. *Walls* composed of a dense spongy tissue. Epitheca complete.

Typ. sp., *Astroides calicularis*, Blainville, Dict. des Sc. Nat., vol. lx. ; *Caryophyllia calicularis*, Lamarck, op. cit., vol. ii, p. 226 ; Milne Edwards, Atlas du Règne Anim. de Cuvier, Zooph., tab. lxxxiii, fig. 2.

65. *Genus* PRIONASTREA.

Milne Edw. and J. Haime, Comptes rend., t. xxvii, p. 495, 1848.

Corallum forming a convex or gibbose mass, the under surface of which constitutes a common plate, covered with a thin, complete epitheca. Gemmation sub-marginal. *Calices* distinct, polygonal ; fossula deep ; margins united so as to form a simple crest between the different corallites. *Columella* spongy. *Septa* thin, crowded, delicately granulated on their sides, and strongly dentated at their apex ; the largest of these teeth are those nearest the columella. *Dissepiments* well developed. *Walls* in general independent towards the basis of the coral, but uniting to the adjacent ones near the calices, so that the visceral chambers appear to be separated only by a single simple lamina.

Typ. sp., *Prionastrea abdita*, Milne Edw. and J. Haime, loc. cit. ; *Astræa abdita*, Lamarck, Hist. des Anim. sans Vert., t. ii, p. 265, 1816 ; *Madrepora abdita*, Soland. and Ellis, t. 50, f. 2.

66. *Genus* SIDERASTREA.

(*In parte*) Blainville, Dict. des Sc. Nat., t. lx, p. 335, 1830 ; *Siderina*, Dana, Zooph., p. 218, 1846.

Corallum incrusting, forming a convex mass of a very dense tissue. Gemmation sub-marginal. *Corallites* united by their walls, which are thin, and sometimes indistinct. *Calices* sub-pentagonal, with a deep fossula, and their margins rendered thick by the pro-longation of the septa. *Columella* papillose, in general not much developed, but having a tendency to become compact. *Septa* very closely set, thin, and regularly denticulated ; their lateral surfaces covered with large granulations, which come in contact with those of the adjoining septa, but are not united to them. *Dissepiments* rudimentary.

Typ. sp., *Siderastrea galaxea*, Blainville, loc. cit. ; *Madrepora galaxea*, Ellis and Solander, Hist. of Zooph., tab. xlvii, fig. 7.

67. *Genus* BARYASTREA.

Milne Edw. and J. Haime, Comptes rend. de l'Acad. des Sc., t. xxvii, p. 495, 1848.

Corallum incrusting ; its tissue very dense and compact. Gemmation marginal or sub-marginal. *Corallites* very intimately united by their walls. *Calices* polygonal and

6

indistinctly separated by superficial, narrow grooves. *Columella* not much developed at its apex; but having a tendency to become compact, and to fill up the visceral chamber towards its basis. *Septa* very thick, closely set, scarcely granulated, and very feebly denticulated. *Dissepiments* little developed.

Typ. sp., *Baryastrea solida*, Milne Edw. and J. Haime, loc. cit.

68. *Genus* ACANTHASTREA.

Milne Edw. and J. Haime, Comptes rend., t. xxvii, p. 495, 1848.

Corallum forming a slightly convex mass, with its upper surface strongly echinulate, and its under surface constituting a plate, covered with a complete, thin epitheca. Gemmation sub-marginal or marginal. *Corallites* united by their walls, which are somewhat cellular. *Calices* sub-polygonal, with broad, spiniferous, simple, common margins. *Columella* rudimentary or septal. *Septa* exsert, strong, and armed with projecting spiniform teeth, the largest of which are situated near the walls, instead of being the central ones, as in the preceding genera. *Dissepiments* very numerous.

Typ. sp., *Acanthastrea spinosa*, Milne Edw. and J. Haime, Comptes rend., t. xxvii, p. 495.

69. *Genus* SYNASTREA.

Milne Edw. and J. Haime, Comptes rend., t. xxvii, p. 495, 1848.

Corallum pediculate, and increasing in breadth more than in height. Gemmation sub-marginal. *Corallites* intimately united by their walls. *Calices* superficial, distinct at their centre, but not so towards their circumference. *Columella* very small. *Septa* confluent, progressing from one calicular centre to another without interruption, exsert, and hiding the walls, over which they extend; their calicular margin almost horizontal, and armed with nearly equal teeth.

Typ. sp., *Synastrea Savignyi*, Milne Edw. and J. Haime, Ann. des Sc. Nat., 3ᵐᵉ série, vol. x, tab. ix, fig. 12.

70. *Genus* THAMNASTREA.

Thamnasteria (in parte), Le Sauvage, Mém. de la Soc. d'Hist. Nat. de Paris, vol. i, p. 241, 1822 ; *Thamnastrea*, ejusd., Ann. des Sc. Nat., 1ʳᵉ série, vol. xxvi, p. 328.

Corallum having confluent septa, and most of the other characters of Synastrea, but forming a fasciculus of columns or thick branches, erect, and of a more or less arborescent aspect.

Typ. sp., *Thamnastrea dendroidea*, Le Sauvage, Mém. de la Soc. d'Hist. Nat., vol. i, tab. xiv.

71. *Genus* GONIASTREA.

Milne Edw. and J. Haime, Comptes rend., t. xxvii, p. 495, 1848.

Corallum always increasing by successive fissiparity, and forming a convex or lobulated mass, of a dense structure. *Corallites* intimately united from top to bottom by their walls, which thus form simple partitions between the visceral cavities, and are thick and compact. *Calices* polygonal; fossula rather deep. *Columella* spongy. *Septa* slightly exsert, their apex arched and denticulated. Well-characterised, denticulated *pali*, corresponding to all the septa, except those of the last cyclum.

Typ. sp., *Goniastrea solida*, Milne Edw. and J. Haime, Ann., 3ᵐᵉ série, t. x, pl. ix, fig. 7; *Madrepora solida*, var. *b*, Forskal, Descr. Anim. in Itin. Orient., p. 131.

72. *Genus* APHRASTREA.

Milne Edw. and J. Haime, Comptes rend., t. xxvii, p. 495, 1848.

Corallum increasing by fissiparity, and forming a convex mass of a light cellular structure, presenting on its under surface a complete, common epitheca. *Calices* intimately united by their margins, which thus assume the appearance of simple partitions. *Columella* spongy. *Pali* or paliform lobes of the septa corresponding to all the cycla, except the last. *Septa* denticulated, slightly exsert. *Dissepiments* vesicular, and highly developed. *Walls* extremely thick, and completely vesicular.

Typ. sp., *Aphrastrea deformis*, Milne Edw. and J. Haime, Ann. des Sc. Nat., 3ᵐᵉ série, vol. x, tab. ix, fig. 11; *Astrea deformis*, Lamarck, Hist. des An. sans Vert., t. xi, p. 264.

73. *Genus* PARASTREA.

Milne Edw. and J. Haime, Comptes rend., t. xxvii, p. 495, 1848.

Corallum increasing by fissiparity, and having the same general form as in the preceding genus, but differing from it by the mode of union of the corallites, which takes place by means of the costæ and their dissepiments, so that the calices, instead of being separated only by a common simple margin, have each a distinct margin independent of those surrounding it. *Septa* exsert, and armed with teeth, the largest of which are placed near the centre of the calice, and often assume the appearance of pali. *Dissepiments* well developed.

Typ. sp., *Parastrea amicorum*, Milne Edw. and J. Haime, Ann., 3ᵐᵉ série, vol. x, pl. ix, fig. 9.

Section V.—ASTREINÆ REPTANTES.

Corallum increasing by the development of buds on stolons, or on membraniform basal expansions. The *corallites* not united by their sides, excepting accidentally by means of their walls, and remaining short. *Septa* feebly denticulated. *Dissepiments* almost rudimentary.

74. *Genus* ANGIA.

Milne Edw. and J. Haime, Comptes rend., t. xxvii, p. 496, 1848.

Corallum composed of short, cylindrical corallites, united by a common gemmiferous basal expansion, and completely free laterally. *Calices* sub-circular; fossula broad and deep. *Columella* papillose, well developed. *Septa* thin, not exsert; the principal ones having their upper margin almost entire, the others strongly dentated. *Walls* covered with a complete epitheca.

Typ. sp., *Angia rubeola*, Milne Edw. and J. Haime, Ann. des Sc. Nat., 3me série, vol. x, tab. vii, fig. 6 ; *Dendrophyllia rubeola*, Quoy and Gaimard, Astrolabe, Zooph., tab. xv, figs. 12-15.

75. *Genus* CRYPTANGIA.

Milne Edw. and J. Haime, Comptes rend., t. xxvii, p. 496, 1848.

Corallum composed of agglomerate, cylindro-turbinate corallites, which appear to multiply by gemmation on a non-persistent, soft stolon, so that they cease to be organically united when in the adult state, but remain imbedded in an extraneous mass composed of Cellepora. *Calices* circular, with a well-formed fossula. *Columella* papillose, well developed. *Septa* thin, not very closely set; the upper edge of all of them dentate. *Walls* covered with a complete epitheca.

Typ. sp., *Cryptangia Woodii*, Milne Edw. and J. Haime, loc. cit. ; *Cladocora cariosa*, Wood, Ann. of Nat. Hist., vol. xiii, p. 12.

76. *Genus* RHIZANGIA.

Milne Edw. and J. Haime, Comptes rend., t. xxvii, p. 496, 1848.

Corallum increasing by the gemmation of stolons, which sometimes become calicified, and are persistent. *Corallites* agglomerate, sub-cylindrical. *Calices* circular ; fossula shallow. *Columella* papillose, and not very distinct from the neighbouring denticulations of the septa. *Septa* thin, scarcely exsert, nearly equal, very closely set, with the upper edge slightly arched, and armed with small, regular teeth. *Walls* covered with a complete epitheca, which extends almost as high as the apex of the septa.

Typ. sp., *Rhizangia brevissima*, Milne Edw. and J. Haime, Ann. des Sc. Nat., 3me serie, vol. x, t. vii, figs. 7, 8 ; *Astrea brevissima*, Deshayes, in Ladoucette, Hist. des Hautes Alpes, tab. xiii, fig. 13.

77. *Genus* ASTRANGIA.

Milne Edw. and J. Haime, Comptes rend., t. xxvii, p. 496, 1848.

Corallum incrusting. *Corallites* very short, produced by gemmation on a thin, common, basal expansion, the surface of which is granulated. *Calices* circular, with a deep fossula. *Columella* papillose, sub-echinulate and not distinctly delimitated. *Septa* thin, exsert, nearly equal, granulate, and armed with teeth much resembling those of the columella ; the tertiary septa bent towards those of the second cycla, and united to them. *Dissepiments* in general simple and distant. *Walls* naked, with broad, delicately-granulated costæ.

Typ. sp., *Astrangia Michelinii*, Milne Edw. and J. Haime, Ann. des Sc. Nat., 3me série, vol. x, t. vii, f. 5.

78. *Genus* PHYLLANGIA.

Milne Edw. and J. Haime, Comptes rend., t. xxvii, p. 497, 1848.

Corallum differing from those of the preceding genus by the structure of the *septa*, the upper edge of which is almost entire in the principal cycla, and slightly denticulated in the others. *Columella* rudimentary.

Typ. sp., *Phyllangia americana*, nob.

79. *Genus* OULANGIA.

Milne Edw. and J. Haime, Comptes rend., t. xxvii, p. 497, 1848.

Corallum composed of very low, cylindrical corallites, which appear to arise by gemmation on a basal incrusting expansion, and having their walls naked and costate, as in the preceding genus, but with a highly-developed, papillose columella. *Septa* very exsert, closely set; those of the principal cycla having their upper edge almost entire.

Typ. sp., *Oulangia Stokesiana*, Milne Edw. and J. Haime, Ann. des Sc. Nat., 3ᵐᵉ série, vol. x, tab. vii, fig. 4.

Aberrant Group.
PSEUDASTREIDÆ.

Corallum composite, thin, and foliaceous, and increasing by extra-calicular gemmation. *Corallites* short, well circumscribed, and dispersed on the surface of a common lamellar plate. Cœnenchyma echinulate. *Septa* well developed, very echinulate. *Dissepiments* not numerous. No synapticulæ. Common basal wall imperforate, sub-costulate, and naked.

Genus ECHINOPORA.

Lamarck, Hist. des An. sans Vert., vol. ii, p. 252, 1816; *Echinastrea*, Blainville, Dict. des Sc. Nat., vol. lx, p. 343, 1830.

Corallum adherent, near the centre, and expanding into large foliaceous, lobated laminæ. *Calices* circular, with an exsert margin.

Typ. sp., *Echinopora rosularia*, Lamarck, loc. cit., p. 253; Milne Edw., Atlas du Règne Anim. de Cuvier, Zooph., tab. lxxxiii ter, fig. 1.

Transitional Group.
PSEUDOFUNGIDÆ.

Corallum composite and foliaceous, having a perforated plate or basal wall (as in Fungidæ) and interseptal dissepiments (as in Astreidæ). *Calices* forming radiating series, separated by lobes or ridges. No synapticulæ.

Genus MERULINA.

Ehrenberg, Corall. des Roth. Meeres, p. 104, 1834. Typ. sp., *Merulina ampliata*, Ehrenberg, loc. cit.

Family IV.
FUNGIDÆ.

Dana, Expl. Exped., Zooph., p. 283, 1846.

Corallum simple or composite, very short and expanding, so as to constitute a disc or foliaceous lamina. *Calice* very shallow, and open laterally in simple species;

confluent, and not circumscribed in the compound species. *Septa* not distinct from the costæ, and formed by complete, imperforate laminæ, with the edge dentate, and the sides covered with styliform or echinulate processes, which, in general, meet so as to constitute numerous *synapaticulæ*, or transverse props, extending across the loculi like the bars of a grate. No dissepiments or tabulæ, so that no part of the visceral chamber is completely closed. *Walls* basal, in general porous. The compound species increasing by sub-marginal gemmation, and not by fissiparity.

First Tribe—CYCLOLITINÆ.

Corallum simple. *Plate* or basal wall having a well-developed *epitheca*, presenting concentric folds.

1. *Genus* CYCLOLITES.

Lamarck, Syst. des Anim. sans Vert., p. 369, 1801.

Corallum circular, or nearly so, and covered with an immense number of very thin septa. Fossula oblong, narrow, and shallow. The small septa in general united to those of the older cycla.

Typ. sp., *Cyclolites elliptica*, Lamarck, loc. cit., p. 234.

2. *Genus* PALÆOCYCLUS.

Milne Edw. and J. Haime, Comptes rend. de l'Acad. des Sc., vol. xxix, p. 71, 1849.

Corallum circular. Fossula deep, very broad, and circular. *Septa* thick and not numerous ; none of them cemented together.

Typ. sp., *Palæocyclus porpita*, Milne Edw. and J. Haime, loc. cit.; *Madrepora porpita*, Fougt, Lin. Amœn. Acad., t. i, tab. iv, fig. 5.

Second Tribe—FUNGINÆ.

Corallum simple or composite. *Plate* or basal wall without an epitheca, in general strongly echinulate, and porous.

3. *Genus* FUNGIA.

(*In parte*) Lamarck, Syst. des An. sans Vert., p. 369, 1801 ; Dana, Zooph., p. 287, 1846 ; Milne Edw. and J. Haime, Comptes rend., t. xxix, p. 71, 1849.

Corallum simple, subdiscoidal. *Septa* very numerous, and united so as to appear ramified. Basal *wall* strongly echinulate, and perforated in an irregular manner.

Typ. sp., *Fungia patellaris*, Lamarck, loc. cit., p. 236 ; Milne Edw. and J. Haime, Ann. des Sc. Nat., t. ix, pl. vi, fig. 1.

4. *Genus* MICRABACIA.

Milne Edw. and J. Haime, Comptes rendus, vol. xxix, p. 71, 1849.

Corallum simple, lenticular, plano-convex. *Septa* not extremely numerous, straight. *Wall* scarcely echinulate, and perforated in a regular manner.

Typ. sp., *Micrabacia coronula*, nob.; *Fungia coronula*, Goldfuss, Petref. Germ., vol. i, tab. xiv, fig. 10.

5. *Genus* ANABACIA.

D'Orbigny MSS.; Milne Edw. and J. Haime, loc. cit., p. 71, 1849.

Corallum simple and lenticular. *Septa* extremely numerous, thin, and projecting on the under side of the corallum without forming a distinct basal wall or plate. Fossula shallow.

Typ. sp., *Anabacia orbulites*, nob.; *Fungia orbulites*, Lamouroux, Exp. Méth., tab. lxxxiii, figs. 1, 2, 3.

6. *Genus* GENABACIA.

Milne Edw. and J. Haime, Comptes rend., t. xxix, p. 71, 1849.

Corallum composite, formed by a parent corallite similar to Anabacia, bearing young calicula arranged circularly.

Typ. sp., *Genabacia stellata*, nob.; *Fungia stellata*, D'Archiac, Mém. Soc. Géol. France.

7. *Genus* HERPOLITHA.

Eschscholtz, Isis, 1825; *Haliglossa*, Ehrenberg, Corall., p. 50, 1834; *Herpetolithus*, Leuckart; Dana, Zooph.,
p. 306, 1846.

Corallum composite, free. *Calicula* sub-radiate, and of two sorts; the central ones multi-lamellate, and arranged in a line; the others pauci-lamellate, and dispersed irregularly. *Septa* strong, and alternately thick and thin. Under surface of the common basal wall very echinulate.

Typ. sp., *Herpolitha limacina*, nob.; *Madrepora pileus*, Ellis and Solander, op. cit., tab. xlv.

8. *Genus* CRYPTABACIA.

Milne Edw. and J. Haime, Comptes rend., t. xxix, p. 71, 1849.

Corallum composite, free and convex above. *Calices* distinctly radiate; the central ones arranged in a line, and more distinct than the others. *Septa* short, and not numerous. Under surface of the common basal walls strongly echinulated.

Typ. sp., *Cryptabatia talpa*, nob.; *Fungia talpa*, Lamarck, Hist. des An. sans Vert., t. ii, p. 237.

9. *Genus* HALOMITRA.

Dana, Zooph., p. 311, 1846.

Corallum composite, differing from the preceding genus by its very long and numerous septa.

Typ. sp., *Halomitra pileus*, Dana, loc. cit., p. 311 ; *Fungia pileus*, Lamarck, Hist. des An. sans Vert., t. ii, p. 237.

10. *Genus* PODOBACIA.

Milne Edw. and J. Haime, loc. cit., p. 71, 1849.

Corallum composite, cyathiform, and adherent by its basis.   *Calices* as in Halomitra.

Typ. sp., *Podobacia cyathoides*, nob.; *Agaricia cyathoides*, Valenciennes MSS., in the Gallery of the Paris Museum.

11. *Genus* LITHACTINIA.

Lesson, Illustr. Zool., 1833.

Corallum composite, free.   *Calices* of one sort only, and not radiate.   *Septa* short, and separated by very thin, transverse laminæ, which appear to be analogous to columellæ.

Typ. sp., *Lithactinia novæhyberniæ*, Lesson, loc. cit., vi, figs. 1, 2.

12. *Genus* POLYPHYLLIA.

Quoy and Gaimard, Voy. de l'Astrolabe, Zooph., p. 184, 1833.

Corallum composite, free, and having calices of two sorts ; the central ones sub-radiate, and arranged in a line.

Typ. sp., *Polyphyllia pelvis*, Quoy and Gaimard, loc. cit., pl. xx, figs. 8—10.

13. *Genus* ZOOPILUS.

Dana, Zooph., p. 318, 1846.

Corallum composite.   *Septa* of two sorts ; the large ones radiately prolonged quite to the margin ; the intermediate much smaller, and those only interrupted by the calicular fossulæ or oririms.

Typ. sp., *Zoopilus echinatus*, Dana, op. cit., p. 319.

Third Tribe—LOPHOSERINÆ.

Plate (or basal wall) not perforate nor echinulate.   No *epitheca*.

14. *Genus* CYCLOSERIS.

Milne Edw. and J. Haime, Comptes rend., t. **xxix**, p. 72, 1849.

Corallum simple, free, and discoidal. *Septa* very numerous, and united by their inner edge. *Wall* completely horizontal.

Typ. sp., *Cycloseris cyclolites,* nob.; *Fungia cyclolites,* Lamarck., Hist. des Anim. sans Vert., t. ii, p. 236.

15. *Genus* DIASERIS.

Milne Edw. and J. Haime, loc. cit., p. 72, 1849.

Corallum simple, free, and discoidal; when young, composed of a certain number of separate, radiating lobes, which, in the adult state, become cemented together. General structure as in Cycloseris.

Typ. sp., *Diaseris distorta,* nob.; *Fungia distorta,* Michelin, in Guerin's Mag. Zool., t. v, Zooph., pl. v, 1843.

16. *Genus* TROCHOSERIS.

Milne Edw. and J. Haime, loc. cit., p. 72, 1849.

Corallum simple, trochoidal, adherent. *Septa* very numerous, and strongly granulated.

Typ. sp., *Trochoseris distorta,* nob.; *Anthophyllum distortum,* Michelin, Icon. Zooph., pl. xliii, fig. 8.

17. *Genus* CYATHOSERIS.

Milne Edw. and J. Haime, loc. cit., p. 72, 1849.

Corallum composite, trochoid, adherent. *Calices* rather distinctly radiate. *Septa* long and thick. Common basal walls, sometimes forming folds, which rise up so as to constitute lobes or ridges on the upper surface of the corallum.

Typ. sp., *Cyathoseris infundibuliformis,* nob.; *Agaricia infundibuliformis,* Michelin, op. cit., tab. xliii, fig. 12.

18. *Genus* LOPHOSERIS.

Milne Edw. and J. Haime, loc. cit., p. 72, 1849; *Pavonia (ex parte),* Lamarck, op. cit., t. ii, p. 238, 1816.

Corallum composite, foliaceous, and adherent, rising in the form of irregular cristæ or of lobes, with confluent, radiate caliculas on each side. *Columella* tubercular.

Typ. sp., *Lophoseris boletiformis,* nob.; *Pavonia boletiformis,* Lamarck, loc. cit., p. 240.

19. *Genus* AGARICIA.

(*Pars*) Lamarck, Syst. des Anim. sans Vert., p. 375, 1801.

Corallum composite, foliaceous, and irregular. *Calices* arranged in concentric series, separated by unequal ridges. *Columella* tubercular.

Typ. sp., *Agaricia undata,* Lamarck, loc. cit.; *Madrepora undata,* Solander and Ellis, Zooph., tab. xl.

20. *Genus* PACHYSERIS.

Milne Edw. and J. Haime, loc. cit., p. 72, 1849.

Corallum similar to Agaricia, excepting that the corallites belonging to the same trench are completely blended together. *Columella* well developed and dense.

Typ. sp., *Pachyseris rugosa*, nob. ; *Agaricia rugosa*, Lamarck, Hist. des An. sans Vert., t. ii, p. 243.

21. *Genus* PHYLLASTREA.

Helioseris, Dana, Zooph., p. 269 ; Milne Edw. and J. Haime, loc. cit., p. 72, 1849.

Corallum composite, composed of frondiform expansions. *Calices* circumscribed, sub-mammillate, and arranged around the parent corallite, which remains larger than the others. *Columella* tubercular.

Typ. sp., *Phyllastrea tubifex*, Dana, loc. cit., tab. xvi, fig. 4.

22. *Genus* HALOSERIS.

Milne Edw. and J. Haime, loc. cit., p. 72, 1849.

Corallum composite, forming foliaceous, crispate, lobulate expansions, the upper surface of which is covered with very long radii, and shows only obsolete calices. *Columella* rudimentary.

Typ. sp., *Haloseris lactuca*, Milne Edw. and J. Haime, loc. cit.

23. *Genus* LEPTOSERIS.

Milne Edw. and J. Haime, loc. cit., p. 72, 1849.

Corallum composite and adherent; the basal walls rising so as to constitute a sub-crateriform disc, in the centre of which is situated a large parent corallite, surrounded by smaller ones. *Calices* very imperfectly circumscribed, but well radiated. *Septa* very long. *Columella* rudimentary.

Typ. sp., *Leptoseris fragilis*, Milne Edw. and J. Haime, loc. cit.

Sub-order 2.

ZOANTHARIA PERFORATA.

Corallum composed essentially of porous sclerenchyma ; with the septal apparatus well characterised; and consisting of six primitive elements, but being sometimes represented only by series of trabiculæ. *Dissepiments* rudimentary ; no tabulæ.

The principal character of this section of Zoantharia is furnished by the structure of the sclerenchyma, which, instead of forming imperforated lamellæ as in the preceding groups, is always porous, or even reticulate. In general the mural apparatus constitutes here the greatest part of the corallum, and does not consist of costal laminæ; the walls are always perforated, and completely or nearly completely naked. It is also to be remarked, that the visceral chamber is almost completely open from top to bottom, and never filled up with dissepiments or synapticulæ, as in most of the *Zoantharia aporosa*, or with tabulæ, as will be seen in the next two sections of this order.

The perforated Zoantharia form three natural families: Eupsammidæ, Madreporidæ, and Poritidæ.

Family V.
EUPSAMMIDÆ.

Milne Edw. and J. Haime, Ann. des Sc. Nat., 3ᵐᵉ série, vol. x, p. 65, 1848.

Corallum simple or complex, with well-developed lamellar septa, a spongiose columella, and perforated, granular, subcostulated walls.

The septa are always numerous, and those of the last cyclum are never situated in the direction of a line drawn from the centre of the calice to its circumference, but are bent towards those of the penultimate cyclum, so as to produce the appearance of a six- or twelve-branched star. The interseptal loculi are completely open from top to bottom, or divided only by a few incomplete trabiculæ. The walls have a granulate vermiculate surface, and become often very thick in advanced age, but never constitute a loose spongy mass, as in Madreporidæ and Poritidæ, or a compact cœnenchyma, as in Oculinidæ.

The star-like arrangement of the septa, which is visible in transverse sections of these corallums, as well as in the calice, is not met with in any other family. The principal septa are sometimes imperforate, but those of the succeeding cycla are more or less porous. It is also to be noted that there are never any pali, and that the costæ are always rudimentary; sometimes there is a rudimentary epitheca.

1. *Genus* EUPSAMMIA.

Milne Edw. and J. Haime, Ann. Sc. Nat., 3ᵐᵉ série, vol. x, p. 77, 1848.

Corallum simple, subturbinate, free, and not presenting any lateral mural expansions. *Calice* oval and rather deep. *Septa* broad, slightly exsert, granulate, closely set, and forming four or five cycla. *Costæ* simple, distinct from the basis of the corallum, nearly equal, slightly vermiculate, and composed of a series of distinct, projecting granulæ.

Typ. sp., *Eupsammia trochiformis*, Milne Edw. and J. Haime, loc. cit., tab. i, fig. 3; *Madrepora trochiformis*, Pallas; *Turbinolia elliptica*, Brongniart.

2. *Genus* ENDOPACHYS.

(*Pars*) Lonsdale, Journ. of the Geol. Soc. of London, vol. i, p. 214, 1845.

Corallum simple, free, and organized as in the preceding genus, but much compressed towards its basis, which is carinate, and continued laterally into two vertical lobiform or cristate expansions. *Calice* arched; fossula long and narrow.

Typ. sp., *Endopachys Maclurii*, Milne Edw. and J. Haime, Ann. Sc. Nat., vol. x, tab. i, fig. 1 ; *Turbinolia Maclurii*, Lea, Contrib. to Geol., tab. vi, fig. 206.

3. *Genus* BALANOPHYLLIA.

Searles Wood, Ann. and Mag. of Nat. Hist., vol. xiii, p. 11, 1844.

Corallum simple and adherent, sub-pediculate, or sub-cylindrical, with a very broad basis. *Columella* well developed, but not projecting at the bottom of the fossula. *Septa* thin, and closely set; those of the last cyclum well developed, and complete in number. *Costæ* narrow, crowded, and nearly equal; no mural expansions.

Typ. sp., *Balanophyllia caliculus*, Searles Wood, loc. cit.

4. *Genus* HETEROPSAMMIA.

Milne Edw. and J. Haime, Ann. Sc. Nat., 3ᵐᵉ série, vol. x, p. 89, 1848.

Corallum simple, adherent, and growing by its basis so as to cover completely the shell on which it is fixed, and to assume the appearance of being free. *Calice* smaller than the basal part of the corallum. *Columella* well developed. *Septa* thick, slightly exsert, and closely set. *Walls* not having distinct costæ, but presenting small striæ or small papillæ, composed of minute granulæ, and arranged in an irregular manner.

Typ. sp., *Heteropsammia Michelinii*, Milne Edw. and J. Haime, loc. cit., p. 89.

5. *Genus* LEPTOPSAMMIA.

Milne Edw. and J. Haime, loc. cit., p. 90, 1848.

Corallum simple, adherent. *Calice* elliptical. *Columella* much developed, and projecting at the bottom of the fossula. *Septa* neither exsert nor crowded, very thin, and presenting scarcely any granulations; those of the fifth order rudimentary. *Walls* thin and translucid. *Costæ* distinct from the basis, and formed by series of small granulæ.

Typ. sp., *Leptopsammia Stokesiana*, Milne Edw. and J. Haime, loc. cit., tab. i, fig. 4.

6. *Genus* ENDOPSAMMIA.

Milne Edw. and J. Haime, loc. cit., p. 91, 1848.

Corallum simple, erect, and adherent. *Calice* circular. *Columella* much developed, but not projecting. *Septa* thick, strongly granulated, and slighty exsert, forming four cycla, the last of which is almost rudimentary. *Walls* covered with an indistinct pellicular epitheca, and having broad, straight costæ.

Typ. sp., *Endopsammia Philippensis*, Milne Edw. and J. Haime, loc. cit., tab. i, fig. 5.

7. *Genus* STEPHANOPHYLLIA.

Michelin, Dict. des Sc. Nat., Suppl., vol. i, p. 484, 1841.

Corallum simple, free, and presenting no trace of adherence. *Wall* discoidal, horizontal. *Calice* circular and open. *Septa* tall, thin, crowded, not projecting laterally beyond the edge of the mural disc, covered with conical granulations on each side, and all, excepting those of the first cyclum, united by the inner edge. *Costæ* delicate, straight, composed of simple series of obscure granulations, and radiating regularly from the centre of the mural disc to its circumference. No epitheca.

Typ. sp., *Stephanophyllia elegans*, Michelin, Icon. Zooph., pl. viii, fig. 2. Milne Edw. and J. Haime, loc. cit., tab. i, fig. 10.

8. *Genus* DENDROPHYLLIA.

Blainville, Dict. des Sc. Nat., vol. lx, p. 319, 1830.

Corallum composite, and in general arborescent. Corallites cylindrical, or cylindrico-turbinate, and formed by lateral gemmation. *Calices* circular, or nearly so; fossula deep. *Columella* well developed, and in general projecting much at the bottom of the fossula. *Septa* not exsert, thin, and closely set; those of the fourth cyclum well developed. *Walls* becoming very thick, and presenting narrow vermiculate *costæ*, formed by series of granulæ.

Typ. sp., *Dendrophyllia ramea*, Blainville, loc. cit.; Milne Edw., Atlas du Règne Animal de Cuvier, Zooph., pl. lxxxiii, fig. 1.

9. *Genus* LOBOPSAMMIA.

Milne Edw. and J. Haime, Ann. des Sc. Nat., 3ᵐᵉ série, vol. x, p. 105, 1848.

Corallum composite, arborescent, increasing by successive fissiparity. *Calices* irregular in form. *Septa* forming four complete and well-developed cycla. In other respects similar to Dendrophyllia.

Typ. sp., *Lobopsammia cariosa*, Milne Edw. and J. Haime, loc. cit.; *Lithodendron cariosum*, Goldfuss, Petref. Germ., vol. i, tab. xiii, fig. 7.

10. *Genus* CŒNOPSAMMIA.

Milne Edw. and J. Haime, loc. cit., p. 106, 1848; *Tubastrea*, Lesson, Voyage aux Indes orient. par Belanger, 1834.

Corallum composite, dendroid, or sub-globose, increasing by lateral or sub-basal gemmation. *Corallites* cylindrical. *Calices* circular, or nearly so. *Columella* tubercular, well developed. *Septa* not exsert, distant, and forming three cycla; those of the fifth order rudimentary. *Costæ* narrow, sub-vermiculate towards the bases, simple, and formed of a series of granulæ near the calice.

Typ. sp., *Cænopsammia coccinea*, Milne Edw. and J. Haime, loc. cit., p. 107; *Tubastrea coccinea*, Lesson, op. cit., Zooph., tab. i; *Astrea calicularis*, Blainville, Manuel d'Actinol., tab. liv, fig. 2.

11. *Genus* STEREOPSAMMIA.

Corallum presenting most of the characters of Cænopsammia, but not having any *Columella*.

Typ. sp., *Stereopsammia humilis*, nob., tab. v, fig. 4.

Family VI.
MADREPORIDÆ.

Corallum composite, increasing by gemmation and not by fissiparity. *Cœnenchyma* abundant, spongy, and reticulate. *Walls* very porous, and not distinct from the cœnenchyma. *Septa* lamellose, and well developed; loculi free.

First Tribe—MADREPORINÆ.

Visceral chambers divided into two equal parts by two of the principal septa, which are more developed than the others, and meet by their inner edge.

1. *Genus* MADREPORA.

Lamarck, Hist. des Anim. sans Vert., t. xi, p. 277, 1816.

Corallum composite, forming ramified, lobate, or fasciculate masses. *Cœnenchyma* loose, and delicately echinulate. *Calices* projecting, with a thick margin. No *columella*.

Typ. sp., *Madrepora muricata*, Ellis and Solander, Zooph., tab. lvii; *Madrepora abrotanoides*, Lamarck, loc. cit., p. 280.

Second Tribe—EXPLANARINÆ.

Visceral chamber presenting at least six equally developed principal septa.

2. *Genus* EXPLANARIA.

(*Pars*) Lamarck, Hist. des Anim. sans Vert., vol. ii, p. 254, 1816; *Gemmipora*, Blainville,
Dict. des Sc. Nat., vol. lx, p. 352, 1830.

Corallum in general foliaceous. *Cœnenchyma* abundant, rather dense, and delicately echinulate. *Septa* almost all of the same size. *Columella* spongy.

Typ. sp., *Explanaria crater*, nob.; *Madrepora crater*, Pallas, Eleuch. Zooph., p. 332.

3. *Genus* ASTREOPORA.

Blainville, Dict. des Sc. Nat., vol. lx, p. 348, 1830.

Corallum massive. *Cœnenchyma* of a loose texture, and strongly echinulated. *Septa* unequally developed. No *columella*.

Typ. sp., *Astreopora myriophthalma*, Blainville, loc. cit.; *Astrea myriophthalma*, Lamarck, op. cit., p. 260.

Family VII.
PORITIDÆ.

Corallum entirely composed of reticulate sclerenchyma. Septal apparatus well developed, but never lamellar, and composed only of series of styliform processes or trabiculæ, constituting by their junction a sort of irregular trellis. *Walls* presenting the same structure, and not distinct from the cœnenchyma. Visceral chamber containing some small dissepiments, but never divided by tabulæ.

First Tribe—PORITINÆ.

Cœnenchyma rudimentary, or not existing.

1. *Genus* PORITES.

(*Pars*) Lamarck, Hist. des An. sans Vert., t. ii, p. 267, 1816 ; Milne Edw. and J. Haime, Comptes rend., t. xxix, p. 258, 1849.

Corallum composed of sclerenchyma, very irregularly reticulated. *Calices* shallow. *Septa* not numerous, rudimentary, and appearing to be represented by a circle of *pali*, the apex of which is papillose.

Typ. sp., *Porites conglomerata*, Lamarck, loc. cit., p. 269.

2. *Genus* LITHARÆA.

Milne Edw. and J. Haime, Comptes rend., t. xxix, p. 258, 1849.

Sclerenchyma very irregularly reticulated. *Calices* not very deep. *Columella* spongy. *Pali* rudimentary, or not existing. *Septa* well developed, particularly towards the wall.

Typ. sp., *Litharæa Websteri*, nob.; *Astrea Websteri*, Bowerbank, Mag. of Nat. Hist., new series, vol. iv, p. 27, figs. A, B, 1840.

3. *Genus* COSCINARÆA.

Milne Edw. and J. Haime, Comptes rend. de l'Ac. des Sc., t. xxvii, p. 496, 1848.

Corallum of a dense structure. *Calices* rather deep ; neither pali, nor distinct walls. *Septa* crowded, very regularly fenestrate, and with crispate edges, passing without any interruption from one visceral chamber to the adjacent one. No *epitheca*.

Typ. sp., *Coscinaræa Bottæ*, Milne Edw. and J. Haime, Ann. des Sc. Nat., 3ᵐᵉ série, vol. ix, tab. v, fig. 2.

4. *Genus* MICROSOLENA.

Lamouroux, Exp. méth., p. 65, 1821.

Corallum differing from the preceding genus by the structure of the septa, the perforations of which are much larger than in Coscinaræa, and by the existence of a strong, common epitheca.

Typ. sp., *Microsolena porosa*, Lamouroux, op. cit., tab. lxxiv, fig. 24.

5. *Genus* GONIOPORA.

Quoy and Gaimard, Voy. de l'Astr., Zooph., p. 218, 1833 ; *Goniopora* and *Porastrea*, Milne Edw. and J. Haime, Comptes rend. de l'Acad. des Sc., t. xxvii, p. 496, 1848.

Corallum having distinct, elevated walls, of a fenestrate structure. *Calices* deep. *Columella* spongy. *Septa* well developed, and fenestrate. No *pali*.

Typ. sp., *Goniopora pedunculata*, Quoy and Gaimard, Voyage de l'Astrolabe, Zooph., tab. xvi, figs. 9-11.

6. *Genus.* RHODARÆA.

Milne Edw. and J. Haime, Comptes rend., t. xxix, p. 259, 1849.

Corallum with thick walls, rather high. *Septa* rudimentary. *Pali* greatly developed, and forming a rosette in the centre of the calice.

Typ. sp., *Rhodaræa calicularis*, nob.; *Astrea calicularis*, Lamarck, Hist. des An. sans Vert., t. ii, p. 266.

7. *Genus* PORARÆA.

Milne Edw. and J. Haime, Comptes rend., t. xxix, p. 259, 1849.

Walls thin, and widely fenestrated. *Septa* formed by a series of spiniform processes, which sometimes ramify towards the centre of the visceral cavity, so as to constitute a sort of spurious columella.

Typ. sp., *Poraræa fenestrata*, nob.; *Pocillopora fenestrata*, Lamarck, Hist. des An. sans Vert., t. ii, p. 275 ; Milne Edw. and J. Haime, Ann. des Sc. Nat., 3ᵐᵉ série, vol. ix, fig. 1.

8. *Genus* HOLARÆA.

Milne Edw. and J. Haime, Comptes rend., t. xxix, p. 259, 1849.

Calices with distinct polygonal margins, rather deep. Septal apparatus composed of irregular trabiculæ, completely blended with the walls, and constituting thus a delicate spongy mass. *Columella* fasciculate and short.

Typ. sp., *Holaræa Parisiensis*, nob. ; *Alveolites Parisiensis*, Michelin, Icon. Zooph., pl. xlv, fig. 10.

Second Tribe—MONTIPORINÆ.

Cœnenchyma abundant and spongy.

9. *Genus* ALVEOPORA.

Quoy and Gaimard, Voyage de l'Astrolabe, Zooph., p. 240, 1833.

Corallum arborescent. *Cœnenchyma* very porous and echinulate, but not bearing large excrescences. Margins of the *calices* scarcely distinct. *Septa* not numerous, and formed by series of spiniform processes. No *columella*.

Typ. sp., *Alveopora rubra*, Quoy and Gaim., loc. cit., Zooph., tab. xix, figs. 11-14.

10. *Genus* MONTIPORA.

Quoy and Gaimard, op. cit., p. 247, 1833; *Manopora*, Dana, Zooph., p. 489, 1846.

Corallum of various forms, differing from Alveopora by the existence of large projections of the cœnenchyma between the caliculas. *Cœnenchyma* much more abundant, and more delicately spongy.

Typ. sp., *Montipora verrucosa*, Quoy and Gaim., op. cit., Zooph., pl. xx, fig. 11.

11. *Genus* PSAMMOCORA.

Dana, Zooph., p. 344, 1846.

Cœnenchyma somewhat compact, of a fasciculate structure, and having its surface papillose. *Calices* very shallow, confluent, and without distinct walls. *Septa* thick, and formed by strong spiniform processes.

Typ. sp., *Psammocora obtusata*, Dana, loc. cit., p. 345; *Pavonia obtusangula*, Lamarck, Hist. des An. sans Vert., t. ii, p. 240.

Sub-order 3.

ZOANTHARIA TABULATA.

Corallum essentially composed of a well-developed mural system, and having the visceral chambers divided into a series of stories by complete transverse tabulæ or diaphragms. Septal apparatus rudimentary.

The principal character of this sub-order is founded on the existence of the lamellar diaphragms that close the visceral chamber of the corallites at different heights, and differ from the dissepiments of the Astreidæ by not being dependent on the septa, and forming as many complete horizontal divisions extending from side to side of the general cavity,

8

instead of occupying only the one or two loculi. It is also to be remembered that the septal apparatus, although more or less rudimentary, has the same general mode of arrangement as in the preceding sub-orders, and never presents the crucial character which we shall find in *Zoantharia rugosa*.

This section comprises four families: Favositidæ, Milleporidæ, Seriatoporidæ, and Thecidæ.

Family VIII.
MILLEPORIDÆ.

Corallum principally composed of a very abundant cœnenchyma, distinct from the walls of the corallites, and of a tubular or cellular structure. *Septa* not numerous; tabulæ numerous, and well formed.

1. *Genus* MILLEPORA.

(*Pars*) Lamarck, Syst. des An. sans Vert., p. 373, 1801; *Palmipora*, Blainville, Dict. des Sc. Nat., t. lx, p. 356, 1830.

Corallum of various forms, but more or less foliaceous. *Cœnenchyma* extremely abundant, of an irregular subtubular structure. *Calices* of very different dimensions in the same corallum. No distinct septa. *Tabulæ* horizontal.

Typ. sp., *Millepora alcicornis*, Lamarck, loc. cit.; Milne Edw., Atlas du Règne Anim. de Cuvier, Zooph., tab. lxxxix, fig. 1.

2. *Genus* HELIOPORA.

(*Pars*) Blainville, Dict. des Sc. Nat., vol. lx,.p. 357, 1830; Dana, Zooph., p. 539, 1846.

Corallum lobulate, somewhat massive, and differing from Millepora by the regular tubular structure of the cœnenchyma, and the existence of small but distinct septa.

Typ. sp., *Heliopora cærulea*, Blainville, loc. cit., p. 357.

3. *Genus* HELIOLITES.

Dana, Zooph., p. 541, 1846; *Palæopora*, M'Coy, Ann. and Mag. of Nat. Hist., 2d series, vol. iii, p. 129, 1849; *Geoporites*, D'Orbigny, Prodr. de Palæont. stratif. Univers., t. i, p. 49, 1849.

Corallum sub-globose. *Cœnenchyma* regularly tubular. Septal radii advancing almost to the centre of the visceral chamber on the upper surface of the tabulæ, which are horizontal.

Typ. sp., *Heliolites pyriformis*, Dana, loc. cit., p. 542; *Heliolite pyriforme*, etc., Guettard, Mém. sur les Sc. et les Arts, vol. iii, pl. xxii, figs. 13, 14.

4. *Genus* FISTULIPORA.

M'Coy, Ann. and Mag. of Nat. Hist., 2d series, vol. iii, p. 130, 1849.

Corallum with vesicular cœnenchyma; thick walls and infundibuliform tabulæ.

Typ. sp., *Fistulipora minor*, M'Coy, loc. cit., figs. *a, b*.

5. *Genus* PLASMOPORA.

Milne Edw. and J. Haime, Comptes rend., t. xxix, p. 262, 1849.

Corallum free, sub-hemispheric, with a basal plate covered with an epitheca presenting concentric folds. *Calices* immersed. *Septa* rudimentary. *Tabulæ* horizontal. *Walls* thin. *Cœnenchyma* composed of large, vertical, radiate laminæ, united by smaller horizontal plates, and resembling much the costal cœnenchyma of the Astreidæ.

Typ. sp., *Plasmopora petaliformis*, nob.; *Porites petaliformis*, Lonsdale, in Murchison, Sil. Syst., pl. xvi, fig. 4.

6. *Genus* PROPORA.

Milne Edw. and J. Haime, Comptes rend., t. xxix, p. 262, 1849.

Corallum differing from the preceding genus by the calices having exsert margins; the septa being more developed, and extending outwards so as to constitute small costæ.

Typ. sp., *Propora tubulata*, nob.; *Porites tubulata*, Lonsdale, Sil. Syst., pl. xvi, figs. 3, 3*ᵃ*, 3*ᵇ* (cæteris exclusis).

7. *Genus* AXOPORA.

Corallum composite, incrusting, and forming thin expansions, which are often superposed. *Cœnenchyma* abundant, and forming irregular ridges between the calices, which are small and deep. *Septa* rudimentary. *Columella* well developed, fasciculate, and expanding at its passage through each of the tabulæ.

Typ. sp., *Axopora pyriformis*, nob.; *Geodia pyriformis*, Michelin, Icon., tab. xlvi, fig. 2.

8. *Genus* LOBOPORA.

Corallum having the same structure as in the preceding genus, but forming large, thick, foliaceous expansions, the two surfaces of which are covered with calices.

Typ. sp., *Lobopora Solanderi*, nob.; *Palmipora Solanderi*, Michelin, op. cit., tab. xlv, fig. 9.

Family IX.
FAVOSITIDÆ.

Corallum essentially formed by lamellar walls, with little or no cœnenchyma.   Visceral chambers divided by numerous and well-developed complete tabulæ.

First Tribe—FAVOSITIDÆ.

Corallum massive.   *Walls* perforated.   *Septa* rudimentary.   No *cœnenchyma*.

1. *Genus* FAVOSITES.

Lamarck, Hist. des An. sans Vert., vol. ii, p. 204, 1816; *Calamopora*, Goldfuss, Petref. Germ., vol. i, p. 77, 1826-30.

Corallum composed of basaltiform corallites, and having a basal plate covered with an epitheca, and no radiciform appendices.   *Calices* at right angle with the axis of the corallite, and in general hexagonal.   *Walls* perforated in a very regular manner.   *Tabulæ* horizontal, and very regularly superposed.   No *cœnenchyma*.

Typ. sp., *Favosites Gothlandica*, Lamarck, loc. cit., p. 206.

2. *Genus* MICHELINIA.

De Koninck, Descr. des Anim. foss. des Terr. houilliers de la Belgique, p. 30, 1842-44.

Corallum having a basal plate with radiciform prolongations.   *Tabulæ* very irregular, and subvesicular.   The other characters as in Favosites.

Typ. sp., *Michelinia tenuisepta*, De Koninck, loc. cit., pl. c, fig. 3 *a, b.*

3. *Genus* KONINCKIA.

Milne Edw. and J. Haime, Comptes rend., t. xxix, p. 260, 1849.

Corallum resembling Favosites, but having the walls larger and less regular, and the septa constituted by series of distinct and spiniform processes, interrupted at certain distances by the tabulæ, which are horizontal.

Typ. sp., *Koninckia fragilis*, Milne Edw. and J. Haime, loc. cit.

4. *Genus* ALVEOLITES.

(*Pars*) Lamarck, Syst. des An. sans Vert., p. 375, 1801 ; Steininger, Mém. Soc. Géol. France, vol. i.

Corallum composed of superposed strata of corallites very similar to those of Favosites, but much shorter, and terminated by an oblique semicircular or subtriangular calice, the edge of which projects on one side.

Typ. sp., *Alveolites spongites*, Steininger, Mém. de la Soc. Géol. de France, vol. i; *Calamopora spongites*, Goldfuss, Petref. Germ., pl. xxviii, figs. 1ᵃ, 1ᵇ, 1ᶜ.

Second Tribe—CHÆTETINÆ.

Corallum massive. *Walls* not perforated. Neither septa nor cœnenchyma.

5. *Genus* CHÆTETES.

Fischer, Oryct. du Gouv. de Moscou, p. 159, 1837.

Corallum glomerate. *Corallites* very long, basaltiform, and in general more or less bent. *Calices* polygonal. *Tabulæ* independent, not connected in the adjoining corallites, nor placed on the same level throughout the corallum.

Typ. sp., *Chætetes radians*, Fischer, loc. cit., pl. xxxvi, fig. 6.

6. *Genus* DANIA.

Milne Edw. and J. Haime, Comptes rend., t. xxix, p. 261, 1849.

Corallum having most of the characters of Chætetes, but with the tabulæ connected through the different corallites so as to constitute a series of common plates, and to divide the whole mass into a great number of parallel strata.

Typ. sp., *Dania Huronica*, Milne Edw. and J. Haime, loc. cit.

7. *Genus* STENOPORA.

(*Pars*) Lonsdale, Geol. of Russia and Ural Mount., vol. i, p. 631, 1845.

Corallum very similar to Chætetes, but having small styliform processes at the angles of the calices.

Typ. sp., *Stenopora spinigera*, Lonsdale, loc. cit., pl. A, fig. 2.

8. *Genus* CONSTELLARIA.

Dana, Zooph., p. 537, 1846.

Third Tribe—HALYSITINÆ.

Corallum composed of corallites constituting vertical laminæ or fasciculi, but more or less free laterally, and united by means of connecting tubes or mural expansions. *Walls* well developed, and not porous. *Septa* distinct, but small.

8. *Genus* HALYSITES.

Fischer, Zoognosia, 3d edit., vol. i, p. 387, 1813; *Catenipora*, Lamarck, Hist. des An. sans Vert., t. ii, p. 206, 1816.

Corallites extremely long, arranged in a single series, and united laterally, so as to constitute large flabelliform expansions, which remain free laterally, but often meet, and thus form a lacunous mass. *Epitheca* very thick. *Septa* almost rudimentary, but very distinct in perfect specimens. *Tabulæ* horizontal.

Typ. sp., *Halysites escharoides*, Fischer; *Catenipora escharoides*, Lamarck, loc. cit., p. 207.

9. *Genus* HARMODITES.

Fischer, Notice sur les Tubipores fossiles, 1828; *Syringopora*, Goldfuss, Petref. Germ., vol. i, p. 75, 1826-33.

Corallum fasciculate. Corallites irregularly cylindrical, very long, and united by horizontal connecting tubes. *Tabulæ* infundibuliform.

Typ. sp., *Harmodites ramulosa*, nob.; *Syringopora ramulosa*, Goldfuss, loc. cit., pl. xxv, fig. 7.

10. *Genus* THECOSTEGITES.

Milne Edw. and J. Haime, Comptes rend., t. xxix, p. 261, 1849.

Corallites cylindrical, short, and united by strong mural expansions situated at various heights. *Tabulæ* horizontal.

Typ. sp., *Thecostegites Bouchardi*, nob.; *Harmodites Bouchardi*, Michelin, Icon. Zooph., pl. xlviii, fig. 3.

Fourth Tribe—POCILLOPORINÆ.

Corallum massive, gibbous, or subdendroid, with thick imperforated walls, forming, towards the surface, an abundant compact cœnenchyma. *Septa* quite rudimentary.

11. *Genus* POCILLOPORA.

(*Pars*) Lamarck, Hist. des An. sans Vert., t. ii, p. 273, 1816; Dana, Zooph., p. 523, 1846.

Calices shallow, and presenting, at their bottom, a transverse, thick, projecting ring, resembling a columella.

Typ. sp., *Pocillopora acuta*, Lamarck, loc. cit., p. 274; Milne Edw., Atlas du Règne Animal de Cuvier, Zooph., pl. lxxxi, fig. 3.

Family X.
SERIATOPORIDÆ.

Corallum arborescent or bushy, with an abundant compact cœnenchyma. Visceral chambers filling up by the growth of the columella and the walls, and showing but few traces of tabulæ.

1. *Genus* SERIATOPORA.

Lamarck, Hist. des An. sans Vert., vol. ii, p. 282, 1816.

Corallum arborescent, with echinulated branches. *Calices* arranged in ascending series. *Septa* scarcely visible. *Columella* large and compact.

Typ. sp., *Seriatopora subulata*, Lamarck, loc. cit., p. 282.

2. *Genus* DENDROPORA.

Michelin, Icon. Zooph., p. 187, 1845.

Corallum arborescent, with very delicate smooth branches. *Calices* distant, and surrounded by a narrow, obtuse margin. *Septa* small, but distinct.

Typ. sp., *Dendropora explicita*, Michelin, op. cit., pl. xlviii, fig. 6.

3. *Genus* RHABDOPORA.

Milne Edw. and J. Haime, Comptes rend., t. xxix, p. 262, 1849.

Corallum with prismatic echinulate branches. *Calices* arranged in series. *Septa* very distinct.

Typ. sp., *Rhabdopora megastoma ; Dendropora megastoma*, M'Coy, Ann. and Mag. of Nat. Hist., 2d series, vol. iii, p. 129.

Family XI.
THECIDÆ.

Corallum massive, with an abundant, compact, spurious cœnenchyma, produced by the septa becoming cemented together laterally. *Tabulæ* numerous.

Genus THECIA.

Milne Edw. and J. Haime, Comptes rend., t. xxix, p. 263, 1849.

Septal system highly developed. *Calices* shallow, with a very small deep fossula.

Typ. sp., *Thecia Swinderniana ; Agaricia Swinderniana*, Goldfuss, Petref. Germ., pl. xxxviii, fig. 3 ; *Porites expatiata*, Lonsdale, ap. Murchison, Sil. Syst., p. 678, tab. xv, fig. 3.

Sub-order 4.

ZOANTHARIA RUGOSA.

Corallum simple or composite, with a septal apparatus never forming six distinct systems, as in all the preceding Zoantharia, but appearing to be derived from four primary elements. Sometimes this disposition is rendered manifest by the existence of four well-characterised primary septa, or of an equal number of depressions occupying the bottom of the calice, and assuming a crucial appearance: in other cases only one of these primary septa or excavations is well developed so as to interrupt the radiate form of the system; and in others, again, no trace of septal groups can be discovered, and the whole apparatus is represented by numerous equally developed radiate striæ rising on the surface of the tabulæ, and extending up the inner side of the walls. The corallites are always perfectly distinct, and are never united by means of a cœnenchyma; nor do they ever form linear series, which is often the case in the preceding sections. They multiply by gemmation, and the reproductive buds are in general developed on the surface of the calices of the parents: this often arrests the growth of the latter, and gives rise to a superposition of generations. It is also to be noted that the septa, although in general very incomplete, are never porous, and never bear synapticulæ, but that the visceral chamber is in general filled up from the bottom by a series of transverse tabulæ, or by a vesicular structure, which often constitutes the principal part of the corallum.

Family XII.

STAURIDÆ.

Corallum with well-developed septa, extending without any interruption from the bottom to the top of the visceral chamber, united by lamellar dissepiments, and arranged in four systems, characterised by an equal number of large primary septa.

1. *Genus* STAURIA.

Corallum composite, massive, astreiform, and increasing by calicular gemmation. *Corallites* united by their walls, or free in part, and not presenting any *costæ*. *Septa* large, and with undivided edges, united along the axis of the visceral chamber. No *columella*.

Typ. sp., *Stauria astreiformis*, nob.

2. *Genus* HOLOCYSTIS.

Lonsdale, in the Quarterly Journal of the Geol. Soc. of London, vol. v, part i, p. 83, 1849.

Corallum composite, massive, astreiform, and increasing by extra-calicular gemmation. *Corallites* united by means of well-developed *costæ*. *Columella* styliform.

Typ. sp., *Holocystis elegans; Cyathophora elegans*, Lonsdale, loc. cit., tab. iv, figs. 12, 13, 14, 15.

Family XIII.

CYATHAXONIDÆ.

Corallum with well-developed, complete *septa*, which extend without interruption from the bottom to the top of the visceral chamber, and not forming a regular radiate circle; those of the primary cyclum not much larger than the others, and not forming a four-branched cross, as in the Stauridæ; one well-characterised septal fossula. No dissepiments nor tabulæ.

Genus CYATHAXONIA.

Michelin, Icon. Zooph., p. 258, 1846.

Corallum simple. *Calice* deep. *Columella* styliform, strong, and very prominent. *Septa* extending to the columella; the place of one of them occupied by a deep depression or septal fossula.

Typ. sp., *Cyathaxonia cornu*, Michelin, loc. cit., p. 258, pl. lix, fig. 9.

Family XIV.

CYATHOPHYLLIDÆ.

Corallum with incomplete *septa*, that do not extend from the bottom to the top of the visceral chamber, in the form of uninterrupted laminæ; those of the primary cyclum similar to the others, and not forming a central four-branched cross. Septal fossulæ varying in number and in size. Visceral chamber divided by a series of superposed tabulæ.

First Tribe—ZAPHRENTINÆ.

A single septal fossula, well developed, or replaced by a sulcus or a crestiform process, and occasioning more or less irregularity in the radiate arrangement of the septal apparatus. The corallum is simple, and free in all the known species.

1. *Genus* ZAPHRENTIS.

Rafinesque and Clifford, Ann. des Sciences physiques de Bruxelles, vol. v, p. 234, 1820; *Caninia*, Michelin, Dict. des Sc. Nat., Supplém., vol. i, p. 485; *Siphonophyllia*, Scouler, in M'Coy's Carbonif. Foss. of Ireland, p. 187, 1844.

Corallum simple and trochoid. *Calice* deep. Septal fossula strongly developed, and occupying the place of one of the septa. No *columella*. *Tabulæ* moderately developed, and bearing on their upper surface a series of septa, which extend from the wall to the centre of the visceral chamber, and are denticulate all along their calicular edge.

Typ. sp., *Zaphrentis patula; Caninia patula*, Michelin, Icon. Zooph., tab. lix, fig. 4.

9

2. *Genus* AMPLEXUS.

Sowerby, Miner. Conchol., vol. i, p. 165 ; *Amplexus* and *Cyathopsis*, D'Orbigny, Prodrome de Paléontol., vol. i, p. 105, 1850.

Corallum resembling Zaphrentis, excepting that the *septa* do not extend to the centre of the visceral chamber, and leave the upper surface of the tabulæ naked and smooth in that part. The *septal fossula* well characterised in the upper portion of the corallum, but not so on the lower floors. *Tabulæ* highly developed.

Typ. sp., *Amplexus coralloides*, Sowerby, loc. cit., tab. lxxii.

3. *Genus* MENOPHYLLUM.

Corallum resembling Zaphrentis, excepting that a small septal fossula is situated on each side of the large one, and that one half of the central part of the calice is occupied by an elevated, smooth portion of the tabula, which resembles a crescent.

Typ. sp., *Menophyllum tenui-marginatum*, nob.

4. *Genus* LOPHOPHYLLUM.

Corallum resembling Zaphrentis, excepting that a crestiform *columella* occupies the centre of the calice, and is in continuity by one of its ends with a small septum, placed in the middle of the septal fossula, and by the other end with the opposite primary septum.

Typ. sp., *Lophophyllum Konincki*, nob.

5. *Genus* ANISOPHYLLUM.

Corallum resembling Zaphrentis, excepting by the great development of three primary *septa*, one of which is placed facing the septal fossula ; this *fossula* extending much towards the centre of the visceral chamber, and ceasing there to be distinct from the bottom of the calice.

Typ. sp., *Anisophyllum Agassizi*, nob.

6. *Genus* BARYPHYLLUM.

Corallum very short. *Calice* quite superficial. A slightly developed septal fossula, corresponding to one of the branches of a cross, the three other branches of which are constituted by well-developed primary septa. The younger septa not arranged in a regular radiate circle, but inclined obliquely towards the primary ones.

Typ. sp., *Baryphyllum Verneuilanum*, nob.

7. *Genus* HALLIA.

Corallum tall, turbinate. *Septa* highly developed, and extending to the centre of the tabulæ. No *columella*. One remarkably large primary septum occupying the place of the septal fossula, and the neighbouring septa directed towards it, so as to assume a pinnate arrangement; the septa belonging to the two other systems presenting the usual regular radiate position.

Typ. sp., *Hallia insignis*, nob.

8. *Genus* AULACOPHYLLUM.

Corallum resembling Hallia by the mode of arrangement of the septa, but having the septal fossula not replaced by a primary septum, and affecting the form of a narrow groove, at the bottom of which the septa of the two adjoining systems meet, and even cross each other.

Typ. sp., *Aulacophyllum sulcatum; Caninia sulcata*, D'Orbigny, Prod. de Paléont., vol. i, p. 105.

9. *Genus* TROCHOPHYLLUM.

Corallum simple, trochoid. *Calice* rather shallow. Septal fossula rudimentary, and occupied by a small septum. The other *septa* thick, not denticulate, presenting a regular radiate mode of arrangement, and extending almost to the centre of the visceral chamber, where a small tabula is visible.

Typ. sp., *Trochophyllum Verneuili*, nob.

10. *Genus* HADROPHYLLUM.

Corallum short. *Calice* superficial. One very large septal fossula, and three small ones, representing a cross. The radiate arrangement of the septa somewhat irregular.

Typ. sp., *Hadrophyllum Orbignyi*, nob.

11. *Genus* COMBOPHYLLUM.

Corallum presenting the general form of a Cyclolites. A single septal fossula. *Septa* exsert, and regularly radiate.

Typ. sp., *Combophyllum osismorum*, nob.

Second Tribe—CYATHOPHYLLINÆ.

Septal apparatus regularly radiate, and uninterrupted, or equally divided into four groups by four superficial septal fossulæ. No true *columella*, but sometimes a spurious one formed by the inner edge of the septa.

12. *Genus* CYATHOPHYLLUM.

Goldfuss (in parte), Petref. Germ., vol. i, p. 54, 1826.

Corallum simple or composite. No *costæ*. *Septa* well developed, extending to the centre of the calice, and twisted together so as to produce the appearance of a small columella. *Tabulæ* occupying only the centre of the visceral chamber; the outer portion of which is filled up with numerous vesicular dissepiments. A single wall, situated exteriorly, and provided with a complete epitheca.

Typ. sp., *Cyathophyllum helianthoides*, Goldfuss, loc. cit., tab. xx, fig. 2.

13. *Genus* PACHYPHYLLUM.

Corallum composite, and increasing by lateral gemmation. *Corallites* united in their lower portion by means of the great development of the costæ and the exotheca, and not delimitated by an individual epitheca. *Tabulæ* well characterised.

Typ. sp., *Pachyphyllum Bouchardi*, nob.

14. *Genus* CAMPOPHYLLUM.

Corallum simple, very tall, and protected by an epitheca. *Septa* well developed. *Tabulæ* very large, and smooth towards the centre. Interseptal loculi filled with small vesiculæ.

Typ. sp., *Campophyllum flexuosum; Cyathophyllum flexuosum*, Goldfuss, Petref. Germ., vol. i, tab. xvii, fig. 3.

15. *Genus* STREPTELASMA.

Hall, Palæont. of New York, p. 17, 1847.

Corallum simple, and differing from Cyathophyllum by the structure of the wall, which is destitute of epitheca, and covered with sublamellar costæ.

Typ. sp., *Streptelasma corniculum*, Hall, loc. cit., tab. xxv, fig. 1.

16. *Genus* OMPHYMA.

Rafinesque and Clifford, in Ann. des Sc. Phys. de Bruxelles, vol. v, p. 234, 1820.

Corallum simple, turbinate. Wall provided with a rudimentary epitheca, and producing radiciform appendices. *Septa* very numerous, equally developed, and divided into four groups by an equal number of shallow septal fossulæ. *Tabulæ* well developed, and smooth towards the centre.

Typ. sp., *Omphyma turbinata; Madrepora turbinata*, Lin. Amœn. Acad., vol. i, tab. iv, fig. 2.

17. *Genus* GONIOPHYLLUM.

Corallum simple, and affecting the form of a quadrangular pyramid. *Calice* deep and square. *Septa* thick and well developed. *Tabulæ* central, and but little developed.

Typ. sp., *Goniophyllum pyramidale; Turbinolia pyramidalis,* Hisinger, Lethæa Suecica, tab. xviii, fig. 12.

18. *Genus* CHONOPHYLLUM.

Corallum simple, and constituted principally by a series of infundibuliform *tabulæ*, superposed and invaginated, the surface of which presents numerous septal radii equally developed, and extending from the centre to the circumference. No *columella* nor *walls*.

Typ. sp., *Chonophyllum perfoliatum; Cyathophyllum perfoliatum,* Goldfuss, tab. xviii, fig. 5.

19. *Genus* PTYCHOPHYLLUM.

Strombodes (pars), Lonsdale, Sil. Syst., p. 691, 1839 (not Schweigger.).

Corallum simple, and organized as in the preceding genus, but having the septal radii strongly twisted towards the centre of the tabulæ, so as to constitute a spurious *columella.*

Typ. sp., *Ptychophyllum Stokesi,* nob.; C. Stokes, Trans. of the Geol. Soc., 2d series, vol. i, tab. xxix, fig. 1. (N.B. The second figure bearing this number, but not the first.)

20. *Genus* HELIOPHYLLUM.

Hall, in Dana, Zooph., p. 396. 1846.

Corallum simple. Septal apparatus well developed, and producing lateral lamellar prolongations, which extend from the wall towards the centre of the visceral chamber, so as to represent ascending arches and to constitute irregular central *tabulæ,* and which are united towards the circumference by means of vertical dissepiments.

Typ. sp., *Heliophyllum Halli,* nob.; *Strombodes helianthoides,* Hall, Geol. of New York, No. 48, fig. 3 (not *S. helianthoides* of Phillips).

21. *Genus* METRIOPHYLLUM.

Corallum simple, turbinate. *Septa* well developed, slightly twisted, and extending to the centre of the visceral chamber, through well-developed *tabulæ.*

Typ. sp., *Metriophyllum Bouchardi,* nob.; *Cyathophyllum mitratum,* Michelin, Icon. Zooph., tab. xlvii, fig. 7 (not *C. mitratum* of Schlotheim).

22. *Genus* CLISIOPHYLLUM.

(*Pars*) Dana, Exploring Exped., Zoophytes, p. 361, 1846.

Corallum simple, turbinate. *Septa* well developed, and rising towards the centre of the calice so as to form a spurious *columella*, but not twisted.

Typ. sp., *Clisiophyllum Danianum*, nob.

23. *Genus* AULOPHYLLUM.

Corallum simple. *Septa* well developed. A double mural investment; the interior wall dividing the visceral chamber into two portions—one central and columnar, the other exterior and annular. No *columella*. *Tabulæ* but little developed.

Typ. sp., *Aulophyllum prolapsum* ; *Clisiophyllum prolapsum*, M'Coy, in Ann. and Mag. of Nat. Hist., 2d series, vol. iii, p. 3.

24. *Genus* ACERVULARIA.

Schweigger, Handb. der Naturg., p. 418, 1820.

Corallum composite, increasing by calicular gemmation. *Corallites* provided with a double mural investment; the inner wall disposed as in the preceding genus. Septal apparatus well developed between the outer and the inner walls, but much less so in the central area. No *columella*. *Tabulæ* not well developed.

Typ. sp., *Acervularia Rœmeri* ; *Astrea Hennahi*, Rœmer, Verst. der Hartzgeb., tab. ii, fig. 13 (not Lonsdale).

25. *Genus* STROMBODES.

(*Pars*) Schweigger, Handb. der Naturg., p. 418, 1820; Goldfuss, Petref. Germ., vol. i, p. 62, 1826; *Acervularia*, Lonsdale, Sil. Syst., p. 689, 1839 ; *Arachnophyllum*, Dana, Zooph., p. 360, 1846 ; *Strombodes* and *Actinocyathus*, D'Orbigny, Prod. de Paléont. stratigr., vol. i, p. 107, 1849.

Corallum composite, increasing by calicular gemmation. *Corallites* constituted principally by a series of superposed, invaginated, infundibuliform *tabulæ*, united by ascending trabiculæ, so as to form a columnar mass. *Calices* pentagonal, well circumscribed, and completely covered with the septal radii. *Outer walls* not well developed ; the inner mural investment rudimentary.

Typ. sp., *Strombodes pentagonus*, Goldf., Petref. Germ., vol. i, tab. xxi, fig. 3.

26. *Genus* PHILLIPSASTREA.

D'Orbigny, Note sur des Polypiers fossiles, p. 2, 1849.

Corallum composite, resembling Strombodes, but differing from them by the septal or costal radii of the neighbouring corallites, being confluent, and consequently the calices not being definitely circumscribed. No exterior *walls;* the interior mural investment well characterised. The centre of the *tabulæ* presenting a columellarian tubercle.

Typ. sp., *Phillipsastrea Hennahi,* D'Orbigny, loc. cit.; *Astrea Hennahi,* Lonsdale, in Geol. Trans., 2d series, vol. v, tab. lviii, fig. 3.

27. *Genus* ERIDOPHYLLUM.

Corallum composite, and increasing by lateral gemmation. *Corallites* tall, cylindroid, and provided with a thick epitheca, which gives rise to a vertical series of short and thick subradiciform productions that extend to the next individual and unite them together. *Tabulæ* well developed, and occupying the central area circumscribed by the inner wall. Septal apparatus occupying the annular area situated between the outer and inner mural investment, but not extending into the inner or central area.

Typ. sp., *Eridophyllum seriale,* nob.

Third Tribe—LITHODENDRONINÆ.

Axis of the visceral chamber of the corallites occupied by a styliform or lamellar columella.

28. *Genus* LITHODENDRON.

Phillips, Geol. of Yorkshire, vol. ii, p. 200, 1835 (but not *Lithodendron* of Schweigger, which is not an admissible genus); *Siphonodendron,* M'Coy, in Ann. and Mag. of Nat. Hist., 2d series, vol. iii, 1849.

Corallum composite, arborescent, or massive. *Corallites* cylindrical or prismatic. *Columella* styliform, compact. *Septa* well developed, but not reaching to the columella. *Tabulæ* well developed. Interior wall rudimentary.

Typ. sp., *Lithodendron irregulare,* Phillips, loc. cit., pl. ii, figs. 14, 15.

29. *Genus* NEMATOPHYLLUM.

Nematophyllum and *Stylaxis,* M'Coy, loc. cit., 1849.

Corallum composite, massive. *Corallites* prismatic, with a well-developed interior *wall.* *Columella* lamellar. *Septa* well developed, and united by transverse dissepiments, which extend to the columella, but do not constitute true tabulæ. Exterior area vesicular.

Typ. sp., *Nematophyllum arachnoideum,* M'Coy, loc. cit., p. 16.

30. *Genus* LITHOSTROTION.

(*Pars*) Fleming, British Animals, p. 508, 1828; *Strombodes* and *Lonsdaleia*, M'Coy, in Ann. of Nat. Hist., 2d series, vol. iii, pp. 10, 11, 1849.

Corallum resembling *Nematophyllum*, but having the *columella* formed by a fasciculus of twisted bands, and the *septa* subvesicular exteriorly, and joining the columella along their inner edge.

Typ. sp., *Lithostrotion floriforme*, Fleming, loc. cit., p. 508.

31. *Genus* AXOPHYLLUM.

Corallum simple, trochoid, and resembling Lithostrotion by its structure.

Typ. sp., *Axophyllum expansum*, nob.

32. *Genus* SYRINGOPHYLLUM.

Sarcinula, Dana, Zooph., p. 363, 1846 (not *Sarcinula*, Lamarck).

Corallum composite, astreiform. *Corallites* provided with strong *walls*, and much developed costæ. *Septa* large. *Tabulæ* but little developed. *Columella* styliform.

Typ. sp., *Syringophyllum organum*; *Madrepora organum*, Linnæus, Syst. Nat., ed. xii, vol. i, p. 1278.

Family XV.

CYSTIPHYLLIDÆ.

Corallum essentially composed of a vesicular tissue, and presenting little or no traces of septa or radiate striæ.

1. *Genus* CYSTIPHYLLUM.

Lonsdale, in Murchison's Silurian Syst., p. 691, 1839.

Corallum simple, turbinate; the visceral chamber filled with small vesicular laminæ. *Calice* shallow. *Walls* vesicular.

Typ. sp., *Cystiphyllum Siluriense*, Lonsdale, loc. cit., tab xvi bis, fig. 1 (but not fig. 2).

Sub-order 5.

ZOANTHARIA CAULICULATA.

Antipathacea, Dana, Zooph., p. 574.

Polypi supported on a *sclerobasis* or epidermic stem-like corallum.

The general form of the corallum is similar to that of the Isis, Gorgonia, &c., in the order of Alcyonaria; but may be distinguished from these by its surface being spinulous or smooth, whereas it is always sulcated in Alcyonaria.

Family ANTIPATHIDÆ.

Gray, Synop. of the Brit. Mus., p. 135, 1842; Dana, Zooph., p. 574, 1846.

1. *Genus* ANTIPATHES.

(In parte) Pallas, Elench. Zooph., p. 209, 1766.

Corallum arborescent; its surface spinulous.

Typ. sp., *Antipathes myriophylla*, Ellis and Solander, Zooph., tab. xix, figs. 11, 12.

2. *Genus* CIRRHIPATHES.

De Blainville, in Dict. des Sc. Nat., vol. lx, p. 475, 1830.

Corallum not arborescent, and having the form of a simple cylindrical stem; its surface spinulous.

Typ. sp., *Cirrhipathes spiralis*, Blainv., loc. cit.; *Antipathes spiralis*, Ellis and Soland., Zooph., tab. xix, fig. 1.

3. *Genus* LEIOPATHES.

Gray, Synops. of the Brit. Mus., p. 135, 1842.

Corallum arborescent; its surface smooth.

Typ. sp. *Leiopathes glaberrima*; *Antipathes glaberrima*, Esper, Pflanz., Antipathes, tab. ix.

ZOANTHARIA INCERTÆ SEDIS.

1. *Genus* HETEROPHYLLIA.

M'Coy, Palæozoic Corals, in Ann. and Mag. of Nat. Hist., 2d series, vol. iii, p. 126, 1849.

Corallum composed of a tall, subcylindrical, irregularly fluted, stem (or tube), containing a few laminæ, irregularly branching and coalescing, but not presenting a radiate appearance.

Typ. sp., *Heterophyllia grandis*, M'Coy, loc. cit., figs. A, B.

10

2. *Genus* MORTIERIA.

De Koninck, Anim. foss. du Terr. carbon. de Belgique, p. 12, 1842.

Corallum having the form of a bi-concave disc, presenting a radiate structure and numerous costæ.

Typ. sp., *Mortieria vertebralis*, De Koninck, loc. cit., pl. B, fig. 3.

3. *Genus* CYCLOCRINITES.

Eichwald, Uber das Silurische Schichten-System in Esthland, p. 192, 1840.

Corallum composite, astreiform. *Calices* hexagonal and shallow. *Septa* well characterised, but not extending to the centre of the visceral chamber, which appears to be occupied by small tabulæ. (?)

Typ. sp., *Cyclocrinites Spaskii,* Eichwald, Die Urwelt Russlands durch abbildungen erlaeutert, p. 48, tab. i, fig. 8, 1842.

Order 11.

ALCYONARIA.

Alcyoniens, Audouin and Milne Edwards, Recherches sur les Anim. sans Vertèbres faites aux iles Chausay, Ann. des Sc. Nat., 1st series,, vol. xv, p. 18, 1828 ; *Zoophytaria,* Blainville, Manuel d'Actinologie, p. 496, 1834 ; *Zoophyta asteroïdea,* Johnston, Brit. Zooph., p. 164, 1838 ; *Alcyonaria* Dana, Exploring Expedition, Zooph., p. 586, 1846 ; *Anthozoa asteroïdea,* Johnston, Brit. Zooph., 2d edit., p. 138, 1847.

Polypi with bi-pinnate tentacula, and only eight perigastric membranaceous laminæ, containing the reproductive organs.

Alcyonaria have, in general, their dermal tissue consolidated by isolated spiculæ or nodular concretions only, and very rarely present a vaginal polypidom similar to that of the Zoantharia; but even when that is the case, the visceral chamber is never subdivided by any longitudinal septa, and consequently the calice never presents any appearance of radii. In general, the corallum is entirely composed of epidermic tissue, (or basal secretion, Dana,) and constitutes a sort of stem or axis in the centre of the compound mass formed by the gemmation of the Polypi. This sclerobasis is always covered by soft dermic tissue, and increases by the addition of concentric layers.

This order is far from being as numerous as the preceding division of Corallaria, and comprises three natural families,—Alcyonidæ, Gorgonidæ, and Pennatulidæ.

Family I.

ALCYONIDÆ.

Polypi adherent and not provided with an epidermic sclerenchyma.

In this family, the dermic tissue is usually consolidated by a great number of sclerenchymous spicula imbedded in its substance, and constitutes sometimes a tubular corallum, but there is never any trace of a central stem or axis, like that which is constituted by the sclerobasis in Gorgonidæ and in most of the Pennatulidæ.

First Tribe—CORNULARINÆ.

Polypi simple or segregate, and produced by gemmation on creeping stolons, or basal membranaceous expansions, and having no lateral buds or connecting appendices.

1. *Genus* CORNULARIA.

Lamarck, Hist. des An. sans Vert., vol. ii, p. 111, 1816.

Polypi rising by gemmation from creeping filiform stolons, and provided with a tough or subcorneous tubiform polypidom, the surface of which is not costulated.

Typ. sp., *Cornularia cornucopiæ*, Cuvier; *Tubularia cornucopiæ*, Cavolini, Mem. per Servire alla Storia de Polipi Marini, tab. ix, figs. 11, 12; *Cornularia rugosa*, Lamárck, loc. cit.

2. *Genus* CLAVULARIA.

Quoy and Gaimard, ap. Blainville, Dict. des Sc. Nat., vol. lx, p. 499, 1830; *Actinantha*, Lesson, Zool. de la Coquille, Zooph., p. 89, 1831.

Polypi resembling Cornularia, but having their tubular polypidoms costulated and incrustated with long spicula.

Typ. sp., *Clavularia viridis*, Quoy and Gaim., Voyage de l'Astrolabe, Zooph., tab. xxi, fig. 10.

3. *Genus* RHIZOXENIA.

Ehrenberg, Corall. Roth. Meer., p. 55, 1834.

Polypi resembling those of the preceding genus, but not retractile.

Typ. sp., *Rhizoxenia thalassantha*, Ehr.; *Zoantha thalassantha*, Lesson, Voyage de la Coquille, Zooph., tab. i, fig. 2.

4. *Genus* SARCODICTYON.

E. Forbes ap. Johnston, Brit. Zooph., 2d ed., p. 179.

Polypi rising from creeping, filiform, anastomosing stolons, distant, uniserial, and appearing verruciform (not tubular) when retracted. Differ from Cornularia by the shortness of the polypidoms.

Typ. sp., *Sarcodictyon catenatum,* Forbes, loc. cit., tab. xxxiii, figs. 4, 7.

5. *Genus* ANTHELIA.

Savigny, ap. Lamarck, An. sans Verteb., vol. ii, p. 407, 1816.

Polypi not retractile, and rising from a thin fleshy incrusting plate.

Typ. sp., *Anthelia glauca,* Savigny, Egypte, Polypes, tab. i, fig. 7.

6. *Genus* SYMPODIUM.

Ehrenberg, Corall., p. 61, 1834.

Polypi resembling Anthelia, but being retractile.

Typ. sp., *Sympodium fuliginosum,* Ehrenb., Savigny, Egypte, Polypes, tab. i, fig. 6.

7. *Genus* AULOPORA.

Goldfuss, Petref. Germ., vol. i, p. 82.

The fossil corals forming this genus greatly resemble Cornularia and Sarcodictyon, but differ from all the preceding genera by their thick, calcareous polypidom.

Typ. sp., *Aulopora serpens,* Goldfuss, loc. cit., tab. xxix, fig. 1.

8. *Genus* CLADOCHONUS.

M'Coy, in Ann. and Mag. of Nat. Hist. 1st series, vol. xx, p. 227.

Corallum resembling Aulopora, but composed of cup-shaped calices, arranged in a regularly alternate manner, and bent in nearly opposite directions.

Typ. sp., *Cladochonus tenuicollis,* M'Coy, loc. cit., tab. xi. fig. 8.

Second Tribe—TUBIPORINÆ.

Polypi fasciculate, and provided with independent tubular polypidoms, united at various heights by means of horizontal connecting plates, the surface of which produces the reproductive buds.

9. *Genus* Tubipora.

Lamarck, Hist. des Anim. sans Verteb., vol. ii, p. 207, 1816.

Typ. sp. *Tubipora musica*, Lamarck, loc. cit., p. 209.

Third Tribe—TELESTHINÆ.

Polypi segregate and multiplying by lateral gemmation, so as to form arborescent tufts.

10. *Genus* Telestho.

Lamouroux, Polypiers Flexibles, p. 232.

Polypidom composed of ramified tubes of a subcalcareous structure.

Typ. sp., *Telestho aurantiaca*, Lamouroux, loc. cit., tab. vii, fig. 6.

Fourth Tribe—ALCYONINÆ.

Polypi aggregate and multiplying by lateral gemmation, so as to constitute a ramified, lobate or simple mass.

11. *Genus* Alcyonium.

Pallas, Elenchus Zooph., p. 342, 1766; *Lobularia*, Savigny, ap. Lamarck, Hist. des Anim. sans Verteb. vol. ii, p. 412, 1816.

Polypi retractile, and united by a thick tough common tissue, so as to form gibbose or subramified masses.

Typ. sp., *Alcyonium digitatum*, Lin. Solander and Ellis, op. cit., p. 175.

12. *Genus* XENIA.

Savigny, Egypte, Atlas and op., Lamarck, op. cit., vol. ii, p. 629, 1816.

Polypi forming subramified masses, as in Alcyonium, but not retractile, and not having a thick coating of spiculæ at the basis of the tentacula.

Typ. sp. *Xenia umbellata,* Savigny, Egypte, Polyp., tab. i, fig. 3.

13. *Genus* NEPHTHYA.

Savigny, Atlas de l'Egypte; Blainville, Manuel d'Actinol, p. 523; *Spoggodes,* Lesson, Illustr. de Zoologie, 1831.

Polypi forming arborescent masses, incompletely retractile, and having the borders of the calice thick and incrustated with large navicular spiculæ.

Typ. sp,, *Nephthya Chabroli,* Audouin, ap. Savigny, Egypte, Pol. tab. ii. fig. 5.

14. *Genus* PARALCYONIUM.

Alcyonidia, Milne Edwards, Ann. des Sc. Nat. 2d series, vol. iv, p. 323.

Polypi resembling Nephthya, but being completely retractile, and having the lower part of the common mass incrustated with a thick coating of long navicular spiculæ, but the upper part membranaceous and retractile.

Typ. sp., *Paralcyonium elegans ; Alcyonidia elegans,* Milne Edwards, loc. cit., tab. xii and xiii.

15. *Genus* SARCOPHYTON.

Lesson, Zoologie du Voyage de la Coquille, Zooph., p. 92, 1831.

Differs from the genus Alcyonium by the great abundance and the peculiar structure of the common tissue, the cells of which are tubular, and arranged with great regularity in fasciculi, perpendicularly to the upper surface of the mass.

Typ. sp., *Sarcophyton plicatum,* Valenciennes MSS.; *Sarcophyton lobulatum,* Lesson, loc. cit. ; *Alcyonium plicatum,* Lamarck, Hist. des An. sans Verteb., vol. ii, p. 395.

16. *Genus* CESPITULARIA.

Valenciennes MSS.

Polypi non-retractile, arranged in fasciculi, and united in the greatest part of their length by a dense, tough, common tissue, as in Alcyonium.

Typ. sp., *Cespitularia multipinnata,* Valen.; *Cornularia multipinnata,* Quoy and Gaimard, Voyage de l'Astrolabe, Zooph., tab. xxii, figs. 1-4.

17. *Genus* DISTICHOPORA?

Lamarck, loc. cit., p. 197.

This singular zoophyte appears to have more affinity to Alcyonium than to any other form of polypi; but the place belonging to it in a natural system of classification is as yet very uncertain. It is characterised by a calcareous, dendroid corallum, composed of long tubular cells, that present no traces of septa or tubulæ, and are disposed in a flabellate manner, so as to constitute a vertical plane, the two sides of which are covered with a thick and compact cœnenchyma, and the edge assumes the appearance of a calicular groove, limited laterally by two rows of circular pores. Nothing is known concerning the structure of the soft parts.

Typ. sp., *Distichopora violacea*, Lamarck, op. cit., p. 305. (For the structure of the Corallum, see Milne Edwards, Atlas du Regne Animal de Cuvier, Zooph., tab. lxxxv, fig. 4.)

Family II.

GORGONIDÆ.

Polypiers corticiferes, Lamarck, Hist. des Anim. sans Verteb. vol. ii, p. 288, 1816; *Polypes corticaux*, Cuvier, Regne Animal, vol. iv, p. 78, 1817; *Corallia*, Blainville, Manuel d'Actinologie, p. 501, 1834; *Cerato-corallia*, Ehrenberg, Corall. des Roth. Meeres, 1834; *Coralliadæ*, Gray, Synop. Brit. Mus. p. 134; *Gorgoniadæ*, Johnston, British Zooph., p. 182, 1838; *Gorgonidæ*, Dana, Exploring Expedition, Zooph., p. 637, 1846; *Gorgoniadæ*, Gray, List of British Anim. of the British Museum, p. 55, 1848.

Polypi provided with a thick, suberous cœnenchyma, surrounding a central stem that is adherent to an extraneous body by its basis, and is formed of epidermic sclerenchyma.

First Tribe—GORGONINÆ.

Gorgonia, Pallas, Elenchus Zoophytorum, p. 160, 1766; Cuvier, Regne Animal, 1st ed., vol. iv, p. 80; Lamarck, Hist. des Anim. sans Verteb., vol. ii, p. 309; *Gorgoninæ*, Dana, Exploring Expedition, Zooph., p. 641, 1846.

Common axis inarticulate, horny or fasciculate, but not calcareous.

1. *Genus* GORGONIA.

Pallas, loc. cit., (in parte.)

Axis corneous. *Calices* disposed irregularly round the ramified cylinders formed by the cœnenchyma, and not encircled by imbricated squammæ. *Polypi* retractile.

Typ. sp., *Gorgonia tuberculata*, Esper. Pflanz., Gorg., tab. xxxvii.

2. *Genus* PTEROGORGIA.

Ehrenberg, Corall. des Rothen Meeres, p. 144, 1834 ; Dana, op. cit., p. 647, 1846.

Differs from Gorgonia by the polypi being bifarious.

Typ. sp., *Pterogorgia anceps*, Ehrenb., loc. cit., p. 145.

3. *Genus* BEBRYCE.

Philippi, Zoologesche Beobachtungen, in Archiv. fur Naturgeschichte, von Erichson, vol. viii, p. 35, 1842.

Arborescent compound polypi, resembling Gorgonia by their corneous sclerobasis, but differing from the preceding genera by not being retractile.

Typ. sp., *Bebryce mollis*, Philippi, loc. cit.

4. *Genus* PHYLLOGORGIA.

Differs from Gorgonia by the cœnenchyma not constituting a cylindrical sheath around the ramifications of the sclerobasis, but extending between them so as to constitute large foliaceous, frondiform laminæ, the two surfaces of which are studded with the calices of the individual polypi.

Typ. sp., *Phyllogorgia dilatata ; Gorgonia dilatata*, Esper, Pflanz. Gorg. tab. xli.

5. *Genus* PHYCOGORGIA.

Sclerobasis flabelliform, divided into digitated lobes, and composed of delicate corneous fibres united into laminæ, the two sides of which are covered with the cœnenchyma, and densely studded with numerous non-prominent calices.

Typ. sp., *Phycogorgia fucata ; Gorgonia fucata*, Valenciennes, Voyage de la Venus, tab. xi, fig. 2.

6. *Genus* MURICEA.

Lamouroux, Exposit. Method. des Polyp. p. 36, 1821.

Differs from Gorgonia by the calices being surrounded with imbricated squammulæ, but not supported on long, verruciform, moveable appendices, as in Primnoa.

Typ. sp., *Muricea spicifera*, Lamouroux, op. cit., tab. lxxi, figs. 1, 2.

7. *Genus* PRIMNOA.

Lamouroux, Hist. des Polypiers Flexibles, p. 440, 1816.

Differs from the preceding genus by the polypi constituting long verruciform subpediculated appendices, which are capable of motion at their bases.

8. *Genus* Solanderia.

Duchassaing and Michelin, in Guerin's Revue Zoologique, June, 1846.

Differs from Gorgonia by the suberous texture of the sclerobasal axis, which resembles the non-calcified joints of Melitæa.

Typ. sp., *Solanderia gracilis*, Duchassaing and Michelin, loc. cit.

9. *Genus* Briareum.

Blainville, Manuel d'Actinologie, p. 520, 1830.

Axis soft, suberous, or composed of spicula. This genus is intermediate between Alcyonium and Gorgonia.

Typ. sp., *Briareum gorgonoideum*, Blainv. ; *Gorgonia briareus*, Lin. ; Ellis and Solander, tab. xiv, figs. 1, 2.

The genus Hyalonema established by M. Gray, (' Proceed. of the Zool. Soc.' 1835, p. 63,) is also referred by some zoologists to the tribe of Gorgoninæ; but the recent observations of M. Valenciennes tend to establish that the fasciculi of siliceous threads, which constitute the axis of this singular production, belong to the class of Spongidæ, and the polypi which we have observed in a dried state on different parts of the axis appear to be parasites, belonging to the order of Zoantharia.

Second Tribe—ISINÆ.

Dana, Exploring Exped., Zooph., p. 677, 1846.

Common axis articulated, or composed of segments, the structure of which differ alternately.

10. *Genus* Isis.

Linnæus, Syst. Nat., 12th ed., p. 1287, 1767.

Axal sclerobasis composed of joints, alternately corneous and calcareous; branches proceeding from the calcareous joints.

Typ. sp., *Isis hippuris*, Lin., loc. cit.

11. *Genus* Mopsea.

Lamouroux, Polyp. Flex., p. 466, 1816.

Axis presenting the same structure as in the preceding genus, but with the branches proceeding from the corneous joints.

Typ. sp., *Mopsea dichotoma*, Lamouroux, loc. cit., p. 467.

11

12. *Genus* MELITÆA.

Lamouroux, Polyp. Flex., p. 461, 1816.

Axis composed of joints, which are alternately calcareous and suberous.

Typ. sp., *Melitæa ochracea*, Lamouroux, loc. cit., p. 462.

Third Tribe—CORALLINÆ.

Dana, loc. cit., p. 639, 1846.

Common axis inarticulate, solid and calcareous.

13. *Genus* CORALLIUM.

Lamarck, Hist. des Anim. sans Vert., t. ii, p. 295, 1816.

Typ. sp., *Corallium rubrum*, Cavolini, Mem. per Servire all. Hist. des Polypi Marini, tab. ii.

Family III.
PENNATULIDÆ.

Pennatula, Linnæus, Syst. Nat., 10th ed., p. 818; Pallas, Elen. Zooph., p. 362, 1766; *Polypi natantes*, Lamarck, op. cit., p. 415; *Pennatulidæ*, Fleming, Brit. Animals, p. 507, 1828; *Pennatularia*, Blainville, Manuel, p. 512, 1830; *Calomides*, Latreille, Fam. du Reg. Anim. p. 543; *Pennatulina*, Ehrenberg, loc. cit., p. 63, 1834; *Pennatulidæ*, Johnston, Brit. Zooph., p. 175; Dana, Explor. Exped. p. 587, 1846.

Polypi aggregate, and having a common peduncle, the centre of which is occupied by a peculiar cavity, and usually contains a solid axis; this sclerobasis styliform, and never expanding at its under extremity, so as to adhere to extraneous bodies. The polype-mass is consequently free.

1. *Genus* PENNATULA.

(In parte.) Linnæus, Syst. Nat., 10th ed., p. 818, 1760; Lamarck, Syst. des An. sans Vert., p. 380, 1801.

Polype mass plume-shaped, with the shaft composed of contractile common tissue, containing a short subosseous axis, and bearing on each side of its upper part a series of large spreading pinnules, on the upper edge of which, the retractile exhalic portion of the polypi protudes. The axis is cylindrical at its upper part, and more or less quadrangular towards its lower end; its structure is somewhat fibrous, and its tissue is not very brittle.

Typ. sp., *Pennatula setacea*, Esper, Pflanz., Pennat. tab. vii.

2. *Genus* VIRGULARIA.

Lamarck, Hist. des Anim. sans Verteb. vol. ii, p. 429, 1816.

Differs from Pennatula by the length of its shaft and the shortness of its pinnules, which assume the form of lunate lobes, or simple transverse striæ. Axis cylindrical, calcareous, very long, slender, tapering, and presenting in its transverse section a radiate structure.

Typ. sp., *Virgularia mirabilis*, Lamk.; *Pennatula mirabilis*, Müller, Zool. Danica, vol. i, tab. xi.

3. *Genus* PAVONARIA.

Cuvier, Regne Animal, vol. iv, p. 85, 1816; *Funicularia*, Lamarck, op. cit., p. 423, 1816.

Polype mass virgate; the polypi not retractile, arranged on one side of the stem. Axis quadrangular, long, and very tapering.

Typ. sp., *Pavonaria quadrangularis*, Cuv.; *Pennatula antennina*, Lin.; Johnston, Brit. Zooph., tab. xxxi.

4. *Genus* GRAPHULARIA.

Corallum styliform, straight, very long, cylindroid towards the lower extremity, sub-tetrahedral at the upper part, and presenting on one side a broad shallow furrow. Transverse section showing the existence of a thin coating, and a radiate structure in the body of the coral.

Typ. sp., *Graphularia Wetherelli*, nob.; *Pennatula*, Sowerby and Wetherell, in Geol. Trans. 2d series, vol. v, part i, p. 136, tab. viii. fig. 2.

5. *Genus* UMBELLULARIA.

Cuvier, Regne Animal, vol. iv, 1807.

Resembling Pavonaria, but having all the polypi collected in a terminal bunch at the extremity of the stem. Axis quadrangular and twisted.

Typ. sp., *Umbellularia Groenlandica*, Cuv.; *Hydra Marina arctica*, Ellis, Corallines, tab. xxxvii.

6. *Genus* VERETILLUM.

Cuvier, Regne Animal, vol. iv.

Resembling Pennatula, but not having any lateral pinnules, with the polypi arranged all round the upper part of the stem. Axis rudimentary, and of a form almost navicular.

Typ. sp., *Veretillum cynomorium*, Cuvier; *Pennatula digitiformis*, Ellis.

7. *Genus* LITUARIA.

Valenciennes MSS., Cat. of the Zoophytes in the Museum of Paris.

Resembling Veretillum, but having a long well-developed axis, quadrangular and tapering towards its lower part, inflated, claviform, pitted and echinulate at its upper end.

Typ. sp., *Lituaria phalloides*, Valenciennes, loc. cit. ; *Pennatula phalloides*, Pallas, Miscel. Zool., tab. xiii.

8. *Genus* CAVERNULARIA.

Valenciennes, loc. cit., MSS.

Resembling Veretillum, but having in its centre a large fibrous tube divided longitudinally into four cavities, and not containing any calcareous or horny axis.

Typ. sp., *Cavernularia obesa*, Valenciennes MSS.

9. *Genus* RENILLA.

Lamarck, Hist. des Anim. sans. Verteb., vol. ii, p. 428, 1816.

Polyp-mass explanate, unifacial, reniform, with a short, slender peduncle, containing a central cavity as in Pennatula, but not having any solid axis.

Typ. sp., *Renilla Americana*, Lamarck ; *Pennatula reniformis*, Ellis and Solander, p. 67 ; Shaw, Miscel. iv, tab. cxxxix.

The genus GRAPTOLITHUS (Linnæus, *Iter Scan.* 1751,) appears to have more affinity with Virgularia than with any other recent zoophyte. The polype mass is slender, virgate, and often becomes bifurcate by the progress of growth. The axis projects at the inferior extremity of the stem, and is often bifurcate.

Example, *Graptolithus ramosus*, Hall, Palæont. of New York, tab. lxxiii, fig. 3.

The genus WEBSTERIA, nob. appears to be very similar to Graptolithus by its general structure, but offers also a certain resemblance to some Sertularidæ and to certain Bryozoa. In the present state of our knowledge, the natural affinities of these fossil zoophytes are indeed so obscure, that we hesitate to place them in any of the preceding zoological divisions, and prefer leaving them in the *incertæ sedis*.

Typ. sp., *Websteria Crisioïdes*, nob., tab. vii, fig. 5.

Order 3.

PODACTINARIA.

Polypi having the gastric cavity surrounded by four vertical membranaceous *septa*, at the upper end of which are placed four pairs of intestiniform reproductive organs. The tentacula discoidal, pedunculated, not tubular as in Zoantharia and Alcyonaria, but organized much in the same way as in Echinoderma. The mouth proboscidiform, and the fauces surrounded by numerous internal, filiform, contractile appendices.

The genus LUCERNARIA is the only known representative of this zoological type, and comprises no coralligenous polypi.

Sub-class 2.

HYDRARIA.

Polypes sertulariens, Audouin and Milne Edwards, Recherches sur les Anim. sans Verteb., faites aux îles Chausay, in Ann. des Sc. Nat., 1st series, vol. xv, p. 18, 1828, ap. Lamarck, Hist. des An. sans Verteb. 2d ed., vol. ii., p. 105; *Sertulariacæa (in parte),* Blainville, Manuel d'Actinologie, p. 465, 1834; *Zoocorallia oligactinia,* Ehrenberg, Coral. Roth. Meeres, p. 67, 1834; *Zoophyta Hydroida,* Johnston, in Mag. of Zool. and Bot., vol. i, p. 447, 1837; *Polyparia,* Gray, Synop. Brit. Mus.; *Nudibranchiata,* Farre, on the Structure of Polypi, Phil. Trans. 1837; *Hydrozoa,* R. Owen, Lectures on the Comp. Anat. of the Inverteb. Animals, p. 82, 1843; *Hydroidea,* Dana, Exploring Expedition, Zooph. p. 685, 1846; *Anthozoa Hydroidea,* Johnston, British Zooph, 2d ed. p. 5, 1847.

Polypi with a simple, non-lamelliferous, digestive cavity. No internal generative organs. Tentacula filiform and subverrucose.

The naked, fresh-water zoophytes of the genus HYDRA constitute the type of this group, and till very lately were considered as being closely allied to Sertularia, Campanularia, &c.; but the recent observations of divers zoologists tend to establish that all the coralligenous animals of this form belong to the class of Medusa. Till this question is decided, it would therefore be idle to make any modifications in the systematic arrangement of these problematic polypi, and it will suffice for us to refer the reader to Dr. Johnston's valuable work on 'British Zoophytes,' for the characters of the generic divisions generally adopted.

DESCRIPTION

OF

THE BRITISH FOSSIL CORALS.

CHAPTER I.

CORALS OF THE CRAG.

THE Crag formation of the East of England is generally reputed very rich in Fossil Corals; and the name given to the lower strata of this system is even derived from the abundance of various organic remains of coralloid appearance which occur in some localities. But this opinion arises from the confusion which has till lately been made between Bryozoa and Polypi; in reality true Corals are far from being common in any of these beds. The four species mentioned by Mr. Searles Wood, in the Catalogue of the Zoophytes of the Crag, published in 1844 in the 'Annals of Natural History,' are the only known Polypidoms belonging to this geological division.

These fossils are found in the Red Crag as well as in the Coralline Crag, and most of them are as yet peculiar to England; only one species has been met with on the Continent, in the Crag of Antwerp, a strata belonging to the same geological horizon; and none of them are known to live in the seas of the present period. The *Sphenotrochus intermedius* has, it is true, been considered as existing on the coast of England as well as in the Crag; but the recent species, which has lately received the name of *Sphenotrochus Andrewianus*,[1] is perfectly distinct from the fossil Coral to which it was at first referred. It is also worthy of remark that the Crag Corals belong to four distinct genera, each of which is represented by different species in the other Miocene formations; that three of these genera are also represented by peculiar species in our actual Fauna, and that none of them have been discovered in strata anterior to the older tertiary formations.

[1] Milne Edwards and J. Haime, Monographie des Turbinolides, Ann. des Sc. Nat., 3me série, vol. ix, p. 245, tab. vii, fig. 4.

ORDER ZOANTHARIA (p. ix).

Family TURBINOLIDÆ (p. xi).

Tribe TURBINOLINÆ (p. xvi).

Genus SPHENOTROCHUS (p. xvi).

1. SPHENOTROCHUS INTERMEDIUS. Tab. I, figs. 1, 1 *a*—1 *i*.

TURBINOLIA INTERMEDIA, *Münster*, ap. Goldfuss, Petref. Germ., vol. i, p. 108, tab. xxxvii, fig. 19, 1826. (This figure is good, excepting that the basis of the Coral appears too truncate.)

 — — *Ch. Morren*, Descrip. Corall. foss., in Belgio Repertorum, p. 52, 1828.

 — . . . . *R. C. Taylor*, in Mag. of Nat. Hist., vol. iii, p. 272, fig. 2, 1830. (A rough figure.)

 — INTERMEDIA, *Milne Edwards*, Notes in the second ed. of Lamarck's Anim. sans Vert., vol. ii, p. 361, 1836.

 — — *Galeotti*, Mém. couron. par l'Acad. de Bruxelles, vol. xii, p. 188, 1837.

 — *Hagenow*, in Neues Jarhb. für Miner. Geol., 1839, p. 291.

 — — *Nyst*, Coquilles et Poly. foss. des Terr. Tert. de la Belgique, p. 631, tab. xlviii, fig. 14, 1843. (This figure is incomplete, and does not show the columella.)

 — MILLETIANA, *Searles Wood*, Ann. and Mag. of Nat. Hist., vol. xiii, p. 12, 1844.

SPHENOTROCHUS INTERMEDIUS, *Milne Edwards* and *Jules Haime*, Monogr. des Turbinolides, Ann. des Sc. Nat., 3me série, vol. ix, p. 243, 1848.

Corallum simple, straight, free, presenting no trace of adherence, cuneiform, strongly compressed in its lower part, and truncate at its basis, which is very broad (fig. 1) ; sometimes even as much so as the calice (fig. 1*a*). This last character exists also in the *Sphenotrochus Milletianus ;* but this Coral, instead of being much compressed in the lower part, is, on the contrary, very thick down to its extremity.

Costæ smooth, rather thick, especially near the calicular edge, closely set, but separated by deep grooves (fig. 1*b*). They all occupy almost the whole length of the corallum ; and it is therefore difficult to recognise their relative age by the height at which they begin. This difficulty is also augmented by the form of those situated near the middle of the flattened sides, which in their lower part are constituted by small, rather irregular papillæ. The median costæ are nearly straight, nearly equal, not very prominent, and narrowing as they approach the base; the lateral costæ, and those situated near them, are, on the contrary, larger, separated by deeper grooves, slightly curved towards their lower end, sometimes rather undulate, and thicker at their base than higher up.

It may also be worth remarking that similar smooth and simple costæ exist in all the species of this genus belonging to the present period or to the Miocene deposits; whereas the older species, found in the Eocene formation, have the costæ crispate, and composed of series of papillæ.[1]

The *calice* (fig. 1a) is regularly elliptic and slightly arched, the extremities of its great axis being lower than those of its small axis. The proportion between the two diameters is nearly constant, and the form of the ellipse, represented by the calicular margin, is intermediate between that of *Sphenotrochus granulosus*, which is much shorter, and that of *Sphenotrochus Andrewianus*, which is more elongated; it is approximately as 100 : 150. The size of the calice is also subject to very slight variations in individuals which have attained their definitive form, whether they be short or tall.

The *fossula* is very shallow.

The *columella* (figs. 1a and 1c) has the form of a rather thin, vertical lamina, situated in the direction of the long axis of the calice and of the basal edge of the corallum. Its upper edge is nearly horizontal, and reaches almost to the level of the apex of the septa; it is obtuse at its angles, and divided into two equal lobes by a small notch; sometimes three of the lobes are visible. The structure of this part of the polypidom may be very well shown by a vertical section corresponding to the small axis of the calice; it is formed by two delicate parallel laminæ, applied together, thickened near its upper edge, and united, towards its base, to the wall, so as to form with the mural sclerenchyma one compact mass.

The *septa*, as in all the other species of this genus, form three complete and well-developed cycla (fig. 1a); they are consequently twenty-four in number, and they are closely set, straight, thick exteriorly, and becoming gradually thinner towards the centre of the calice, exsert, arched at their apex, truncate at the upper end of their inner edge, and granulated on their surface. These granulations are easily brought to view by a vertical section of the corallum (fig. 1c); they are small, unequal in size, pointed, not numerous, and not disposed in a regular manner, excepting near the upper edge of the septa, where they form a curved line nearly parallel to the edge. The septa of the first and second cycla are nearly similar; and, as is often the case in Corals with an elliptic calice, the two primary septa, corresponding to the long axis of the calice, are a little smaller than the four others of the same cyclum, and the six secondary ones. The tertiary septa are

---

[1] The fossil Coral figured by Mr. Isaac Lea, under the name of *Turbinolia nana*, and mentioned by that author as belonging to the Eocene strata of Alabama, would appear to be an exception to this rule, for it resembles much the *Sphenotrochus Milletianus*, and seems to have smooth costæ; but the figure given by Mr. Lea is not sufficiently explicit for us to be able to decide the question, or even to be quite sure that this Turbinolida really belongs to the genus Sphenotrochus, and in the text the author says that he could see no trace of a columella (Lea, Contrib. to Geol., p. 195, tab. vi, fig. 209). In the present state of palæontology, we may, therefore, consider the above-mentioned observation as still holding good; and the distinction between the Eocene species of Sphenotrochus and the more recent representatives of the same generic type is a result not devoid of interest for geologists as well as for zoologists.

not quite as much developed as the older ones, but are broad enough to reach the columella, to which they are united, at least towards their base. The union between the septa and the columella is not complete, but is effected by means of a double series of trabiculæ extending from the inner edge of the septa, bent alternately to the right and to the left; so that in a vertical section of the visceral chamber a series of pores is seen along the line of junction of each septa with the columella (fig. 1c). This mode of arrangement of the marginal trabiculæ gives also to the septa, when viewed from above (fig. 1d), or by means of a horizontal section, the appearance of bifurcation along their inner margin, and may easily be mistaken for a disjunction of their two constituent laminæ, an error which has been committed by Goldfuss and by ourselves in our first observations.

Sphenotrochus intermedius is the largest known species of this genus; sometimes, however, *S. Milletianus* and *S. granulatus* are almost as long. Its usual length is about three lines, but there are individuals half an inch long. The long axis of the calice is about two lines and a half.

Mr. Searles Wood, to whose kindness we are indebted for the specimens here described, has collected an interesting series of these Corals, showing the changes of form which they experience before arriving at the adult state, and has thus enabled us to study their mode of growth, as we had already done for Fungia in a preceding memoir.[1] We have not met with any of these young Turbinolidæ with only six septa and the same number of costæ; the youngest in Mr. Searles Wood's collection (fig. 1e) has twelve well-marked costæ, distinct from the top to the bottom of the corallum; but the six primary septa are the only ones which are pretty well developed, and those of the second cyclum are still in a rudimentary state. There is no trace of the columella, which appears at a later period and the general form of the corallum is almost cylindrical; its height is then not more than two thirds of a line, and its calice is circular. The base of the corallum is circular; it is truncate, but not spread out, and its adherence must have been of very short duration.

Before the tertiary costæ make their appearance, the calice begins to enlarge in one direction more than in the other, so as to assume an oval form; a slight coarctation becomes visible towards the middle of the corallum, its upper part swells out laterally, and the peduncle enlarges and becomes smooth. Soon after this the tertiary costæ begin to be formed (fig. 1f), and the calice becomes completely elliptical, but is still quite horizontal. The coarctation above the peduncle still exists, and we at first supposed that the upper part of the corallum became free by rupture, as is the case in Flabellum;[2] but the series of specimens collected by Mr. Searles Wood shows that such is not the case, and that the peduncle does not lose its vitality, but is gradually absorbed. Its truncate extremity is first

[1] Observations sur la Structure et le Mode de Développement des Polypiers, Ann. des Sc. Nat., 3ᵐᵉ série, vol. ix, p. 76, tab. vi.

[2] Loc. cit.

cicatrized, and becomes rounded, at the same time that it expands laterally, as does the rest of the corallum, which ceases to be cylindrical, and assumes a compressed form (figs. 1*g* and 1*h*). Soon after the peduncle begins to become thin, and to shorten (fig. 1*i*); the absorption continues till it disappears completely, and the under edge of the corallum becomes long and obtuse. While these modifications are going on, other changes are produced in the internal structure of the corallum. As soon as the tertiary septa appear, the columella begins to rise, and the primary and the secondary septa, which have become rather broad, give off some spiniform trabiculæ, that unite with the columella. The simultaneous development of the twelve tertiary costæ also determines considerable change in the general form of the corallum; the calice, instead of being horizontal, becomes arched (figs. 1*g*, 1*h*, 1*i*), and the sides of the wall corresponding to the long axis of the calice not having yet expanded towards the basis, the corallum has the form of a small battledore; but when the tertiary costæ increase in size, the convexity of the calice diminishes, and the base of the corallum spreads out, till it assumes the form of a broad, obtuse wedge (figs. 1, 1*a*), which it retains in the adult state.

The *Sphenotrochus intermedius* is easily distinguished from *Sphenotrochus crispus*, *S. mixtus*, *S. pulchellus*, *S. granulosus*, and *S. semigranosus* (species which all belong to the Eocene period), by the costæ being smooth, and not formed by a series of large granules. It resembles *S. granulosus* by its general form, and *S. crispus* by its calice. We are acquainted with only three other species, which have also smooth costæ, and have often been confounded with *S. intermedius*. One of these lævicostate species is the *S. Andrewianus*, which lives on the coasts of Cornwall and of the Isle of Arran, but is easily distinguished by its narrow subconical base, and the slight elongation of its calice, the two diameters of which are as 100 : 120. The second lævicostate species, which we designate by the name of *Sphenotrochus Rœmeri*,[1] differs also from *S. intermedius* by its narrow base. The third species, *S. Milletianus*, bears great resemblance to the latter, and belongs to strata occupying the same geological formation, a circumstance that has also contributed to create confusion between them. But the *S. Milletianus* found in the Faluns of Anjou is characterised by its lateral costæ being much less prominent, and its base being more rounded and less compressed than in the *S. intermedius*.

This fossil is common in the Coralline Crag, and the Red Crag at Sutton. We have ascertained its identity with the species found in the Crag of Antwerp, by comparing it with the specimens belonging to the collection of M. H. Nyst, at Louvain, and with that of Goldfuss, in the Poppelsdorf Museum, at Bonn. Specimens of this species exist in the

[1] This undescribed species has most of the characters of *S. mixtus*, but the costæ are all similar and smooth. The lateral ones are not notably larger than the others, and those adjacent are slightly curved near their lower end, and sometimes interrupted. The primary and secondary septa are equal, and those of the third cyclum are narrow; all are thick towards the outer edge, and but slightly granulate. Calice twice as long as it is broad. Length two lines; breadth one line and a half; thickness one line. A fossil of the Miocene strata of Cassel and Hildesheim, belonging to the Museum of Bonn. M. Nyst possesses a specimen of the same species found in the Crag of Antwerp.

collections of the Geological Society of London, and of MM. Searles Wood, Bowerbank, and Frederick Edwards, in London; of the Museum of Natural History, and of MM. d'Archiac, Michelin, and Milne Edwards, in Paris; M. Nyst, at Louvain; M. de Koninck, at Liége, &c.

Genus FLABELLUM (p. xviii).

1. FLABELLUM WOODII.   Tab. I, figs. 2, 2 *a*, 2 *b*.

FUNGIA SEMILUNATA,[1] *Searles Wood*, Ann. and Mag of Nat. Hist., vol. xiii, p. 12, 1844.

FLABELLUM WOODII, *Milne Edwards* and *J. Haime*, Monogr. des Turbinolides, Ann. des Sc. Nat., 3me série, vol. ix, p. 267, 1848.

Corallum simple, erect, rather short, much compressed, especially towards its base, cuneiform, subdeltoid, with a peduncle short and rather thick, and lateral edges straight, and diverging at an angle of rather less than 90°.   All the *costæ*, even the lateral ones, simple, flat, equal, indistinct, and crossed by scarcely developed rugæ and slight folds of the epitheca, which is very thin.   The surface of the wall is also marked by small longitudinal sulci, corresponding to the outer edge of the septa; those referable to the small septa but slightly marked.

Calice having the form of a very long ellipse, and rather arched.   In one specimen the proportion of its two axes was as 100 : 280, and in another as 100 : 300; the extremities of the ellipse corresponding to the great axis are obtuse, and on a level rather lower than that of the small axis.   The *fossula* is long, narrow, and deep.

The *columella* represented only by few large granulæ adhering to the inner edge of the septa, and assuming the form of short, thick trabiculæ.

The *septa* constitute five complete cycla, very well developed, and a sixth cyclum incomplete, more or less rudimentary in some parts, but most apparent in the systems situated near the long axis of the calice.   The septa of the first three cycla are nearly of the same size, and the septal apparatus is therefore divided into twenty-four groups or apparent systems, containing each seven septa, or only five, as is often the case when those of the sixth cyclum are missing in half of these groups.   In general, these minor septa are most developed in the half of the lateral groups adjoining the extremities of the long axis of the calice, and at the same time the septa of the fourth cyclum enlarge in these groups so as to resemble the neighbouring ones of superior orders, and produce an appearance of there being twenty-six or twenty-eight systems; but in these lateral groups the number of septal elements never exceeds three.

The septa are straight, thin, closely set, and do not rise quite so high as the mural

---

[1] The *Fungia semilunata* of Lamarck, to which this fossil was referred by Mr. Searles Wood, belongs to the genus *Diploctenium* of Goldfuss; hence the necessity of giving a new name to the above-mentioned species. (See our Monograph of Astreidæ, Ann. des Sc. Nat., 3me série, vol. x, p. 248.)

epitheca; their upper edge is slightly sinuous, and their surface covered with projecting granulæ of various sizes, disposed rather irregularly in rows nearly parallel to the upper edge. These granulæ are much larger along the inner and inferior part of the edge of the septa of the superior orders, where they assume the appearance of alternate trabiculæ or spines. It is also to be noted that the principal septa are slightly emarginated near the border of the calice, and that their free edge is thin and arched above, thick, subflexuous, and obliquely truncate towards the columella. A horizontal section of the corallum, made a little below the edge of the calice, shows the thickness of the walls, and of the inner part of the large septa; it also renders evident the bifoliate structure of these septa. Height twelve lines; long axis of the calice from twelve to sixteen lines; the short axis from four to six lines.

The genus Flabellum contains a great number of species, and has been subdivided into three sections, according to the state of the basis of the corallum, which is sometimes pedicellate or truncate, and in others widely adherent. The *Flabellum Woodii* is easily distinguished from the fixed *Flabellum* and the truncated *Flabellum*, by the permanence of its narrow peduncle, and differs from most of the pedicellated *Flabella* by its simple non-cristate, non-spinous costæ. Seven species, *F. Gallapagense*, *F. Michelinii*, *F. Thouarsii*, *F. cuneatum*, *F. subturbinatum*, *F. majus*, and *F. Sinense*, have the same character; but *F. subturbinatum* and *F. Michelinii* are recognisable by their horizontal calice and their lateral costæ, almost vertical. *F. Gallapagense* also resembles *F. Woodii* by the rudimentary state of its columella, but is of a more elongated form, and is much less compressed laterally. *F. cuneatum* and *F. majus* are still nearer allied to *F. Woodii*, their characters, however, are not yet completely known; but the first of these fossil species has the septa much thicker than in the above-described Coral, and *Flabellum majus* is remarkable by its great size, its highly-compressed calice, and the peculiar structure of its principal septa.[1]

The *Flabellum Woodii* has been found in the Coralline Crag at Iken, and appears to be very rare; for in 1844, when Mr. Searles Wood published his 'Catalogue of the Zoophytes of the Crag,' only two specimens, one belonging to Mr. Bunbury, and the other to Mr. W. Colchester, were known, and we believe that since that time only two more specimens, now in the possession of Mr. Searles Wood, have been found. Those figured and described in this Monograph were communicated to us by Mr. Searles Wood.

[1] See our Monograph of Turbinolidæ, loc. cit., p. 260.

Family ASTREIDÆ (p. xxiii).

Tribe ASTREINÆ (p. xxxi).

Genus CRYPTANGIA (p. xliv).

1. CRYPTANGIA WOODII.   Tab. I, figs. 4, 4 *a*, 4 *b*, 4 *c*, 4 *d*, 4 *e*.

> CLADOCORA CARIOSA, *Lonsdale;* in Searles Wood's Catal. Ann. of Nat. Hist., vol. xiii,
> p. xii, 1844.[1]
> CRYPTANGIA WOODII, *Milne Edwards* and *J. Haime,* Mém. sur les Astreides, Comptes rend.
> de l'Acad. des Sciences, vol. xxvii, p. 496, 1848.

This singular fossil Coral is always found immersed in a mass of Cellepora, a peculiarity which is also met with in another species of the same genus, belonging to the Faluns of Touraine.   At first sight, the vesicular mass formed by these Bryzoa may easily be mistaken for a cellular epithecal cœnenchyma, resembling that of Sarcinula ; but an attentive examination of the cells will lead to a recognition of their real nature, and similar masses of Cellepora, not containing any Cryptangia, are often found in the same localities.   It is however remarkable, that Corals of this genus should never be found adhering to other extraneous bodies, and should always take up their abode on a cluster of Cellepora, which, increasing as they themselves grow up, imbeds them so completely, that the calices alone remain free on the surface of the common mass.

The mode of multiplication of Cryptangia is also worthy of notice.   These Corals always form clusters, and must be produced by gemmiferous stolons, but the radiciform expansions from which they must proceed do not become sclerenchymatous, and leave little or no trace of their existence ; so that when the soft parts are destroyed, as is always the case in fossils, the different corallites appear to be quite independent, and would be free, were it not for the extraneous cellular mass in which they are so deeply immersed. It is therefore easy to perceive that these Corals differ widely from *Cladocora*, to which they were referred by Mr. Lonsdale, and are equally distinct from the generic forms to which the name of *Lithodendron*, applied by M. Michelin to the Touraine species, had been previously given.   They are nearly allied to the *Astreinæ reptantes*, for which we have established the genera *Angia* and *Rhizangia*, but must constitute a separate generic group, which we have proposed calling *Cryptangia*.

[1] The *Madrepora cariosa* of Goldfuss, to which this fossil was referred by the above-mentioned author, is a true Madrepora, and neither the one nor the other can be placed in Ehrenberg's genus Cladocora, The typic specimen of *M. cariosa*, figured and described by Goldfuss, is preserved in the Museum of Bonn, where it was attentively examined by one of us ; it is a fossil of the Parisian basin, having a spongy cœnenchyma, and the visceral cavity of the corallites divided into two parts in consequence of the great development of two opposite primary septa.

The corallites penetrate almost perpendicularly to the surface of the celleporous mass, and, when isolated from this extraneous body, present the appearance of small, subturbinate cylinders, the walls of which are covered with a thick epitheca; there is no trace of costæ visible, and the epitheca forms round the calice a small exsert rim. The *calice* is circular, and its fossula large, but not deep. The *columella* is well developed, papillose, and not projecting, nor is it placed exactly in the axis of the visceral chamber, the *septa* being more developed on one side of the corallite than on the other. The *septa* of different orders are nearly equal in size, and do not form well-characterised systems; they vary in number from sixteen to twenty, and consequently must belong to three cycla, the first two of which are probably complete, and the third developed only in two or four of the six systems normal in all Astreidæ. It is also to be noted, that all these septa are very thin excepting near the wall, closely set, slightly bent inwardly, and terminated by an oblique edge, armed all along with strong dentations, the size of which increases towards the columella. A few large granulæ are seen on the lateral surfaces of the septa, and the loculi are divided by very thin dissepiments, placed at a distance of about two thirds of a line from each other.

The length of these corallites, when adult, is about four lines; the diameter of the calice, one line and a half; and the depth of the fossula, two lines.

Cryptangia parasita[1] of the Faluns of Touraine, is very nearly allied to the above-described species, but differs from it by the small dimensions of its calices, and the constant existence of eight principal septa.

Cryptangia Woodii is found in a good state of preservation in the Coralline Crag at Ramsholt. Specimens which appear to belong to the same species, but are not well preserved, are met with in the Red Crag of Sutton.

These fossils are to be seen in the collections of the Geological Society of London, and of Messrs. Searles Wood, Bowerbank, D'Archiac, and Milne Edwards.

Family EUPSAMMIDÆ (p. li).

Genus BALANOPHYLLIA (p. lii).

BALANOPHYLLIA CALYCULUS. Tab. I, figs. 3, 3 *a*, 3 *b*, 3 *c*, 3 *d*.

. . . . . . *R. C. Taylor*, Mag. of Nat. Hist. vol. iii, p. 272, fig. D, 1830. (Very rough figure.)

BALANOPHYLLIA CALYCULUS, *Searles Wood*, Ann. of Nat. Hist., vol. xiii, p. 12, 1844.

— — *Milne Edwards* and *Jules Haime*, Annales des Scien. Nat., vol. x, p. 84, 1848.

Corallum simple, cylindrico-turbinate, adherent by a large basal surface, erect, and in general not very tall. The *walls*, of a spongy tissue and rather thin, are covered in most

---

[1] *Lithodendron parasitum* Michelin Icon. Zooph., pl. lxxix, fig. 3.

parts by a pellicular epitheca, which Mr. Searles Wood has designated by the name of *periostracum*, and presents some slight transverse folds. In the parts where the epitheca is worn off, the *costæ* become visible. These are narrow, equal, closely set, and composed of a single series of indistinct, obtuse granulæ. The intercostal spaces present a series of small mural perforations, disposed with some regularity. None of the numerous specimens of this fossil which we have examined had the calice well preserved, and consequently we have not been able to ascertain as yet whether its margin is crenulated or entire, the fossula deep or shallow, and the columella projecting or not; but it is evident that the calice must be sub-circular, or slightly elliptic, with its two axes in the proportion of 100 : 120, and that the *columella* is spongy, not greatly developed, and spread in the direction of the long axis of the calice.

The *septa* are well developed, and always form five cycla, but do not appear ever to constitute a sixth cyclum. The mode of arrangement of these laminæ, which is characteristic in the family of Eupsammidæ, is very evident in this species : the septa of the first four cycla are straight, but those of the fifth cyclum deviate a great deal from the direction of the radii of the circle represented by the calice, and are bent. In this last cyclum the septa of the sixth order are placed very close to the primary septa, and are united to them to a certain extent, near the wall, but diverge strongly from them as they advance towards the centre of the visceral chamber, and join the ternary septa near the columella; those of the seventh order are disposed in the same way near the secondary septa, and are also united to the ternary septa by their inner edge, but do not advance quite so near the centre of the visceral chamber ; the septa of the eighth and ninth orders, which complete the fifth cyclum, are smaller than the preceding ones, and are strongly bent, so as to join the septa of the sixth and seventh orders ; and the septa of the fourth and fifth orders, which constitute the fourth cyclum, remain free, and advance in the middle of the sort of irregularly circular depressed area, formed by the coalescence of the septa of the eighth and ninth orders with those of the sixth and seventh. All the septa are very closely set and thin, but the primary and secondary ones enlarge a little towards their inner edge, and are almost equally developed, so that the adult corallum assumes the appearance of having twelve septal systems instead of six, which is the real number. We must also add, that the laminæ constituting all these septa are cribriform, and not very granulate.

The length of this corallum is commonly about eight lines, but the individual represented by fig. 3 is more than twice as tall, without being broader than usual. The calice is in general about seven or eight lines broad in one direction, and six lines in the other.

The greater development of the epitheca might suffice to distinguish *Balanophyllia calyculus* from all the other species belonging to the same genus, but it differs also from *B. prælonga*[1] (a fossil species belonging to the Miocene deposits of Turin) by its broad basis,

---

[1] *Turbinolia prælonga*, Michelin, Icon., pl. ix, fig. 1.

whereas the *B. prælonga* and the *B. Gravesii*[1] of the Paris basin have a narrow peduncle; from *B. cylindrica*[2] (a Miocene species found at Turin and Verona), *B. geniculata*[3] (a fossil belonging to the Nummulitic formation of Port des Basques), and *B. Cumingii*[4] (a recent species from the Philippine Islands), by the existence of the fifth cyclum of septa; and from *B. tenuistriata* (fossil of the Paris basin), *B. desmophyllum*[5] (fossil of the London Clay), *B. italica*[6] (fossil of the Pliocene deposits of Asti), and *B. Bairdiana* (a recent species), by its general form, and the slight elongation of its calice. The species which it most resembles is *B. verrucaria*, which exists at the present period on the coast of Corsica; but in the latter the columella is less developed, and the arrangement of the septa of the last orders is less regular.[7]

Balanophyllia calyculus is common in the Red Crag of Sutton, but has not, to our knowledge, been met with in other localities. Mr. Searles Wood considered it as identical with some fossil Coral found in the Faluns of Touraine, but the latter are young specimens of the *Dendrophyllia amica*; they resemble *B. calyculus* by their epitheca, but are easily recognisable by the mode of arrangement of their septa, and their multiplication by gemmation when in the adult state.[8]

We have examined numerous specimens of this species in the collections of the Geological Society of London, of Messrs. Bowerbank and Searles Wood, of the Museum of Natural History, and of MM. D'Archiac, Michelin, and Milne Edwards, in Paris.

[1] *Turbinolia Gravesii*, ibid., pl. xliii, fig. 7.

[2] *Turbinolia cylindrica*, Michelin, ibid. pl. viii, fig. 19.

[3] *Caryophyllia geniculata* d'Archiac, Mém. Soc. Géol. France, 2d series, vol. ii, pl. vii, fig. 7 a.

[4] See our Monograph of Eupsammidæ, Ann. des Sc. Nat., 3me série, vol. x, pl. i, fig. 8.

[5] See tab. vi, fig. 1.

[6] *Caryophillia italica*, Michelin, Icon., pl. ix, fig. 19.

[7] See our Monograph of Eupsammidæ, Ann. des Sc. Nat., 3me série, vol. x, p. 85, tab. i, fig. 6.

[8] Loc. cit., tab. i, fig. 9.

CHAPTER II.

CORALS OF THE LONDON CLAY.

THE Eocene deposits, known by the name of London Clay, contain various Corals, most of which belong to the two subordinate forms predominant among the Polypi of the present period, *Zoantharia aporosa* and *Zoantharia perforata;* but none of these organic remains can be considered as appertaining to species now in existence, or even to those found in the more recent tertiary formations. The general aspect of this portion of the fossil Fauna of England resembles very much that of the Corals imbedded in the " Calcaire grossier" of the Parisian basin. Some species, such as *Turbinolia sulcata, Stylocœnia emarciata, Stylocœnia monticularia,* and *Holarœa Parisiensis,* are common to both these localities; but most of those found in the Eocene strata of the environs of Paris have not been met with in the London Clay, and many of the Corals belonging to these last-mentioned deposits have not been discovered elsewhere. Thus the London Clay appears not to contain any Milleporidæ, Madreporidæ, or Lophoserinæ, families which have various representatives in the Fauna of the Calcaire grossier, and the only Parisian fossil Coral referable to the order of Alcyonaria is a Distichopora; whereas both Pennatulidæ and Gorgonidæ have been met with in the London Clay. At the present period similar differences exist at small distances in the same zoological region, and appear to depend principally on the depth of the sea and the nature of the bottom; by analogy we are therefore led to suppose that in the Eocene marine Fauna they are only indicative of some such local peculiarities. Indeed, most of the Corals of the London Clay belong to Polypi nearly allied to species which are now found in very deep water, and seem to be particularly organized for living on a loose, muddy, or sandy ground; whereas many of the fossil Corals of the Calcaire grossier resemble those which now inhabit rocky shores, and are seen very near the surface of the sea.

The principal localities from which our London Clay Corals have been obtained are, Haverstock Hill, Highgate and Holloway, near London; Barton, Sheppy, Bracklesham Bay, on the coast of Sussex, and Alum Bay (Isle of Wight); most of the species were found by Mr. Bowerbank, Mr. Frederick Edwards, and Mr. Frederick Dixon, to whom we are indebted for the specimens figured in this Monograph.

Order 1.—ZOANTHARIA.

Family TURBINOLIDÆ (p. xi).

Tribe TURBINOLINÆ (p. xvi).

1. *Genus* TURBINOLIA (p. xvi).

1. TURBINOLIA SULCATA.   Tab. III, figs. 3, 3 *a*, 3 *b*, 3 *c*.

TURBINOLITE DE DEUXIÈME GRANDEUR, *Cuvier* and *Alex. Brongniart*, Géogr. Minéral. des
    Environs de Paris, pl. ii, fig. 3, 1808.

TURBILONIA SULCATA, *Lamarck*, Hist. des An. sans Vert., t. ii, p. 231, 1816 ; 2d edit., p. 361.

—   —   *Lamouroux*, Expos. méth. des Genres de Polypiers, p. 51, tab. lxxiv,
    figs. 18-21, 1821. (Very bad figures.)

—   —   *Cuvier* and *Brongniart*, Descript. Géol. des Environs de Paris, p. 33,
    tab. viii, fig. 3, 1822.

—   —   *Deslongchamps*, Encyclop. méthod. Zooph., p. 761, 1824.

—   —   *Goldfuss*, Petref. Germ., vol. i, p. 51, tab. xv, fig. 3, 1826. (This
    figure is very good, excepting that the columella is not conical
    enough.)

—   —   *Fleming*, Hist. of British Animals, p. 510, 1828.

—   —   *Defrance*, Dict. des Scien. Nat., vol. lvi, p. 93, 1828. (The Coral
    figured under this name in the Atlas of the Dictionnaire des Sci-
    ences Naturelles, tab. xxxvi, fig. 2, and in the Manuel d'Actinologie,
    by M. de Blainville, is not a Turbinolia, and appears to belong to
    the genus Trochocyathus.)

—   —   *Holl*, Handb. der Petref., p. 415, 1829.

—   —   *Bronn*, Lethæa Geognostica, vol. ii, p. 899, tab. xxxvi, fig. 4, 1838.
    (This figure is good, but the columella is rather too thick.)

—   —   *Nyst*, Descript. des Coquilles et Polypiers fossiles de la Belgique,
    tab. xlviii, fig. 11, 1843. (This figure is copied from Goldfuss ;
    the description is referable to the *Turbinolia Nystiana*.)

—   —   *Michelin*, Iconogr. Zooph., p. 151, pl. xliii, fig. 4, 1844.

—   —   *Graves*, Topogr. Géogn. de l'Oise, p. 701, 1847.

—   —   *Milne Edwards* and *Jules Haime*, Annales des Scien. Nat., 3ᵐᵉ série,
    vol. ix, p. 236, 1848.

This corallum has the form of a cylindroid, elongated cone, and is not contracted just
above its basis, nor inflated near the calice (figs. 3, 3 *b*) ; sometimes only the cone is
somewhat shorter in proportion to its length (fig. 3 *a*). The *costæ* are very thin, sharp,
straight, and very prominent from top to bottom, but particularly so near the basis of the
corallum. The secondary costæ are nearly as long as the primary ones ; they do not,
however, originate quite at the same level. The tertiary costæ begin to appear about half
way up the wall in young specimens, and occupy two thirds of the height of the Coral in

the adult state. The intercostal grooves are deep and broad; near the calice a small longitudinal line is visible in each of them, and indicates the existence of a fourth cyclum of rudimentary costæ, which do not correspond to any of the septa on the inner side of the wall. These vertical furrows also present a double series of small dimples, which are prolonged laterally on the sides of the costæ, so as to constitute a sort of transverse fluting, and are arranged alternately; they are very closely set, and about fifteen occupy the space of a line. The *wall* is very thin. The *calice* is circular, and its fossula is not very deep. The *columella* is terminated by a conical, pointed apex, which rises higher than the septa, and is delicately granulated. The *septa* are thin and very exsert, but not quite so much so as in the *Turbinolia Dixonii*;[1] their upper edge is strongly arched, and their lateral surfaces present small granulations, which form short submarginal, radiate lines near the apex, and are arranged in nearly horizontal rows towards the lower part of the visceral chamber. The inner edge of the apical portion of the septa is slightly concave, and soon becomes horizontal, so as to meet the columella, to which it unites. A projecting line extends from each of the six primary septa up the apical portion of the columella; the secondary septa join the columella much lower down, but they are broader than the primary ones. The tertiary septa are narrower at the apex, and less exsert than the preceding ones; they converge towards the intermediate primary septa, and become united to them all along their inner edge, at about two thirds of the breadth of the latter. The height of the corallum is usually about three or four lines, and the diameter of the calice about one line and a half. In young specimens the calice is larger in proportion.

This fossil is the only species belonging to the genus Turbinolia as now circumscribed, which was known at the time when Lamarck first established the group bearing that name. Shortly afterwards, Mr. Defrance discovered a second species, and Mr. Isaac Lea has since then found a third. In our Monograph of Turbinolidæ, published about a year ago, six species were described, and we now know double that number of true Turbinoliæ, but they all belong to the same geological period, and are imbedded in Eocene deposits. They appear to be more abundant in England than elsewhere; the London Clay contains eight species, only one of which (the fossil just described) has been met with in the synchronous formation of the Parisian basin.

Turbinolia sulcata differs from *Turbinolia dispar*,[2] and from *Turbinolia costata*,[3] by the number of the septa which in these two last-mentioned species form four cycla. An additional cyclum of costæ distinguishes *Turbinolia Fredericiana*[4] from it; in *Turbinolia Prestwichii*,[5] *T. minor*,[6] and *T. firma*,[7] the costæ are not so thin, prominent, and wide apart as in this species, and the last of these characters separates it also from *Turbinolia pharetra*[8]

[1] See plate iii, fig. 1.   [2] Michelin, Icon., pl. xliii, fig. 5.

[3] These species, as well as the others only quoted here, have been described at full length in our Monograph of Turbinolidæ, published in the Annales des Sciences Naturelles, 3ᵐᵉ série, vol. ix.

[4] See pl. iii, fig. 2.   [5] See tab. iii, fig. 5.   [6] See tab. ii, fig. 5.

[7] See tab. ii, fig. 4.   Contrib. to Geol., tab. vi, fig. 210.

and *T. Nystiana*,[1] to which it is, however, closely allied. The species which it resembles most are, however, *Turbinolia Dixonii*, *T. humilis*,[2] and *T. Bowerbankii*.[3] The last of these differs from *T. sulcata* by its form, which is more elongate and conical, by the thickness of the lower part of its primary costæ, and by its very slender columella. *Turbinolia Dixonii* is easily distinguished by its compressed columella, by the enlargement of its walls near the calice, and by the great prominence of its costæ. To conclude this brief comparison, we must add, that *Turbinolia sulcata* differs from *T. humilis* by its size, by its form, which is not near so cylindrical as in the latter, and by the normal number of its septa.

This species is extremely abundant in certain localities of the environs of Paris, such as Grignon, Parnes, and Auvert; it is also found in the tertiary strata of Hauteville, in Normandy, and in the London Clay at Bracklesham Bay, but it is not common in this last-mentioned deposit. We are indebted to Mr. Frederick Edwards and to Mr. F. Dixon for the specimens figured in this Monograph.

2. TURBINOLIA DIXONII. Tab. III, figs. 1, 1 *a*, 1 *b*, 1 *c*, 1 *d*.

> TURBINOLIA DIXONII, *Milne Edwards* and *J. Haime*, Monogr. des Turbinolides, Ann. des Sc. Nat., 3me série, vol. ix, p. 238, tab. iv, figs. 2, 2 *a*, 2 *b*. 1848.[4]
> — SULCATA, *Lonsdale*, in the MS. work of M. Dixon on the Chalk Formations and Tertiary Deposits of Sussex.

Corallum slightly contracted just above its basis, and rather inflated near the calice. *Costæ* very thin, and projecting very much, especially towards the lower part of the wall; those of the third cyclum beginning very near the basis, and those of the first and second cycla beginning almost at the same height. Intercostal furrows nearly of the same size, very broad, and very deep; intercostal dimples very distinct, separated by small transverse laminæ, disposed as usual, in two vertical rows, and prolonged laterally, so as to produce the appearance of transverse fluting on the sides of the costæ; about ten of these dimples

---

[1] We have given this specific name to the Turbinolia described by M. Nyst, and considered by that author as being referable to the *Turbinolia sulcata* (see Coquilles et Polyp. des Ter. tert. de la Belgique, p. 629; but not the corresponding figure, which is copied from the work of Goldfuss, and belongs to *T. sulcata*). In order to facilitate the comparison between the British Turbinolia and the species found in other countries, we think it may be useful to point out the characteristic features of the *T. Nystiana*, which were not known to us when we published our Monograph of Turbinolidæ.

Turbinolia Nystiana, nob. (*T. sulcata*, Nyst, loc. cit.) Corallum elongated, slightly contracted a little above its basis, and somewhat inflated near the calice. Costæ very slightly prominent, and rather thick; the primary and secondary ones very broad towards the basis; the dimples of the intercostal furrows very small, but distinct, and those of one series alternating with those of the other. No rudiments of a fourth cyclum of costæ. Columella small, and almost cylindrical. Septa rather thick, slightly granulated, and forming three complete cycla. Length 3⅓ lines; diameter of the calice 1⅓ line. Fossil from the environs of Brussels. (Cabinet of M. Nyst at Louvain.)

[2] See tab. ii, fig. 4. [3] See tab. ii, fig. 3.

[4] In fig. 2 *a* of this plate, the principal septa are not broad enough towards the calice, and the concavity of their inner edge is placed rather too high.

occupy a line in length. No rudiments of a fourth cyclum of costæ, and a well-marked depression in calicular edge of the wall, corresponding to each of the intercostal spaces. *Calice* with a very narrow, but rather deep fossula. *Columella* compressed, arched at the apex, granulated on the surface, reaching in general to the same height as the septa, or even higher, and presenting, in the part where it begins to become isolated, six vertical striæ, which are in continuity with the inner edge of the six primary septa; rather lower down, the columella is slightly contracted, and a vertical section of the corallum (fig. 1*b*) shows that its tissue is compact, and that towards the bottom of the visceral chamber it becomes united with the septa, so as to form a solid mass. *Septa* thin, unequal, very exsert, having their upper edge strongly arched near the outer margin, but concave near the centre of the calice, slightly granulated laterally, and forming three cycla; those of the first and the second cycla nearly of the same height, but the secondary ones much broader at the apex than the primary ones, and not extending so far up the columella. The tertiary septa much narrower and shorter than the older ones; very thin towards their inner edge, and cemented to the primary septa, as in the preceding species (fig. 1*b*). Interseptal loculi large. Height of the corallum about four lines. Diameter of the calice, two lines and one third. The form and the proportions not differing in the young and in the adult specimens.

Turbinolia Dixonii is the largest known species of the genus, but *T. dispar* and *T. Prestwichii* are almost of the same size. This species is very closely allied to *T. sulcata*, from which it differs principally by its form (rather more inflated near the calice), by its compressed columella, by its septa being more exsert, and its costæ more projecting and more distant. The breadth of the intercostal furrows distinguishes both *T. Dixonii* and *T. sulcata* from *T. Prestwichii*, *T. minor*, *T. firma*, *T. pharetra*, *T. Nystiana*, and *T. Bowerbankii*. The existence of only three cycla of costæ does not allow of its being confounded with *T. Fredericiana*, *T. dispar*, and *T. costata*, and, finally, *T. humilis* is easily distinguished from it by its cylindroid form, non-compressed columella, and glabrous septa.

Turbinolia Dixonii is a fossil very abundant in the London Clay at Bracklesham Bay, and has probably been confounded with *T. sulcata* by Mr. Fleming, and some other geologists, who mention the latter as being found in that locality, where it appears to be very rare. In Mr. Dixon's work, now passing through the press, Mr. Lonsdale has also described it as a variety of the *T. sulcata* of Lamarck.

The specimens, the examination of which has enabled us to recognise this new species, were given to us by Mr. Dixon and by Mr. Frederick Edwards.

3. TURBINOLIA BOWERBANKII. Tab. II, figs. 3, 3 *a*, 3 *b*.

Corallum almost conical, rather short. *Costæ* not very prominent; those of the first two cycla inflated near the basis; the tertiary ones beginning at less than a quarter of the distance from the basis to the calicular edge of the wall; slight rudiments of a fourth cyclum of costæ appearing near the calice, and consisting in very short, prominent, thin lines, most developed between the primary and the tertiary septa. Intercostal furrows rather narrow, but presenting very clearly a double row of small dimples. *Calicular fossula* not deep. *Columella* cylindrical, prominent, and very slender in proportion to the size of the calice. *Septa* very thin, exsert, rather unequal, and forming six regularly-developed tertiary systems (fig. 3 *b*). No traces of a fourth cyclum of septa corresponding to the rudimentary costæ of the fourth cyclum. The tertiary septa joining the primary ones very near the columella. The lateral surfaces of all the septa presenting delicate granulations. Height two lines; diameter of the calice one line and a third.

This species bears great resemblance to *Turbinolia Fredericiana;* it differs from it by the rudimentary state of the fourth cyclum of costæ, and by its slender, round columella. The existence of well-formed intercostal dimples distinguishes it from *T. minor, T. costata, T. Prestwichii,* and *T. firma;* the costæ are much less prominent than in *T. sulcata,* and *T. Dixonii,* from which this Coral may also be distinguished by its form; the costæ are thinner than in *T. pharetra* and *T. Dixonii,* and the complete development of its six systems of septa does not admit of its being confounded with *T. humilis.*

We have seen but one specimen of this species; it was found at Barton, and belongs to the collection of the fossils of the London Clay formed by Mr. Frederick Edwards. We have dedicated it to our friend Mr. J. S. Bowerbank, whose active researches have much contributed to the extension of our knowledge relative to this portion of British palæontology.

4. TURBINOLIA FREDERICIANA. Tab. III, figs. 2, 2 *a*, 2 *b*.

Corallum of a regular conical form, not much elongated, and rather broad towards the calice. *Costæ* numerous, forming four cycla, closely set, unequal, and projecting very little; the secondary ones beginning a little above those of the first cyclum, but very near the basis of the corallum, and being, as well as the former, much thicker near their lower end than higher up, where they become very delicate (fig. 2 *a*). The tertiary costæ begin also at a short distance from the basis, but those of the fourth cyclum appear only in the upper half of the corallum; they are also rather thinner than the others. The intercostal furrows very narrow, and not very deep; the mural dimples not very apparent, small, closely set, and forming towards the calice, if not from top to bottom, only a single series in each intercostal furrow. *Calicular fossula* very narrow and shallow. *Columella* thick, compressed, granulated, rising higher than the septa, and presenting well-marked prolongations of the principal septa. Three cycla of septa, and no vestiges of a fourth cyclum corresponding to the quaternary costæ (fig. 2 *b*). The septa are much like those of the two preceding species, but they are a little thicker, and not so exsert; the primary ones are, as usual, narrower than the secondary ones, and these reach higher up along the columella; the tertiary septa are small, and join the primary ones, but appear to be cemented to them. The sides of all the septa present granulations arranged in radiate lines, but not very prominent. Height of the corallum two lines and a half; diameter of the calice one line and a third.

Mr. Frederick Edwards, to whom we dedicate this species, has submitted to our investigation a series of young individuals, showing the changes of form produced by age. The young Corals are rather shorter in proportion, to their breadth, than the adult ones, and consequently never resemble *Turbinolia humilis*, whatever their size may be, for the latter species is always much more cylindrical. The tertiary costæ make their appearance in *T. Fredericiana* when very young, but those of the fourth cyclum exist only in individuals that are nearly adult,

These quaternary costæ, occupying at least one third of the height of the corallum, and not corresponding to any rudiments of septa, distinguish *Turbinolia Fredericiana* from all the other species of the same genus; in some others, such as *T. sulcata* and *T. Bowerbankii*, the rudiments of similar costæ can be seen with the help of a strong lens, but these intercostal lines never become cristiform, as is the case here. The great development of quaternary costæ and the general form of the corallum make this species have some resemblance to *Turbinolia dispar*; but in the latter, as well as in *T. costata*, there is always a fourth cyclum of septa corresponding to the last cyclum of costæ. It is to *T. Bowerbankii* that *T. Fredericiana* approximates most; but in the former the columella is perfectly cylindrical and extremely slender, whereas in the latter it is large and compressed.

Turbinolia Fredericiana has as yet been found only in the London Clay, at Barton, and the specimen figured in this Monograph belongs to the collection of Mr. Frederick Edwards.

5. TURBINOLIA HUMILIS. Tab. II, figs. 4, 4 *a*, 4 *b*.

This little Turbinolia is of a much more cylindroid form than preceding species, and is not so slender at its basis. The *costæ* are thin, prominent, and not closely set; the secondary ones begin very near the basis, and those of the third cyclum at about a quarter of the way up the wall. The intercostal furrows are broad, and present each a double row of small dimples, separated by transverse or oblique bars (fig. 4 *a*). The *columella* is prominent, round, and conical. The *septa* belong to three cycla, the last of which is always incomplete, and is wanting in two of the systems;[1] but it is nevertheless evident that the number of systems is as usual six, and not five, as would at first appear, for the secondary costæ corresponding to the two incomplete systems begin near the basis of the corallum, at the same level as those of the other systems, and are as much developed as these, whereas they would have been much shorter, and would have began much higher

[1] By an inadvertency of our artist, the third cyclum is represented in fig. 4 as if it were perfect; but the specimen did not in reality present tertiary septa in more than four of the systems.

up, if they had corresponded to septa belonging to the third cyclum. All the septa are very thin, almost glabrous, exsert, and terminated by a regularly arched apex, rising more or less, according to the cyclum to which they belong. Height not quite a line and a half; diameter of the calice two thirds of a line.

This Coral is one of the smallest of the genus Turbinolia, and is indeed usually even smaller than the species designated by the name of *Turbinolia minor*, for which it may very easily be mistaken; its characteristic features can only be seen with the help of a lens, but when sufficiently magnified, the appearance of its walls will make it immediately recognisable; for in *T. minor* the costæ are very thick, crenulated laterally, and the intercostal furrows do not present any dimples, whereas in *T. humilis* these dimples are well marked, and the costæ are thin. These two species are the only ones of the genus that have apparently but five tertiary systems, and their diminutive size contributes also to make them not easily recognisable.

Turbinolia humilis is found in the London Clay at Barton, where it appears to be abundant. We are indebted to Mr. F. Dixon and Mr. Frederick Edwards for the specimens in our possession.

6. Turbinolia minor. Tab. II, figs. 5, 5 *a*, 5 *b*.

Turbinolia minor, *Milne Edwards* and *J. Haime,* Annales des Sc. Nat., 3ᵐᵉ série, vol. ix, p. 239, 1848.

Corallum of a cylindrico-conical form, rather short, and very obtuse at its basis. *Costæ* very thick, closely set, and not very prominent; those of the first and second order particularly thick near the basis; the outer edge of all very obtuse, and their sides delicately crenulated. Intercostal furrows very narrow, linear, and presenting no trace of the dimples, which are so apparent in the preceding species. The form of the costæ is particularly well marked near the calice (fig. 5 *b*), the lateral crenations of which are sometimes so developed near the basis, that they assume a crispate appearance; in other specimens they are scarcely visible, but the variations met with in the form of these parts are never such as to make them resemble the costæ of *T. humilis*. *Calice* very deep. *Columella* cylindrical, slender, and exsert. *Septa* belonging to three cycla, and appearing to form only five systems, although there are in reality six systems as usual; only in two of these there are no tertiary septa, and the secondary septa are of the size of the other tertiary ones, but correspond to secondary costæ, the development of which are normal (fig. 5 *b*). All the septa are thin, exsert, and slightly granulated on their lateral surfaces; the secondary ones are nearly as large as those of the first order, but those of the third cyclum are much smaller. This species is always remarkably small; it is not more than a line and a half high, and two thirds of a line in diameter.

Turbinolia minor differs from *T. sulcata, T. pharetra, T. Nystiana, T. Dixonii, T. Fredericiana,* and *T. humilis,* by not having the intercostal furrows ornamented with a double row of dimples, a character which in these can always be ascertained with the aid of a good lens. The imperfect development of two of the systems of septa, and the apparent existence of only five systems which is thus produced, is also sufficient to distinguish *T. minor* from *T. costata, T. dispar, T. Prestwichii,* and *T. firma.*

This fossil has been found only in the London Clay, at Alum Bay, in the Isle of Wight. The specimen figured in this Monograph belongs to the cabinet of Mr. J. S. Bowerbank.

7. TURBINOLIA FIRMA. Tab. II, figs. 4, 4 *a*, 4 *b*.

Corallum subturbinate, and elongated; narrow at the basis. *Costæ* thick, obtuse, closely set, and prominent; those of the first two cycla very broad below the under end of the tertiary ones. Intercostal furrows narrow, and presenting neither mural dimples nor well-marked lateral transverse flutings or costal crenations. *Columella* compressed, and not very large. *Septa* rather thin, delicately granulated, and forming three complete cycla; the tertiary ones less developed than the secondary ones, and cemented to the primary ones at a small distance from the columella. Height three lines and a half; diameter of the calice, one line and a half.

Turbinolia firma differs from *T. costata, T. dispar, T. Bowerbankii,* and *T. Fredericiana,* by the non-existence of a fourth cyclum of more or less developed costæ; from *T. minor* and *T. humilis,* by the complete development of the tertiary septa in the six systems, and from *T. sulcata, T. pharetra, T. Nystiana, T. Dixonii,* and *T. humilis,* by the non-existence of dimples in the intercostal furrows. It resembles very much *T. Prestwichii,* but differs from it by its general form and by its thick obtuse costæ.

We have as yet seen but one specimen of this species; it was found at Barton, and given to us by Mr. Dixon : unluckily the artist in whose hands it was placed in order to have it figured, has broken it so much that it is no longer recognisable.

8. TURBINOLIA PRESTWICHII. Tab. III, figs. 5, 5 *a*, 5 *b*.

Corallum of a cylindroid form, much elongated, and very obtuse at the basis. *Costæ* strong, rather thick, and very prominent, especially towards the basis; those of the third cyclum beginning much lower down than in most species (figs. 5 *a*), and contributing to form the convex star seen at the basis of the corallum (fig. 5 *b*). Some slight vestiges of a fourth cyclum of costæ at the bottom of the intercostal furrows near the calice. These furrows very deep, becoming very narrow near the wall, and not presenting any mural dimples but irregularly crenulated laterally, especially towards the basis. Height four lines; diameter one line and two thirds.

The only specimen of this species which we have seen belongs to the collection of Mr. Frederick Edwards, and is so much filled up with clay at its upper end, that we have not been able to ascertain well the form of the columella and the septa; we are, however, inclined to think that the columella is slightly compressed, and the septa rather thick.

Turbinolia Prestwichii differs from all the preceding species by its cylindrical form and broad convex basis; it differs also from *T. sulcata, T. Dixonii, T. pharetra, T. Nystiana, T. humilis,* and *T. Bowerbankii,* by not presenting any vertical rows of intercostal dimples; from *T. dispar, T. costata,* and *T. Fredericiana,* by having only three cycla of costæ, instead of four, and from *T. minor* and *T. firma,* in which the intercostal dimples are equally wanting, by its sharp-edged costæ.

This remarkable species was found at Haverstock Hill, and appears to be very scarce, for Mr. Frederick Edwards, whose collection of London Clay Fossils is extremely rich, has only one specimen of it, and we are not aware of its existing in the cabinet of any other palæontologist.

SUB-FAMILY OF THE CYATHININÆ (p. xii).

1. *Genus* LEPTOCYATHUS (p. xiv).

LEPTOCYATHUS ELEGANS. Tab. III, figs. 6, 6 *a*, 6 *b*, 6 *c*.

Corallum extremely short, nearly discoidal, and presenting, in the adult state, no trace of adherence. *Costæ* distinct from the centre of the under part of the corallum to the calice, strong, projecting externally, cristiform, closely set, rather unequal, separated by rather deep radiate furrows, and rendered echinulate by the presence of a multitude of granulations crowded together (figs. 6 *a*, 6 *b*). *Calice* circular, and regularly excavated in the centre. *Columella* not much developed, and delicately papillose. *Septa* constituting four complete cycla, closely set, broad, projecting much above and externally; very thin near the columella, but remarkably thick towards the circumference, and rather unequal (fig. 6 *c*); those of the first two cycla nearly equal, and larger than the others; the tertiary ones broader than those of the fourth cyclum, but not so high; all are straight, and none adhere together by their inner edge; their sides are covered with granulations, which are obtuse towards the circumference of the calice, but become spiniform in the inner part, where the septa themselves are slender. *Pali* corresponding to all the septa (even to those of the last cyclum, a mode of structure which is very rare), very thin, slightly echinulated, becoming broader as they correspond to younger septa, and appearing to be lobated, as in the genus Paracyathus. Height of the corallum, one line; diameter, three lines and a half.

The genus Leptocyathus, which we have established for this fossil, is nearly allied to Trochocyathus, but differs from it by its subdiscoidal form, the absence of all sign of adhesion at the basis, and the existence of pali corresponding to all the septa. The genus Ecmesus of Philippi[1] appears to present most of the same characters, but, as far as we can judge by the very short description, and by the rough figure given by that author, the calice appears to be eccentric, a mode of structure which is quite exceptional, and very remarkable.

The fossil Coral from the environs of Biaritz, mentioned by Viscount d'Archiac[2] under the name of *Turbinolia atalayensis*, belongs probably to the same generical division as our *Leptocyathus elegans*, but differs from it by its large size, by the existence of a fifth cyclum of septa, and by the smooth surface of the central portion of its wall.

Leptocyathus elegans was found in the London Clay, at Haverstock Hill, by Mr. Frederick Edwards, who obligingly communicated to us the only specimen that has as yet been seen.

2. *Genus* TROCHOCYATHUS (p. xiv).

TROCHOCYATHUS SINUOSUS.

> TURBINOLIA TURBINATA (*pars*), *Lamarck*, Hist. des An. sans Vert., t. ii, p. 231, 1816.
> — . . . . *Parkinson*, Organic Remains, vol. ii, tab. iv, fig. 11, 1820.
> — SINUOSA, *Alex. Brongniart*, Mém. sur les Terr. du Vicentin, p. 83, pl. vi, fig. 17, 1823.
> — — *Bronn*, Syst. des Urweltlichen Pflanz., tab. v, fig. 12, 1825. (Bad figure.)
> — DUBIA, *Defrance*, Dict. des Sc. Nat., vol. lvi, p. 92, 1828.
> — SINUOSA, *Bronn*, Lethea Geognostica, vol. ii, p. 897, 1838.
> — — *Leymerie*, Mém. de la Soc. Géol. de France, 2ᵐᵉ série, pl. xiii, figs. 7, 8, 1845.
> — — *Michelin*, Icon. Zooph., p. 270, pl. lxiii, fig. 1, 1846.
> TROCHOCYATHUS SINUOSUS, *Milne Edwards* and *J. Haime*, Ann. des Sc. Nat., 3ᵐᵉ série, vol. ix, p. 314, 1848.

We have not met with this fossil in any collection of the British Corals, but Parkinson has figured it in a very recognisable way, and mentions it as having been found in the Isle of Sheppy; we must therefore recall its specific characters in this monograph, although we deem it advisable not to have it figured from a foreign specimen.

Corallum subturbinate, rather compressed, and having its inferior extremity slightly curved in the direction of the small axis of the calice. *Costæ* distinct from the basis, very narrow, numerous, closely set, simple, unequal, delicately granulated, and projecting very little. *Calice* oval, and contracted in the middle, so as to assume the form of an 8.

---

[1] Neues Jahrb. für Mineral. Geol., vol. ix, p. 665, tab. xi, fig. B 1, 1841.
[2] Bulletin de la Soc. Géol. de France, 2ᵐᵉ série, vol. ii, p. 1010, 1847.

Columella fasciculate, with very slender elements. *Septa* forming six complete cycla, closely set, very thin and broad; those of the first three cycla nearly equal. *Pali* rather large, and thin, scarcely thicker than the septa, and presenting laterally spiniform granulations; those corresponding to the penultimate cyclum of septa being the most developed, the others nearly equal.

This Coral soon acquires all its septa and its final diameter, but continues growing up, so that it becomes sometimes very tall, without expanding proportionally; we have seen specimens three or four inches high, or even still longer. It has been found in the lower tertiary deposits of several localities in the south of France and the north of Italy.

3. *Genus* PARACYATHUS (p. xiv).

1. PARACYATHUS CRASSUS. Tab. IV, figs. 1, 1 *a*, 1 *b*, 1 *c*.

Corallum subturbinate, short, fixed by a very broad basis, slightly contracted just above the lower end, and rather inflated at the upper part. *Costæ* well marked from top to bottom, closely set, nearly equal in breadth, but alternately more or less prominent, especially near the calice, and covered with very delicate granulations. *Calice* nearly circular when young, but becoming soon more or less oval; fossula deep. *Columella* concave, papillose, thick, and not distinctly separated from the inner lobes of the pali. *Septa* forming four complete cycla, and an incomplete rudimentary fifth cyclum, in one half of the systems corresponding to the long axis of the calice; closely set, straight, slightly exsert, thin towards the centre of the visceral chamber, rather thick externally, granulated laterally, and unequally developed according to relative age. *Pali* corresponding to the septa of the first three cycla, thick, tall, strongly granulated, and denticulated along the inner edge, which is rather oblique; those corresponding to the tertiary septa larger than the others, and those that correspond to the primary septa being the smallest of all. Height, five or six lines; long axis of the calice, four lines; short axis, three lines; depth of the fossula, three lines.

This Paracyathus is easily distinguished from the other species of the same genus by the number of the septa, which in *P. procumbens*,[1] *P. Stokesii*,[2] and *P. Desnoyersii*, form an additional cyclum; by the size of the pali, which are much thicker than in *P. caryophyllus*, and *P. brevis*,[4] and by the lobulate edge of these same organs, and the oval form of the calice, from *P. æquilamellosus*, *P. Pedemontanus*,[5] and *P. Turonensis*.

Paracyathus crassus has as yet been found only in the London Clay of Bracklesham Bay, and has been communicated to us by Mr. Dixon and Mr. Frederick Edwards.

[1] Milne Edwards and J. Haime, Monogr. of Turbinolidæ, in Ann. Sc. Nat., 3d ser., vol. ix, pl. x, fig. 6.
[2] Idem, loc. cit., pl. x, fig. 7.   [3] See tab. iv, fig. 2.   [4] See tab. iv, fig. 3.
[5] *Caryophyllia pedemontana*, Michelin, Icon., pl. ix, fig. 16.

2. PARACYATHUS CARYOPHYLLUS. Tab. IV, figs. 2, 2 *a*, 2 *b*, 2 *c*, 2 *d*, 2 *e*.

> TURBINOLIA CARYOPHYLLUS, *Lamarck.* Hist. des Anim. sans Vertèb., t. ii, p. 232, 1816;
> 2d edit., p. 362.
> —   —   *Deslongchamps.* Encyclop. méthod., Zooph., p. 761, 1824.
> —   —   *Lamarck.* Tableau encyclop. et méthod. des trois Règnes,
> t. iii, p. 483, fig. 3, 1827.
> —   —   *Defrance.* Dict. des Sc. Nat., t. lvi, p. 92, 1828.
> PARACYATHUS CARYOPHYLLUS, *Milne Edwards* and *J. Haime.* Monogr. des Turbinolides,
> Ann. des Sc. Nat., 3ᵐᵉ série, vol. ix, p. 322, 1848.

Corallum turbinate, elongated, almost cylindrical in the tallest specimens, usually straight, or very slightly curved, and adhering by a moderately developed basis. *Costæ* well marked, distinct from top to bottom, nearly equal, rather narrow, not much more prominent towards the calice than near the basis, separated by broad, deep furrows, and covered with small granulations, which exist also in the intercostal furrows (fig. 2 *b*). *Calice* circular; fossula not very deep. *Columella* concave, delicately papillose, and not distinctly separated from the pali (fig. 2 *g*). *Septa* forming four cycla, the last of which is wanting in half of one or two systems; closely set, not very exsert, thin, strongly granulated laterally, and rather unequal in accordance with their relative age; the primary and secondary ones rather thick externally. *Pali* very thin, rather tall, lobated, with the inner edge oblique, and gradually larger as the septa to which they correspond are younger. Height, varying from five to eight lines; diameter of the calice, three or four lines; depth of the fossula, one line and a half.

This fossil is in general found in a bad state of preservation, being much worn away, with its basis obtuse, its wall almost entirely destroyed, and the visceral chamber filled with a carboniferous substance, the black tint of which contrasts with the white colour of the septa. Lamarck, who had only seen specimens in this state, was thus led to suppose that the corallum was free, and to consider it as appertaining to the genus *Turbinolia*. But, through the kindness of Mr. Bowerbank and Mr. Dixon, we have been enabled to examine a great number of specimens, some of which presented a complete calice, well-preserved costæ, and a basis that had evidently been adherent, so that no uncertainty could remain as to their belonging to our genus *Paracyathus*. This species differs from most other nearly allied species by the thinness of the pali, a character which is to be seen only in one other species; the *Paracyathus brevis*, from which *T. caryophyllus* is easily distinguished, by its septa being also much thinner. The number of the septa can equally serve as a character, for in *Paracyathus Stokesii*, *P. Desnoyersii*, and *P. procumbens*, there is a cyclum more than in the species here described.

Paracyathus caryophyllus is a very common fossil in the London Clay at the Isle of Sheppy; specimens of it are preserved in the collections belonging to the Geological Society of London, Mr. Bowerbank, Mr. Dixon, Mr. Fredcrick Edwards, the Museum of Paris, and M. Milne Edwards.

3. PARACYATHUS BREVIS, Tab. IV, figs. 3, 3 *a*, 3 *b*, 3 *c*.

> PARACYATHUS BREVIS, *Milne Edwards* and *J. Haime*. Ann. des Scien. Nat., 3ᵐᵉ série, vol. ix, p. 323, 1848.

The fossil remains of this species which we have had an opportunity of examining, are all more or less imperfect, and could not give us a complete knowledge of its character, but are sufficient to show that it belongs to the genus Paracyathus, and differs from all the other species of the same group. The general form of the corallum appears to be usually subturbinate and short (as in fig. 3) ; but if, as we have some reason to think, the natural interior cast represented in fig. 3 *c* belongs to this species, the proportions of height and breadth must be very variable. The *costæ* are closely set, nearly equal, thick, and covered with dense granulations. The *calice* is circular, and the fossula deep. The *columella* is concave, large, and delicately papillose. The *septa* are but very slightly exsert, closely set, thin towards the centre, and very thick towards the outer edge, strongly granulated laterally, and almost equally developed. The *pali* correspond to the septa of the first three cycla, and are very thin, lobulated, and, as usual, developed in an inverse ratio with the septa, in the prolongation of which they are placed. Height, from four to seven lines ; diameter of the calice, six lines ; depth of the fossula, one line and a half.

The existence of only four cycla of septa distinguishes this species from *Paracyathus Stokesii*, *P. Desnoyersii*, and *P. procumbens*, in which there are five of these cycla ; the tenuity of the pali distinguishes it from *P. æquilamellosus*, *P. Pedemontanus*, *P. Turonensis*, and *P. crassus ;* it is nearest allied to *P. caryophyllus*, in which the pali are also very thin, and lobulated, but in the latter the septa are much thinner, and the general form is very different.

Paracyathus brevis is found at the Isle of Sheppy, and the specimens figured in this Monograph belong to the cabinet of Mr. Bowerbank.

ABERRANT GROUP OF THE PSEUDOTURBINOLIDÆ (p. xix).

Genus DASMIA (p. xix).

DASMIA SOWERBYI. Tab. IV, figs. 4, 4 *a*, 4 *b*.

> DESMOPHYLLUM, *J. Decarle Sowerby*. Trans of the Geol. Soc. of London, vol. v, p. 136, tab. viii, fig. 1, 1834.
>
> DASMIA SOWERBYI, *Milne Edwards* and *J. Haime*. Ann. des Sc. Nat., 3ᵐᵉ série, vol. ix, p. 329, tab. vii, figs. 8, 8 *a*, 1848.

Corallum subturbinate, straight, or slightly curved, and subpedicellate or adherent by a very narrow basis. *Costæ* extremely broad, separated by deep, narrow furrows, obtuse towards the basis, rather prominent; and subcristate near the calice, covered with granulations, which become much larger towards the calice, and varying in number (17 in one specimen, 18 in another, and 22 in a third); about two thirds of them begin at the basis of the corallum, and the others about half way up towards the calice, but all are of the same breadth; the position of these younger costæ does not appear to be constant, for some are separated by three longer ones, and others by two, or only one; in general, however, two long ones are placed between two short ones, so that the latter are only about half as numerous as the former. The *calice* is nearly circular, or slightly elliptical, and the fossula appears to be deep; we are also inclined to think that there is no columella, and that the septa are free all along their inner edge, but the calice being clogged up with carboniferous matter in all the specimens that we have seen, we have not been able to determine these points with any degree of certainty. The mode of arrangement of the septa is quite abnormal; three vertical plates advance from each of the costæ towards the centre of the visceral chamber; they are all extremely thin, broad, somewhat flexuous, free from all adherence among themselves, and rendered echinulate laterally by a few prominent granulations; the plate placed in the middle of each of these groups is rather thicker than the others, and the space existing between it and the latter is rather larger than that comprised between the lateral laminæ of two neighbouring groups. Height of the corallum, about four lines; long axis of the calice, three lines and a half; short axis, two lines and a half; breadth of the costæ, more than half a line.

The three specimens of this species, from which we have drawn up the preceding description, belong to Mr. Bowerbank's palæontological collection, and were found at Highgate; Mr. Prestwich has met with it also at Clarendon Hill.[1]

Dasmia Sowerbyi is the only known species of this genus which by its general characters appears to be closely allied to the family of Turbinolidæ, but differs from it, and even from all the other Zoantharia, by the abnormal structure of the septal apparatus: when our attention was first called to this point, we endeavoured to explain the mode of radiation of the calice by supposing that each of the laminæ corresponding to the middle of the costæ belonged to one cyclum, and that the two lateral laminæ of two neighbouring groups, corresponding to the two sides of each intercostal furrow, represented the two halves of septa belonging to another cyclum;[2] the slight difference in the thickness of the middle and the lateral laminæ, as well as the facility with which the two constituent plates of the septa separate from each other in some Corals, had induced us to admit that this structure was only an exaggerated form of that which is frequently met with in certain Turbinolidæ, in many species of Flabellum, for example, where the line of junction of the two laminæ that constitute each septum is indicated externally by a single costal ridge. But a more attentive study of this singular fossil has made us change our opinion, and

---

[1] Journ. of the Geol. Soc. of London, vol. iii, p. 368.

[2] Monogr. des Turbinolides, loc. cit.

has induced us to think that each group composed of three laminæ, and corresponding to one costa must be the homologue of a single normal septum. It also appears evident that the first-mentioned hypothesis is incompatible with the mode of development of the younger septa corresponding to the short costæ; for wherever one of these younger costæ interposes itself between two older ones, a new group of three septal laminæ makes its appearance in the visceral chamber, between two of the old lateral plates, which, according to this view, would belong to one septum, and a young septum, accompanied by two half septa, would thus be included in the interior of an elder septum. Nothing of the sort is ever met with in any known corallum, and would be contrary to the general laws which appear to regulate the formation of the septal apparatus; but if we admit that each group of these vertical laminæ corresponds to a single septum in the ordinary Polypidoms, all serious difficulties disappear, and a circumstance that tends to corroborate this view of the subject, is, that in some Turbinolidæ an intermediate tissue is seen between the two lateral plates constituting each of the larger septa, so that if these three vertical strata of sclerenchyma, instead of being in contact, and intimately united, were separated by a membranous fold or duct, each septum would no longer have the appearance of a simple partition, but would resemble the trilaminate septal groups of the Dasmia. In the present state of our knowledge concerning the structure and the mode of development of this curious fossil, we must be cautious in our speculations concerning the signification of the parts just described; but it is to be hoped that a complete solution of the question will be obtained by the study of a greater number of these Corals. At all events, the development of the septal apparatus must be very abnormal in Dasmia, and appears to warrant the establishment of a separate zoological division for the reception of this extinct genus.

Family OCULINIDÆ (p. xix).

1. *Genus* OCULINA (p. xix).

OCULINA CONFERTA. Tab. II, figs. 2, 2 *a*, 2 *b*.

Corallum composite, incrusting, forming an irregular, subglobose, or lobated mass, and appearing to have always grown on some extraneous stem, which has disappeared during the process of fossilization. The corallites are not arranged in a regular way, but are usually very closely set, and the calices are unequally prominent on the surface of the cœnenchyma, which is compact, and moderately thick; its surface is covered with round, unequal, crowded granulations, and presents no distinct costæ. The *calices* are in general quite circular, excepting when preparing to multiply by fissiparity, which is very seldom the case; the edge is rather thin, and the fossula large, but not very deep. The *columella* is sub-papillose. The *septa* (fig. 2 *b*) constitute three complete cycla, besides which some

vestiges of an incomplete fourth cyclum often exists.   The six systems, independently of
these rudimentary septa, are equally developed; the septa are thin, narrow towards the
apex, strongly granulated laterally, of unequal size according to their relative age, scarcely
exsert, and terminated by a slightly arched, almost undivided edge.   The *pali* are thick,
narrow, and crispate; they form two coronets, and those corresponding to the secondary
septa are rather larger and more distant from the columella than those corresponding to the
primary septa.   Diameter of the calice, two thirds of a line; depth of the fossula, half a line.

A vertical section of one of these corallites (fig. 2*a*) shows that the walls, as well as the
cœnenchyma, are of a very compact structure, and are covered with minute granulations;
that the small tubercles arising on the lateral surface of the septa are much less crowded;
that the columella is constituted by small, irregular, filiform, ascending trabiculæ, and that
the loculi are devoid of dissepiments, or only contain very few of them.

The genus Oculina, reduced to the limits here assigned to that zoological division,
appears to have very few fossil representatives, for this is as yet the only known species
belonging to it that is not exclusively recent; and it might be almost considered as
constituting a distinct generic type, for it differs from all the recent species of Oculina by
the mode of arrangement of the corallites.   In the latter the corallites affect a spiral order
in the ascending branches constituted by their union, and the cœnenchyma presents near
the calices some slight indication of radiate costæ, whereas in this fossil the corallites, as
we have already remarked, are quite irregularly grouped, and the surface of the cœnenchyma
is not at all striated.

Oculina conferta appears to be abundant in the London Clay at Bracklesham Bay.
We have received specimens of this fossil from Mr. Dixon and Mr. Frederic Edwards.

2. *Genus* Diplhelia (p. xxi).

Diplhelia papillosa.   Tab. II, figs. 1, 1*a*, 1 *b*.

Corallum composite, subdendroid, and rather tall.   *Corallites* usually disposed alternately
in contrary directions, but appearing sometimes irregularly grouped, in consequence of
two series becoming united so as to form a single branch, or of a few individuals multiplying
by fissiparity.   The *calices* placed far apart, quite circular, scarcely prominent, if at all so,
and united by a highly-developed mural cœnenchyma, the surface of which is covered with
closely-set, unequal, minute granulations, rather oblong, especially in the vicinity of the
calices (fig. 1 *a*).   Calicular margin very thin; fossula large, and very deep.   *Columella*
very large, of a spongiose texture, and sub-papillose at the apex.   *Septa* forming three
complete cycla, and six equally-developed systems; very narrow at the upper end, not
exsert, thin, granulated on their lateral surface, and presenting along their inner edge
delicate denticulations, which become larger towards the columella, but do not assume the

appearance of rudimentary pali. The secondary septa are almost as large as the primary ones, and thus give the appearance of twelve systems (fig. 1 *b*). Sometimes septa of the fourth order exist in one of the real systems, and in that case the neighbouring tertiary septa become at the same time as large as the elder ones, so that the septal apparatus becomes divided into fourteen almost equal parts. Diameter of the calice, one line; depth of the fossula, one line, or more.

The great development and the compact structure of the cœnenchyma are rendered manifest by a vertical section of one of these corallites (fig. 1*a*); this preparation is also necessary to show the denticulations of the inner edge of the septa, and if continued to a certain distance from the calice, brings to view a few irregular, incomplete, locular dissepiments.

The new generic division, to which we have given the name of *Diplhelia*, comprises the Oculinidæ that multiply by alternate gemmation, and have denticulated, non-exsert, unequal septa, no pali, and a large columella. It differs from Astrhelia by the mode of arrangement of the corallites dependent on the alternate position of the reproductive buds, by the existence of a well-developed columella, and the absence of costal striæ near the calices. The mode of gemmation is the same in Amphelia and Enallhelia, but in these Oculinidæ the septa are entire and exsert, the columella is rudimentary, and the costal striæ are well marked near the calices.

Four species compose at present this small group; they are all fossil, and belong to the Eocene Fauna. Two of these Corals have been described by M. Defrance, under the names of *Oculina raristella* and *O. Solanderi*; the third is the *Caryophyllia multostellata* of M. Nyst; the fourth is our *Diplhelia papillosa*. *Diplhelia Solanderi*, of which a pretty good figure has been given by M. Michelin,[1] differs from the latter by the existence of numerous delicate, vermiculated, longitudinal sulci on its surface. *Diplhelia raristella*[2] differs from *D. papillosa*, by the calices being smaller and not so deep, by the septa being thicker, and the columella less developed. *Diplhelia multostellata*[3] is principally characterised by the approximation of the calices, and their dilated form.

Diplhelia papillosa has as yet been found only at Bracklesham Bay, where it appears to be abundant. The specimens here described have been communicated to us by Mr. Bowerbank, Mr. Dixon, and Mr. Frederic Edwards.

---

[1] Icon. Zooph., tab. xliii, fig. 19.   [2] Michelin, loc. cit., tab. xliii, fig. 16.
[3] Nyst, Coq. et Pol. foss. des Terr. tert. de la Belgique, tab. xlviii, fig. 10.

Family ASTREIDÆ (p. xxiii).

Tribe EUSMILINÆ (p. xxiii).

(*Eusmilinæ aggregatæ*).

1. *Genus* STYLOCŒNIA (p. xxix).

STYLOCŒNIA EMARCIATA.　Tab. V, figs. 1, 1 *a*.

ASTROITE DEMI-CYLINDRIQUE, *Guettard*, Mém. sur les Arts et les Sciences, t. iii, p. 480,
　　　　　　tab. xxxi, figs. 40, 41, 42, 1770.
ASTREA EMARCIATA, *Lamarck*, Hist. des Anim. sans Vertèb. t. ii, p. 266, 1816 ; 2ᵐᵉ edit.
　　　　　　p. 417.
— 　　— 　　*Lamouroux*, Encyclop. Zooph., p. 127. 1824.
— 　　— 　　*Defrance*, Dict. des Scien. Nat., t. xlii, p. 389, 1826.
— 　CYLINDRICA, Ejusd., loc. cit., p. 379. (From a worn specimen.)
— 　STYLOPORA, *Goldfuss*, Petref. Germ., vol. i, p. 71, tab. xxiv, fig. 4, 1826. (From a
　　　　　　frustrate specimen.)
CELLASTREA EMARCIATA, *Blainville*, Dict. des Sc. Nat., vol. lx, p. 342, 1830 ; and Manuel
　　　　　　d'Actinologie, p. 377. (The fossil figured in the atlas of this
　　　　　　work, pl. liv, fig. 5, under the name of *Cellastrea hystrix*,
　　　　　　belongs to this species.)
ASTREA EMARCIATA, *Michelin*, Icon. Zooph., p. 154, tab. xliv, fig. 6, 1844.
— 　CYLINDRICA, Ejusd., op. cit., tab. xliv, fig. 4.
— 　DECORATA, Ejusd., op. cit., p. 161, tab. xliv, fig. 8.
STYLOCŒNIA EMARCIATA, *Milne Edwards* and *J. Haime*, Monogr. des Astreides, Ann. des
　　　　　　Sciences Naturelles, 3ᵐᵉ série, vol. x, p. 293, tab. vii, figs. 2, 2 *a*,
　　　　　　1848.

It is only in the Eocene deposits of the Parisian basin at Grignon and at Parnes that
this species has as yet been met with in a good state of preservation, but its existence in
the London clay is sufficiently established by two small fossils found at Bracklesham Bay,
by Mr. Frederick Edwards, which do not appear to differ from the worn specimens found,
together with the well-characterised ones in the first-mentioned localities. The following
description is consequently derived principally from the Parisian specimens ; but in order
to avoid introducing into this Monograph any uncertain elements, we have figured the
British specimens in preference to more perfect foreign fossils with which we consider them
as being specifically identical.

Astrea emarciata is a composite Coral, of an oval, gibbous, or subramose form, which at
first sight appears to be completely free, but was in all probability primitively fixed on
some soft, globular, extraneous body, which after having been completely covered by the
incrusting Coral, disappeared by the progress of putrefaction, and has only left a central
cavity in the middle of the irregular globose mass thus produced : it consists of a thick

lamellar expansion, bent so as to shut up completely an irregular cavity, and to have all the calices of its constituent corallites turned outwards. The basal or inner surface of this lamellar corallum is coated with a thin, membranous epitheca, in which circular striæ, indicative of its mode of growth, are perceptible. The *calices* are polygonal, and rather unequal in size; they are separated by a simple edge, which is common to the two adjoining corallites, and is thin where these corallites are crowded together, but rather thick where the reproductive process has been less active; in the latter case these mural edges are covered with numerous well-marked granulations, but in the former, no appearance of granulations is to be seen. Sometimes these two states are met with in different parts of the same specimen, but in others the whole mass presents one or the other of these forms, and may then be easily mistaken for distinct species. It is thus that M. Michelin has been led to consider the thick-walled variety as constituting a new species to which he has applied the name of *Astrea decorata*. The calicular margins present also at each corner a well-formed cylindro-conical columnar tubercle or process, which is not very thick at the basis, and is usually fluted by six or eight vertical furrows. In specimens that have been much rolled by the sea, these mural processes are often worn away, and these dilapidated Corals have also been described by palæontologists as a distinct species; they constitute the *Astrea cylindrica* of M. Defrance. The *columella* is slender, cylindrical, and free down to a great distance from its apex, but presents at the bottom of the fossula vertical striæ, which are produced by the prolongation of the principal septa along its sides, and are particularly manifest in some worn-down specimens, such as those found at Bracklesham, and figured in this Monograph (fig. 1 *a*). The *septa* form two complete well-developed cycla; a third cyclum is rudimentary in four of the systems, but well developed in two systems, the secondary septa of which become nearly as large as the primary ones, so as to give to the calice the appearance of having eight systems instead of six, which is the fundamental number. The eight large septa thus formed are broad, very thin, almost glabrous, not exsert, and terminated by regularly-arched, undivided edges; the other intermediate septa are very small. The interseptal dissepiments are simple, somewhat concave, slightly raised towards the columella, and placed at the distance of about one third of a line apart. The breadth of the calice is about one line and a third; the height of the mural processes two thirds of a line.

The British fossils which we refer to this species, and which we have figured in the annexed plates, have evidently been modified by the long-continued action of the sea; the septa are much broken, and the granulations of the calicular margins are not visible; it appears not improbable that the polypi to which they belonged did not live in the locality where these remains have been found, and that the Corals were brought there by some marine current. They are very rare at Bracklesham Bay, but extremely common in the Calcaire grossier of the environs of Paris. M. Michelin states that the same species is met with at La Palarea, and the Stylocœnia found in this locality is certainly very similar to *S. emarciata*, but all the specimens that we have been able to examine, were in such a bad condition that we cannot give any decided opinion as to their specific identity with the Corals described above.

Stylocœnia emarciata differs from *S. monticularia*,[1] and from *S. Taurinensis*[2] by the number of the large septa which in the latter species is only six ; it differs much by its general form from *S. Lapeyrousiana*,[3] and resembles most *S. lobato-rotundata*,[4] from which it may be distinguished by a greater development of the mural tubercles, the tenuity of the septa, and the general form of the mass.

2. STYLOCŒNIA MONTICULARIA.   Tab. V, figs. 2, 2 *a*, 2 *b*.

> STYLOPORA MONTICULARIA, *Schweigger*, Beob. auf Naturg. reisen, tab. vi, fig. 62, 1819.
> (Correct figure.)
> ASTREA HYSTRIX, *Defrance*, Dict. des Sc. Nat., vol. xlii, p. 385, 1826.
> CELLASTREA HYSTRIX, *Blainville*, Dict. des Sc. Nat., vol. lx, p. 342, 1830 ; Manuel d'Actin.,
> p. 377.
> ASTREA HYSTRIX, *Michelin*, Icon., p. 160, tab. xlv, fig. 1, 1845.
> STYLOCŒNIA MONTICULARIA, *Milne Edwards* and *J. Haime*, Ann. des Sc. Nat., 3ᵐᵉ série,
> vol. x, p. 294, 1848.

Corallum composite, elongated, and nearly cylindrical when young, but becoming, by the progress of growth, oval, sub-globose, and free, (fig. 2) ; with an empty central cavity, the parietes of which are coated with a thin epitheca, wrinkled circularly (fig. 2 *b*).   The exterior surface of this hollow mass is covered with the *calices*, the borders of which vary in form according to the age of the compound Coral ; when the colony of polypi is young, the margins of the calices are circular, prominent, and separated from each other by a striated surface ; but in older groups, they become polygonal and united, so as to form a single thin ridge common to the two neighbouring corallites.   The mamilliform processes that rise at the angles of the calices are stout, conical, broad at the basis, and covered with numerous prominent, sub-lamellar, finely-denticulated, vertical striæ.   The *columella* is very slender, prominent, free a great way down, cylindrical towards the apex, and slightly compressed at the basis.   The *septa* form only two complete cycla ; the primary ones are much larger than those of the second cyclum, broad, not exsert, granulated laterally, and terminated by an undivided convex edge.   Sometimes the striæ of the mural processes are continued down along the parietes of the visceral chamber, and assume the appearance of rudimentary septa.   Diameter of the calice, two thirds of a line ; height of the mural processes, half a line.

This species differs from *Stylocœnia emarciata*, and *S. lobato-rotundata*, by the regular and equal development of the six systems of septa, whereas, in the latter, two of these

---

[1] See tab. iv, fig. 2.   [2] Michelin, Icon., tab. xiii, fig. 2.
[3] Michelin, op. cit., tab. lxx, fig. 3.   [4] Michelin, op. cit., tab. xiii, fig. 2.

systems are apparently double, so that the septal apparatus is divided into eight nearly equal groups. *Stylocœnia Lapeyrousiana* differs from it by its conical form, and by the existence of three complete cycla of septa. *Stylocœnia Taurinensis* has equally but two septal cycla, but the mural processes are much smaller than in *S. monticularia;* the calices are larger, and the primary septa are united to columella very high up.

Stylocœnia monticularia has been found at Bracklesham Bay, by Mr. Frederick Edwards, but appears to be rare in that locality; it is, on the contrary, very common at Grignon, and in many other places near Paris.

2. *Genus* Astrocœnia (p. xxx).

Astrocœnia pulchella. Tab. V, figs. 3, 3 *a*, 3 *b*, 3 *c*.

Corallum composite, astreiform, massive, or subcolumnar, and presenting at its under surface a common plate, covered with a complete epitheca, delicately wrinkled by concentric striæ. Corallites approximating more or less, according to the age of the Coral and the degree of activity with which gemmation has been carried on. When the corallites are not crowded together, the calices are circular, and have a distinct though not prominent edge; they are also separated by a pseudo-cœnenchyma, the surface of which is covered with small costal ridges, that are usually denticulated, so as to assume the appearance of rows of round, obtuse granulæ (fig. 3*c*). When the calices approximate, they become somewhat polygonal, and their margins are separated only by a narrow furrow, or united so as to appear simple. The calicular fossula is very shallow. *Columella* cylindrical, obtuse, and free to a considerable extent, but not rising quite so high as the septa. Three complete cycla of *septa*, and six equally-developed systems; the septa of unequal size, according to their relative age, straight, slightly exsert, closely set, feebly granulated, rather thick externally, and having their upper edge entire and convex. Breadth of the calices, two thirds or three quarters of a line.

This species belongs to the division of the irregular Astrocœnia,[1] for independently of the slight inequality perceptible in the size of the calices, it is evident that gemmation takes place in this Coral simultaneously at various parts; but it differs from the other species of the same section, by the number of the septa; here, as we have already said, the six systems are equally developed, whereas in *Astrocœnia Koninckii, A. Orbignyana, A. reticulata, A. ornata, A. ramosa,* and *A. decaphylla,* there is always apparently eight or even ten systems.

We know of only three small specimens of this species, which were found at Bracklesham Bay, and belong to the cabinet of Mr. Frederick Edwards.

---

[1] See our Monograph of the Astreidæ, Ann. des Sc. Nat., 3ᵐᵉ série, vol. x.

Family EUPSAMMIDÆ (p. li).

1. *Genus* STEPHANOPHYLLIA (p. liii).

STEPHANOPHYLLIA DISCOIDES.   Tab. VI, figs. 3, 3 *a*, 3 *b*.

STEPHANOPHYLLIA DISCOIDES, *Milne Edwards* and *J. Haime*, Ann. des Sc. Nat., 3ᵐᵉ série,
vol. x, p. 93, 1848.

Corallum simple, extremely short, and discoidal; its under surface almost horizontal, somewhat prominent in the middle, and showing no trace of adhesion. *Costæ* very narrow, radiate, alternating with the septa, corresponding to these in number, and nearly equal in breadth, but differing much in length, according to the cycla to which they belong; the smaller ones often united to the larger ones at their basis, and thus giving to the latter a dichotomous appearance (fig. 3*b*). All these costal striæ are composed of a single row of rather indistinct granulæ, and are united by small intercostal trabiculæ, thus constituting the tissue of the discoidal wall, and the radiate rows of pores that exist in this part of the corallum, and give to it the appearance of a microscopical sieve. The upper or *calicular surface* somewhat convex, and presenting in the centre a small, narrow fossula, at the bottom of which there appears to be a rudimentary papillose columella. *Septa* forming five cycla, of very unequal size, thin, very slightly granulated, not projecting laterally beyond the edge of the wall, and having the upper edge rather angular. Those of the first and second order large, straight, and free at their inner end; all the others bent towards one another, and cemented along their inner edge, so as to constitute a series of slightly undulated arches, superposed and increasing in size from the circumference of the calice towards the centre; the largest are formed by the septa of the fourth cyclum, which unite two by two, along the inner edge of the tertiary ones (which are very short), and thus constitute on each side of the secondary septa a single lamina, that advances still further towards the centre of the calice, and joins the neighbouring secondary septum opposite the point of junction of its homologue, so as to give to the central portion of the calice the appearance of a six-branched cross of Malta; the septa of the fifth cyclum very small and marginal. Diameter, two lines or two lines and a half; height, about half a line.

This fossil Coral differs from *Stephanophyllia Suecica*[1] and *S. Bowerbankii* (Tab. IX, fig. 4), by the form of the septa, which in the latter two species are terminated by an arched edge, and are spinulose laterally; it is distinguished from *Stephanophyllia astreata*[2] by the distance that separates the large septa near the columella, and from *S. elegans*,[3] *S. imperialis*,[4] and

---

[1] See our Monograph of Eupsammidæ, Ann. des Sc. Nat., 3ᵐᵉ série, vol. x, p. 94.

[2] *Fungia astreata*, Goldfuss, Petref. Germ., vol. i, tab. xiv, fig. 1.

[3] Ann. des Sc. Nat., 3ᵐᵉ série, vol. x, tab. i, figs. 10, 10 *a*.

[4] Michelin, Icon. Zooph., tab. viii, fig. 1.

S. Nystii,[1] by its diminutive size, and by its low, very feebly-granulated septa. It is worthy of notice that *S. discoides* is the only species of this genus that has as yet been found in the Eocene formations. We have seen four specimens of this Coral; they were all met with at Haverstock Hill, and belong to the cabinet of Mr. Frederick Edwards.

2. *Genus* BALANOPHYLLIA (p. lii).

BALANOPHYLLIA DESMOPHYLLUM. Tab. VI, figs. 1, 1 *a*, 1 *b*, 1 *c*.

BALANOPHYLLIA DESMOPHYLLUM, *Milne Edwards* and *J. Haime*, Monog. des Eupsammides, Ann. des Sc. Nat., 3me série, vol. x, p. 86, 1848.

Corallum simple, adherent by a broad, incrustating basis, subturbinate, straight, rather elongated, and slightly compressed. *Costæ* almost straight, closely set, and formed of one or more rows of irregular granulæ; the primary and secondary ones much taller and much larger than the others, especially near the calice, and usually separated by five small ones, two of which begin to appear at about two thirds down the wall, whereas the others extend to the basis. *Calice* slightly arched, and almost elliptical; its long axis being to the short one in the proportion of 100 : 160. Calicular fossula deep and narrow. *Columella* spongy, not highly developed, flat, and not prominent at the bottom of the calice. *Septa* forming five cycla, usually complete; those of the fifth cyclum more developed than those of the third order, and becoming cemented together two by two beyond the inner edge of those of the fourth cyclum, and constituting thus in each half system two septal laminæ, that in their turn unite between the inner edge of the tertiary septa and the columella. In the neighbourhood of the wall, the septa of the fifth cyclum that are situated next the primary and the secondary ones are cemented to them, and do not usually correspond to any distinct costæ; so that in each half system there are only five costæ corresponding to seven septa. The large septa are terminated by an undivided edge, and are much more exsert than the others; all are thin, granulated laterally, very porous, and closely set; those of the younger orders are delicately denticulated. Height of the corallum about

---

[1] We have given this name to a Stephanophyllia of the Antwerp Crag that we have seen in M. Nyst's cabinet at Louvain, and had been referred by that author to the *S. imperialis* (Coquilles et Polyp. foss. de Belgique, p. 633, tab. xlviii, fig. 17). This figure is pretty good, but does not show the small septa. Not having described it in our Monograph of Eupsammidæ, we point out here its characteristic features. The under surface of *Stephanophyllia Nystii,* nob., is somewhat concave. The *costæ* are of almost equal thickness, and do not appear distinctly composed of rows of granulæ; they alternate with the well-developed septa, but correspond to rudimentary septa of the sixth cyclum; the younger ones are, as usual, united by their base to the elder ones, but this apparent bifurcation takes place only very near the centre of the corallum; the intercostal furrows become gradually wider from the centre towards the circumference of the wall, and are bored with pores, that increase in size in the same manner. The *calicular fossula* is very deep. The *septa* are disposed in the same way as in *S. discoides* and *S. elegans,* but are much taller, thinner, and more angular; they are denticulated externally, and present on their lateral surfaces radiate striæ, which resemble incomplete synapticulæ; those of the last cyclum are very small. Diameter nearly an inch; height, 5½ lines.

seven lines; long axis of the calice, three lines and a quarter; short axis, two lines; depth of the fossula, two lines.

This species belongs to the section of the fixed Balanophylliæ, and is consequently easily distinguished from *B. prælonga*[1] and *B. Gravesii*,[2] which are only sub-pedicellate. The nakedness of the wall, and quite rudimentary state of the epitheca, distinguishes it also from *B. calyculus*,[3] *B. verrucaria*,[4] and *B. cylindrica*.[5] In *B. geniculata*[6] and *B. Cumingii*[7] there are but four cycla of septa, whereas in the above-described fossil there are five cycla. It differs from *B. italica*[8] by its elongated and compressed form, from *B. Bairdiana* by its exsert septa, and from *B. tenuistriata* by the compressed form of its lower part, and the thickness of its principal costa. It resembles most this last-mentioned species, which belongs to the Calcaire grossier of the Parisian basin.

Balanophyllia desmophyllum is found at Bracklesham Bay, and has been communicated to us by Mr. Dixon and Mr. Frederick Edwards.

3. *Genus* DENDROPHYLLIA (p. liii).

DENDROPHYLLIA DENDROPHYLLOIDES. Tab. VI, figs. 2, 2 *a*, 2 *b*, 2 *c*.

> OCULINA DENDROPHYLLOIDES, *Lonsdale*, in Mr. Dixon's manuscript work on the Chalk Formations and Tertiary Deposits of Sussex.
> DENDROPHYLLIA DENDROPHYLLOIDES, *Milne Edwards* and *J. Haime*, Ann. des Sc. Nat., 3ᵐᵉ série, vol. x, p. 102, 1848.

Corallum composite, appearing usually to have incrusted the stem of some marine plant which has been destroyed during the process of fossilization. Gemmation irregular. Corallites short, very unequal in size, rather closely set, united by their basis, and free down to a variable distance from the calice, so as to project more or less on the surface of the common mass, or even to form a certain number of somewhat ramified branches (fig. 2). *Costæ* delicate, numerous, closely set, almost equal in breadth, composed of a row of irregular, conical granulæ, having a sub-vermiculate appearance, and becoming more irregular and more flexuous in the parts where they unite with those of neighbouring corallites (fig. 2 *b*). Mural pores large, and very distinct near the calice, but ceasing to be so lower down, where the tissue of the wall becomes very compact. *Calices* regularly circular, with the edge rather thin, and the fossula infundibuliform, but not deep. *Columella* spongiose, not much developed, and appearing to be but slightly prominent at the bottom of the fossula. *Septa* forming four complete cycla, and sometimes a rudimentary incomplete fifth cyclum; very thin, unequal, not exsert, or only very slightly so, and granulated

---

[1] Michelin, Icon., tab. ix, fig. 2.   [2] Michelin, Icon., tab. xliii, fig. 7.   [3] See tab. i, fig. 3.
[4] Milne Edwards and J. Haime, Ann. des Sc. Nat., 3ᵐᵉ série, vol. x, tab. i, figs. 6, 6 *a*.
[5] Michelin, op. cit., tab. viii, fig. 15.
[6] D'Archiac, Mém. de la Soc. Géol., 2ᵐᵉ série, vol. ii, tab. vii, fig. 7.
[7] Milne Edwards and J. Haime, loc. cit., fig. 8.   [8] Michelin, loc. cit., tab. ix, fig. 15.

laterally. The six systems very distinct; the primary septa much broader and taller than the others; the secondary ones smaller than those of the fourth order, which unite to those of the fifth order opposite the almost rudimentary tertiary septa, and continuing to bend towards the secondary ones, unite two by two along the inner edge of these, and so constitute six laminæ, that advance almost to the columella, and appear at first sight to be prolongations of the secondary septa (fig. 3 a). The largest of these corallites are about two lines in diameter at the calice, and project little more than a line above the common mass; the depth of the fossula is about one line and a quarter. The young individuals very soon acquire all their septa.

The genus Dendrophyllia has many representatives in the seas of the present period, and in the upper tertiary formations, but the species here described is the only one that has as yet been found in Eocene deposits, and is the oldest known, for the various fossil Corals appertaining to remoter geological periods that have been referred to this generic division by M. Michelin, and by some other authors, do not in reality belong to it. *Dendrophyllia dendrophylloides* differs from *D. ramea*,[1] *D. Taurinensis*,[2] and *D. digitalis*,[3] by its irregular gemmation; from *D. Cornigera*,[4] *D. irregularis*,[5] *D. amica*,[6] and *D. axifuga*, by having fewer septa (a cyclum less); and from *D. gracilis*[7] by not being arborescent. It resembles most our *Dendrophyllia Cecilliana*, but this last-mentioned species, which lives in the Chinese seas, is sufficiently characterised by its broad costæ, formed by double or triple rows of small granulations, and by its large prominent columella.

This fossil Coral is found at Bracklesham Bay, and has been communicated to us by Mr. Bowerbank, Mr. Dixon, and Mr. Frederick Edwards.

4. *Genus* STEREOPSAMMIA (p. liii).

STEREOPSAMMIA HUMILIS. Tab. V, figs. 4, 4 a, 4 b.

Corallum composite, incrusting, glomerulate, remaining low, and increasing by means of an irregular basal gemmation. Corallites crowded together, cylindrical, short, united by the basis, and free down at least two thirds of their length from the calice. The costal tissue which unites them at their basis, is in general but little developed, and does not deserve the name of cœnenchyma. Gemmation seldom takes place laterally; sometimes, however, reproductive buds are formed on the side of a parent corallite at a certain distance from its basis, and will thus produce a slight appearance of ramification. *Costæ* very

---

[1] *Madrepora ramea*, Solander and Ellis, Zooph., tab. xxxviii.

[2] Michelin, op. cit., tab. x, fig. 8.

[3] Michelin, loc. cit., tab. x, fig. 10; and tab. lxxiv, fig. 4.

[4] Esper, Pflanz. Madrep., tab. x.

[5] Michelin, op. cit., tab. lxxiv, fig. 3.

[6] Milne Edwards and J. Haime, Ann. des Sc. Nat., 3me série, vol. x, tab. i, fig. 9.

[7] Milne Edwards and J. Haime, loc. cit., fig. 13.

slender, closely set, projecting very little, equal, sub-vermiculate, and assuming the appearance of vertical striæ irregularly broken at short distances, but not distinctly composed of granulations, as is the case in most Eupsammidæ (fig. 4 *a*). *Walls* perforated, as usual in this family, near the calice, but becoming compact lower down. *Calices* circular, infundibuliform, rather deep, and having an obtuse edge. *Columella* quite rudimentary, or not existing at all. Four *septal cycla*, the last of which is quite rudimentary, whereas the others are well developed proportionably to their age. The septa are very thin, closely set, not remarkably exsert, terminated by an oblique, nearly entire edge, and proceed in a straight direction towards the axis of the corallum, but present an undulate appearance, due principally to the existence of a few large lateral granulæ. The six systems are equally developed, and very distinct; the primary and secondary septa meet along their inner edge in the middle of the visceral chamber. Height of the corallites, about two or three lines; diameter of the calices, about two thirds of a line.

This fossil, of which we have seen but one specimen, that was found at Bracklesham Bay, and belongs to the cabinet of Mr. Frederick Edwards, is the only known species of the genus Stereopsammia. The regular radiate structure of its calice distinguishes it from most of the Eupsammidæ, and in the genus Cœnopsammia, where the same character is met with, the columella is essential and well developed,[1] whereas in Stereopsammia it does not exist, or is quite rudimentary.

Family PORITIDÆ (p. lv).

Tribe PORITINÆ (p. lv).

1. *Genus* LITHARÆA (p. lv).

LITHARÆA WEBSTERI. Tab. VII, figs. 1, 1 *a*, 1 *b*, 1 *c*.

ASTREA WEBSTERI, *J. S. Bowerbank*, on the London Clay Formation, in Charlesworth's Mag. of Nat. Hist., new series, vol. iv, p. 23, figs. A, B, 1840.
SIDERASTREA WEBSTERI, *Lonsdale*, in Mr. Dixon's unpublished work on the Chalk Formations and Tertiary Deposits of Sussex.

Corallum composite, incrusting, adhering in general to large pebbles, and forming a thick convex mass, on the edge of which some traces of a rudimentary epitheca are sometimes perceptible. Multiplication by gemmation in the spaces comprised between the calices. Corallites sometimes united by a spongy cœnenchyma; in other parts crowded together so as to render the calices almost polygonal, and separated only by a thin, simple, common margin. *Calices* infundibuliform, but not deep. *Columella* well developed,

---

[1] Annales des Sc. Nat., 3ᵐᵉ série, vol. x, tab. i, figs. 11, 12.

of a spongy texture, not projecting at the bottom of the fossula, and terminated by a somewhat papillose surface. *Septa* thin towards the columella, thick externally, strongly echinulated laterally, broad, closely set, not exsert, terminated by an oblique crenulated edge, and forming three complete cycla, besides which there is sometimes a very incomplete fourth cyclum. The secondary septa differ but little from those of the first order; the tertiary ones also well developed, bent towards those of the second cyclum, and cemented to them along the inner edge near the columella. A horizontal section shows that the visceral chamber is cylindroid at some distance from the calice, and that the spongy tissue of the walls and the columella becomes much developed (fig. 1 c). The fenestrate structure of the septa is seen in a vertical section of the corallum, represented at fig. 1 b.

Breadth of the calices, nearly two lines; depth, half a line.

This fossil is very abundant at Bracklesham Bay. The specimens from which we have drawn up the preceding description belong to the collections of the Geological Society, of Mr. Bowerbank, Mr. F. Dixon, and Mr. Frederick Edwards.

Some other Corals that have been described under the names of *Astrea* or of *Porites*, and that belong to the Calcaire grossier of the Parisian basin, are also referable to our genus Litharæa, but all differ specifically from *L. Websteri*. In *L. Deshayesiana*,[1] *L. Heberti*,[2] and *L. bellula*,[3] the calices are smaller, and the septa less numerous; the third cyclum, which is always complete in *L. Websteri*, is incomplete in the last of these three species, and does not exist in the first two. In *Litharæa ameliana*[4] and *L. crispa*,[5] which resemble most the London Clay fossil, the walls are thinner and more prominent, and the septa more echinulate.

---

[1] *Porites Deshayesiana*, Michelin, Icon. Zooph., tab. xlv, fig. 4.

[2] *Litharæa Heberti*, nob. This undescribed species presents the following characters: Corallum composite, convex, massive, and often formed of superposed layers. Common epitheca moderately developed. Walls scarcely distinct. Calices polygonal and shallow. Columella not well developed, and appearing to be formed only by the inner marginal *denta* of the septa. Septa not exsert, very thick, especially outwardly, strongly echinulated laterally, terminated by an almost horizontal, spinular edge, and forming only two cycla. The twelve septa are nearly equal in size, and of a very porous structure; the spiniform granulations which cover their lateral surfaces are so highly developed, that they often become united to those of the neighbouring septum. This fossil has been found in an excellent state of preservation at Auvert, by M. Hebert, and appears to be specifically identical with same dilapidated corals met with at Valmondois.

[3] *Astrea bellula*, Michelin, op. cit., tab. xliv, fig. 2.

[4] *Astrea ameliana*, Defrance; *Astrea muricata*, Goldfuss, Petref. Germ., vol. i, tab. xxiv, fig. 3.

[5] *Astrea crispa*, Michelin, loc. cit., tab. xliv, fig. 7; (but not the *Astrea crispa* of Lamarck).

2. *Genus* HOLARÆA (p. lvi).

HOLARÆA PARISIENSIS.   Tab. VI, figs. 2, 2 *a*.

ALVEOLITES PARISIENSIS, *Michelin*, Icon. Zooph., p. 166, tab. xlv, fig. 10, 1845.

Corallum composite, and appearing to have lived fixed to the stem of some Fucus, which it incrusted all round, so as to constitute, after the destruction of this extraneous body, a hollow cylinder, open at both ends.   The lamellar expansion thus rolled round is very thin, and its inner or basal surface is covered by an extremely delicate epitheca.   The *calices* which occupy the opposite surface, and are consequently placed all round the exterior of the above-described cylinder, are infundibuliform, deep, irregularly polygonal, surrounded by a prominent margin, and sometimes slightly turned towards one of the extremities of the corallum, which was probably its upper end.   The fossula is small and circular; its centre is occupied by a fasciculated *columella*, composed of delicate vertical processes, which are quite separated from each other, excepting towards the apex (fig. 2 *a*). The vertical section of the corallum, by means of which the composition of the columella is seen, shows also that the tissue of the whole mass is uniformly and delicately spongy; no appearance of costæ, of septa, or of any radiate structure is perceptible.   The diameter of the specimen that we have figured is about a line and a half, and the thickness of the lamellar expansion that constitutes this cylinder, about half a line; the calices are also about half a line in breadth.

This species has been found both in the London Clay at Barton and the Calcaire grossier of the environs of Paris.   The British specimen represented in our plates belongs to the cabinet of Mr. Frederick Edwards.   We have examined many of these fossils, but owing to the very small size of the corallites, and the extremely delicate structure of their constituent parts, we fear that some of their characteristic features may have escaped from observation, and we feel much uncertainty respecting the natural affinities of the generic division of which it is as yet the only representative.   We have not been able to ascertain the existence of any tabulæ in the interior of the visceral cavity, and therefore it would appear to be allied to Poritidæ rather than to Milleporidæ; but it bears great resemblance to the latter, and we are inclined to think that, when better-preserved and older specimens become known, it will prove to be a tabulated Zoantharia, and if that be the case, there will no longer be any reason for distinguishing Holaræa from our genus Axopora (p. lix).   It is therefore only provisionally that we place it in the family of the Poritidæ.

ORDER 2.—ALCYONARIA.

Family PENNATULIDÆ.

Genus GRAPHULARIA.

GRAPHULARIA WETHERELLI.   Tab. VII, figs. 4, 4 *a*, 4 *b*, 4 *c*, 4 *d*, 4 *e*.

PENNATULA, *J. Decarle Sowerby* and *Wetherell*, in Geol. Trans., 2d series, vol. v, part 1, p. 136, tab. viii, fig. 2 *a*, *b*, 1834.

Corallum styliform, straight, very long, cylindroid towards the lower extremity, sub-tetrahedral at the upper part, and presenting on one side a broad shallow furrow. Surface appearing smooth, when examined by the naked eye, but showing, when placed under the microscope, a multitude of small, longitudinal, closely-set, striæ, that seem to indicate a fibrous structure. Transverse section showing the existence of a thin coating and a radiate structure in the body of the Coral. Diameter of the thickest part, two thirds of a line; probable length, more than a foot.

We have only seen small fragments of this styliform Coral, that evidently constituted the central stem of some aggregate polypi of the family of the Pennatulidæ. Some of these broken remains are almost cylindrical, and usually thicker than others that are imbedded in the same mass of clay, and have a sub-tetrahedral form; others, again, are intermediate between the former, both by their size and their form, and have the same radiate structure and striated surface. It is therefore probable that they all belonged to the same species, and constituted a long, slender, sclerenchymatous axis, somewhat similar in form to that of Pennatula, but resembling that of Virgularia by its structure. The characters of the corallum thus reconstructed are also nearly allied to those of Pavonaria and Umbellularia, but differ from those of all the known recent genera of Pennatulidæ. It is brittle, and presents a radiate section, as in Virgularia, but is not cylindrical from one end to the other, as is the case in the latter, nor is it from top to bottom of a tetrahedral form, as in Pavonaria; it never appears to be twisted like the stem of Umbellularia; it united in its different parts the two forms that are found separately in the two first-mentioned genera, and so far resembles Pennatula; but in the latter the square portion is situated towards the lower end, and the apex is cylindrical, whereas in the above-described fossil, it is the upper slender part that presents a square section, and the thick basal part is cylindrical; it must also be remembered that the axis of Pennatula is not very brittle, and does not present a radiate structure when cut transversely, but appears rather of a fibrous structure. In the recent genus Lithuaria, the styliform axis is tapering towards the lower end, and inflated, pitted, and even somewhat echinulate at its upper extremity. It is also impossible to refer the elongate stem of our London Clay Pennatulida to the genus

Veretellum, for in the latter the axis is quite rudimentary. We therefore considered it advisable to designate this fossil Coral by a peculiar generic name, but we are not as yet sufficiently well acquainted with its characters to be able to form a complete idea of the polypi to which it belonged.

Graphularia Wetherelli is the only known species of this genus of Pennatulidæ; it was discovered about twenty years ago by Mr. Wetherell in the London Clay at Hampstead Heath and at Highgate. Mr. Frederic Edwards has also found fragments of it at Barton and at Haverstock Hill; and it is to his kindness that we are indebted for the specimens described in this Monograph.

Family GORGONIDÆ.

Tribe ISINÆ.

1. *Genus* MOPSEA.

1. MOPSEA COSTATA.   Tab. VII, figs. 3, 3*a*.

Corallum arborescent, dichotomous, and composed of epidermic basal sclerenchyma, the ossification of which is intermittent, so as to constitute a series of calcified cylinders, separated by non-ossified discs. The branches appear to spread out in one plane; they are thin, elongated, cylindrical, and deeply fluted longitudinally; each of them, immediately above its separation from the parent branch or stem, is bent outwards and upwards, so as to represent an inverted ogival angle. The corneous articulations, which have been destroyed during the process of fossilization, are very thin, and do not appear to have existed in any of the non-bifurcated branches. The longitudinal costæ are straight, thick, prominent, denticulated on their sides, and separated by deep furrows. All those belonging to the same joint are in general of the same size; but in some of the thickest branches, where they are the most numerous (about twelve), some very small ones are sometimes visible between the larger ones. The diameter of the thickest branches in the specimens here described is about half a line; that of the smallest not more than a tenth of a line.

The genus Mopsea was established by Lamouroux, but more correctly characterised by Ehrenberg, who refers to it four recent species—the *Mopsea dichotoma* of Lamouroux, the *M. gracilis*, *M. erythræa*, and *M. encrinula*, Ehrenberg; but it is doubtful whether the last does in reality appertain to this division of the Isinæ. Mr. Dana is of opinion that the *Isis coralloides* of Lamarck is also a Mopsea; but in all of these species the corallum is but slightly striated, and never presents anything like the strong costæ which exist in the above-described fossil.

We have seen two specimens of this Gorgonida, found in the London Clay at Holloway, by Mr. Frederic Edwards.

2. *Genus* WEBSTERIA.

WEBSTERIA CRISIOIDES. Tab. VII, figs. 5, 5 *a*.

Corallum composite, slender, and dichotomous, with its branches straight, flat, of the same dimensions as the stem, spreading out on one plane, and forming very acute angles with each other. Corallites subverruciform, disposed in opposite order, and forming two vertical series, the outer edge of which is occupied by a row of small, oblique, circular calices. These two lateral parts are separated by a median line, which usually has the appearance of a small furrow; sometimes they disunite, and so bring to view a small, styliform, central axis. Breadth of the branches about a fifth of a line; height of the corallites, a quarter of a line.

It is not without much uncertainty that we refer this delicate fossil to the family of the Gorgonidæ. By its general aspect, and by the mode of arrangement of the verruciform individuals of which it is composed, it resembles the genus Pterogorgia of Ehrenberg, and the existence of a central axis furnishes a strong argument in favour of the opinion which we have adopted provisionally; but, on the other hand, the structure of the individuals is very similar to that of some Sertularidæ, and still more so to divers Bryozoa, such as *Crisia denticulata*. The specimens that we have examined have not enabled us to decide the question concerning the natural affinities of the animals to which these organic remains belonged; but whether they be Polypi or Bryozoa, they appear to differ from all the known generic forms, and to constitute the type of a new genus, that we dedicate to Mr. Webster, whose observations on the formation in which they are found have been so serviceable to geology.

Websteria crisioides is the only species belonging to this zoological division. It was discovered in the London Clay at Haverstock Hill, by Mr. Frederic Edwards, to whose kindness we are indebted for the communication of the specimens here described.

CHAPTER III.

CORALS OF THE UPPER CHALK.

THE fossil Corals found in the Upper Chalk of England are not numerous; they belong principally to the section of simple Eusmilinæ, and appear to be peculiar to the British Fauna. One species, it is true (the *Parasmilia centralis*), has been mentioned by different geologists as existing also in the Chalk of Beauvais and in the north-west of Germany, but we have great reason to think that these fossils are not specifically identical. It is also worthy of notice, that even no species corresponding to those met with among the Corals of the Upper Chalk of England have as yet been seen in the Chalk of Meudon, and that a great difference exists between the predominant generical forms in the first of these formations, and in the Chalk of Maestricht. In the latter some Cyathininæ nearly allied to that of England are met with; but the Diploctenium, the Cyclolites, and the aggregate Astreidæ of Maestricht are represented by no corresponding forms in this part of the British fossil Fauna, and the organic remains found in these two cretaceous deposits have consequently a very different aspect. We must add, that the fossil Corals of the Chalk of Faxoe are equally distinct from the British species, and that none of the latter have been met with in the Lower Chalk Formations of England.

ORDER ZOANTHARIA (p. ix).

Family TURBINOLIDÆ (p. xi).

Tribe CYATHININÆ (p. xii).

Genus CYATHINA (p. xii).

CYATHINA LÆVIGATA. Tab. IX, figs. 1, 1*a*, 1*b*, 1*c*, 1*d*.

CYATHINA LÆVIGATA, *Milne Edwards* and *J. Haime*, Monogr. des Turbinolides; Ann. des Sc. Nat., 3ᵐᵉ série, vol. ix, p. 290, 1848.
MONOCARYA CENTRALIS (*pars*), *Lonsdale*, in Dixon's unpublished work on the Chalk Formations and Tertiary Deposits of Sussex, tab. xviii, figs. 12, 12*a*, (perhaps also fig. 5, but not the other figures bearing the same name, which are *Parasmilia* and probably *Cœlosmilia*).

Corallum simple, elongated, adherent, cylindro-turbinate, straight, and in general much contracted just above the basis, which is broad. *Walls* quite smooth, and polished towards the basis, but presenting towards the calice slightly-marked *costæ*, which are closely set, glabrous, or very delicately granulated, and almost equal in size. *Calice* circular, or sometimes rather oval, shallow. *Columella* moderately developed, not projecting in the centre of the calicular fossula, composed of six or eight twisted, vertical processes, and terminated by an equal number of papillæ. *Septa* forming four cycla, the last of which is in general incomplete; the septa of the fourth and fifth orders not existing in one half of three of the systems or even of all six of these, so that the number of these radiate laminæ is reduced to 42, or even to 36; sometimes, however, four of the systems are complete, and the number of the septa then amounts to 48. These septa are well developed, closely set, thin, but slightly granulated, rather exsert, and almost equal; the principal ones are, however, a little thicker than the others. The *pali* are narrow, but very thick, prominent, aud terminated by a curved edge; they all correspond to the tertiary septa, and in the specimens where the fourth cyclum is complete, they exist in front of all the septa of the penultimate cyclum, and are therefore twelve in number; but they are never so numerous when the fourth cyclum remains incomplete, and never correspond to tertiary septa that are not separated by septa of the fourth cyclum. As mentioned above, these latter septa are often wanting in one half of three of the systems, and in that case there are consequently no pali corresponding to the tertiary septa of these incomplete half systems, so that the number of pali is reduced to nine; two belonging to each of the three complete systems, and one to each of the incomplete ones. The same rule also holds good when all the six systems are deficient of the septa of the fourth cyclum in one of their halves; the tertiary septa of the complete halves are the only ones having corresponding pali, so that the number of these organs is only six. The height of the corallum varies between one inch and one inch and a half; in the tall specimens the diameter of the calice is about four lines; in the short and broad ones it is sometimes five lines.

This species is easily distinguished from the *Cyathina Cyathus*,[1] *C. Smithii*,[2] and *C. pseudoturbinolia*,[3] by its never having a fifth cyclum of septa. *C. Guadulpensis*[4] and *C. arcuata*[5] differ from the above-described fossil by the existence of distinct costæ down to the basis of the walls, and by the large size of the pali. *C. lævigata* most resembles *C. Koninckii*,[6] *C. Bowerbankii*,[7] *C. Debeyana*, *C. Bredæ*, and *C. cylindrica*,[8] and it may

[1] See our Monograph of Turbinolidæ, Ann. des Sc. Nat., 3ᵐᵉ série, vol. ix, p. 287, tab. iv, fig. 1.

[2] Loc. cit., p. 288.   [3] Loc. cit., p. 289, tab. ix, fig. 1.   [4] Loc. cit., p. 290.

[5] Loc. cit., p. 290.   [6] Loc. cit., p. 290.   [7] Loc. cit., p. 292.

[8] The three last-mentioned species were not known to us when we published our Monograph of Turbinolidæ, and in order to render the comparison between the *T. lævigata* and the rest of the genus more complete, it appears to us advisable to give a description of them here.

CYATHINA CYLINDRICA, nob. Corallum fixed by a broad basis, regularly cylindrical, straight, and not very tall. Costæ equal, flat, straight, closely set, not very broad, and very indistinct, especially towards the basis. Calice circular, having a thick edge, and the fossula rather shallow. Columella very small, and reduced to two or three small, almost indistinct, tubercles. Septa forming four complete cycla, very closely

not be unworthy of notice that these five species are as yet the only representations of the genus Cyathina that have been met with in the Chalk Formation. At first sight they appear very similar, but by an attentive examination, constant and well-defined characteristic differences are found between all. In *C. Koninckii*, the corallum is always shorter, and more regularly turbinate; the pali are thicker, and the columella is reduced to two or three thick, twisted processes. In *C. Bowerbankii*, on the contrary, the pali are much thinner, and the surface of the walls appears granulous. *C. cylindrica* and *C. Bredæ* differ from it by a very peculiar character, which exists also in *C. Koninckii*, but which is not met with in any other species of the same genus, and is indeed quite an exception to the family of Turbinolidæ, the pali being only six in number, although the four cycla of septa be complete, and corresponding to the septa of the antepenultimate cyclum, whereas they usually correspond to those of the penultimate cyclum. The thin, elongate form of *C. Bredæ* and the quite cylindrical form of *C. cylindrica*, will also help to distinguish them from *C. lævigata*, which differs also from *C. Debeyana*, by the latter having a well-marked epithecal fold near the calice, a small columella, and thinner pali.

Cyathina lævigata is found in the Upper Chalk at Dinton, in Wiltshire; specimens may be seen in the collections of the Geological Society, of Mr. Bowerbank, and of the Museum at Paris.

set, and having stronger lateral granulations near the inner edge. The primary ones larger and rather thicker than the others, but differing very little from the secondary ones; the tertiary ones are thinner and smaller; those of the fourth cyclum are distinct, but very small. Pali prominent, extremely thick, narrow, strongly granulated laterally, and corresponding to the secondary septa. Height of the corallum about six lines; diameter of the calice three lines. Fossil from the Chalk of St. Peter's Mountain, at Maestricht; specimens exist in the Museum of Natural History of Paris, and in the Tylerian Museum at Haarlem.

CYATHINA BREDÆ, nobis. This fossil corallum, which we dedicate to Professor Van Breda, is adherent by a rather broad basis, contracted immediately above, elongate, slender, much bent, and cylindrical towards its upper part. The costæ are not well marked, and the walls are almost smooth, but present sometimes slight horizontal folds. Calice circular; fossula shallow. Columella but little developed, and sometimes reduced to a single twisted process. Septa forming four complete cycla; but those of the last cyclum rudimentary though distinct; the primary ones rather thick, especially towards the inner edge; the secondary ones resembling those of the first cyclum, but rather narrower; the others very thin. The granulations on the sides of the septa are conical, and very prominent. The pali corresponding to the secondary septa, well developed, prominent, narrow, and appearing very thick, because they are flexuous. Height, seven or eight lines; diameter of the calice, two lines and a half. This species is also found in the fossil state in the Chalk of St. Peter's Mountain, at Maestricht; specimens exist in the collections of MM. Van Riemsdyck and Bosquet, at Maestricht; of M. Van Breda, at Haarlem; and of the Museum at Paris.

CYATHINA DEBEYANA nob. Corallum cylindrical, elongate, slightly curved, and presenting near the calicular margin a small but well-marked circular band, representing an incomplete epitheca. Calice circular; fossula not deep. Septa unequal, closely set, somewhat exsert, rather thick externally, but thin towards the inner edge, and forming four complete cycla; the secondary ones almost as large as those of the first cyclum. Pali rather narrow, and not very thick. Height, one inch; diameter of the calice, three lines; depth of the fossula, one line. Fossil from the Chalk of Aix-la-Chapelle, discovered by M. Debay.

Family ASTREIDÆ (p. xxiii).

Tribe EUSMILINÆ (p. xxiii).

1. *Genus* PARASMILIA (p. xxiv).

1. PARASMILIA CENTRALIS. Tab. VIII, figs. 1, 1*a*, 1*b*, 1*c*.

MADREPORITE, *Parkinson*, Organ. Remains of a FormerWorld, vol. ii, tab. iv, figs. 15, 16, 1820.

MADREPORA CENTRALIS, *Mantell*, Geol. of Sussex, p. 159, tab. xvi, figs. 2, 19, 1822. (Correct figures.)

CARYOPHYLLIA CENTRALIS, *Fleming*, British Animals, p. 509, 1828.

— — *Mantell*, Trans. of the Geol. Soc., 2d series, vol. iii, p. 204, 1829.

— — *Phillips*, Illust. of the Geol. of Yorkshire, part i, p. 119, tab. i, fig. 13, 1829 ; 2d edit., p. 91.

— — *S. Woodward*, Synoptic Table of Brit. Org. Remains, p. 6, 1830.

CARYOPHYLLIA, *R. C. Taylor*, in Mag. of Nat. Hist., vol. iii, p. 271, fig. *f*, 1830.

LITHODENDRON CENTRALE, *Ch. Keferstein*, Die Naturgeschichte des Erdkorpers, vol. ii, p. 789, 1824.

TURBINOLIA EXCAVATA (?), *Hagenow*, in Leonard's und Bronn's Jahrbuch für Mineral., p. 229, 1839.

— CENTRALIS, *Fred. Adolph Rœmer*, Verstein. des Norddeutschen Kreidegebirges, p. 26, 1840.

— — *Bronn*, Index Paleontologicus, p. 314, 1848.

PARASMILIA CENTRALIS, *Milne Edwards* and *J. Haime*, Monogr. des Astreides, Ann. des Sc. Nat., 3me série, Zool. vol. x, p. 244. 1848.

MONOCARYA CENTRALIS (*in parte*), *Lonsdale*, in Dixon's unpublished work on the Chalk Formations of Sussex, tab. xviii, figs. 1, 3, 7, 7 *a*, 9 (cæteris exclusis).

Corallum simple, cylindrico-turbinate, fixed by a rather broad basis, above which it is much contracted, elongate, irregularly bent in various directions, and presenting a series of unequal contractions and circular dilatations. *Costæ* closely set, and distinct from the calicular margin down to the basis, where they are the most prominent ; those corresponding to the primary and secondary septa are rather larger than the others towards the basis ; but the tertiary ones soon become almost similar to the former, and at the upper part of the wall all these large costæ alternate with smaller ones belonging to the fourth cyclum. All are covered with delicate granulations, which are most prominent towards the lower part of the costæ of the fourth cyclum, where they form simple series. *Calice* circular, with the fossula less shallow than usual in this genus. *Columella* well developed, somewhat prominent and crispate. *Septa* forming six equally developed systems and four complete cycla ; closely set, very unequal, broad, thin, slightly exsert, straight, or very slightly flexuous, and presenting laterally a few large granulations. *Dissepiments* simple, almost horizontal, and few in number; about three from the top to the bottom of each principal septum, as may be seen by means of a vertical section. Height varying from one to two inches; diameter of the calice, four lines; depth of the fossula, two lines.

The genus Parasmilia, circumscribed within the limits assigned to it in the Introduction to this Monograph, only contains seven species, all of which belong exclusively to the upper beds of the Chalk Formations. Three of them (*P. centralis, P. Gravesiana,* and *P. elongata*) have already been described in our Monograph of the Astreidæ,[1] and the four others (*P. Mantellii, P. Fittonii, P. cylindrica,* and *P. serpentina*) will be made known in the present work. They all differ but little from each other, and in order to recognise them, it is necessary that they should be compared together with attention. *P. centralis,* which may be considered as the type of this small generic group, differs from *P. Gravesiana, P. elongata, P. cylindrica,* and *P. Mantellii,* by its costæ being always straight, rather thick and never sub-lamellous, and rather flexuous, as in the four last-mentioned species; it is also to be remarked, that its costæ are rather more prominent near the basis than higher up, whereas the contrary is seen in the *P. Gravesiana,* and that the loculi are never subdivided by small dissepiments, as is the case in *P. elongata, P. cylindrica,* and *P. Mantellii.* *P. serpentina,* which bears more resemblance to it, is characterised by the septa forming only three cycla, and the costæ being very delicate, and rather indistinct towards the basis. But it is with *P. Fittonii* that *P. centralis* is most closely allied; the former, however, is of a thicker form, its tertiary costæ are more developed and more delicately granulated, and its columella is much larger, and terminated by a sub-papillose surface.

The specimens of *P. centralis* which we had the opportunity of examining were found in the Upper Chalk at Northfleet, near Gravesend, and at Norwich. Mr. Phillips mentions the existence of the same fossil at Dane's Dike;[2] and Dr. Mantell has met with it at Brighton, Lewes,[3] Steyning, and Heytesbury.[4] Mr. Graves also alludes to it as being found in the Chalk Formation of the Parisian basin at Beauvais;[5] but we have much reason to think that the species observed by that geologist is not the one here described, and must be referred to our *P. Gravesiana.* M. Rœmer and other authors equally apply the name of *C. centralis*[6] to a fossil found in the north-west of Germany, but we have not been able as yet to verify the propriety of this determination, not having seen any of the specimens discovered in that part of the Continent.

---

[1] The species described in that work under the names of *Parasmilia poculum, P. Faujasii,* and *P. punctata,* must now be referred to our genus *Cœlosmilia,* which is characterised by the entire absence of the columella.

[2] Op. cit., part i, p. 119.

[3] Illust. of the Geol. of Sussex, p. 160.

[4] Geol. Trans., 2d series, vol. iii, p. 204.

[5] Geogn. de l'Oise, p. 701.

[6] Versteinerungen des Norddeutschen Kreidebirges, p. 26.

We must also remark that, in a note just published, our able friend M. Alcide d'Orbigny[1] refers to the *Caryophyllia* or *Parasmilia centralis* as the type of a new genus, designated under the name of *Cyclosmilia*, and characterised in the following terms: " Cyclosmilia are Parasmilia, in which the loculi are but very little divided by dissepiments, the growth of the corallum is intermittent, the calice circular instead of being oval, and the external costæ distant from each other." Now, with the exception of this last peculiarity, which is not even met with in *P. centralis*, all these characters may be seen in every species belonging to our genus Parasmilia, and we therefore can find no reason for separating from it this new generic division.

2. PARASMILIA MANTELLI. Tab. VIII, fig. 2, 2 *a*.

We have as yet seen but one specimen of this species, which appears to be very distinct from all others. It is a small corallum, nearly straight, adherent by a broad basis, regularly turbinate, and not very tall; but being in all probability susceptible of increasing much in height by progress of age, as is the case with the other species belonging to the same genus. The *costæ* are narrow, sublamellar, closely set, distinct down to the basis of the corallum, very echinulate, and somewhat crispate. Those of the primary and secondary cycla are equally developed, and rather more prominent than the others, especially towards the basis and the calicular margin; the tertiary ones also extend on the basal expansion of the corallum, but are smaller; and those of the fourth cyclum begin at a short distance above the basis, and are very narrow at their lower part. The intercostal furrows are broad, deep, and divided by small transverse dissepiments, formed by rudiments of an exotheca. *Calice* circular; fossula not deep. *Columella*, as far as we can judge by the specimen here described, very similar to that of *P. centralis*. *Septa* forming four complete cycla; well developed, thin, straight, closely set, rather unequal, and presenting well-marked striæ on their lateral surfaces. Height, seven or eight lines; diameter of the calice, nearly five lines.

This fossil differs from the other species belonging to the same genus, and more especially from *P. centralis* and *P. Fittoni*, by its costæ, which are equally prominent and subcrispate, whereas in the latter they are smooth and never sublamellar. It resembles more closely *P. Gravesiana*, *P. cylindrica*, and *P. elongata;* but it differs from them by the strong granulations of the costæ. In *P. serpentina* the basis is almost smooth, and the septa do not form so many cycla.

Parasmilia Mantelli was met with in the upper chalk at Bromley in Kent, by our friend Mr. J. S. Bowerbank.

[1] This paper, bearing the title of ' Note sur des Polypiers Fossiles,' and published on the 10th of October, 1849, contains the exposition of the characters of a series of new genera proposed by M. d'Orbigny. The author assigns to most of these divisions the date of 1847, a period at which he appears to have adopted them in the arrangement of his private collection; but in referring to them here or elsewhere, we have considered it proper to quote the year of their publication, which is the only authentic date that could be made use of if any question of priority should arise concerning them.

3. Parasmilia Cylindrica.   Tab. VIII, fig. 5.

It is not without some hesitation that we inscribe this species in the list of our Parasmilia, for the specimen about to be described is extremely incomplete; but it does not present the specific characters of any other species, and although very nearly allied to *P. elongata* and *P. Mantelli*, it appears to differ from both in some essential points.   The fragment here alluded to is deficient both in the basis and in the calice, but it appertained to a tall, nearly cylindrical corallum, that was somewhat bent.   The *costæ* are almost equal, extremely thin, sublamellar, but not very prominent, subflexuous, very slightly granulated, and divided at short intervals by circles of small dissepiments, formed by rudiments of an exotheca.   The intercostal furrows are broad, rather shallow, almost destitute of granulations, and presenting sometimes in the middle a small rudimentary costa.   *Columella* well developed.   *Septa* forming four complete cycla, not very closely set, somewhat flexuous, and slightly granulate; those of the first and second cycla equally developed, and rather thick; the tertiary ones smaller and thinner; those of the fourth cyclum very small, although the costæ corresponding to them are as large as those of the other cycla.   Length, above two inches; diameter, about six lines.

This fossil much resembles the *Parasmilia elongata* found at Ciply, but differs from it by the unequal development of the septa belonging to the first two and to the last two cycla, a mode of structure which does not exist in *P. elongata*.   It differs from *P. centralis*, *P. Fittoni*, and *P. serpentina*, by the delicacy and almost lamellar form of the costæ, and bears greater resemblance to *P. Mantelli* and *P. Gravesiana*, from which it may, however, be easily distinguished by the breadth of its intercostal furrows.

The specimen here described belongs to the Palæontological collection of Mr. J. S. Bowerbank, and was found in the upper chalk at Norwich.   Another fossil, which we consider as belonging to the same species, exists in the Poppelsdorf Museum at Bonn, and was found in the upper chalk at Darup, in Westphalia.

4. Parasmilia Fittoni.   Tab. IX, fig. 2, 2 *a*, 2 *b*.

Corallum stout, adherent by a somewhat broad basis, immediately above which it in general becomes very narrow; elongate, much bent, and presenting at intervals circular constrictions.   *Costæ* broad, closely set, not very prominent, excepting near the basis, down to which they are quite distinct, rather unequal alternately, and covered with very numerous and small granulations.   *Calice* circular; fossula large and rather shallow.   *Columella* well developed, but very slightly prominent, of a spongy structure, and terminated by a broad subpapillose surface.   *Septa* forming four complete cycla, rather thin, straight, not very closely set, slightly exsert, and having but few granulations on their lateral surfaces.

Those of the second order are nearly as large as those of the first set, and thus produce the appearance of twelve tertiary systems. Height, from one to two inches; diameter of the calice five lines; its depth, two lines.

This species is easily distinguished from all the other Parasmilia by the great development of its columella, which occupies nearly half the diameter of the calice, and by the spongy structure of this organ. It most resembles *P. centralis*, from which, however, it differs also by its thick form and the greater development of the tertiary costæ towards its basis. The breadth and delicate granulations of the costæ may equally serve to distinguish it from *P. Gravesiana*, *P. elongata*, *P. cylindrica*, and *P. Mantelli*. It differs from *P. serpentina* by having an additional cyclum of septa, and by its basis not being smooth, as is the case in the latter.

This fossil is found in the upper chalk of Norwich, and exists in the collections of the Geological Society, of the Geological Survey, and of the Museum at Paris. It appears probable that it has often been confounded with *P. centralis*, and that some of the figures referred to that species may in reality belong to it; but the engravings here alluded to are not correct enough to enable us to decide this question.

5. PARASMILIA (?) SERPENTINA. Tab. VIII, fig. 3, 3 *a*, 3 *b*.

It is not without some doubts that we place this fossil in the genus Parasmilia, for in the unique specimen that has come under our observation, the calice was in so bad a state of preservation that it was impossible to decide whether the papillæ seen near the centre of that part were fractured septa or remains of pali, or even trabiculæ belonging to the columella. However, the first hypothesis appears most probable, and the general appearance of the corallum is also very similar to that of all the other Parasmilia.

This fossil is almost cylindrical, slender, much elongated, and bent; it presents some strongly marked circular constrictions, indicative as usual of a certain intermittence in the progress of its growth. The *costæ* are narrow, straight, rather unequal alternately, scarcely distinct near the basis, but more prominent towards the upper part of each inflated ring and near the calice; the *calice* is circular. The *septæ* form three complete cycla and are rather closely set, exsert, and somewhat dilated exteriorly. The *columella* is well developed. Length, one inch, seven lines. Diameter of the calice, two lines and a half.

This coral, belonging to Mr. Bowerbank's collection, was found in the upper chalk at Bromley, in Kent.

It is the only species of Parasmilia in which the fourth cyclum of septa does not exist; it is also characterised by its basis not being costulated.

2. *Genus* CŒLOSMILIA (p. xxv.)

CŒLOSMILIA LAXA.   Tab. VIII, fig. 4, 4 *a*, 4 *b*, 4 *c*.

Corallum simple, turbinate, slightly bent, rather intermittent in its growth, and appearing to have been adherent. *Costæ* distinct from the basis to the calice, very distant from each other; those belonging to the first three cycla subcrestiform; those of the last cyclum flat and scarcely visible, delicately granulated and crossed by small horizontal striæ. *Calice* circular; fossula narrow and rather deep. No *columella. Septa* forming four complete cycla; but those of the last cyclum almost rudimentary. The six systems equally developed. The septa very unequally developed, broad, very exsert; thin, but rather less so near the inner margin, presenting a few round granulations on their lateral surfaces. Those of the first and second cycla united along the lower part of their inner edge. Height, from one inch to one inch and a half; diameter of the calice, seven lines.

We have given the generic name of *Cœlosmilia* to a certain number of Eusmilinæ which we formerly placed in our genus *Parasmilia*, but which are characterised by the absence of the columella and the rudimentary state of the endotheca. *Parasmilia poculum, P. Faujasi,* and *P. punctata*[1] belong to this group, and differ from *C. laxa* by their costæ being flat and granulated near the calice, whereas in the above-described fossil these parts are subcrestiform. It is also to be remembered that in *Cœlosmilia poculum* and *C. Faujasi* the septa form five complete cycla, and that in the last-mentioned species, as well as in *C. punctata,* the principal septa are much thicker than in *C. laxa.* M. Alcide d'Orbigny has lately discovered in the white chalk of Césanne a new species which he designates by the name of *Cœlosmilia Edwardsiana,* and which differs from *C. laxa* by its costæ being rudimentary and its septa thinner.

---

[1] See our Monograph of the Astreidæ (Ann. des Scien. Nat. 3me série, vol. x). It is possible that our *Cœlosmilia punctata* may be only a young form of *C. Faujasi,* but we have not as yet seen a sufficient number of specimens to be able to decide the question.

CHAPTER IV.

CORALS FROM THE LOWER CHALK.

THE number of British Corals known to belong to this formation is as yet so very small, that it would be premature to speculate on their mode of distribution. We have seen but two species, one appertaining to the family of Oculinidæ, the other to that of Eupsammidæ; both appear to be peculiar to the lower chalk of England.

Family OCULINIDÆ (p. xix).

Genus SYNHELIA (p. xx).

SYNHELIA SHARPEANA. Tab. IX, fig. 3, 3 *a*.

Corallum composite, dendroid, with thick, erect branches, forming acute angles with each other, and presenting on their surface large, non-exsert, circular calices, which are not closely set, and are united by rather indistinct, small costal striæ. *Calices* quite superficial, and presenting scarcely any central depression. *Columella* assuming the appearance of a small, obtuse tubercle. Three complete cycla of *septa*, and in one half of each system two quaternary septa, of which no homologues exist in the other half. The septa are thick, very closely set, almost straight, and unequally developed, but those of the second order differ but little from the primary ones. The upper edge of all is horizontal, and closely denticulated; towards the columella the denticulations are rather larger than towards the calicular margin, and we have not been able to decide whether some of them do not constitute pali. The lateral surfaces of the septa present oblong transverse granulations, which much resemble incomplete synapticulæ, but they are not prominent enough to meet those of the adjoining septa, and to subdivide the interseptal loculi. The height of the specimen here described is about two inches and a half, and the diameter of the calices two lines.

We are as yet acquainted with but two other species that can be referred to our genus *Synhelia;* one is the *S. gibbosa,* which was first described by Goldfuss under the name of

Lithodendron gibbosum,[1] and which belongs also to the lower Chalk formation, but is found at Bochum, in Westphalia, and at Blaton, near Mons, in Belgium. It differs from *S. Sharpeana,* by its calices being more closely set; rather oblong, with a more prominent margin, and twenty-four nearly equal, very thick septa, separated by an equal number of rudimentary ones. The other is the *Madrepora Meyeri,* found by MM. Koch and Dunker in the Jurassic formation at Elligser-Brinke; it has deep calices.[2]

The unique specimen here described appears to have been found in the lower chalk near Dover, and was kindly communicated to us by Mr. Daniel Sharpe.

Family EUPSAMMIDÆ (p. li).

Genus STEPHANOPHYLLIA (p. liii).

STEPHANOPHYLLIA BOWERBANKII. Tab. IX, fig. 4, 4 *a,* 4 *b,* 4 *c.*

STEPHANOPHYLLIA BOWERBANKII, *Milne Edwards* and *J. Haime,* Monogr. des Eupsammides,
in Annales des Sciences Naturelles, 3ᵐᵉ série, Zool.
vol. x, p. 94, 1848.

Corallum simple, resembling, in its general form, a plano-convex lens. *Wall* discoidal and horizontal. *Costæ* numerous, delicate, nearly equal, closely set by pairs, and formed by a simple series of granulations, which become the most distinct near the outer edge of the mural disc. Twenty-four of these costæ begin near the centre of the corallum, and soon after bifurcate; the forty-eight costæ thus produced soon divide again, in the same manner, and near the edge of the disc the number of these radiate ridges amounts to ninety-six. The mural pores are small, not very distinct, and arranged in series in the intercostal furrows. *Calice* quite circular, and appearing to be regularly convex, excepting towards the centre, where there is a slightly-marked, shallow fossula. *Columella* almost rudimentary, and formed only by two or three trabiculæ, which are often scarcely distinct from the edges of the *septa.* These last-mentioned organs arise from the upper surface of the mural disc, and are thin, especially outwards, closely set, and covered laterally with large, prominent granulations. They form five complete cycla, and represent six well-characterised and equally-developed systems. The primary and secondary septæ are straight, and extend to the columella; their upper edge is arched, or slightly angular. The tertiary septa are also much developed, and bend towards the secondary ones, to which they become united by their inner edge, near the columella. The septa of the fourth and fifth orders, constituting the fourth cyclum, are united in a similar way to the tertiary septa, at about half way from the margin of the mural disc to the columella, but not exactly at the same point, those of

---

[1] Petref. Germ., vol. i, tab. xxxvii, fig. 9.
[2] Beiträge zur Kenntniss des Norddeutschen oolithgebildes, p. 55, tab. vi fig. 11, 1837.

the fifth order being rather longer than those of the fourth order. The septa of the fifth cyclum are small, thin, low, and unite to the neighbouring principal septa; those of the sixth order join the primary ones; those of the seventh order adhere by their inner and upper edge to the secondary ones, and those of the eighth and ninth orders to the tertiary septa; or, in other words, each element of this fifth cyclum joins the eldest of the two septa between which it is placed. Independently of these junctions, which are normal, and always take place along the inner edge of the septa, the interseptal loculi are irregularly divided in some places by the projecting lateral granulæ of two neighbouring septa meeting, and becoming cemented together. By this character, as well as by its general form, this species tends to unite the family of Eupsammidæ with the Fungidæ.

Height of the corallum, one and a half or two lines; diameter, three or four lines. Some specimens, which were probably not adult, were only two lines and a half in diameter.

This delicate little Coral differs from *Stephanophyllia elegans*, *S. imperialis*, and *S. discoides*[1] by the form of the septa, which do not appear to be angular and lacerated, as in the three latter species. *Stephanophyllia astreata*[2] differs from it by having a large fossula and a well-developed columella. It most resembles *S. suecica*;[3] but in this species the two tertiary septa of each system unite below the columella and the secondary septa, which consequently do not extend to the centre of the calice; whereas in *S. Bowerbankii* these tertiary septa, as we have already stated, adhere to the secondary septa, and these last-mentioned septa extend to the columella. The Fossil Coral figured by M. von Hagenow, under the name of *Fungia clathrata*,[4] and found by that geologist in the chalk formation of Rugen, is evidently very nearly allied to the British species here described; but as far as we can judge of it by M. von Hagenow's engraving, it appears to differ from it by its more elevated form, by the strongly-marked concentric striæ visible on the mural disc, and by its basis being more prominent.

We must also remark, that the section of the genus Stephanophyllia, to which this species belongs, and to which we applied the name of *Lenticular Stephanophyllia*,[5] has of

[1] See our Monograph of the Eupsammidæ, Ann. des Sc. Nat., 3ᵐᵉ série, vol. x.

[2] *Fungia astreata*, Goldfuss, Petref. Germ., vol. i, p. 47, tab. xiv, fig. 1 (where it is by mistake designated under the name of *Fungia radiata*). This species not having been, as yet, well characterised, we think it may be useful to give a short description of it here. Corallum simple, very short, and having the form of a plano-convex lens. Calicular fossula circular, and well developed. Costæ very delicate and not closely set. Septa forming five complete cycla, and appearing to be thin and strongly granulated. Size very variable; in the adult, diameter three lines, height about one line. Fossil found at Aix-la-Chapelle, in Westphalia, and existing in the Museums of Bonn and Paris. All the specimens yet found are in a very bad state of preservation.

[3] Monogr. of the Eupsammidæ, loc. cit., p. 94.

[4] In Leonhard and Bronn's Jahrbuch für Mineralogie, 1840, p. 684, tab. ix, fig. 3.

[5] Monogr. of the Eupsammidæ, loc. cit., p. 94, 1848.

late been considered by M. Alcide d'Orbigny as deserving to be elevated to the rank of a genus, and has been named by that author *Discopsammia;*[1] but M. d'Orbigny has not pointed out any new characters in addition to those on which this separation was primitively established in our Monograph, and consequently we see no reason for altering the classification previously adopted.

Stephanophyllia Bowerbankii is found in the lower chalk near Dover, and does not appear to differ from some corals which one of us[2] has lately met with in a bed of chlorited chalk at Orcher, near le Havre. The specimens here described belong to the collections of Mr. Bowerbank, Mr. D. Sharpe, and the Geological Society.

[1] Note sur les Polypiers Fossiles, p. 10, 1849.
[2] M. Jules Haime.

CHAPTER V.

CORALS FROM THE UPPER GREENSAND.

THE class of Polypi had not, in all probability, numerous representatives in the seas where the Upper Green Sand was deposited, for we have as yet seen only four British species belonging to that formation, and the English geologists do not appear to have met with many more. Most of these fossils belong to the family of Astreidæ, and have been found at Haldon, at Blackdown, or at Warminster. One of these British species appears to be identical with a coral described by Goldfuss, and found in the chalk formation of Essen; and Mr. Morris has pointed out two others as being referable to species found in the chalk of Maestricht, but we have not had an opportunity of recognising the specific identity of these last-mentioned fossils.

Family ASTREIDÆ (p. xxiii).

Tribe EUSMILINÆ (p. xxiii).

1. *Genus* PEPLOSMILIA (p. xxv).

PEPLOSMILIA AUSTENI.   Tab. X, fig. 1, 1 *a*, 1 *b*.

Corallum simple, fixed by a broad basis, cylindrical, and surrounded from top to bottom by a membraniform epitheca, presenting some slight transverse folds. *Calice* circular, or somewhat oval; fossula shallow, narrow, and elongated. *Columella* well developed and lamellar. *Septa* appearing to form four well-developed cycla, and a fourth rudimentary one. The primary and secondary ones equal, and differing but little from the tertiary ones; they are all thick, broad, closely set, slightly exsert, not quite straight, those on one side inclining to the right near the columella, and those of the other side bending in an opposite direction. A vertical section of this Coral (fig. 1 *b*) shows that the septa are granulated on their lateral surfaces, especially near their inner edge, which joins the columella, and that these granulations form closely-set radiate rows. *Dissepiments* vesicular, and rather abundant. Height of the coral, one inch and a half; diameter of the calice, above an inch.

8

This species is as yet the only known representative of our genus Peplosmilia,[1] and is easily distinguished from the other true Eusmilinæ, either by its lamellar columella or its complete epitheca; it may be considered as a Montlivaltia, having a lamellar columella. We have seen but one specimen of this fossil; it was found in the Greensand at Haldon, and presented to the Geological Society by Mr. R. H. C. Austen.

2. *Genus* TROCHOSMILIA (p. xxiv).

TROCHOSMILIA (?) TUBEROSA. Tab. X, fig. 2, 2 *a*.

TURBINOLIA COMPRESSA, (?) *Morris*, Cat. of Brit. Foss., p. 46, 1843.

Corallum simple, compressed, even at its basis, cuneiform, subpedunculated, and presenting on each of its lateral edges, at a short distance above the basis, a broad but not very prominent tuberosity. *Costæ* delicate, straight, not prominent, but very distinct from the basis upwards, closely set and somewhat unequal. *Calice* elliptic and horizontal; its small axis only half the length of the long axis. Fossula narrow, rather shallow, and elongated. No *columella*. *Septa* forming five complete cycla; very thin, straight, closely set, and delicately granulated laterally; those of the first and second cycla nearly equal in size and larger than the others, so as to produce the appearance of twelve systems; those of the fifth cyclum very small. Height, seven lines; diameter of the calice, eight lines by four.

The above-described specimen was found in the Greensand of Blackdown by our able friend Mr. J. S. Bowerbank. We have not, as yet, been able to ascertain the existence of dissepiments in the interseptal loculi, and consequently are not quite sure that it belongs to the genus Trochosmilia; if these parts do not exist it must be referred to the family of the Turbinolidæ, but we have not had the materials necessary for deciding that question. We shall therefore only add here, that this coral differs from the other species of Trochosmilia described in our ' Monograph of the Astreidæ' by the existence of the lateral tuberosities, and the basis presenting scarcely any traces of adherence.

It is probably this fossil which Mr. Morris referred to the *Turbinolia compressa* of Lamarck, and mentioned as existing in the Greensand of Blackdown. *T. compressa* belongs also to our genus Trochosmilia, and is found in the Greensand at Uchaux in the South of France, but differs from *T. tuberosa* by its general form.

---

[1] The fossil described by M. Michelin under the name of *Anthophyllum detritum* (Icon. Zooph., tab. x, fig. 1) might at first sight be supposed to belong to this genus, for it presents some appearance of a lamellar columella; but that is owing to the presence of some extraneous matter adhering to the specimen figured by M. Michelin, and although the epitheca does no longer exist in this fossil, we have no doubt that it is in reality a Montlivaltia.

Tribe ASTREINÆ (p. xxxi).

Genus PARASTREA (p. xliii).

PARASTREA STRICTA. Tab. X, fig. 3, 3 *a*.

Corallum composite, forming a mass not very tall, and slightly convex on its upper surface. *Calices* seldom circular, in general oblong or irregularly polygonal, projecting very little, and having always distinct margins. *Costæ* delicate, closely set, nearly equal, almost horizontal, nearly straight or slightly bent, and united by their extremity to those of the neighbouring corallites, which, however, remain circumscribed by a small furrow. *Calicular fossula* shallow. *Columella* of a dense tissue, subpapillose, and not much developed. *Septa* thin, broad, closely set, terminated by a series of calicular dentations, the last of which (towards the columella) appears to be more developed than the others; the number of these septa seldom exceeds forty, and they are rather unequal. *Walls* thin, but well developed. Diameter of the calices, usually between two lines and two lines and a half; distance between the calices, at least half a line.

This species, found in the Greensand at Blackdown, is characterised from a specimen belonging to the Geological Society; it differs from all the previously described Parastrea by the approximation and delicate structure of the septa.

Mr. Morris mentions, in his ' Catalogue of British Fossils,'[1] two other species which have been found by M. Austen in the Greensand at Haldon, and which belong to the family of Astreidæ. M. Austen considers the one as being identical with the Maestricht fossil coral described by Goldfuss under the name of *Astrea elegans*,[2] and he refers the other to the *Astrea escharoides*[3] of the same author[4]. We regret not having had an opportunity of examining these fossils.

[1] Loc. cit., p. 31.

[2] Petref. Germ., vol. i, tab. xxiii, fig. 6.

[3] Goldfuss, op. cit., tab. xxiii, fig. 2; fossil from Maestricht.

[4] Austen, on the Geol. of the South-east of Devonshire, Trans. of the Geol. Soc., Second Series, vol. vi, p. 452.

Family FUNGIDÆ (p. xlv).

Genus MICRABACIA (p. xlvii).

MICRABACIA CORONULA.   Tab. X, fig. 4, 4 *a*, 4 *b*, 4 *c*.

> CYCLOLITES, *W. Smith,* Strata identified by Organic Fossils, p. 12; Greensand, p. 15, 1816.
> FUNGIA CORONULA, *Goldfuss,* Petref. Germ., vol. i, p. 50, tab. xiv, fig. 10, 1826.
>   —   —   *F. A. Rœmer,* Die Verstein. des Norddeutschen Kreidegebirges, p. 25, 1840.
>   —   —   *Morris,* Cat. of Brit. Fossils, p. 38, 1843.
> FUNGIA CLATHRATA (?) *Geinitz,* Grundriss der Versteinerungskunde, tab. xxiii, fig. 2, 1849.

Corallum simple, lenticular, short; its under surface horizontal or slightly concave; its upper surface somewhat convex. Mural disc completely naked and regularly perforated by small intercostal pores. *Costæ* closely set, almost straight, equally narrow, not prominent, and but slightly echinulated; only twelve of them arise in the centre of the disc, but these soon bifurcate, and the twenty-four costæ so formed soon divide again; at about half the distance from the centre to the circumference of the disc each costa bifurcates once more, and the two terminal costæ so formed are grouped two by two towards the periphery of the disc. The granulations which form all these costæ are not very distinct, and are arranged in single lines. *Calicular fossula* small and not very deep, but well marked and rather elongated laterally. *Columella* very small, oblong, and subpapillose. *Septa* forming five complete cycla, and corresponding to the intercostal spaces; those of the last cyclum quite rudimentary; the others tall, thin, straight, and united by sub-spiniform trabiculæ. Those of the first cyclum larger than the others, and augmenting slightly in thickness towards the middle; the secondary ones almost as large; all delicately denticulated along their upper edge, and much thinner towards their outer and inferior angle than in any other part. Diameter, three or sometimes four lines; height, one line and a half.

The above-described fossils were found in the Greensand at Warminster, in Wiltshire, and according to William Smith, who was the first author that mentions this fossil, are also met with at Chute Farm and Puddle Hill, near Dunstable.

By an attentive comparison with the specimens described by Goldfuss, and belonging to the Poppelsdorff Museum at Bonn, we have ascertained the specific identity of this British Coral with the *Fungia coronula* found in the chalk of Essen. Specimens exist in Mr. Bowerbank's cabinet, and in the collections belonging to the Geological Society, the Museum of Paris, the Museum of Bonn, and M. Defrance at Sceaux, who has designated it by the unpublished name of *Fungia dubia.*

CHAPTER VI.

CORALS FROM THE GAULT.

THE Fossil Corals contained in the Gault are more numerous than those imbedded in the upper greensand and the lower chalk. Most of them belong to the family of Turbinolidæ, and the principal localities where they have been met with in England are Folkstone and Cambridge.

Family TURBINOLIDÆ (p. xi).

Tribe CYATHININÆ (p. xii).

1. *Genus* CYATHINA (p. xii).

CYATHINA BOWERBANKII. Tab. XI, fig. 1, 1 *a*, 1 *b*.

CYATHINA BOWERBANKII, *Milne Edwards* and *J. Haime*, Monogr. Turbin., in Ann. des Sc. Nat., 3ᵐᵉ série, vol. ix, p. 292, 1848.

Corallum simple, elongated, turbinate, very narrow, and slightly bent near the basis, which does not appear to have expanded much. *Wall* quite naked. *Costæ* almost flat, distinct from the basis, or nearly so, covered with small granulations, nearly equal, and showing a slight tendency to form binary groups. *Calice* circular. *Columella* not much developed, and composed of twisted blades. *Septa* forming four complete cycla; very thin, but slightly granulated, and rather unequal. Those of the last cyclum very little developed, and the tertiary ones rather thickened towards the inner edge. *Pali* corresponding to the penultimate cyclum of septa, and rather broad. Height of the coral, eight or nine lines; diameter of the calice, three lines and a half.

This fossil was found in the Gault at Folkstone, by our friend Mr. Bowerbank. All the specimens that we have seen were very incomplete, but some showed all the principal characters represented in the figures which we have given.

C. Bowerbankii is easily distinguished from *C. Smithii* and *C. pseudoturbinolia*, by not having a fifth cyclum of septa. It differs also from *C. arcuata* by the delicacy of its septa, and from *C. Guadulpensis* by the circular form of its calice, and its round columella.

C. cylindrica, *C. Bredæ*, and *C. Koninckii*, have only six large *pali*, whereas in *C. Bowerbankii* the number of these organs amounts to twelve. *C. lævigata*[1] differs from the above-described species, by the pali being narrow, and very thick, and *C. Debeyana* by the existence of a well-marked epithecal band near the calice.

M. Alcide d'Orbigny has, in a recent publication,[2] referred to this species as the type of his new genus *Amblocyathus*, which he defines as being *Cyathina*, with a circular calice and a round columella. He adds that Amblocyathus is a lost genus, and contains three fossil species belonging to the Neocomian and Albian[3] strata. We must, however, beg leave to remark, that the two above-mentioned characters are met with in almost every species of our Cyathina, and most especially in *C. cyathus*, which is the type of the genus *Cyathina*, and is actually living in the Mediterranean sea. Only two of the species referred to the genus Cyathina in our 'Monograph of the Turbinolidæ' present a slightly oval calice and a transversal columella—*C. pseudoturbinolia* and *C. Guadulpensis*. In *C. Smithii* the columella is oblong, but the calice is circular, or nearly so. If it be con_ sidered necessary to separate the Cyathina with a circular calice from those that have an oval calice, it would therefore be more proper to give a new generic name to the latter, and not to change the denomination of the group containing the very species for which Ehrenberg first established the genus Cyathina. But this innovation, proposed by M. d'Orbigny, appears to us as being in every respect unnecessary, for the slight deformation of the calice and the columella which forms the sole basis of the new generic division, can hardly be considered as characters of sufficient value; species that differ in no other respect are often found to vary in this way, and even specimens belonging to the same species sometimes differ much in the form of the calicular margin. Thus, although the calice is circular, or nearly so, in most specimens of *C. cyathus* and *C. Smithii* that are met with, we have seen some that were compressed, and had the calice as oval as in *C. pseudo-turbinolia* and *C. Guadulpensis;* similar deviations from the normal form are also to be met with in the columella; in *C. Smithii*, for example, this organ is sometimes quite circular, although it is in general oblong. Differences of this kind, when not more marked than is the case among the various species of Cyathina, can therefore scarcely be deemed important enough to characterise generic divisions; and, as in the present case, they do not appear to coexist with any other structural peculiarity, we see no reason for admitting the new genus *Amblocyathus*.

---

[1] Tab. ix, fig. 1.

[2] Note sur des Polypiers Fossiles, Paris, 1849.

[3] M. d'Orbigny employs the name of *Albian* formation to designate the *Gault*.

2. *Genus* CYCLOCYATHUS (p. xiv).

CYCLOCYATHUS FITTONI.   Tab. XI, fig. 3, 3 *a*, 3 *b*.

Corallum simple, discoidal, short; mural disc horizontal, or slightly concave, and presenting in its centre a small, irregular cicatrix, indicative of its primitive adherence. *Epitheca* very thin, presenting some slight concentric striæ, and not preventing the radiate costæ from being visible.   These are straight, and not very prominent; but those of the first and second order are well marked.   The edge of the mural disc is thin, and slightly prominent.   The upper or *calicular surface* of the corallum is rather convex externally, and concave towards the centre.   The fossula is shallow, but large, and well marked. *Columella* fasciculate, well developed, and terminated by a broad, papillose surface.   *Septa* forming four complete cycla.   The six fundamental septal systems distinct, but the septa of the second order not differing much from those of the first order.   All the septa well developed, straight, rather thick exteriorly, arched above, and granulated laterally; their outer edge somewhat crenulated, granulose, slightly concave near the mural disc, and projecting a little towards the upper part.   *Pali* well developed, very distinct from the septa, and corresponding to those of the third cyclum.   Height of the corallum, two or three lines; diameter, in general not more than five or six lines.

This fossil is the only known species of the genus *Cyclocyathus;* its form renders it very remarkable.   It has been found in the Gault at Cambridge, Drayton, West Malling, and Folkstone, but appears to be most abundant in the last-mentioned locality.   The specimens here described belong to the collections of the Geological Society, of Mr. Bowerbank, and of Mr. D. Sharpe.

3. *Genus* TROCHOCYATHUS (p. xiv).

1. TROCHOCYATHUS CONULUS.   Tab. XI, fig. 5, 5 *a*.

> CARYOPHYLLIA CONULUS, (?) *Phillips*, Illust. of the Geol. of Yorkshire, tab. ii, fig. 1, 1829.
> (A rough figure without any description.)
> —   —   *Michelin*, Mém. de la Soc. Géol. de France, vol. iii, p. 98, 1838.
> TURBINOLIA CONULUS, *Michelin*, Icon. Zooph., p. i, pl. i, fig. 12, 1840.
> TROCHOCYATHUS CONULUS, *Milne Edwards* and *J. Haime*, Monogr. des Turbin., Ann. des
> Sc. Nat., 3me série, vol. ix, p. 306, 1848.

Corallum simple, turbinate, rather elongate, straight or slightly bent, and pedicellated. *Wall* presenting in general some slight traces of an incomplete epitheca.   *Costæ* simple, distinct from the basis, closely set, delicately granulated, not very prominent, and alternately of unequal size towards the calicular edge. *Calice* almost circular, or somewhat oval and shallow. *Columella* fascicular, well developed, not prominent at its apex, and terminated by ten or fifteen papillæ of equal size. *Septa* forming four complete cycla and six well-marked, equally developed systems, in which, however, the secondary ones differ but little from those of the first cyclum. The septa are slightly exsert, closely set, unequal, and rather thicker outwards than towards the columella. *Pali* narrow and unequal; those corresponding to the tertiary septa broad and rather stout; the others, and most especially those corresponding to the primary septa, narrow and thinner. Height of the corallum, seven or eight lines; diameter of the calice, almost seven lines.

This species belongs to the first section of the genus Trochocyathus (*T. simplices*), and differs from *T. impari-partitus*[1] and *T. Bellingherianus* by not having a fifth cyclum of septa; its general form distinguishes it from *T. mitratus*,[2] *T. crassus*, *T. simplex*, and *T. costulatus*, which are all short, broad, and curved; and from *T. elongatus*, *T. Koninckii*, and *T. gracilis*, which are much elongated, curved, and very narrow towards the basis. It appears to resemble most, especially by its general form, *T. cupula*,[3] which is also conical and straight, but this last-mentioned species differs from it by the thickness and strong granulations of the septa, and by the breadth of the basis.

Trochocyathus conulus appears to have been very widely spread in the seas where the Gault formations were deposited. The specimens which we most particularly studied were

[1] See our Monograph of Turbinolidæ, loc. cit., p. 307.

[2] Since the publication of our Monograph of the Turbinolidæ (in 1848) we have recognised that the fossils from Tortona, which M. Michelotti designates under the name of *Turbinolia plicata*, do not differ specifically from the specimens existing in the Poppelsdorf Museum under the name of *Turbinolia mitrata*, Goldfuss. As we already expected, the latter specific name must therefore be substituted for the one employed by M. Michelotti, and M. Michelin.

[3] This new species, designated under the name of *Turbinolia cupula*, by M. Alex. Rouault, (Bulletin de la Soc. Géol. de France, 2ᵐᵉ série, vol. ix, p. 206, 1848), was found by that geologist at Bos d'Arros, in the department of the Lower Pyrennees, and does not appear to differ from a fossil which exists in the collection of M. Nyst, and was found in the Eocene formation at Lacken, near Brussels. *Trochocyathus cupula* belongs to the first section of our genus Trachocyathus, and presents the following characters :

Corallum straight, or almost so, subturbinate, but short, and having a broad peduncle, but not remaining adherent in the adult state. *Costæ* distinct from the basis, straight, unequally developed alternately, rather prominent, especially near the calice, granulated and striated transversely; rudimentary costæ, that do not correspond to any septa, are seen in the intercostal furrows. *Calice* circular; fossula not deep. *Columella* crispate, well developed. *Septa* forming three complete cycla, and in general a fourth incomplete cyclum in one half of three of the systems; exsert, rather unequal, strong, and presenting on their lateral surfaces large prominent granulations, which are arranged in lines nearly parallel to the upper edge. *Pali* thick, strongly granulated, and unequal; those corresponding to the tertiary septa the largest in the half systems where the septa of the fourth cyclum exist, and those corresponding to the secondary septa most developed in the other part of the calice. Height of the corallum, three lines; diameter of the calice almost as much. By the strong granulations of the septa, and the breadth of its basis, this species tends to establish a transition between the genus *Trochocyathus* and the genus *Paracyathus*.

found near Cambridge, in England; at Gatis de Gerodot, Dienville, near Brienne (department of the Aude), and Etrepy (department of the Marne), in France. Other specimens, which in all probability belong also to this species, are designated in M. Michelin's collection as having been found at Novion-en-Porcien; at Macheromenil, in the Ardennes, and at the Perte du Rhone, in the department of the Ain; but we suspect that some mistake may have been made in the labelling of the specimen which is designated in the same collection as belonging to the chalk of Tournay, in Belgium. We must also add, that the fossil designated by Professor J. Phillips under the name of *Turbinolia conulus* was found by that eminent geologist at Speeton, in Yorkshire; but its characters are not sufficiently well known for us to be able to identify it with the above-described species, specimens of which exist in the collections of the Geological Society, of the Museum at Paris, and of MM. d'Orbigny, Michelin, and Milne Edwards.

M. Al. d'Orbigny has lately given the name of *Aplocyathus*[1] to those species of our genus Trochocyathus in which the calice is circular. If this new generic division was adopted, the species here described would be referred to it; but that is not, in our opinion, advisable. The calice, which is quite circular in a great many species of our genus Trochocyathus, becomes slightly elongated in some, quite elliptical in others, and not only would the line of separation be difficult to establish between these different forms, but certain species which are evidently most closely allied by all their other organic characters, would be separated generically in the classification proposed by M. d'Orbigny. We cannot, therefore, adopt his views in this respect; but, in justice to that distinguished palæontologist, we must remark that the species[2] chosen by him as the type of his genus *Aplocyathus* differs much in its general aspect from most species of our genus Trochocyathus, and, when more completely known, may be found to present characters of sufficient value to authorise the establishment of a separate generic group, which must then be so defined as not to comprehend *T. conulus*, nor most of the other species that have a circular calice.

2. TROCHOCYATHUS HARVEYANUS. Tab. XI, fig. 4, 4 *a*, 4 *b*.

TROCHOCYATHUS HARVEYANUS, *Milne Edwards* and *J. Haime*, Monogr. des Turbinolides, in Ann. des Sc. Nat., 3ᵐᵉ série, vol. ix, p. 314, 1848.

Corallum simple, straight, short, almost hemispherical, and terminated by a very short peduncle, the basal surface of which is concave. *Costæ* distinct from the basis, and delicately striated transversely; the primary and secondary ones very prominent and sharp; those of the third cyclum well developed along the upper half of the wall, but those of the fourth cyclum very small and obscure. *Calice* circular and flat; fossula shallow. *Columella* well developed and papillose. *Septa* forming four complete cycla; exsert, thin,

[1] Note sur des Polypiers Fossiles, p. 5, 1849. [2] The *Trochocyathus armatus*.

broad, straight, granulated laterally, unequally developed, but not differing much in the first and second cycla. *Pali* corresponding to the septa of the first three cycla, rather narrow, and unequally developed in an inverse ratio to the corresponding septa; no pali in the radii of the septa belonging to the fourth cyclum. Height of the corallum, three lines; diameter of the calice, four lines.

This species belongs to the fourth section of our genus Trochocyathus (*T. breves*), and consequently its characters need not be compared with those of the various species belonging to the sections of the *T. simplices*, *T. cristati*, and *T. multistriati*, the description of which may be found in our 'Monograph of the Turbinolidæ.' It differs from *T. obesus*, *T. armatus*, and *T. perarmatus*,[1] by not having any costal spines, and from *T. Michelini* by the costæ being distinct down to the basis, and by its general form being less depressed. It appears to be most closely allied to the fossil which we shall next describe under the name of *Trochocyathus* (?) *Konigi*, but is of a more slender form.

T. Harveyanus was found in the Gault at Folkstone, the birthplace of the illustrious physiologist to whom we have dedicated this species. The specimens here described belong to the collections of Mr. Bowerbank and Mr. D. Sharpe.

3. Trochocyathus (?) Konigi.

> Turbinolia Konigi, *Mantell*, Illust. of the Geol. of Sussex, p. 85, tab. xix, figs. 22, 24, 1822.
> — — *Fleming*, British Animals, p. 510, 1828.
> — (Trochocyathus?) Konigi, *Milne Edwards* and *J. Haime*, Monogr. des Turb., in Ann. des Sc. Nat., 3ᵐᵉ série, vol. ix, p. 335, 1848.

The specimens of this fossil figured by Mr. Mantell, and those which we have seen in the collections of MM. d'Orbigny and Michelin, are in a very bad state of preservation,

---

[1] This species, which has been lately designated under the name of *Turbinolia perarmata* by M. Talavignes, but has not yet been described, and has been given to us by that geologist, was discovered at Fabresan, in the department of the Aude. M. Alex. Rouault has since then met with the same species at Bos d'Arros in the Lower Pyrennees. (See Bull. Soc. Géol., 2ᵐᵉ série, vol. v, p. 206.) It may be recognised by the following characters:

Corallum very short, subdiscoidal; its under surface flat and almost smooth; sometimes adhering to a small shell. *Costæ* distinct near the calice, projecting very little, closely set, almost equal, and delicately granulated; those of the first cyclum not differing much from the others, but bearing, at a short distance from the calicular edge, a strong spiniform appendix, which is rather compressed, extends outwards, and presents, on its under edge, a small pointed tubercle. *Calice* circular. *Septa* forming four complete cycla and six equally developed systems; closely set, rather exsert, thin, and unequally developed; but those of the second cyclum differing very little from the primary ones. *Pali* narrow and rather thick. Height of the corallum, one line and a half; diameter of the calice, two lines and a half. Fossil from the Nummulitic formation at Fabresan and Bos d'Arros.

and have lost their walls; we are, therefore, unable to characterise the species with any degree of precision, and it is with much doubt that we refer it to the genus Trochocyathus, for we are not as yet sufficiently satisfied as to the existence of pali. M. Michelin is of opinion that these fossils are merely specimens of *Trochocyathus conulus* with their basis worn away. They are of a conico-convex form, and are broader in proportion than *T. Harveyanus*, to which they bear, however, great resemblance. Their height is about four lines, and their diameter a little more. We have not considered it necessary to give a new figure of these corals, for the specimens in our possession do not show anything more than those represented in Dr. Mantell's plates.

The specimens that we have had an opportunity of examining were found in the Gault at Folkstone, in the environs of Boulogne-sur-Mer, at Wissant, at Les Fiz, near Chamounix, and at the Perte du Rhone, in the department of the Ain. According to Dr. Mantell the same species is met with at Lewes in Sussex, and Godstone in Surrey, at Malling in Kent, in Cambridgeshire,[1] at Ringmer, and at Bletchingley.[2]

TROCHOCYATHUS (?) WARBURTONI.

We are inclined to think that a cast found in the Gault of Cambridgeshire by Mr. H. Warburton, and presented by that gentleman to the Museum of the Geological Society, must belong to a distinct species of Trochocyathus. It is about six lines in height, and seven in diameter; the number of septa is forty-eight. For the sake of convenience we have given a specific name to it, but we are not able to characterise it.

4. *Genus* BATHYCYATHUS (p. xiii.)

BATHYCYATHUS SOWERBYI. Tab. XI, fig. 2, 2 *a*.

BATHYCYATHUS SOWERBYI, *Milne Edwards* and *J. Haime*, Monogr. des Turbinolides, Ann. des Sc. Nat., 3me série, vol. ix, p. 295, 1848.

Corallum simple, adherent by a broad basis, straight, tall, compressed, and having its lateral edges somewhat prominent. *Wall* delicately granulated. *Costæ* not very distinct in the lower half of the corallum, but becoming rather prominent higher up, especially those of the first and second orders. *Calice* elliptical and horizontal, the relative length of its long and short axis varying much (in one specimen = 100 : 170, and in another = 100 : 250). *Fossula* narrow, and appearing to be deep, but completely filled up with extraneous matter in all the specimens that we have seen, so as not to enable us to obtain any knowledge respecting the columella and the pali. It is therefore with some uncertainty that we refer this species to the genus Bathycyathus, and in doing so we have been guided

[1] Geol. of Sussex.   [2] Trans. of the Geol. Soc., s. 2, vol. iii, p. 210.

only by characters of secondary value, which agree, however, very well with those of the other Corals belonging to the same generical division. *Septa* forming four complete cycla; exsert, thick exteriorly, but thin inwardly, and presenting but few granulations on their lateral surfaces. Those of the second cyclum almost as large as the primary ones; the tertiary ones but little developed, although they correspond to large costæ, and not as tall as those of the last cyclum, which are grouped very closely on each side of the primary and secondary ones. Height of the corallum, one inch two or three lines; great diameter of the calice, six or seven lines.

The genus *Bathycyathus* contains two other species, which are both recent: *B. Chilensis*[1] and *B. Indicus,*[2] which differ from *B. Sowerbyi* in having an additional cyclum of septa, the calice arched, and the costæ more developed near the basis. We have seen but two specimens of this fossil; one, belonging to the collection of Mr. D. Sharpe, is catalogued as having been found in the Gault near Folkstone; the second, belonging to the museum of the Geological Society, is referred with doubt to the upper greensand of Kidge, in Wiltshire.

Family ASTREIDÆ (p. xxiii).

Tribe EUSMILINÆ (p. xxiii).

1. *Genus* TROCHOSMILIA (p. xxiv).

TROCHOSMILIA SULCATA. Tab. XI, fig. 6, 6 *a*, 6 *b*.

Corallum simple, turbinate, straight, tall, much compressed, subpedicellate, and appearing to be free. *Wall* presenting on each side two deep longitudinal furrows. *Costæ* distinct from the basis, slightly prominent, closely set, and unequal, especially towards their upper end. *Calice* elliptical, sublobulated, and slightly arched; its long and short axis in the proportion of 100 : 200. *Fossula* very narrow, elongated, and not very deep. No *columella*. *Septa* forming four cycla or more, rather unequal, closely set, thin, and slightly exsert. *Dissepiments* not numerous. Height of the corallum, nearly one inch; diameter of the calice, six or seven lines by three; depth of the fossula, two lines and a half.

We have seen but one specimen of this fossil, which, although somewhat weather-worn, appeared sufficiently distinct from all other species to authorise us in giving it a peculiar specific name. It differs from *Trochosmilia didyma*[3] by its calice being straight, and not

---

[1] See our Monograph of Turbinolidæ, tab. ix, fig. 5.

[2] Loc. cit., tab. ix, fig. 4.

[3] *Turbinolia didyma*, Goldfuss, Petref. Germ., vol. i, tab. xv, fig. 11.

bent in two; from *T. Boissyana*, *T. Patula*,[1] *T. cernua*, and *T. crassa*,[2] by being sub-pedicellated, and not adherent in the adult state; from *T. irregularis*, *T. corniculum*, *T. Faujasii*, *T. Gervillii*, and *T. uricornis*,[3] by being strongly compressed quite down to the basis; and from *T. Saltzburgiensis*, *T. cuneolus*, *T. compressa*, *T. complanata*, *T. Basochesii*, and *T. tuberosa*, by the existence of the above-mentioned four deep mural furrows. By their general form, all these corals much resemble many species belonging to the division of Cyathininæ, but differ from them, and from all other Turbinolidæ, by having interseptal dissepiments.

This fossil was found in the Gault at Folkstone, by Mr. Bowerbank.

The LITHODENDRON GRACILE, Goldfuss,[4] is mentioned by Mr. Morris[5] as having been found in the Gault of Kent, but as yet we have not met with any specimens of that species in any of the British palæontological collections.

[1] See our Monogr. des Astreides, Ann. des Sc. Nat., 3ᵐᵉ série, vol. x, p. 236.

[2] We here designate, under the name of *Trochosmilia crassa*, the fossil described by M. Michelin under the name of *Turbinolia cernua*, Goldfuss, and by ourselves as *Trochosmilia cernua* ; for, on comparing it with the specimens previously described by Goldfuss under the name of *Turbinolia cernua*, we have ascertained that they are not specifically identical.

The species which must retain the name first applied by Goldfuss presents the following characters :

Corallum pedicellated and strongly compressed quite from the basis. *Costæ* thin, alternately unequal ; the larger ones rather prominent and somewhat lamellar. *Calice* arched and elongated in the proportion of 100 : 230. *Septa* thin, very closely set, and presenting on their lateral surfaces a great number of granulations arranged somewhat regularly in convex lines parallel to the upper edge. Forty-eight principal septa, separated by an equal number of small ones ; some indications of an additional rudimentary cyclum. Height of the corallum, one inch and a half; long diameter of the calice, twelve lines ; short axis, five lines. (The figure given by Goldfuss, tab. xv, fig. 8, is not quite accurate.)

[3] Monogr. des Astreides, loc. cit.

[4] Petref. Germ., vol. i, tab. xiii, fig. 2.

[5] Catalogue of British Fossils, p. 40.

CHAPTER VII.

CORALS FROM THE LOWER GREENSAND.

THE remains of true Polypi are very rare in this part of the British geological strata; the fossil which Mr. Lonsdale has lately described under the name of *Choristopetalum impar*,[1] and which was found in the lower greensand at Atherfield, does not appear to us to belong to this class, and is, in our opinion, a Bryozoon. We have as yet met with but one species of Zoantharia, which can be referred with any degree of certainty to this formation.

Family STAURIDÆ (p. lxiv).

Genus HOLOCYSTIS (p. lxiv).

HOLOCYSTIS ELEGANS.   Tab. X, fig. 5, 5 *a*, 5 *b*.

ASTREA, *Fitton*, On the Strata below the Chalk, in Geol. Trans., s. 2, vol. iv, p. 352, 1843.
ASTREA ELEGANS, *Fitton*, in Quarterly Journ. Geol. Soc., vol. iii, p. 296, 1847.
CYATHOPHORA (?) ELEGANS, *Lonsdale*, Proceed. of the Geol. Soc., vol. v, part i, p. 83, tab. iv,
fig. 12,‑15, 1849.

Corallum complex, astreiform, constituting a convex mass, and augmenting by extra calicular gemmation; the young individuals being produced at the point of junction of the surrounding calices. *Corallites* somewhat prismatic, and cemented together laterally, either by the direct union of their walls, or by means of the costæ, which are thick, and in general pretty well developed. *Calices* subpolygonal, separated in general by a simple but thick mural ridge; sometimes by walls that remain distinct, and are in their turn separated by a small intermural furrow. *Fossula* deep. *Columella* very small, and appearing to be styliform. *Septa* forming three complete cycla, and four well-characterised systems. The four primary ones much more developed than the others, reaching almost to the centre of the fossula, and giving to the calice a crucial character, which is never met with in Astreidæ, Oculinidæ, Turbinolidæ, &c. The septa are slightly exsert, closely set, thick exteriorly, and very slightly granulated laterally; they appear to have undivided edges, and they differ much in size, according to the cycla to which they belong. The interseptal *dissepiments* are simple, horizontal, or slightly convex, and placed at the same level in the different loculi,

[1] Proceedings of the Geol. Soc., vol. v, part i, p. 69, tab. iv, figs. 5 to 11, 1849.

so as to constitute by their union a series of complete tabulæ, subdivided by the primary septa, and distant from each other about one fifth of a line. Exothecal dissepiments much resembling the preceding ones. Diameter of the calices, and depth of the fossula, about one line and a fourth.

Fossil from the lower greensand at Redhill cutting, Atherfield, in the Isle of Wight, and at Peasemarsh.

The specimens here described belong to the Museum of the Geological Society, and had been named by Mr. Lonsdale. The propriety of establishing a new generical division for this remarkable coral, was very judiciously pointed out by that indefatigable palæontologist; but, guided by reasons which we do not quite understand, he refers, with a sign of doubt, this same species to the *Cyathophora* of M. Michelin, a genus which, in our opinion, does not differ from true *Stylina*. The genus *Holocystis* differs from our genus *Stauria* by its extra calicular gemmation, and its costulated walls. It is the most modern representative of the great division of Zoantharia rugosa, which becomes predominant in the Palæozoic formations, and is principally characterised by the tendency to a quadrate arrangement of the constitutive parts of the Corallites, whereas in the other sections of Zoantharia, six is the fundamental number of the radiate organs.

THE

PALÆONTOGRAPHICAL SOCIETY.

INSTITUTED MDCCCXLVII.

LONDON:

MDCCCLI.

A MONOGRAPH

OF THE

BRITISH FOSSIL CORALS.

BY

H. MILNE EDWARDS,

DEAN OF THE FACULTY OF SCIENCES OF PARIS; PROFESSOR AT THE MUSEUM OF NATURAL HISTORY;
MEMBER OF THE INSTITUT OF FRANCE;
FOREIGN MEMBER OF THE ROYAL SOCIETY OF LONDON, OF THE ACADEMIES OF BERLIN, STOCKHOLM, ST. PETERSBURG,
VIENNA, KONIGSBERG, MOSCOW, BRUXELLES, BOSTON, PHILADELPHIA, ETC.

AND

JULES HAIME.

SECOND PART.

CORALS FROM THE OOLITIC FORMATIONS.

LONDON:

PRINTED FOR THE PALÆONTOGRAPHICAL SOCIETY.

1851.

C. AND J. ADLARD, PRINTERS, BARTHOLOMEW CLOSE.

DESCRIPTION

OF

THE BRITISH FOSSIL CORALS.

CHAPTER VIII.

CORALS FROM THE PORTLAND STONE.

THE most recent Oolitic Formations appear to contain very few corals. We have seen only three species which can, with any degree of certainty, be referred to the continental deposits belonging to this group, and but one British species. The former were found in strata which are considered as corresponding to the Kimmeridge Clay, and we are not aware of any having been met with in that Formation in Great Britain. The latter is contained in the Portland stone, and appears to be peculiar to that upper portion of the superior Oolite.

Family ASTREIDÆ, (p. xxiii.)

Genus ISASTREA.

ISASTREA OBLONGA, (Tab. XII, fig. 1, 1*a*, 1*b*, 1*c*, 1*d*.)

> CORALLOIDEA COLUMNARIA, etc., *Parkinson*, Org. Remains, vol. ii, p. 60, tab. vi, figs. 12, 13, 1808.
>
> SILICIFIED CORAL, *W. Conybeare* and *W. Phillips*, Outlines of the Geol. of England and Wales, p. 176, 1822.
>
> LITHOSTROTION OBLONGUM, *Fleming*, British Animals, p. 508, 1828.
>
> MADREPORA, SILICIFIED, *E. Benett*, Catal. of the Organic Remains of the County of Wilts, p. 7, 1831.
>
> ASTREA TISBURIENSIS, *Fitton*, On the strata below the chalk; Geol. Trans., 2d series, vol. iv, p. 347, 1843.
>
> LITHOSTROTION OBLONGUM and ASTREA TISBURIENSIS, *Morris*, Cat. of Brit. Fossils, pp. 31, 40, 1843.
>
> ISASTREA OBLONGA, *Milne Edwards* and *J. Haime*, Polypiers Fossiles des Terrains Palæoz., etc. Archives du Museum, vol. v, p. 103, 1851.

10

Since the publication of our Monograph of the Family of AstreidÆ (Annales des Sciences Naturelles, s. 3, vols. xi and xii, 1849), and that of the First Part of this Work, we have been led to consider, as being of generical value, a group composed of most of the species which we formerly placed in the Second Section of our genus *Prionastrea* (Introd., p. xii), and in the Conspectus forming the Introduction to our Monograph of the Palæozoic Corals, we have designated this new division by the name of Isastrea. In the genus *Prionastrea*, as now circumscribed, the walls are double in the lower part of the corallum; in *Isastrea* they are always simple. In the latter the columella is rudimentary or does not exist, and the *septa* are terminated by a crenulated edge, the denticulations of which are of nearly equal size; in *Prionastrea* the columella exists, and the marginal denticulations of the *septa* increase in size from the circumference of the corallum towards its centre. These differences are shown in some well-preserved specimens which we have but lately had the opportunity of examining, and it may be worth noticing, that the two generical groups thus separated appear to have each a distinct geological range; all the true *Prionastrea* being either recent or tertiary species, whereas *Isastrea* have as yet been met with only in secondary deposits.

The fossil here designated by the name of *Isastrea oblonga* is very common in the Portland beds of Tisbury, Wiltshire; it is a massive corallum, completely silicified, and when polished shows its characters in a very distinct manner. By means of a horizontal section (fig. 1c) it is easy to see that the corallites are circumscribed by simple, thick *walls* of a pentagonal or hexagonal form; that the columella is quite rudimentary, if not completely deficient; and that the principal *septa* reach quite to the centre of the visceral cavity, but do not join together by their inner edge, and are united only by means of small trabiculæ, which occupy the place usually filled by the columella. The six septal systems are in general very distinct, in consequence of the primary *septa* being much more developed than those of the following cycla, and two of these systems are much larger than the four others. The *septa* form four complete cycla, but those of the last cyclum are rudimentary in the four small systems above mentioned; they are all nearly straight, somewhat thick, and strongly granulated laterally; they are very unequal in size in the different cycla, and it often happens that those of the fourth cyclum are more developed in one half of each system than in the other, and in that case the tertiary septa situated between the former, are also somewhat more developed than those of the other half systems; but the secondary *septa* are all nearly of equal size, and even those of the two large systems are never as much developed as the primary ones, which alone reach to the centre of the visceral chamber, and become rather thicker internally.

A vertical section (fig. 1f) shows that the inner edge of the *septa* is delicately and almost regularly denticulated. The *dissepiments*, which in many specimens have disappeared completely, or have been more or less modified in form by the process of fossilisation, are well developed, arched, somewhat decline inwards, and situated at one third or one fourth of a line apart; some remain simple, but most of them become bifurcate inwards.

The corallites are very tall, and the calices vary somewhat in size in the different specimens which have come under our observation; in small adult individuals their width (at the great diagonal) is about two and a half lines, and in the large ones scarcely more than three lines.

The genus *Isastrea* comprises a considerable number of species, the list of which has been given in our above-mentioned work. In order to avoid unnecessary details, we will therefore only add, that *T. oblonga* is easily distinguished from the other British fossils belonging to the same generical group, by the thickness of its walls, and the strong lateral granulations of the *septa*. By its general aspect this coral bears some likeness to the *Isastrea polygonalis*,[1] a fossil of the Muschelkalk, of which the cast only has as yet been found; but in this last-mentioned species all the *septa* of the fourth cyclum are well developed.

Isastrea oblonga has been met with only at Tisbury, Wiltshire. The specimens which we have examined belong to the collections of the Geological Society of London, the Museum of Bristol, the Museum of Paris, Mr. Bowerbank, and Mr. Stokes.

CHAPTER IX.

CORALS FROM THE CORAL RAG.

The Coral Rag, as may be inferred from its name, is a deposit most abundant in fossil corals; but the number of species found in this Formation is by no means proportionate to that of the specimens met with. The British species are indeed very limited; and, although we have had access to all the richest palæontological collections in England, we have only seen fourteen species of true corals belonging to this portion of the oolitic series. Twelve of these are *Astreidæ*, and two *Fungidæ*; five of these species are also found in the Coral Rag of France and Germany; the nine others are, as yet, peculiar to England. One (*Thamnastrea concinna*) appears to exist also in the Great Oolite, and probably even in the Inferior Oolite; but most have not been met with in any other Formation.

The principal fossiliferous beds, from which these corals have been obtained, are situated at Steeple Ashton, in Wiltshire, and Malton, in Yorkshire. Some species have also been found at Hackness, in Oxfordshire; at Osmington, near Weymouth; at Upware, in Cambridgeshire, &c.

---

[1] *Astrea polygonalis*, Michelin, Iconogr., tab. iii, fig. 1.

Family—ASTREIDÆ, (p. xxiii.)

Genus STYLINA, (p. xxix.)

1. STYLINA TUBULIFERA. Tab. XIV, figs. 3, 3*a*, 3*b*, 3*c*.

CORALLOID BODY? *J. Morton*, Nat. Hist. of Northamptonshire, p. 184, tab. ii, fig. 10, 1712.

ASTREA TUBULIFERA, *Phillips*, Illustr. of the Geol. of Yorkshire, vol. i, p. 126, tab. iii, fig. 6, 1829; and second edition, p. 98. (The specimen figured was much worn away.)

HYDNOPHORA FRIESLEBENII? *Fischer*, Oryctographie de Moscou, pl. xxxiii, fig. 2, 1837.

STYLINA TUBULOSA, *Michelin*, Icon. Zooph., p. 97, pl. xxi, fig. 6, 1843.

ASTREA TUBULOSA and AGARICIA LOBATA, *Morris*, Cat. of Brit. Foss., pp. 20, 31, 1843.

DENTIPORA GLOMERATA, *M'Coy*, Ann. of Nat. Hist., s. 2, vol. ii, p. 399, 1848.

STYLINA TUBULOSA, *Milne Edwards* and *J. Haime*, Monogr. des Astreides, Ann. des Sc. Nat., s. 3, vol. x, p. 289, 1848. (Wrongly referred to the *Astrea tubulosa* of Goldfuss, whose figure is inexact.)

DECACŒNIA MICHELINI, *D'Orbigny*, Prodr. de Paléontol., v. i, p. 33, 1850.

STYLINA TUBULIFERA, *Milne Edwards* and *J. Haime*, Polyp. des Terrains Palæoz., etc., p. 59, 1851.

Corallum massive, more or less elevated, convex on the upper surface, and somewhat gibbose. Common basal plate or wall with a very thin epitheca, which is most distinct on the accretion ridges, and is always more or less worn away, but was probably continuous in the natural state. In the parts thus denuded, the *costal* striæ became visible; they are very delicate, closely set, quite straight, and equally developed. The corallites are almost cylindrical, and diverge in fasciculi from the common basis (fig. 3). The upper surface of the corallum is occupied by the calices, which are placed at some distance from each other, and very unequally exsert; the terminal portion of the corallites which thus protrudes has the form of a short truncate cone, and is surrounded by straight, delicate, closely-set, well-marked, and equally developed *costæ*. These are composed of a single row of granulations, and meet at the bottom of the intercalicular depressions, where those of two adjoining corallites often become completely blended together (fig. 3*a*). The *calices* are perfectly circular and somewhat unequal in size; the fossula is circular, rather narrow, and not deep; the *columella* is styliform, small, and slightly prominent; it is somewhat compressed, and its transverse section is suboval. The *septa* form three complete cycla, and in four of the six systems there are *septa* belonging to a fourth cyclum. In these highly-developed systems the secondary *septa* are almost as large as the primary ones, and thus give to the calice the appearance of having ten equal systems (fig. 3*a*). In each system the *septa* of the last-formed cyclum (that is to say, the tertiary septa in the small ones and those of the fourth cyclum in the large ones) are quite rudimentary on the inside of the wall, but correspond to well-developed costæ outwardly. The principal *septa* are strong, somewhat exsert, granulated laterally, thick exteriorly, but thin towards their middle part, and their upper edge, which is strongly arched, becomes thicker again towards the central part of the visceral chamber, but does not quite reach to the columella, so that this last-mentioned organ remains quite free to some distance from its upper end. In some of these composite Corals one or two corallites may be found, having the fourth septal cyclum complete, and all the system equally developed.

A transverse section shows that the *walls* of the corallites are very thick, and are principally formed by the corresponding part of the *septa*. A vertical section brings to light a structure which appears to belong to all the species of the genus *Stylina*. The tissue, which occupies the spaces existing between the cylindrical walls of the corallites, is not formed solely by the costæ and the exothecal laminæ, as in *Astrea*, but is divided into superposed layers, by means of prolongations from the walls which bend down in the intercalicular spaces.

Diameter of the calices, 1½ line; distance between them, 1 or 2 lines, or even more.

This fossil is found at Steeple-Ashton, Wiltshire, and at Malton, in Yorkshire. The British specimens submitted to our investigations belong to the collections of the Museum of Practical Geology, the Geological Society, the Bristol Museum, the Cambridge Museum, the Paris Museum, Mr. Phillips, and Mr. Bowerbank. Some fossil Corals, which we have seen in M. Michelin's Cabinet, and which were found in the Coralline Oolite of St. Mihiel, and some other localities in France, belong to the same species.

The genus *Stylina*, as defined in the Introduction to this Monograph, contains a considerable number of species, and corresponds to no less than eleven genera, lately proposed by M. D'Orbigny. These new generical divisions are founded on the differences existing: 1st, in the general form of the corallum, which is well known to be very variable; 2d, in the depth of the interseptal loculi which that author measured by means of casts, and which decreases gradually from one species to another; 3d, in the number of the principal septa, which is sometimes six, in other instances eight, ten, or even twelve, but can always be easily explained by slight modifications in the development of the same number of septal systems; and 4th, in the axis of the corallites, where the columella is sometimes most evident, and in other cases cannot be seen. The absence of a columella in some species of *Stylina* would certainly be a character of sufficient importance to justify the establishment of a generical division, were it not merely an accident dependent on the process of fossilisation, or some other cause independent of the structure of the corallite; but in many instances that is the case. Sometimes, however, we have not sufficient grounds for explaining in this manner the absence of the columella, and we therefore have provisionally adopted the genus *Cyathophora* of M. Michelin, containing the *Stylina* that show no traces of that organ;[1] but the divisions founded on the various combinations of

---

[1] See our above-mentioned Memoire in the 'Archives du Museum,' vol. v, p. 58.

the other above-mentioned differences do not appear to have sufficient value, and the genera *Lobocænia, Conocænia, Adelocænia, Tremocænia, Cryptocænia, Dendrocænia, Aplosastrea, Octocænia, Decacænia,* and *Pseudocænia* of M. d'Orbigny, may still remain united in a single generical group, under the old name of *Stylina*. This genus belongs exclusively to the secondary period, and most of its representatives are found in strata of the Jurassic formation.

Stylina tubulifera, having 10 principal *septa*, belongs to the genus *Decacænia* in M. D'Orbigny's method of classification; and this peculiarity, which is met with but in a few other species of *Stylina,* distinguishes it from all those in which the calice is divided into 6, 8, or 12 equal parts. All the *Stylina* which have this number of principal *septa* are very nearly allied to each other, and most of those which are at present considered as being specifically distinct, may very likely prove to be nothing more than varieties of one species; but we have not been able to examine a sufficient number of well-preserved specimens in order to decide this question. Thus, *Stylina lobata*[1] may perhaps be a young specimen of *S. tubulifera,* with short corallites, and very prominent, widely-set calices; and *Stylina octonis*[2] only differs from the above-described species by the calices being more closely set, somewhat unequal in size, and about $1\frac{1}{4}$ lines in diameter. The specimens on which these two species were established are both in a very bad state of preservation. *S. tubulifera* also resembles very much another fossil which was found in the Great Oolite near Bath, and will be described in a subsequent chapter of this Monograph, under the name of *S. Ploti;* the latter, however, has a smaller *columella,* thinner *septa,* and less prominent *calices.* The fossil coral mentioned by M. D'Orbigny, under the name of *Decacænia magnifica*[3] is more easily distinguished from *S. tubulifera,* and in some calices shows only 8 large *septa* instead of 10, as is the case in most. It is a slightly convex mass, with calices of unequal size, but slightly prominent, and of $2\frac{1}{2}$ lines diameter; the costæ are nearly equal, and delicate; three very small but well-characterised septa exist between each of the principal septa. This new species appears to have been found in the Coral Rag of Chatel Censoir and of Wagnon, Ardennes. We must also add that, by its general aspect, *S. tubulifera* resembles very much *S. tubulosa,*[4] described by Goldfuss; and in the figure given by that able Palæontologist, this latter species is represented as having 10 principal *septa;* but that is not in reality the case, for in the original specimen belonging to the Poppelsdorf Museum, at Bonn, we ascertained the existence of 12 of these *septa.*

---

[1] *Explanaria lobata,* Goldfuss, Petref. Germ., tab. 38, fig. 5.

[2] *Pseudocænia octonis,* D'Orbigny, Prodrome, vol. ii, p. 34.

[3] Prod. de Paléontol., vol. ii, p. 33.

[4] *Astrea tubulosa,* Goldfuss, Petref. Germ., vol. i, tab. xxxviii, fig. 15.

2. STYLINA DELABECHII. Tab. XV, figs. 1, 1*a*, 1*b*, 1*c*. 1*d*.

Corallum massive, convex, seldom subgibbose, and sometimes composed of a series of thick superposed layers; common basal plate or wall covered with an epitheca presenting concentric folds, and appearing to have been complete originally. *Calices* not projecting much, nor closely set, and placed at very unequal distances from each other. *Costæ* sub-granulose, slightly prominent, rather closely set, straight, or slightly curved towards their lower end, and alternately larger and smaller; the former corresponding to the *septa* of the last cyclum: those of adjoining corallites meet at the bottom of the intercalicular spaces, but remain in general distinct. *Calices* quite circular, but rather unequal in size, especially in different specimens; fossula large, open, and rather shallow. *Columella* styliform, slightly prominent, somewhat compressed, and quite distinct from the *septa*. Three complete septal cycla, and the elements of a fourth cyclum in two of the six systems. The secondary *septa* very little developed in the four small systems, but becoming as large as the primary ones in the two other systems, so as to form with these eight principal *septa*, which are somewhat exsert, thicker at their inner and outer edge than in the middle, and quite straight. The secondary *septa* are small and delicate; those of the last cyclum are rudimentary, but are represented externally by well-developed mural costæ, which are even larger than those corresponding to the *septa* of superior orders. A vertical section of the corallum shows that the intermural spaces are principally filled up with exothecal tissue and costal laminæ, but present also some horizontal prolongations of the walls forming ill-defined strata. The *septa* are composed of non-perforated laminæ, and the dissepiments, which are horizontal, correspond to each other in the different interseptal loculi, so as to divide the visceral chamber into a regular series of superposed spaces, somewhat as in the *Cyathophyllidæ*.

The diameter of the calices varies from 1 to 2 lines, and the breadth of the inter-calicular spaces is often double that size.

The specimens which we have examined were found at Steeple-Ashton, and belonged to the collections of the Museum of Practical Geology, the Geological Society of London, the Bristol Museum, Mr. Phillips, Mr. Bowerbank, Mr. Stokes, Mr. Walton, Mr. Sharpe, Mr. Pratt, M. de Koninck, and the Paris Museum.

Stylina Delabechii is easily distinguished from most of the other species belonging to the same genus, by the existence of eight apparent systems. The same character is met with only in *S. ramosa*[1] and in *S. Lugdunensis*.[2] The first of these species is of a subdendroid form, its calices are unequal, rather distant, and a little more than a line in diameter; the total number of the septa is only sixteen. As to *S. Lugdunensis*, the specimen from which the characters were taken is in a very bad state of preservation, and we are not able to add

---

[1] *Pseudocænia ramosa*, and *P. digitata*, D'Orbigny, Prod., vol. ii, p. 34.

[2] *Octocænia Lugdunensis*, D'Orbigny, Prod., vol. i, p. 222.

any details to the brief indications given by M. D'Orbigny. That author gives the following definition :—A fine species, with large cells, somewhat elevated above the common surface.

Genus MONTLIVALTIA, (p. xxv.)

MONTLIVALTIA DISPAR. Tab. XIV, figs. 2, 2*a*.

> FUNGITE, *Knorr* and *Walch*, Rec. des Monum. des Catastr., vol. ii, p. 23, tab. i, *i*, fig. 3, 1775.
> TURNIP-SHAPED MADREPORA, *G. Young*, Geol. Surv. of York, p. 195, tab. iv, fig. 8, 1828.
> TURBINOLIA DISPAR, *Phillips*, Illustr. of the Geol. of York, part i, p. 126, tab. iv, 1829;
> (a very incomplete figure.)
> ANTHOPHYLLUM OBCONICUM, *Goldfuss*, Petref. Germ., vol. i, p. 407, tab. xxxvii, fig. 14, 1829.
> LITHODENDRON DISPAR, *Goldfuss*, MS., name in the collection of the Poppelsdorf Museum
> at Bonn.
> MONTLIVALTIA (?) DISPAR and M. OBCONICA, *Milne Edwards* and *J. Haime*, Monogr. des
> Astreides, Ann. des Sc. Nat. 3ᵐᵉ série,
> vol. x, pp. 256, 259, 1848.
> MONTLIVALTIA DILATATA, M. MOREAUSIACA, and M. OBCONICA (?), *M'Coy*, Ann. and Mag.
> of Nat. Hist., s. 2, vol. ii, p. 419, 1848.
> THECOPHYLLIA ARDUENNENSIS, *D'Orbigny*, Prod. de Paléont., vol. i, p. 384, 1850; (a young
> specimen.)
> LASMOPHYLLIA RADISENSIS, *D'Orbigny*, op. cit., vol. ii, p. 30, 1850; (adult.)
> MONTLIVALTIA DISPAR, *Milne Edwards* and *J. Haime*, Polyp. Foss. des Ter. Palæoz., etc.,
> p. 73, 1851.

Corallum turbinate, straight, or slightly curved, somewhat elongated, and presenting in some specimens thick circular accretion wrinkles. *Calice* circular or suboval, with the fossula rather shallow, and but slightly compressed transversely. *Septa* thin, quite straight, not presenting many granulations, very closely set, and forming six complete cycla. Those of the first four cycla almost equal, and reaching nearly to the centre of the visceral chamber, where they meet along their inner edge. Those of the fifth cyclum almost as thick as the principal ones, but not extending as far inwards, and quite free along their inner edge. Those of the sixth cyclum extremely thin, and not joined to the neighbouring *septa* of the superior orders. *Dissepiments* well developed, and appearing to be very oblique, for, in a horizontal section of the corallum, a considerable number of them are shown, especially near the wall, and are situated at about one line apart.

Height 3 or 4 inches; breadth of the calice 2 or 2½.

We have remarked in the Poppelsdorf Museum a specimen of this species, which Goldfuss had catalogued under the name of *Lithodendron dispar*, and which presents a fissiparous calice; but we must consider this anomaly as being quite accidental, for we know of no specimen of a compound *corallum* which can be referred to the same species, and we have sometimes met with similar cases of monstrosity in corals which are evidently simple, and incapable of fissiparous generation : *Sphenotrochus crispus* for example.

All the specimens which we have met with were worn, and had lost their wall as well as their basis ; we are, therefore, unable to decide whether this species was free or adherent, and had or had not a complete epitheca. There remains, therefore, some uncertainty relative to the zoological affinities of this fossil, but we have referred it to the genus *Montlivaltia* rather than to the genus *Trochosmilia*, on account of its great resemblance to some other corals which undoubtedly belong to the genus *Montlivaltia*, and also because we have as yet not met with any species of *Trochosmilia* in deposits formed before the cretaceous period.

The genus *Montlivaltia*, established by Lamouroux, contains a great number of species; more than thirty have been described in our Monograph of the family of Astreidæ, but many of them are as yet but imperfectly known. In a note published a short time ago,[1] M. D'Orbigny has considered it advisable to form a separate generic division for the species which are of a compressed form, and he has given the name of *Perismilia* to the group thus characterised. But this innovation is not, in our opinion, judicious, for, independently of there being instances of every intermediate degree between the species with a calice perfectly circular (such as *Montlivaltia brevissima*), and those in which the great axis of the calice is to the short axis as 260 : 100, we see no reason for establishing generical divisions on a character which, although to a certain degree constant in some cases, is in others variable in the different individuals belonging to the same species. We must add that no important difference in other parts of the corallum corresponds with the modifications in the form of the calice.

Since the publication of the first part of the Monograph, we have been enabled to examine a great number of well preserved specimens of *Montlivaltia*, and have been thus led to rectify an error which the study of imperfect specimens had led us into ; we have ascertained, in many species, that the edge of the *septa* is not entire, as we formerly supposed, but is crenulate or regularly denticulated. There is, therefore, no longer any reason for separating from the genus *Montlivaltia* the group which we established some years ago under the name of *Thecophyllia*,[2] and the genus *Montlivaltia*, thus extended, must no longer be placed in the section Eusmilinæ (p. xxiii), but be referred to the tribe of the Astreinæ (p. xxxi).

Montlivaltia dispar differs from *M. deltoides*,[3] *M. rudis*,[4] *M. cornucopia*,[5] *M. bilobata*,[6]

---

[1] Note sur des Polypiers Fossiles, 1849.

[2] Compt. Rend. de l'Acad. des Sc., t. xxvii, p. 491, 1848.

[3] Annales des Sc. Nat., s. 3, vol. x, pl. 6, fig. 3.

[4] *Cyathophyllum rude*, Sowerby, Trans. of the Geol. Soc., s. 2, vol. iii, pl. xxxvii, fig. 2.

[5] Milne Edwards and J. Haime, loc. cit., p. 298. M. D'Orbigny places this species in his genus *Ellipsosmilia*, (Note sur les Polypiers Fossiles, p. 5,) which is composed of our compressed *Trochosmilia* ; all the specimens known are in a very bad state of preservation, but we are inclined to think that this fossil had a complete epitheca, as is the case with *Montlivaltia* ; if not, it must be referred to our genus *Trochosmilia*, for the subdivision of which, proposed by M. D'Orbigny, does not appear to rest on sufficient grounds.

[6] *Turbinolia bilobata*, Michelin, Icon., pl. lxii, fig. 1 ; (not pl. lxi, fig. 7.)

and *M. irregularis*,[1] by the somewhat compressed form of its calice. It differs from *M. caryophyllata*,[2] *M. brevissima*,[3] *M. pateriformis*,[4] *M. Guerangeri*,[5] *M. Lotharinga*,[6] *M. Goldfussana*,[7] *M. hippuritiformis*,[8] *M. plicata*,[9] *M. pictaviensis*,[10] *M. decipiens*,[11] *M. Guettardi*,[12] *M. cyclolitoides*,[13] *C. trouvillensis*,[14] by its numerous *septa*, whereas in the above-mentioned species there are only five cycla. The same character distinguishes it from *M. detrita*,[15] *M. inæqualis*,[16] *M. sycodes*,[17] *M. Stutchburyi*,[18] *M. luciensis*,[19] and *M. striatulata*,[20] which have only four cycla. On the contrary, the *septa* are more numerous in *M. truncata*[21] and in *M. Lesueuri*[22] than in *M. dispar*; instead of six cycla, they constitute in general seven complete cycla.

Six cycla are also met with in *M. trochoides*,[23] *M. ponderosa*,[24] *M. Beaumonti*,[25] *M. patellata*,[26] *M. subtruncata*,[27] and *M. dilatata*[28]; but *M. dispar* may be distinguished from *M. trochoides*[29] and *M. Beaumonti* by its general form, which is not so regularly conical as in these, by the apparent shallowness of its fossula and its large size. It is proportionally much taller than *M. subtruncata* and *M. patellata*, and its basis is far from being as broad as the calice, as is the case with the latter. *M. ponderosa* is also much shorter than *M. dispar*, and its basis is quite convex.

The general form of the corallum is also very different in most of the other species of this numerous group. Thus, *M. numismalis*,[30] *M. depressa*,[31] *M. lens*,[32] *M. Delabechii*,[33]

---

[1] Milne Edwards and J. Haime, Ann., l. c., p. 298; *Anthophyllum dispar*, Michelin, Icon., pl. xc, fig. 6.

[2] Lamouroux, Exp. Meth. des Genres de Pol., pl. lxxix, figs. 8, 9, 10.

[3] Milne Edwards and J. Haime, l. c., p. 293.

[4] *Anthophyllum pateriforme*, Michelin, Icon., pl. xc, fig. 3.

[5] Milne Edwards and J. Haime, l. c., p. 293.   [6] Ib., p. 294.   [7] Ib.

[8] *Turbinolia hippuritiformis*, Michelin, Icon., pl. lxix, fig. 7.

[9] *Ellipsosmilia plicata*, D'Orbigny, Prod., vol. ii, p. 30.

[10] D'Orbigny, Prod., t. i, p. 292.

[11] *Anthophyllum decipiens*, Goldfuss, Petref., tab. lxix, fig. 3.

[12] Blainville, Dict. Sc. Nat., t. lx, p. 302.

[13] *Thecophyllia cyclolitoides*, Milne Edwards and J. Haime, l. c., p. 242.

[14] D'Orbigny, Prod., t. i, p. 384.

[15] *Anthophyllum detritum*, Michelin, Icon., pl. x, fig. 1.

[16] *Anthophyllum inæquale*, Michelin, Icon., pl. xc, fig. 4.

[17] Milne Edwards and J. Haime, l. c., p. 299.

[18] Tab. xxvii, figs. 3, 9.

[19] D'Orbigny, Prod., t. i, p. 321.

[20] *Caryophyllia striatulata*, Michelin, Icon., pl. xc, fig. 9.

[21] *Caryophyllia truncata*, Defrance, Dict. Sc. Nat., vol. vii, fig. 198.

[22] Milne Edwards and J. Haime, l. c., p. 297.   [23] Ib., p. 299.   [24] Ib., p. 242.

[25] *Thecophyllia Beaumonti*, Milne Edwards and J. Haime, p. 243.

[26] *Anthophyllum patellatum*, Michelin, Icon., pl. xc, fig. 2.

[27] *Caryophyllia truncata*, Lamouroux, Exp. Meth., pl. lxxviii, fig. 9.

[28] *Caryophyllia dilatata*, Michelin, Icon., pl. xvii, fig. 4.   [29] Tab. xxvi, figs. 2, 4.

[30] D'Orbigny, Prod., t. i, p. 321.   [31] Tab. xxix, fig. 5.   [32] Tab. xxvi, fig. 7.   [33] Tab. xxvi, fig. 5.

M. orbitolites, and *M. deformis*, are discoid or subdiscoid ; *M. tenuilamellosa*[1] and *M. dilatata*[2] are conical, but very short and broad ; *M. Waterhousii*[3] is rather tall, but almost cylindrical and regularly convex at its basis, and free ; *M. Smithi*[4] and *M. cupuliformis*[5] are fixed by a very broad basis ; and *M. contorta*[6] is very tall and irregularly bent.

The fossils described by M. Michelin under the names of *Anthophyllum excavatum*,[7] *Caryophyllia subcylindrica*,[8] *C. elongata*,[9] *C. cornuta*,[10] *C. vasiformis*,[11] and *Lobophyllia incubans*[12] ; by Goldfuss under the name of *Anthophyllum turbinatum*,[13] and by Roemer under the name of *Anthophyllum explanatum*,[14] as well as some small specimens from St. Cassian, figured by Munster, appear also to belong to this genus, and differ from *M. dispar* by their form, but are not sufficiently characterised when compared with some of the preceding species. We must also add that the *Caryophyllia Moreausiaca* of M. Michelin is very imperfectly known, and we are not acquainted with the characters that distinguish it from *M. dispar*.

This fossil is found at Malton, Yorkshire, and at Bridport. A specimen belonging to the Museum of Natural History, in Paris, and obtained at Damvilliers, in the department of the Meuse, appears to belong to the same species. There is also in M. Michelin's collection a coral from Is-sur-Thil in the department of La Coté d'Or, which may be referred to the *M. dispar* ; and M. D'Orbigny mentions the same species as having been met with in the island of Ré, on the west coast of France. The British specimens which we have examined belonged to the collections of the Geological Society of London, the Cambridge Museum, Professor J. Phillips, Mr. Bowerbank, and the celebrated cabinet formed by Goldfuss in the Poppelsdorf Museum, at Bonn.

[1] Tab. xxvi, fig. 11.

[2] *Caryophyllia dilatata*, Michelin, Icon., p. 96, tab. xvii, fig. 4. Although this figure does not show the epitheca which is so highly developed in Montlivaltia, we referred the species to that generical division in our Monograph of the Astreidæ, ' Ann. des Sc. Nat.,' s. 3, vol. x, p. 260, and since the publication of that work, we have been able to ascertain the propriety of so doing. We have seen, in the collection of M. Buvignier, some specimen found in the Coral Rag of Chaumont, and having the wall completely covered with a thick epitheca. M. D'Orbigny has mentioned this species as the type of his genus *Lasmophyllia*, the characters of which do not differ from those of our genus *Trochosmilia*, and it appears very probable that all the fossils which that palæontologist refers to his new generical group, are in fact species of *Montlivaltia*, the epitheca of which has been accidentally worn off, as is the case with the *Caryophyllia dilatata* described by M. Michelin.

[3] Tab. xxvii, fig. 7.   [4] Tab. xxi, fig. 1.   [5] Tab. xxvii, fig. 1.

[6] D'Orbigny, Prod., vol. ii, p. 30.

[7] Michelin, Icon., pl. xvii, fig. 10, (non Roemer.)   [8] Ib., figs. 2, 3.   [9] Ib., fig. 7.

[10] Ib. fig. 19.   [11] Ib., fig. 9.   [12] Ib., fig. 2.

[13] Goldfuss, Petref. Germ., vol. i, tab. xxxvii, fig. 13.

[14] Verst. des Norddeutschen ool. geb., tab. xvii, fig. 21.

Genus THECOSMILIA, (p. xxvi.)

THECOSMILIA ANNULARIS.   Tab. XIII, figs. 1, 1*a*, 1*b*, 1*c*, 1*d*; and Tab. XIV, figs. 1, 1*a*, 1*b*, 1*c*, 1*d*.

MADREPORA, *W. Smith*, Strata identified by organic remains, p. 20, figs. 1, 2, 3, 1816. (Good figures.)

—   *Parkinson*, Organic remains, vol. ii, tab. v, fig. 5, 1820.

CARYOPHYLLIA, *Conybeare* and *W. Phillips*, Geol. of England, p. 188, 1822.

—   ANNULARIS, *Fleming*, British Animals, p. 509, 1828.

—   CYLINDRICA, *J. Phillips*, Illustr. of the Geol. of York., part i, p. 126, tab. iii, fig. 5, 1829; and 2d edition, p. 98.   (Incomplete figure.)

CARYOPHYLLŒA, *R. C. Taylor*, Mag. of Nat. Hist., vol. iii, p. 271, fig. *g*, 1830.   (Rough figure.)

CARYOPHYLLIA CYLINDRICA and C. ANNULARIS, *S. Woodward*, Synopt. Table of Brit. Org. Remains, p. 6, 1830.

LITHODENDRON ANNULARE, *Keferstein*, Naturg. des Erdkörpers, vol. ii, p. 785, 1834.

CARYOPHYLLIA ANNULARIS and LITHODENDRON TRICHOTOMUM, *Morris*, Cat. of Brit. Fossils, pp 32, 40, 1843.

THECOSMILIA CYLINDRICA and T. TRILOBATA, *Milne Edwards* and *J. Haime*, Monogr. des Astreides, Ann. des Sc. Nat., s. iii, vol. x, pp. 271-2, 1848.

LOBOPHYLLIA TRICHOTOMA, *M'Coy*, Ann. and Mag. of Nat. Hist., s. ii, vol. ii, p. 419, 1848.

THECOSMILIA ANNULARIS, *Milne Edwards* and *J. Haime*, Polyp. des Terr. Palæoz., &c., loc. cit., p. 77, 1851.

Corallum composite, dendroid, very tall, its branches in general cylindrico-turbinate and not spreading much.   In most instances of fissiparous multiplication, the calice becomes divided only into two parts, one of which rather abruptly bends out and remains short, whilst the other continues ascending and grows much more; so that the calices which take their origin from the same point are placed at very unequal heights, and thence a general form, the aspect of which is very different from that of most species of the same genus, and more especially of *Thecosmilia trichotoma*, where all the twin corallites grow to the same height, and the corresponding calices are situated on the same level.   But this peculiar disposition does not become well marked till the corallum has attained a certain size, and in young specimens not only the first parent calice often becomes multilobate, but those of the second generation thus formed grow up in a uniform manner, and often in their turn give birth to more than two individuals.   The general aspect of the small compound coral so formed, is, therefore, very different from that of the adult specimens, and in order to recognise their specific identity, it is necessary to compare a great number of these fossils.   We have of late been able to make this comparison, but when we published our Monograph of the Astreidæ we had not before us sufficient materials for so doing, and we were therefore unable to recognise that identity.   Thus the fossil which we designated by the name of *Thecosmilia trilobata*,[1] is one of the varieties. Sometimes the young corallites, produced by a simple parent polyp, instead of forming a fascicular group, arrange themselves so as to constitute a short row, and do not separate immediately from each other; it may even happen that a few of these small series of corallites remain in contact laterally, and thus assume the form of *Symphyllia*. But these variations in the general form are only met with in young specimens, and have never been met with in the older, large Corals.

The epitheca is well developed, and extends from the basis of the corallum almost to the edge of the calices, but the *septa* are exsert. Sometimes this coating continues to envelop two neighbouring corallites after these have become quite distinct internally, and it presents numerous strong circular wrinkles or folds, which are closely set and very unequally developed. When the epitheca has been in part, or totally, worn away, as is often the case, the costæ or outer edge of the *septa* become visible, and appear delicately denticulated, not very closely set, and alternatively somewhat more or less thick. There does not appear to be any true walls, and the spaces situated between the costo-septal laminæ are occupied by dissepiments.

The calices are seldom circular, (as in fig. 1, Tab. XIII;) they usually become very soon oval, subtriangular, or lobated, and it often happens that two fossulæ become perfectly distinct some time before any corresponding change takes place in the margin, and are united by common *septa*. The fossulæ are small and rather shallow; there is no appearance of a columella, and the septa meet in the centre of the visceral chamber at a very short distance from the surface of the calice.

The number of the septa is extremely variable, and differs most especially according as the calice belongs to a newly-formed corallite, or is more or less ready to multiply by a fissiparous development. Similar modifications are always met with in fissiparous corals, and renders it very difficult to come at the knowledge of the real specific characters of the septal apparatus. But as far as that can be made out by the examination of the most perfectly circular calices which must be supposed to belong to individuals that have not begun to multiply in this way, it appears that the normal number of cycla is five; the last cyclum being more or less imperfect. The *septa* are thin, closely set, straight or slightly flexuous, exsert, and terminated by an oblique arched edge, which is armed with delicate, nearly equally developed, denticulations. Those of the first three cycla are almost of the same size; those of the fourth cyclum not as thick towards their inner edge, and those of the fifth cyclum are very thin; all present on their sides slight granulations, arranged in radiate series.

This fine coral often forms large arborescent masses, one or two feet in height. The specimen figured in Pl. 13 is eight inches high, and Mr. Charlesworth showed us in the Museum of York a specimen, which, although incomplete, was more than one foot and a half

---

[1] Milne Edw. and J. Haime, Ann. Sc. Nat., s. iii, vol. x, p. 272.

high. Each corallite usually attains the length of about one inch and a half before dividing, and the diameter of the simple calices is in general about eight lines, but the compound calices are often double that size.

We have examined a great number of specimens of *Thecosmilia annularis* found at Steeple Ashton, Wiltshire., Most of these fossils belonged to the collections of the Museum of Practical Geology, the Geological Society, the Bristol Museum, the Paris Museum, Mr. Bowerbank, Mr. Phillips, Mr. Stokes, and Mr. Pratt.

The same species is met with at Slingsby in Yorkshire, and we also refer to it a fossil found at Malton, and belonging to Mr. Bowerbank's collection, which resembles much the specimen figured by Mr. Phillips, but is not in a state of preservation sufficiently good to enable us to be certain as to its specific characters. W. Smith, who gave some good figures of this coral in his remarkable work on 'Organic Remains,' mentions it as having been found in the following localities : Longleat Park, Stratton, Ensham Bridge, Wotton Basset, Banner's Ash, Well near Swindon, Wilts and Berks Canal, Shippon, Bagley Wood Pit, and Stanton, near Highworth. Mr. Phillips also points out its existence at Seamer. We must add that some specimens found at Radcliff by the collectors of the Geological survey, and communicated to us by Sir H. De la Beche, do not appear to differ from the above-described species, but some other fossils from the same locality appear to be more similar to the *Montlivaltia Lesueuri* from the Kimmeridge clay near Havre.

The well-preserved specimens which we have met with in some of the English collections enable us to rectify, concerning *Thecosmilia*, an error of the same kind as that we formerly fell into with respect to *Montlivaltia*. The *septa* are not terminated by an undivided edge, as in the tribe of *Eusmilinæ*, where we placed this genus when the first part of this Monograph was published; they are denticulated in a regular manner, and the *Thecosmilia* may be briefly defined "compound *Montlivaltia*."

This generical division contains fossils belonging to cretaceous as well as jurassic formations, and we are also inclined to admit in it some fragments of corals found in the celebrated fossiliferous deposit at St. Cassian. We have given the list of all these species in the synopsis joined to our Monograph of the Palæozoic corals, but we regret not having had an opportunity of examining some of them, and others that have of late been submitted to our investigation were in a very bad state of preservation; much uncertainty, therefore, still exists respecting the specific character of many *Thecosmilia*. However, the fossil above described is easily recognisable by the unequal size of the calices, which take their origin on the same stem and are of the same age, and by the form of the *septa*. Thus, *T. Konincki*,[1] a species of which we have only seen a young specimen, differs from *T. annularis* by its *septa* being thinner, and more equal in size. In *T. trichotoma*[2] and *T. lobata*[3] the calices are almost circular, smaller, and placed on the same level. *T. semi-*

---

[1] Milne Edwards and J. Haime, Ann. Sc. Nat., 3ᵐᵉ sér. vol. x, p. 272.

[2] *Lithodendron trichotomum*, Goldfuss, Petref. Germ., pl. xiii, fig. 6.

[3] *Lobophyllia lobata*, Michelin, Icon., pl. lxvii, fig. 3.

nuda,[1] which is very much like *T. trichotoma*, by its general aspect, has also thinner and more numerous *septa*. *T. ramosa*[2] may also be distinguished from *T. annularis* by the regular form and small size of its calices. *T. gregaria*[3] differs still more from the preceding species by the corallites remaining in general grouped in fasciculi to a considerable distance from the parent calice, on which they were formed by fissiparous generation, a mode of arrangement which we have not met with in other corals of the same genus.

M. D'Orbigny has recently given the name of *Lasmosmilia*[4] to a certain number of fossil corals, which appear to us to be species of *Thecosmilia* that have been accidentally deprived of their epitheca. The genus *Amblophyllia* of the same author[5] is founded on the existence of a rudimentary epitheca, and is probably composed only of specimens of the same genus less completely weatherworn. If the different species mentioned under these two generical names were well characterised, it would be necessary for us to compare them with the British species described here above; but that is far from being the case.

Genus RHABDOPHYLLIA.[6]

RHABDOPHYLLIA PHILLIPSI. Tab. XV, figs. 3, 3*a*, 3*b*, 3*c*.

> CARYOPHYLLIA, *Phillips*, Illustr. of the Geol. of Yorkshire, vol. i, p. 126, 1829.
> LITHODENDRON EDWARDSII, *M'Coy*, Ann. and Mag. of Nat. Hist., s. 2, vol. ii, p. 419, 1848;
> (but not *Lithodendron Edwardsii* of Michelin, as supposed by that author.)
> RHABDOPHYLLIA PHILLIPSI, *Milne Edwards* and *J. Haime*, Monogr. des Polyp. Palæoz., &c.,
> loc. cit., p. 83, 1851.

Corallum composite, dendroid; corallites tall, almost cylindrical; slightly tumified at short distances, and becoming larger and somewhat compressed where they dichotomise. This division takes place frequently, and the newly formed branches diverge at an angle of about 50°. *Costæ* very distinct, rather thick, granulose, almost equally developed, closely set, and often dichotomose (fig. 3*a*). The *calices*, when young, are regularly circular, as may be inferred from the form of the corallum, but the terminal portion of the branches was broken off in the specimens we have seen. A horizontal section, made at some distance

---

[1] D'Orbigny, Prod., t. i, p. 389.   [2] Ib., p. 291.

[3] Tab. xxviii, fig. 1.   [4] Note sur des Pol. Foss., p. 6.

[5] Note sur des Pol. Foss., p. 8. *Amblophyllia rupellensis* (D'Orbigny, Prod. de Paléont., vol. ii, p. 30,) is a species established for a cast, which does not appear to us susceptible of being characterised.

Amblophyllia obtusa (D'Orbigny, Op. cit., vol. i, p. 285,) is known only by a very young specimen, in which the three calices are not yet become distinct, and present each about sixty *septa* belonging to three or four different cycla, and delicately dentate on the edge; but this species appears to differ from all others previously described by the loosely set prominent radiate striæ that cover the sides of the *septa*.

[6] Milne Edwards and J. Haime, Polyp. Foss. des Terr. Palæoz., &c., p. 83.

from the calice, shows a *Columella* of a spongy texture (fig. 3 *b*), and three complete cycla of *septa*, independently of the rudiments of a fourth cyclum in two of the systems; so that it appears probable that there may be four cycla in the calices which are ready to multiply by fissiparity; the *septa* are thin, not closely set, slightly tumified near their inner edge, sometimes flexuous, and but slightly granulated laterally; the secondary ones in the small systems, and even the tertiary ones in the most developed systems, are almost as large as the primary ones, but those of the last cyclum are much smaller and often even rudimentary. The *wall*, although not very thick, is well formed. The *dissepiments* appear to be rudimentary.

We do not know to what height this coral may grow; the calices are from two to three lines in diameter, and the distance between the successive fissiparous generations varies from seven lines to an inch.

We have seen three specimens of this species: two were found at Malton, and belong, the one to our friend Mr. Bowerbank, the other to the Cambridge Museum; the third was presented to the Geological Society by Sir Roderick Murchison, and had been found in the Coral Rag of Cumnor Hill. If, as we are inclined to think, the fossil coral, mentioned by Mr. Phillips as resembling the *Madrepora flexuosa* of Ellis and Solander, belongs to this species, we must also add to these localities Hackness, in Yorkshire.

The genus *Rhabdophyllia*, which we have recently established[1] for a certain number of arborescent Astreidæ that multiply by fissiparity, and have naked costulated walls, differs from *Calamophyllia* by the absence of mural rings, the existence of a well-characterised columella, and the rudimental state of the interseptal dissepiments. This group is essentially composed of a small number of species belonging to the Coral Rag, and we also include in it an ill-defined species found at St. Cassian, and described by Count Munster.

Mr. M'Coy refers this British species to the *Rhabdophyllia Edwardsi*,[2] which, as far as can be seen by the figure given by M. Michelin, is certainly very much like it; but we are of opinion that these fossils are not identical; the latter appears to differ from *R. Phillipsi* by the corallites being more regularly cylindrical and having thicker costæ. *R. undata*[3] and *R. nodosa*[4] differ from the above-described species by the alternate constrictions and

---

[1] Polypiers Fossiles des terr. Palæoz., p. 83.

[2] *Lithodendron Edwardsii*, Michelin, Icon. Zooph., tab. xxi, fig. 2.

[3] *Calamophyllia undata*, D'Orbigny, Prod., vol. ii, p. 31. This species having been very briefly noticed by M. D'Orbigny, it may be useful to point out its most essential characters:

Corallum arborescent; branches almost cylindrical, dichotomous or trichotomous; *costæ* straight, nearly equal, projecting but little, closely set, and formed by a single series of granulations. The corallites presenting a series of alternate constrictions, and thick, circular, obtuse ridges. *Septa* thin and numerous. Margin of the calice irregular. Diameter 7 or 8 lines. From the Coral Rag of Wagnon, Ardennes.

[4] *Calamophyllia nodosa*, D'Orbigny, Prod., vol. ii, p. 32. Species very nearly allied to the preceding one, but with the circular tumefactions of the walls placed with less regularity and more prominent. Costal striæ very delicate, and of unequal size alternately. Diameter, 5 lines. From the Coral Rag at Oyonnax and Landeyron, Departement de l'Ain, France.

swellings of the walls As to *Lithodendron subdichotomum,*[1] *Calamophyllia simplex,*[2] and *C. Bernardana,*[3] and *Lithodendron Morcausiacum,*[4] which we are inclined to refer to the same genus, we are not sufficiently well acquainted with them to be able to point out their characteristic features.

Genus CALAMOPHYLLIA, (p. xxxiii.)

CALAMOPHYLLIA STOKESI. Plate XVI, figs. 1, 1*a*, 1*b*, 1*c*, 1*d*.

Corallum composite, fasciculate, and composed of very tall subcylindrical or sub-prismatic corallites, which present a considerable number of annular expansions. These circular ridges are placed at a short distance from each other, and appear to be formed by the inferior edge of a series of laminæ lapping over each other. The corallites dichotomise at short distances, and under very acute angles; the new branches thus formed continue ascending parallel to each other, and are in general somewhat constricted immediately above the point of origin. The *costæ* are quite straight, very delicate, and closely set, but separated by deep, well-marked, narrow furrows; they do not project much, and are composed of a series of granulations more or less confounded together (fig. 1*a*). In general they are all nearly of the same size; but in some parts they are alternately a little thicker. The form of the *calice* is somewhat irregular (fig. 1*c*, 1*d*, 1*e*); it is seldom quite circular, and usually more or less oval or subpolygonal. The fossula is shallow, the *columella* null or rudimentary, and the *septa* numerous; in the large calices there are about seventy of these radiate laminæ, and we must, therefore, suppose that there are four complete cycla and a fifth cyclum incomplete; but it is very difficult to distinguish the different systems, and there exists, in all probability, much irregularity in their mode of growth. The *septa* are very thin, broad, closely set, and exsert; their upper edge is slightly arched and regularly crenulated (fig. 1*b*), their sides granulated. Those of the first three cycla differ but little, and those of the fourth cyclum are also highly developed; but those of the fifth cyclum are much smaller. A vertical section of one of these corallites shows that the laminæ which form the *septa* are very cribrate, and by means of a horizontal section numerous small dissepiments are exposed to view; there are seven or eight of these in each interseptal loculæ (fig. 1*d*).

The corallites are very tall—those figured in this work, although broken at the end, were from six to seven inches high, and in general about five or six lines in diameter.

---

[1] *Lithodendron subdichotomum,* Munster, Beitr. zur Petref., 4th part, tab. ii, fig. 3. *Rhabdophyllia?* *subdichotoma,* Milne Edwards and J. Haime, Polyp. Palæoz., &c., p. 83.

[2] D'Orbigny, Prod., v. ii, p. 32.

[3] *Calamophyllia Bernardina,* D'Orbigny, loc. cit., vol. ii, p. 32.

[4] Michelin, Icon., tab. xxi, fig. 3.

Calamophyllia Stokesi is found at Steeple-Ashton; specimens may be seen in the Collection of the Geological Society and of the Paris Museum; that figured in plate 16 belongs to our friend Ch. Stokes, Esq.

The genus *Calamophyllia*, as defined in the Introduction to this Monograph, contained all the fasciculated astreinæ with naked, costulated walls; but as it has been already mentioned here above, we have of late been induced to subdivide that group, and to reserve the name of *Calamophyllia* for the species which present mural annular laps; this characteristic feature coincides with the existence of numerous dissepiments and an irregular cylindrical form, whereas in the species which do not present such mural appendages, and which constitute our genus *Rhabdophyllia*, the endothecal structures are quite rudimentary, and the columella is much more developed. It is also necessary to remark, that the genus *Calamophyllia* thus rectified, must no longer be distinguished from the genus *Eunomia* of Lamouroux, for having had of late the opportunity of examining some specimens of *Eunomia radiata*, in an excellent state of preservation,[1] we have been enabled to ascertain that the walls of this fossil are not covered with an epitheca, as we formerly supposed. By right of priority, Lamouroux's name of *Eunomia* ought, therefore, to be substituted for that of *Calamophyllia*, introduced more recently by Blainville; but the former having been previously employed for a genus of *Lepidoptera*, it seems preferable to abandon it here, and to adopt the latter.

The genus *Calamophyllia* is composed of three of the species described under that name in our Monograph of the Astreidæ: *C. striata*, *C. pseudostylina*, and *C. articulosa*; of *Calamophyllia radiata*, (or *Eunomia radiata*, Lamouroux,) the British fossil which we have called *C. Stokesi*, and a few other fossils mentioned by M. D'Orbigny in his 'Prodrome.'

Calamophyllia Stokesi bears great resemblance to *C. striata*,[2] and differs from it only by the mural laps being more developed and closer set, the *septa* more numerous, and the costæ broader, and separated by deeper furrows. *C. pseudostylina*[3] and *C. articulosa*[4] are easily distinguished from *C. Stokesi* by the large size of their corallites and of the mural annular laps; *C. radiata*,[5] on the contrary, differs from all the preceding species by the slender form of the corallites, and is also recognisable by the small number of its *septa*.

---

[1] These corals were kindly communicated to us by M. D'Orbigny, in whose fine Palæontological collection we have also been enabled to examine many other interesting fossils.

[2] Blainville, 'Manuel d'Actinologie,' p. 346, tab. lii, fig. 4. We have of late been able to obtain a complete confirmation of the views we alluded to in a former work, relative to the identity of this species, and of the *Calamophyllia flabellum*; the fossils described under the latter name by Blainville, and figured by M. Michelin, (Iconogr., tab. xxi, fig. 4,) are specimens of *C. striata*, the costæ of which have been worn away accidentally, and, in some specimens, we have seen on the same corallite the two forms which were considered as characteristic of the two nominal species.

[3] *Lithodendron pseudostylina*, Michelin, Icon., pl. xix, fig. 9.

[4] Milne Edw. and J. Haime, Ann. Sc. Nat., 3d sér., t. xi, fig. 26 *b*.

[5] Tab. xxii, fig. 1.

As to the various fossils which M. D'Orbigny considers as new species referable to this group, they have not been as yet characterised with sufficient minuteness to be recognisable.[1]

Genus CLADOPHYLLIA.[2]

CLADOPHYLLIA CONYBEARII.   Tab. XVI, figs. 2, 2a, 2b, 2c.

> CARYOPHYLLIA CESPITOSA, *Conybeare* and *W. Phillips*, Geol. of England, p. 188, 1822.
> CORAL, LIKE CARYOPHYLLIA CESPITOSA, *Phillips*, Illustr. of the Geol. of Yorkshire, vol. i, p. 126, 1829.
> LITHODENDRON DICHOTOMUM, *M'Coy*, Ann. and Mag. of Nat. Hist., s. ii, vol. ii, p. 418, 1848.

Corallum composite, irregularly cespitose.   Its branches obliquely erect, placed at unequal distances, and bifurcating under a very open angle : the two corallites that rise thus from the same parent resemble young individuals that might be produced by calicular gemmiparity rather than by fissiparity.   The branches are cylindrical, equal in diameter, alternately somewhat constricted or tumefied, and covered from top to bottom by a complete epitheca.   In some parts where this external coating has been worn away, the *costæ* are visible, and assume the form of delicate obtuse, closely set, and equally developed lines.   The *calices* are circular, or nearly so, and the fossula narrow and deep. There appears to exist no indication of a *columella*.   The *septa* form in general three complete cycla, and are broad, thin, not exsert, terminated by an arched, delicately denticulated edge, and granulated laterally.   The dissepiments appear to be numerous.

Diameter of the corallites, $1\frac{1}{2}$ lines; depth of the calicular fossula almost as much.

This fossil is found at Steeple Ashton, and specimens are in the collections of the Museum of Practical Geology, of the Geological Society, of the Cambridge and Paris Museums, of Mr. Bowerbank, and of Mr. Pratt.

The genus *Cladophyllia* which we have recently proposed, comprises the *Astreinæ* which resemble *Calamophyllia* by most of their general characters, but differ from these by the existence of a complete epitheca.   The definition of this group is therefore almost the same as that which we formerly gave to *Eunomia*, but which is not in reality applicable to the species for which Lamouroux established that genus.   These corals are remarkable by

---

[1] *Calamophyllia corallina*, D'Orbigny, Prod., vol. ii, p. 31, *C. Luciensis*, D'Orb., Op. cit., vol. i, p. 321, and *Eunomia contorta*, D'Orb., Op. cit., v. ii, p. 32, are species established on specimens, which appear to us undeterminable.   *C. lumbricalis* and *C. rugosa*, D'Orb., Loc. cit., belong most likely to our genus *Cladophyllia* ; *Eunomia grandis*, D'Orb., Loc. cit., is a *Thecosmilia*, and we are inclined to think that *C. inæqualis*, D'Orb., Loc. cit., belongs to the family of the *Cyathophyllidæ*.   We have not seen the other species of *Calamophyllia* or *Eunomia* mentioned by that author.

[2] Polyp. Palæoz., &c. ; in the Archives du Museum, vol. v, p. 81, 1851.

the regularity of their structure and the circular form of their calices (figs. 2 *b*, 2 *c*). Some of them are met with in the Lias, and the same generic form appears to have existed in the Triasic period, at St. Cassian; but those of which the characters are best known all belong to the Oolitic formation. *Cladophyllia Conybearii* resembles very much *C. dichotoma*,[1] to which Prof. M'Coy has referred it; but we think they are not specifically identical, for the folds of the epitheca appear to be more developed and more irregular in the Giengen coral than in the Steeple-Ashton fossil; but the former has only been found in such a bad state of preservation that it is as yet difficult to decide the question. *C. Babeana*[2] differs also but little from *C. Conybearii*, but has the tertiary *septa* less developed and the folds of the epitheca quite horizontal, whereas they are somewhat oblique in the latter species. *Eunomia rugosa*, D'Orbigny,[3] which appears to be also very nearly allied to the above described species, but may be distinguished by the great obliquity of the epithecal folds. *Cladophyllia articulata*,[4] and *C. lævis*[5] differ from the former by their thick accretion tumefactions; and *C. funiculus*,[6] by the surface of its corallites being quite even and presenting no such swellings. *C. lumbricalis*[7] has a much thicker epitheca than *C. Conybearii*, and its calices are much larger. Some other fossils mentioned by different authors under various specific names appear to belong to the same group, but have not as yet been satisfactorily characterised and it would, therefore, be useless to dwell upon them here.[8]

Genus GONIOCORA.[9]

GONIOCORA SOCIALIS. Tab. XV, figs. 2, 2 *a*, 2 *b*.

LITHODENDRON SOCIALE, *F. A. Roemer*, Versteiner. des Norddeutschen oolithen gebirges Suppl., tab. xvii, fig. xxiii, 1839.[10]

DENDROPHYLLIA PLICATA, *M'Coy*, on some new Mesozoic Radiata, in Ann. and Mag. of Nat. Hist., s. ii, vol. ii, p. 403, 1848.

GONIOCORA SOCIALIS, *Milne Edwards* and *J. Haime*, Polyp. Palæoz., &c., p. 96, 1851.

Corallum composite, dendroid, and presenting in general one or more principal erect stems bearing lateral branches, each of which also gives birth to a series of smaller branches.

---

[1] *Lithodendron dichotomum*, Goldfuss, Petref., tab. xiii, fig. 3.

[2] *Eunomia Babeana*, D'Orbigny, Prod., vol. i, p. 292.

[3] Prodr., vol. ii, p. 32.

[4] *Lithodendron articulatum*, Michelin, Icon., pl. xxi, fig. 1.

[5] *Lithodendron læve*, Michelin, Icon., pl. xix, fig. 8.

[6] *Lithodendron funiculus*, Michelin, Icon., pl. xix, fig. 7.

[7] *Calamophyllia lumbricalis*, D'Orbigny, Prodr., t. ii, p. 3.

[8] The list of these fossils is given in the Introduction to our Memoir on Palæozoic Corals, loc. cit., p. 82.

[9] Milne Edwards and J. Haime, Polyp. Palæoz., &c., p. 96.

[10] But not the figure given under the same name in the first plate of that work (fig. 3), which appears to be a Rhabdophyllia, and does not differ from the *Lithodendron nanum* of the same author.

The branches are situated at a small distance apart, and very often they are arranged two by two opposite each other; they separate from the parent stem at an angle of about 50°, and grow to some distance in a straight direction before they begin to become erect. All are quite cylindrical, and the young ones are almost as thick as those of a superior order. The *walls* appear to be completely naked, and present closely set *costæ*, which are narrow, delicately granulated, alternately small, or larger and more prominent, quite straight, and uninterrupted from the basis to the extremity of each branch, but becoming more developed near their upper end. The *calices* are perfectly circular, and contain no columella; the principal septa meet in the centre of the visceral chamber, and become united together all along their inner edge, or by means of a few thick trabiculæ. There is no appearance of any *pali*. The *septa* are twenty-four in number, and therefore belong to three complete cycla; but there are twice that number of costæ; the fourth cyclum of costæ not having any corresponding septa on the inside of the wall. The septa are well developed, straight, and closely set. Those of the first cyclum are thick, especially near the wall; the secondary ones are almost as strong, but those of the third cyclum are thin; they are all but slightly granulose, and constitute almost perfect laminæ; there appears to be but very few dissepiments, and the walls are thick. The diameter of the branches varies between half a line and two lines and a half.

This fossil is found at Steeple Ashton, and is in the collections of the Museum of Practical Geology, the Geological Society, the Cambridge Museum, Mr. Stokes, Mr. Pratt, and M. D'Archiac. M. Roemer mentions it as being met with also in the Coral Rag, at Speckenbrinke and Knebel, in Germany.

The genus *Goniocora*, which we have established since the publication of the first part of this Monograph, is closely allied to *Cladocora* and *Pleurocora* (p. xxxviii), by its mode of generation, which always takes place by means of lateral gemmation, and not by fissiparity, as in *Calamophyllia*, *Rhabdophyllia*, and *Cladophyllia*. It differs from the above-mentioned dendroid *Astreinæ* by the rudimentary state of the *columella* and the entire absence of *pali*. *G. socialis* is the only well characterised species of this new generical division; but we also refer to it a small fragment found in the trias of St. Cassian, and figured by Count Munster under the name of *Lithodendron verticillatum*.[1] This imperfectly known species differs from that here described by unequal size of the costæ and the verticillate arrangement of the corallites.

[1] Beitr. zur Petref., part iv, tab. xi, fig. 22.

Genus Isastrea.[1]

1. Isastrea explanata.   Tab. XVIII, figs. 1, 1*a*, 1*b*, 1*c*, 1*d*.

> Madrepora, *W. Smith,* Strata identified by organic fossils, p. 20, Coral Rag, fig. 1, 1816.
> Compound madrepora, *G. Young,* Geol. Survey of York, tab. iv, fig. 2, 1828. (Very rough figure.)
> Astrea, approaching to A. Favosa, *W. D. Conybeare* and *W. Phillips,* Geol. of England, p. 188, 1822.
> Astrea explanata, *Goldfuss,* Petref. Germ, vol. i, p. 112, tab. xxxviii, fig. 14, 1829.
> — favosioides, *Phillips,* Illustr. of the Geol. of Yorkshire, vol. i, p. 126, tab. iii, fig. 7, 1829.
> Siderastrea explanata, *Blainville,* Dict. des Sc. Nat., v. lx, p. 337, 1830: and Manuel d'Actinologie, p. 371.
> Astrea explanata, *Milne Edwards,* Annot. to Lamarck, vol. ii, p. 420, 1836.
> — — *Bronn,* Lethea Geognostica, vol. i, p. 299, 1837.
> — helianthoides, *M'Coy,* Ann. of Nat. Hist., s. ii, vol. 2, p. 408, 1848.
> Prionastrea explanata, *Milne Edwards* and *J. Haime,* Monogr. des Astreides, Ann. des Sc. Nat., s. iii, vol. xii, p. 136, 1849.

Corallum composite, massive, and convex. The common basal plate covered with a complete epitheca, which is often partially worn away, and then leaves exposed to view the costal striæ. They are narrow, somewhat unequal in size, and arranged in fasciculi that radiate from the basis to the circumference of the basal plate, so that the outer striæ of each group meet those of the neighbouring fasciculi under a very acute angle (fig. 1), The *calices* are in general polygonal and very unequal in size, especially in large specimens. They are shallow, and present in their centre a small round fossula, at the bottom of which is a rudimentary columella. Sometimes these small fossulæ become filled up with extraneous stony matter, that assumes the appearance of a prominent columella (fig. 1*d*). The edges of the calices are convex, and intimately united together. Sometimes the septa of one calice appears even to extend without interruption into the adjoining calice; but in general the corallites are circumscribed by a very delicate mural line or a narrow furrow. The septal systems are rather irregular; the first three cycla are complete; the fourth cyclum more or less incomplete, and the total number of septa thus varies from twenty-eight to forty-four. The septa are broad, thin towards their outer edge as well as inwards, closely set, often flexuous, and but slightly exsert; their upper edge is almost straight, descends obliquely towards the fossula, and is divided into a series of small, closely set, and nearly equal denticulations, each of which corresponds to a series of granulations situated on the lateral surfaces of the septum. The secondary septa are almost as large as the primary ones, but their inner edge does not ascend so high; the tertiary ones are much smaller, and those of the fourth cyclum still less. The greatest diagonal of the adult individuals is in general about four lines, but varies much in the different parts of the same specimen.

[1] See page 74.

A horizontal section made some way down from the calicular surface shows that the walls remain simple, and very thin or even rudimentary, but the different corallites united in a common mass are always very well delimitated, and the same septa never extend into two adjoining visceral cavities, as might be supposed to be sometimes the case by the aspect of the calices. No columella exists in most corallites, and in those where some traces of a similar organ are met with they consist only in a few small trabiculæ. It must also be noted that although the small *septa* often bend somewhat towards the neighbouring larger ones, they always remain quite free at their inner edge, and that the calicular gemmation takes place at a considerable distance from the centre of the calice.

The British specimens of this species which we have examined had been found at Steeple Ashton, Malton, and Hackness, and belonged to the collections of the Museum of Practical Geology, the Geological Society, the Museums of Bristol and Paris, Mr. Bowerbank, and Mr. J. Phillips. The same species is mentioned by Mr. Smith as having been met with at Stanton near Highworth, Shippon, Bagley Wood Pit, Banner's Ash, Well near Swindon, and Wilts and Berks Canal, South of Bayford. It is found also in abundance at Lifol, in the departement des Vosges, Stenay, in the departement des Ardennes, and at Heidenheim, in Germany; specimens from these localities are in the collections of the University of Bonn, the Paris Museum, M. Michelin, &c.

The genus *Isastrea*, as already stated (p. 74), has been established for a certain number of corals that we formerly placed in the genus *Prionastrea* (p. xli), but that differs from the species considered as the types of this latter group, by the total absence or rudimentary state of the columella, and by the corallites being separated only by a single mural lamina, whereas as in *Prionastrea* they are independent of each other in their lower part, and become intimately cemented together only near the calice.

I. explanata is one of the best characterised species of this generical division. It differs from *I. limitata*,[1] *I. Guettardana*,[2] *I. explanulata*,[3] and *I. Richardsoni*,[4] by the size of the calices and the number of the *septa*. *I. Munsterana*[5] may be distinguished from it by the principal *septa* being thinner outwards, and becoming somewhat thicker towards the two thirds of their breadth inwards. In *I. polygonalis*,[6] *I. oblonga*,[7] and *I. Michelini*,[8] the walls are much thicker. In *I. lamellosissima*[9] the *septa* are more distant, and in *I. tenuistriata*[10] their number is twice as great. In *I. Conybearii*[11] the *septa* are, on the contrary, less numerous, and in *I. serialis*[12] they are very unequal and

---

[1] Tab. xxiii, fig. 2.

[2] *Prionastrea Guettardiana*, Milne Edwards and J. Haime, Ann. Sc. Nat., 3d ser., t. xii, p. 137.

[3] Tab. xxiv, fig. 3.   [4] Tab. xxix, fig. 1

[5] *Prionastrea Munsteriana*, Milne Edw. and J. Haime, loc. cit., p. 136.

[6] *Astrea polygonalis*, Michelin, Icon., pl. iii, fig. 1.

[7] Tab. xii, fig. 1.

[8] *Montastrea Michelini*, Blainville, Dict. Sc. Nat., t. lx, p. 339.

[9] *Astrea lamellosissima*, Michelin, Icon., pl. vi, fig. 1.

[10] Tab. xxx, fig. 1.   [11] Tab. xxii, fig. 4.   [12] Tab. xxiv, fig. 2.

closely set.   The above-described species bears the closest resemblance to *I. helianthoides*,[1] *I. Bernardana*,[2] *I. ornata*,[3] and *I. Greenoughi*,[4] and it is often difficult to distinguish them.   In *I. helianthoides* the calices are in general more regular, and the *septa* are less numerous, and not so strongly granulated; in *I. Bernardana* and *I. ornata* the *septa* are more distant and less granulated on their lateral surfaces; and in *I. Greenoughi* the calices are larger, the *septa* thinner, and the marginal dentations of these less developed. As to the new species which M. D'Orbigny has lately mentioned as appertaining to the Oolitic formation,[5] they are mostly established on specimens that are in a very bad state of preservation, and do not seem to us susceptible of being satisfactorily characterised.[6]

2. ISASTREA GREENOUGHI.   Tab. XVIII, fig. 2.

It is not without much hesitation that we have separated specifically this fossil from *I. explanata*, for we have seen but a small fragment of it in the collection of the Geological Society.   It is in fact very similar to the preceding species, but differs from it by the calices being all larger, and the *septa* being on the contrary thinner in proportion and less denticulated.   The *corallum* appears to have been almost flat at its upper surface, and gemmates at a considerable distance from the fossulæ.   The *calices* are shallow, and the edge is obtuse; the fossula is circular, well defined, and appears to be closed by dissepiments without containing any columellarian trabiculæ.   The *septa* form four complete cycla and an incomplete fifth cyclum; there are often as many as fifty-six; they are thin, broad, slightly flexuous, but little prominent, and not very closely set.   Those of the second cyclum are as large as the primary ones, and reach to the central fossula; the tertiary ones are also large, but those of the fourth cyclum are much smaller.   They all appear to be very delicately denticulated at their upper edge, and slightly striated laterally.   The dissepiments seem to be highly developed.   The great diagonal of the calices varies from six to seven lines.

This fossil was found at Botley Hill by Mr. G. B. Greenough.

---

[1] *Astrea helianthoides*, Goldfuss, Petref., pl. xxii, fig. 4*a*. (Cæt. excl.)

[2] *Prionastrea Bernardina*, D'Orbigny, Prodr., vol. i, p. 293.

[3] *Prionastrea ornata*, D'Orbigny, Prodr., vol. i, p. 293.   [4] Tab. xviii, fig. 2.

[5] Prod. de Paléont., vols. i, ii.

[6] We must also beg leave to remark that the *Astrea dissimilis*, (Michelin, Icon., tab. xciv, fig. 12,) for which M. D'Orbigny has made the new general division *Dendrastrea*, (Prod., v. i, p. 322,) differs only from *Isastrea* by its general form, and that the same species is again entered in that palæontologist's Catalogue under the name of *Dendrastrea Langrunensis*, D'Orb. (Prod., v. i, p. 322.)

Genus THAMNASTREA, (p. lxii.)

THAMNASTREA ARACHNOIDES. Tab. XVII, figs. 1, 1*a*, 1*b*, 1*c*, 1*d*, 1*e*, 1*f*, 1*g*, 1*h*, 1*i*, 1*j*, 1*k*.

MADREPORA ARACHNOIDES, *Parkinson*, Org. Rem., vol. ii, p. 54, tab. vi, figs. 4, 6; and tab. vii, fig. 11, 1808.

ASTREA approaching to A. ANNULARIS, *Conybeare* and *W. Phillips*, Geol. of England, p. 188, 1822.

EXPLANARIA FLEXUOSA, *Fleming*, Brit. Animals, p. 510, 1828.

ASTREA ARACHNOIDES, *Ejusdem*, loc. cit., p. 510.

—   —   *J. Phillips*, Illustr. of the Geòl. of Yorkshire, vol. i, p. 126, 1829.

EXPLANARIA FLEXUOSA and ASTREA ARACHNOIDES, *S. Woodward*, Syn. Table of Brit. Org. Rem., p. 6, 1820.

—   —   *Morris*, Cat. of Brit. Fossils, pp. 31—36, 1843.

SIDERASTREA AGARICIAFORMIS, *M'Coy*, Ann. of Nat. Hist., s. ii, vol. 2, p. 401, 1848.

THAMNASTREA ARACHNOIDES, *Milne Edwards* and *J. Haime*, Polyp. Palæoz., etc., p. 111, 1851.

Corallum composite, massive, and varying in its general form, but appearing in most instances to have been fixed by the central part of its under surface, and to have spread out as it grew up (fig. 1, 1*c*); in other specimens it is composed of foliaceous expansions, which are sometimes superposed, so as to produce thick subdiscoidal masses, more or less lobated towards the margin (fig. 1). The upper surface of these corals is in most specimens slightly convex, but is sometimes very strongly so, or on the contrary quite flat, or even concave, and by the figures given in Parkinson's Work it appears that there are in other instances foliaceous lobes arising from it. In some young specimens the general form is regularly turbinate.

The basal plate is somewhat lobated, and presents some transverse swellings, which are produced by a certain intermittance in the progress of growth. The epitheca appears to be rudimentary, and the costal striæ, which are very distinct, are straight, regularly crenulated, very narrow, of equal size, and very closely set. The *calices* are shallow, unequally developed, and vary considerably in their degree of approximation; in general they are rather closely set, and when most so, often become arranged in concentric series (fig. 1*c*), and produce the appearance observed on the specimens, which Prof. M'Coy has described under the name of *Siderastrea agariciaformis*. When well preserved, the calice shows a slight annular elevation round the fossula, which is well characterised, but very shallow, and contains a *columella* composed of a various number of papillæ (two to eight). The *septa* form three complete cycla, and in the largest calices they also represent an incomplete fourth cyclum in one half of one or two systems, so that their total number

13

amounts often to twenty-six, twenty-eight, or even thirty-two. They are very closely set; their upper edge is almost horizontal, and delicately denticulated; their size differs but little (especially amongst those of the first two cycla), and most of them extend in an almost straight line from the fossula of one calice to that of the neighbouring one. The secondary septa in general bend towards the primary ones near their extremities, and often become united to them by their inner edge (fig. 1*f*). The size of the calices varies somewhat in the same specimen, but presents much greater variations in different specimens; their diameter varies from two to three lines, and even more; the distance between the calicular fossulæ is in general three lines, but sometimes four lines.

The aspect of this Coral differs very much, according to its mode of fossilization and state of preservation. Thus when the upper part of the *septa* has been broken down to a certain extent, as is often the case with the specimens found at Steeple Ashton, the calices appear deep, almost polygonal, and much like those of *Isastrea* (see figs. 1*g* and 1*k*). It is owing to a change of this kind that M. Michelin was led to suppose that a species very nearly allied to this, and found at Le Mans, in France, was composed of two distinct Corals, the one much resembling the specimen represented in fig. 1*g*, and the others extremely thin, enveloping the first, resembling fig. 1*j*, and "disappearing when rubbed with a hard brush," that is to say, when the delicate terminal portion of the septal apparatus had been worn away by the operator and the subpolygonal walls of the corallites denudated. The various appearances here alluded to sometimes exist on different parts of the same specimens.

Thamnastrea arachnoïdes is very common in the Coral rag of Steeple Ashton, and is met with also at Upware, near Cambridge; at Malton; and, according to Parkinson, at Chatelor. We have seen numerous specimens of this species in the collections of the Museum of Practical Geology, of the Geological Society, of the Bristol, Cambridge, Paris, and Bonn Museums; and of Messrs. Bowerbank, Walton, Phillips, d'Archiac, and Michelin.

The genus *Thamnastrea* was established in 1822 by Dr. Lesauvage, for a dendroïd Coral found near Caen, and it was on account of the general form of this fossil that it was thus distinguished from the other Astreidæ. In the Introduction to this Work, as well as in our Monograph of the family of Astreidæ, we adopted this genus, and assigned to it characters furnished by structural peculiarities, which appeared to warrant its separation from our genus *Synastrea*, as well as from the other divisions of the same tribe. But having been enabled of late to examine some more perfect specimens of Lamouroux's *Thamnastrea*, we have ascertained that the differences between these and our *Synastrea* are not by far as great as we at first supposed; thus the *septa* are in reality dentate in the first as well as in the latter, and the columella varies almost as much in different calices of the same species as from one species to the other, being, in some, composed of only one styliform tubercle, and in others of two, three, or more papillæ; the general form of the compound mass is evidently not here a character of generical value; we, therefore, deem it advisable to do away with the nominal distinction between *Thamnastrea* and *Synastrea*, and to designate all the species appertaining to them under the oldest of the two generical names, which is that of *Thamnastrea*.[1]

The group thus composed contains a great number of species, most of which belong to the jurassic or cretaceous formations. *Thamnastrea arachnoides* differs from most of them by the existence of a basal plate destitute of epitheca; but it bears great resemblance to some cretaceous corals, and more especially to *T. agaricites*,[2] *T. cistela*,[3] *T. conica*,[4] and *T. decipiens*.[5] It differs, however, from the first of these by the calices being smaller and the septa more numerous; in *T. cistela* the septa are, on the contrary, more numerous than in *T. arachnoides;* in *T. conica* the septa are thinner, besides the general form being very different; and in *T. decipiens* the septa are thicker and form only three cycla. Compared to the jurassic species, *T. arachnoides* may be distinguished by similar peculiarities; thus in *T. concinna*[6] the calices are much smaller, and the septal systems more simple; in *T. scita*[7] the septa are more delicate and closer set; *T. Lyelli*,[8] *T. dendroidea*,[9] *T. mammosa*,[10] *T. Waltoni*,[11] *T. cadomensis*,[12] and *T. affinis*,[13] are of a dendroid or mammilose form; in *T. Defranciana*[14] and *T. mettensis*[15] the septa are again thinner, and in *T. Terquemi*[16] they are, on the contrary, more robust; in *T. fungiformis*[17] they are more numerous and more strongly denticulated. As to most of the new species mentioned by M. D'Orbigny,[18] their characters have not been pointed out with sufficient minuteness to enable us to distinguish them from the fossil described in this chapter.

[1] In so doing, we must, however, remark that one of the fossil corals of the Neocomian period, the *Astrea micrantha* of Roemer appears to have a real styliform columella highly developed, as well as *septa* with entire edges, characters which we erroneously attributed to *Thamnastrea*. If that be really the case, this species must constitute the type of a distinct genus, to which the name of *Holocœnia* may be given. The genus *Centrastrea* of M. D'Orbigny, (Note sur des Polyp. Foss., p. 9,) does not contain *Astrea micrantha*, and the species referred to this division in that naturalist's 'Prodrome' do not, in reality, differ from *Thamnastrea*, their supposed prominent columella being adventitious. We also see no sufficient grounds for adopting the genus *Polyphyllastrea* of M. D'Orbigny, the species for which it was established differing from *Thamnastrea* only by having a greater number of *septa* than is commonly the case in those corals.

[2] *Astrea agaricites*, Goldfuss, Petref. Germ., t. i, pl. xxii, fig. 9.

[3] *Astrea cistela*, Defrance, Dict. Sc. Nat., t. xlii, p. 388.

[4] *Astrea conica*, Defrance, Dict. Sc. Nat., vol. xlii, p. 387.

[5] *Astrea decipiens*, Michelin, Icon. Zooph., pl. xc, figs. 12, 13.

[6] Tab. xviii, fig. 3. [7] Tab. xxiii, fig. 4. [8] Tab. xxi, fig. 4.

[9] *Astrea dendroidea*, Lamouroux, Exp. meth., pl. lxxviii, fig. 6.

[10] Tab. xxiii, fig. 3. [11] Tab. xxix, fig. 4.

[12] *Astrea cadomensis*, Michelin, Icon., pl. xciv, fig. 4.

[13] Milne Edw. and J. Haime, Ann. Sc. Nat., 3me ser., vol. xii, p. 198.

[14] Tab. xxix, figs. 3, 4. [15] Tab. xxx, fig. 3. [16] Tab. xxx, fig. 2. [17] Tab. xxx, fig. 4.

[18] Tab. xxx, fig. 3.

2. THAMNASTREA CONCINNA.   Tab. XVIII, figs. 3, 3*a*, 3*b*, 3*c*.

ASTREA CONCINNA, *Goldfuss,* Petref. Germ., vol. i, p. 64, tab. xxii, fig. 1*a,* 1826.   (It
            appears doubtful whether the figures 1*a* and 1*b* ought to be
            referred to this species.)
 — MICRASTON ? *Phillips,* Illustr. of the Geol. of Yorkshire, vol. i, p. 126, 1829.
 — CONCINNA, *Holl,* Handb. der Petref., p. 402, 1830.
 — VARIANS, *F. A. Roemer,* Vers. des Norddeutschen Oolithengeb., p. 23, tab. i,
            figs. 10, 11, 1836.
AGARICIA LOBATA, *Morris,* Cat. of Brit. Fossils, p. 36, 1843.
ASTREA VARIANS, *M'Coy,* Ann. and Mag. of Nat. Hist., s. ii, v. xi, p. 418, 1848.
SYNASTREA CONCINNA, *Milne Edwards* and *J. Haime,* Ann. des. Sc. Nat., s. iii, v. xi, p. 135.
STEPHANOCŒNIA CONCINNA and TREMOCŒNIA VARIANS, *D'Orbigny,* Prod., vol. i, p. 386.
THAMNASTREA CONCINNA, *Milne Edwards* and *J. Haime,* Polyp. Palæoz., etc., p. 111.

Corallum massive and varying in form; in some specimens very thick, in others thin
and almost foliaceous, or composed of superposed layers.   Basal plate covered with a
complete epitheca, presenting numerous circular folds (fig. 3*a*).   Upper surface convex and
gibbose.   *Calices* closely set, but unequally so in different parts of the same mass, and,
when not much crowded together, presenting round the fossula a small elevation which
corresponds to a very delicate or even rudimentary wall, as may be seen in corallites that
are worn down.   The fossula is shallow (fig. 3*b*), and contains in general only one small
columellarian tubercle; sometimes there are two.   The *septa* always constitute two
complete cycla; sometimes a third cyclum begins to appear in some of the systems where
the secondary septa become almost as large as the primary ones; so that the apparent
number of systems, composed each of a single septum, increases to 8, 10, or even 12.
This tertiary cyclum is very seldom complete, and in general appears only in four of the
fundamental systems.   The septa are alternately very strong and thin; the thickest are
the most prominent, and all are well denticulated along their upper edge; those of the
first cyclum often present near the fossula a denticulation, which is placed more apart than
the others, and bears some resemblance to a palum.   A horizontal section (fig. 3*c*) shows
that the tertiary septo-costal radiæ are much more numerous outside the walls of the
corallites than in the visceral cavity, and in their costal portion these laminæ bend so as to
join those of the surrounding corallites; they pass thus without interruption from one
fossula to another, but usually change abruptly their direction towards the middle of the
space existing between these.

This fossil is common at Steeple Ashton, Upware, and Malton; the corals briefly
described by Professor J. Phillips, under the name of *Astrea micraston,* was found at
Hackness, Ebberston, and in the south of England, and probably belongs to the same
species.[1]   Prof. M'Coy mentions having found it also in the Great Oolite at Minchinhampton,

---

[1] The only characters assigned by that author to *A. micraston,* are "calices small and equal."

and Mr. Morris had entered it in his Catalogue as having been met with in the Inferior Oolite of Cheltenham. We have also seen in Mr. Walton's collection a fossil from the Inferior Oolite of Coomb Hay, which does not appear to differ specifically from the *T. concinna* of the coral rag. Some other specimens of the same species, belonging to M. Michelin's collection, were found in the coralline formation at Stenay, in the Department des Ardennes, and those which we have seen in the Museum of Bonn were from Giengen and Natheim.

The British specimens which we have examined were communicated to us by the Geological Society, Sir H. De-la-Beche, Mr. Bowerbank, Mr. Walton, Mr. Sharpe, and Mr. Pratt. The one figured in Plate XVIII belongs to the Paris Museum.

This species is remarkable for the small size of the calices, and by that character alone can be easily distinguished from most Thamnastrea; most especially from *T. arachnoides*,[1] *T. fungiformis*,[2] *T. Defranciana*,[3] *T. Terquemi*,[4] and *T. mettensis*;[5] it also differs much by its general form from *T. dendroidea*,[6] *T. affinis*,[7] *T. Lyelli*,[8] *T. mammosa*,[9] *T. Waltoni*,[10] and *T. cadomensis*,[11] which are all much taller and more mamillose. In *T. scita*[12] the septa are much more delicate and more numerous. By its general aspect it bears a great resemblance to *T. tenuissima*,[13] but in the latter the septa are thinner and less unequal in size.

Family FUNGIDÆ, (p. lxv.)

Genus COMOSERIS.[14]

COMOSERIS IRRADIANS, Tab. XIX, figs. 1, 1*a*, 1*b*, 1*c*, 1*d*.

SIDERASTREA MEANDRINOIDES, *M'Coy*, Ann. of Nat. Hist., s. ii, vol. xi, p. 419, 1848.

Corallum massive, thick, orbicular, or sublobate, and free or fixed by a small portion of its basal plate, which is covered with a complete epitheca, presenting circular thick wrinkles or accretion folds. The upper surface convex, uneven, and usually divided into a certain number of irregular radiating valleys, by elevated ridges, which much resemble those of *Meandrina*, and more especially those of *Aspidiscus*. Most of the ridges are straight or slightly flexuous, and often meet towards the centre of the corallum, but become more

[1] Tab. xvii, fig. 1. [2] Tab. xxx, fig. 4. [3] Tab. xxix, figs. 3, 4.

[4] Tab. xxx, fig. 2. [5] Tab. xxx, fig. 3.

[6] *Astrea dendroidea*, Lamouroux, Exp. meth., pl. lxxviii, fig. 6.

[7] Milne Edw. and J. Haime, Ann. Sc. Nat., 3me sér., vol. xii, p. 198.

[8] Tab. xxi, fig. 4. [9] Tab. xxiii, fig. 3. [10] Tab. xxix, fig. 4.

[11] *Astrea cadomensis*, Michelin, Icon., pl. xciv, fig. 4.

[12] Tab. xxiii, fig. 4.

[13] *Synastrea tenuissima*, Milne Edw. and J. Haime, Ann. Sc. Nat., 3me sér., vol. xii, p. 191.

[14] D'Orbigny, Note sur des Polyp. Fossiles, p. 12, 1849.

regularly centrifugous towards the margin of the common mass. In some specimens they are separated by very large, shallow depressions, containing numerous calices confusedly arranged (fig. 1*a*); in most they become more numerous, and approximate so as to be separated only by the breadth of three or four calices (fig. 1), and in others they multiply so much, especially towards the circumference of the corallum, that each valley contains only a single series of calices (fig. 1 *b*). The calices are not originally arranged, either in concentric or radiate series, but irregularly grouped together (fig. 1, 1*d*). The centre of each calice is rendered very distinct by the existence of a small, well-defined fossula; but they are completely confluent by their circumference, and the septa pass without any interruption from one visceral chamber to another. The septal radii thus disposed ascend the above-mentioned ridges, and there become parallel; those of the opposite sides meet at the apex of these cristiform productions, and unite there without ever presenting any trace of a furrow or other separation between them (fig. 1*c*). The *Columella* is rudimentary, and represented only by one or two papillæ, which appear to be merely the inner denticulations of some of the septa. There are only two complete septal cycla; sometimes, but rarely, a few tertiary septa also exist, and the total number of these radiate laminæ is therefore twelve, fourteen, and sometimes sixteen. They are all rather thick; their edge is strongly crenulated, and they are united together laterally by numerous isolated *synapticulæ*. The secondary septa are not as long as the primary ones, and often become united to them by their inner edge. Some of the septa are straight, but most are more or less bent at the place where they pass from one corallite to another. The breadth of the calices does not much exceed a line.

This fossil is abundant at Steeple Ashton, and exists in the collections of the Museum of Practical Geology, Mr. Bowerbank, Mr. Walton, Mr. D. Sharpe, and the Museum of Paris. We are inclined to think that the coral mentioned by Mr. J. Phillips under the name of *Meandrina*,[1] but not described by that geologist, is referable to this species; it was found at Malton. We have also seen in Mr. Sharpe's collection a fossil from this locality, which appears to be a *Comoseris irradians*, but is too ill preserved to be recognised with any degree of certainty.

The genus *Comoseris* has been established by M. D'Orbigny[2] since the introduction to this Monograph was printed. It was formed with a species that had been figured by M. Michelin under the name of *Pavonia Meandrinoides*.[3] It differs from all the meandriniform astreinæ by the mode of union of the septa, having synapticulæ, as in *Fungiæ*, instead of dissepiments; it, at the same time, differs from *Agaricia*,[4] *Oroseris*, *Protoseris*, and *Lophoseris*, by its massive form, and the existence of a complete epitheca on the surface of the basal plate.

[1] Geol. of Yorkshire, vol. i, p. 126.
[2] Note sur des Polypiers Fossiles, p. 12, 1849.
[3] Iconogr. Zooph., tab. xxii, fig. 3.
[4] Introduction, p. xlix.

Comoseris meandrinoides,[1] which we know only by M. Michelin's figures, appears to differ from the British fossil here described, by the serpentine form of the ridges, which pass without interruption from one edge to the other, and by the septa being more unequal. In *Comoseris vermicularis*[2] the ridges are thinner, the septa much more delicate and closer set.

Genus PROTOSERIS.[3]

PROTOSERIS WALTONI. Tab. XX, figs. 1, 1*a*, 1*b*, 1*c*.

Corallum composite, foliaceous, subcrateriform, sometimes lobate and invaginated. The outer surface is formed by a common basal plate covered with delicate, granulated costal striæ, which are well marked, and project somewhat unequally alternately; they are almost straight, and extend from the central basal point to the edge of the corallum, but dichotomise sometimes. This basal plate presents also some transverse constrictions and tumefactions, which form ill-defined accretion ridges, but are not strongly characterised. The upper (or inner) surface of the corallum is almost smooth, and presents neither valleys nor ridges, nor cristæ, but is covered with shallow calices irregularly disposed. These *calices* are individualised by the existence of a well characterised but shallow central depression or fossula, but are not distinct at their circumference where they are completely blended together, the septa passing, without any interruption, from one visceral chamber to another (figs. 1*a* and 1*b*). In the centre of each fossula there exists a small papillose *columella*, formed by the inner septal denticulations. There are from thirty to forty septa round each calice, but not more than half of these extend to the fossula; they are small, delicate laminæ, with a crenulate edge, and are all nearly equal in thickness; some are straight, others more or less bent, or even flexuous, and many of them become cemented to one of the adjoining ones at their end, so as to assume the appearance of bifurcation.

We have seen but one specimen of this remarkable fossil; it was found in the Coral Rag at Osmington, near Weymouth, by Mr. Walton, and communicated to us by that active palæontologist.

This coral cannot be referred to any of the generical divisions established in the introduction to this Monograph, but must form the type of a new genus to which we have given the name of *Protoseris*. This genus is closely allied to *Agaricia*,[4] *Leptoseris*,[5] *Cyathoseris*,[6] *Oroseris*,[7] and *Comoseris*,[8] but differs from all by its frondescent lamellar

---

[1] *Pavonia meandrinoides*, Michelin, loc. cit.; *Comoseris meandrinoides*, D'Orbigny, Prodr., vol. ii, p. 40.

[2] Tab. xxiv, fig. 1.

[3] See our Mémoire sur les Polyp. Palæoz., etc., loc. cit., p. 129.

[4] Introduction, page xlix.   [5] Ib., p. xlx.   [6] Ib., p. xlix.

[7] Milne Edwards and J. Haime, Polyp. Palæoz., etc., p. 131.   [8] See p. 131.

form, its costulæ, the mode of arrangement of its superficial calices, its papillose columella, and the complete absence of intercalicular ridges.

———————

Some other fossil corals are mentioned by geologists as having been met with in the Coral Rag of England, but we have not been enabled to ascertain the character of these species, and can at present only recall what our predecessors have said concerning them, without pretending to distinguish them from those described in this chapter.

  1. ASTREA INÆQUALIS, *Phillips*, Geol. of Yorkshire, vol. i, p. 126. (This fossil was found at Malton, and has been only characterised by the very unequal size of its cells.)

  2. ASTREA, with cells circumscribed, *Phillips*, op. cit., Malton.

  3. TURBINOLIA DIDYMA, *Morris*, Cat. of Brit. Fossils, p. 46. Steeple Ashton. (Referred by that author to the *T. didyma* of Goldfuss, a species which has as yet been found only in cretaceous formations.)

  4. ISIS, *Miss Bennet*. (Fossil from Steeple Ashton and Bradford; no description given.)

———————

CHAPTER X.

CORALS FROM THE GREAT OOLITE.

The Corals from the Great Oolite of England, which have been submitted to our investigations, belong to twenty-two distinct species, and, as well as those met with in the upper deposits of the same geological group, are for the most part *Astreidæ;* together with eighteen species belonging to this family, we have only seen two species of *Fungidæ* and two of *Poritidæ.*

Three of these fossils (*Cladophyllia Babeana, Stylina solida,* and *Anabacia orbulites*), are found in the Inferior Oolite as well as in the Great Oolite; one (*Thamnastrea concinna*) appears to exist in both of these formations as well as in the Coral Rag, and six species have been met with in the Great Oolite in France as well as in England; but the British palæontological Fauna of this period presents sixteen species that have not as yet been discovered on the Continent.

The principal localities at which the species here described were found are the environs of Bath, Minchinhampton, Bradford, and Stonesfield.

Family ASTREIDÆ, (p. xxiii.)

Genus STYLINA, (p. xxix.)

1. STYLINA CONIFERA, Tab. XXI, figs. 2, 2*a*.

GEMMASTREA LIMBATA? *M'Coy*, loc. cit., p.419, 1848.

Corallum composite, massive, tall, very convex, and subgibbose. Corallites free at their upper end, where they assume the form of a truncate cone; more or less prominent, and sometimes rather crowded. *Costæ* closely set, straight, almost lamellar, thick, projecting more or less alternately, and terminated by an almost entire obtuse edge. *Calices*, in general, regularly circular; sometimes slightly compressed, elevated, and containing a small, shallow, round fossula. *Columella* styliform, but very small, and seldom projecting enough to be easily seen at the bottom of the fossula. *Septa* forming six equally developed systems, and two complete cycla; in general, no indications of septa corresponding to the third cyclum of costæ. The six primary septa are well developed; the secondary ones less so; they are all exsert, thick, straight, terminated by a strongly arched edge, and slightly granulated laterally. The calices often project above the common surface of the corallum to the distance of one line and a half; their diameter is not quite one line.

S. conifera was found near Bath: we have seen two well-preserved specimens of this species; one was given to us by Mr. Pratt, the other belongs to the Geological Society.

We are inclined to think that a fossil found at Minchinhampton, and considered by Prof. M Coy[1] as specifically identical with the *Astrea limbata* of Goldfuss[2], may belong to this species; but *A. limbata* differs from it, by being a ramose *Stylina*, and having a longer columella and larger calices.

S. conifera may easily be distinguished from most species of the same genus by its strongly prominent calices, and by their very unequal size.

2. STYLINA SOLIDA, Tab. XXII, figs. 3, 3*a*, 3*b*.

STYLOPORA SOLIDA, *M'Coy*, Ann. of Nat. Hist., s. 2, vol. ii, p. 399, 1848.
STYLINA BABEANA, *D'Orbigny*, Prod., t. i, p. 292, 1850.
—   —   *Milne Edwards* and *J. Haime*, Polyp. Foss. des Terr. Palæoz., etc., p. 59, 1851.

---

[1] *Gemmastrea limbata*, M'Coy, Ann. of Nat. Hist., s. 2, vol. ii, p. 419.
[2] *Petref. Germ.*, t. i, tab. viii, fig. 7, and tab. xxxviii, fig. 7.

Corallum massive, subspheroidal. *Calices* rather distant, not remarkably prominent, somewhat unequal in size, and very open. *Columella* strong and slightly compressed. *Septa* forming three complete cycla, unequally developed, straight and thin. Diameter of the calices one line, or one line and a half.

The specimen which Mr. M'Coy described, and which we have seen in the Cambridge Museum, was found in the Inferior Oolite at Dundry. M. D'Orbigny has shown us some other fossils which belong to the same species, and were met with in the corresponding strata at Morey, Departement de la Haute Saone. Mr. Terquem has also found it in the Inferior Oolite, near Metz; but the cast which we have figured here, and which does not appear to differ specifically from the above-mentioned specimens, was found by Mr. Bowerbank in the Great Oolite, near Bath. It is therefore probable that this species exists in both these formations, which are considered as distinct by some authors, but are placed in the same group of strata by our celebrated friend, M. Elie de Beaumont, under the general denomination of Lower Oolite.

Stylina solida resembles *S. conifera*,[1] *S. echinulata*,[2] *S. Deluci*,[3] and *S. limbata*,[4] in having six simple equally developed septal systems, and three cycla of the costo-septal radii. It differs from *S. limbata* by its massive and almost spherical form; and from *S. conifera*, by its less prominent calices and small tertiary septa. It bears great resemblance to *S. echinulata* and *S. Deluci;* but in the first of these, the columella is perfectly cylindrical, and the primary septa more delicate; and in the latter the calices are more crowded, and have thinner margins; the septa are rather thicker, and the columella is larger.

We consider it necessary to add that, in our figure, 3 *b*, the artist has represented the calices as being much more prominent than they really are, and we regret not having been able to correct that error.

3. STYLINA PLOTI, Tab. XXIII, fig. 1.

ASTROITES (?) *Robert Plot*, Nat. Hist. of Oxfordshire, tab. viii, fig. 2, 1676.

Corallum massive, convex, and somewhat gibbose. *Calices* rather closely set, unequal in size, projecting but little, and widely open. *Columella* small. Septal systems unequally developed; ten principal septa of equal size, thin, straight, reaching to the columella, and alternating with an equal number of very small ones. Diameter of the large calices, two thirds of a line.

The specimens of this species which we have examined were all much weather-worn,

---

[1] Tab. xxi, fig. 2.   [2] Lamarck, Hist. des Anim. sans Vert., vol. ii, p. 221.
[3] *Astrea Deluci*, Defrance, Dict. Sc. Nat., vol. xlii, p. 386.
[4] *Astrea limbata*, Goldfuss, Petref. Germ., t. i, pl. viii, fig. 7, and pl. xxxviii, fig. 7.

and we have not been able to ascertain satisfactorily whether the form of the calices is not due to erosion, and whether there be not another cyclum represented at least by costal striæ. If that were the case, *S. Ploti* would differ very little from *S. tubulifera*,[1] which is found in the Coral Rag; its septa being only thinner, and its columella a little smaller. The same peculiarities distinguish it from *S. lobata*[2] and *S. octonis*,[3] in which the calices equally present ten apparent systems. *S. magnifica*[4] is easily distinguished from it by the size of the calices and its more developed septa.

S. Ploti, figured in this Monograph, was found in the Great Oolite at Comb-Down, and belongs to Mr. Walton's collection.

Genus CYATHOPHORA.[5]

1. CYATHOPHORA LUCIENSIS. Tab. XXX, figs. 5, 5*a*.

CRYPTOCŒNIA LUCIENSIS, *D'Orbigny*, Prod., vol. i, p. 322, 1850.
STYLINA? LUCIENSIS, *Milne Edwards* and *J. Haime*, Polyp. Palæoz., etc., p. 60, 1851.

Corallum massive, convex. *Calices* circular, projecting very little above the common surface, not much crowded, and circumscribed by a very thin wall. Two cycla of septa well developed, and a third rudimentary. *Septa* straight, very unequal in size, thick externally, and continuing to extend outwards under the form of costæ; those of the first cyclum reaching almost to the centre of the calice. No appearance of a columella. Diameter of the calices somewhat more than one line.

This fossil was found in the Bradford Clay at Pound hill, and belongs to Mr. Walton's collection.

The same species has been met with in France, at Luc, and at Ranville, near Caen.

The genus *Cyathophora* of M. Michelin was established on a very imperfect specimen, in which that geologist thought that the visceral chambers were divided at short distances by a series of horizontal tabulæ, as is often the case in the *Cyathophyllidæ*. But having had an opportunity of examining this fossil in M. Michelin's collection, we recognised its specific identity with a better preserved coral that we had before seen in M. Defrance's collection, where it bore the name of *Astrea Bourgueti*, and that presented well developed septa extending almost to the centre of the calices, and united by contiguous dissepiments somewhat resembling tabulæ. We could, therefore, entertain no doubt as to the existence of great affinity between these fossils and *Stylina*; these even lose accidentally their columella in many specimens where the septa remain unimpaired, and as it appeared to us possible to account for the absence of that central axis in *Cyathophora* by similar circumstances, we

[1] Tab. xiv, fig. 3.
[2] *Explanaria lobata*, Goldfuss, Petref. Germ., t. i, pl. xxxviii, fig. 9.
[3] *Pseudocœnia octonis*, D'Orbigny, Prodr. de Paléont., vol. ii, p. 34.
[4] *Decacœnia magnifica*, D'Orbigny, Prodr. de Paléont., vol. ii, p. 33.
[5] Michelin, Iconogr., p. 104.

did not deem it necessary to maintain the distinction between the two generical divisions thus characterised. In our Monograph of the Astreidæ we, therefore, described the typical species of *Cyathophora* under the denomination of *Stylina Bourgueti*. But since the publication of that work we have examined a greater number of specimens of this species without ever finding in them any trace of a columella; other species have shown the same peculiarity; we must consequently feel less confident in the justness of our former views on the subject, and, till further data be procured, we do not feel authorised to abolish the genus *Cyathophora*. We have provisionally replaced it in our synopsis of the classification of corals lately published,[1] and we include in it four species: *C. Bourgueti*, already mentioned, *C. monticularia*,[2] *C. Pratti*,[3] and *C. Luciensis*.[4]

The latter is easily distinguished from the three others by the small size of its calices, and its septa being less numerous.

The genus *Cryptocœnia*, to which M. D'Orbigny refers this fossil, is a subdivision of the genus *Stylina* as delimitated in the system of classification adopted in this Monograph.

2. CYATHOPHORA PRATTI.   Tab. XXI, figs. 3, 3*a*.

Corallum massive, very convex, and fixed by a broad basis. *Calices* unequally distant, quite circular, and not very prominent. *Costæ* thin, straight, or slightly flexuous where they join those of a neighbouring corallite, alternately more or less prominent, but all nearly of the same breadth, and closely set. They belong to four cycla. Calicular fossula not very deep. *Septa* very thin, broad, slightly granulated, very unequal in size, and forming three well developed cycla, besides one rudimentary cyclum. Those of the first cyclum do not extend quite to the centre of the calice, where a small vacant space is visible in all the corallites that we have examined, and no trace of a styliform columella could be discovered. Diameter of the calices two lines or more.

We have seen only three specimens of this species, and all were in a bad state of preservation; two were communicated to us by Mr. Pratt, and had been found at Comb-Down, near Bath; the other forms part of the collection that Mr. Walton has had the kindness to place at our disposal for description.

Cyathophora Pratti differs from *C. Luciensis* by its calices being much larger and multiseptate. In *C. monticularia* the septa are thicker, and in *C. Bourgueti*[5] the septal systems, instead of being uniformly developed, are always unequal, the septa of the fourth cyclum existing only in four of these groups.

---

[1] See Milne Edwards and J. Haime, Polyp. Foss. des Terr. Palœoz., p. 62.

[2] *Cyclocœnia monticularia*, D'Orbigny, Prodr. de Pal., vol. ii, p. 204.

[3] Tab. xxi, fig. 3.   [4] Tab. xxx, fig. 9.

[5] Having examined in the Poppelsdorf Museum the typical specimen of the *Astrea alveolata* of Goldfuss, we have ascertained that it is not, as we formerly supposed, specifically identical with *C. Bourgueti*. The two species are quite distinct; but *Astrea alveolata* does not differ from the Coral which we

Genus CONVEXASTREA.[1]

CONVEXASTREA WALTONI. Tab. XXIII, figs. 5, 5*a*, and 6.

Corallum composite, massive, convex, more or less gibbose, or even dendroid. *Calices* small, rather unequal, and more or less closely set. The parts that have been worn down show that the walls are circular and the visceral chambers very narrow; but in the parts that remain entire the edge of this investment is completely hidden by the septo-costal lamella which are exsert and somewhat cristiform. These radii are twelve in number, and constitute, therefore, two complete cycla; they are very thick, of unequal length alternately, and in general separated from those of the surrounding corallites by narrow subpolygonal furrows, but sometimes one or two of them join these, and others establish an imperfect confluence between the adjoining individuals. The septa become rather thin inwards, and present small spiniform granulations on their lateral surfaces. There appears to be no columella. Diameter of the corallites almost a line; diameter of the mural investment half a line.

This species was found in the Great Oolite at Hampton Cliffs, near Bath, by Mr. Walton, in whose collection are placed the specimens here described. The fossil which Mr. M'Coy refers to the *Astrea reticulata* of Goldfuss, and was found at Minchinhampton, may probably belong to this species. By its general aspect *C. Waltoni* bears some resemblance to the Gosau fossil figured by Goldfuss, but the latter belongs to the genus *Astrocœnia*, and differs from the above-described species by its polygonal walls, its non-exsert septa, and its styliform columella.

The genus *Convexastrea* has been recently established by M. D'Orbigny, for a species found at St. Cassian, and very well figured by M. Klipstein under the name of *Astrea regularis*.[2] This new division is very nearly allied to *Stylina* (p. xxix), and may be defined by the following characters:

Corallum massive, astreiform, increasing by extracalicular gemmation; calices circular, and separated from each other by circumvallating furrows, through which the costal laminæ do not pass (excepting sometimes low down in the corallum); no columella; septa not numerous, and slightly exsert.

described in our Monograph of the Astreidæ, under the name of *Stylina astroides*, and which must now be called *Stylina alveolata*. (See our ' Memoir on the Palæoz. Corals,' &c., p. 59.) It is also to be noted, that the same species has been referred by Blainville to an unrecognisable fossil mentioned by Schlotheim, and has been named by that zoologist, *Siderastrea cavernosa*. (Dict. des Sc. Nat., vol. lx., p. 336, and Manuel d'Actinologie, p. 371.)

[1] D'Orbigny, note, ' Sur des Polypiers Fossiles,' p. 9, 1849.

[2] Beitr., etc. tab. 20, fig. 11.

By an attentive examination of Goldfuss's corals in the Poppelsdorf Museum, and in the cabinet of Professor Bronn, at Heidelberg, we have been able to ascertain that the *Astrea sexradiata*, Goldfuss,[1] belongs to this small group, but the specimen figured is in a very bad state of preservation. The genus *Convexastrea* contains, therefore, at present three species, and the one here described differs principally from the two others by the less regular form and the thickness of the septa, and by these laminæ being less exsert.

Family ASTREIDÆ, (p. xxiii.)

Genus MONTLIVALTIA, (p. xxv.)

1. MONTLIVALTIA SMITHI. Tab. XXI, figs. 1, 1*a*, 1*b*.

> MADREPORA TURBINATA, *Smith*, Strata identified by Organic Fossils, p. 84, tab. Upper Oolite, fig. 3, 1816, (appears to be a specimen, the wall and the basis of which have been worn away.)

Corallum simple, short, fixed by a large basis, which is somewhat expanded. Wall a little constricted near the basis, and covered with a thick epitheca which extends to a short distance from the calicular margin. *Calice* regularly circular; fossula not very deep, but well defined. In general five complete cycla of septa, but sometimes the last cyclum is quite rudimentary in one half of each system. The *septa* are exsert, very strong, broad, quite straight, and very closely set. Those of the first three cycla are almost of the same size, and extend to the centres of the visceral chamber; those of the fourth cyclum are also large, and those of the last cyclum are much smaller, but remain always free from any adherence at their inner edge. The lateral surfaces of the septa appear to be strongly striated, and their upper edge was probably denticulated originally, but had become quite smooth by wear in the two specimens here described. Diameter of the calice one inch three lines; height, seven lines in one specimen, and more than an inch in the other.

These corals were found near Bath and communicated to us by Mr. Pratt. The fossil, which appears to belong to the same species, and is figured in Mr. Smith's work, was found at Farley.

Montlivaltia Smithi differs from most of the other species belonging to the same genus by the thickness of its *septa* and its broad, short form. It resembles most our *M. Wrighti*,[2] but this fossil does not appear to have been fixed by a large basis, and its *septa* are more numerous and more unequal in size.

---

[1] Petref. Germ., vol. i, tab. 24, fig. 9; *Convexastrea sexradiata*, Milne Edwards and J. Haime, Polyp. Palæoz., etc., p. 63.

[2] Tab. xxvi, fig. 12.

2. MONTLIVALTIA WATERHOUSEI.   Tab. XXVII, figs. 7, 7 a.

Corallum straight, erect, cylindroid, and convex at the basis, where no appearance of adherence is perceptible. *Epitheca* very strong. *Calice* circular, or nearly so; fossula well defined and somewhat oblong. *Septa* rather thin, unequal in size alternately, and in general sixty-six in number; sometimes an equal number of rudimentary ones situated between the former. Many of the septa appear to have their upper edge somewhat arched towards the centre of the calice. Height one inch and a half; diameter one inch two lines.

This fossil was found in the Oolite of Minchinhampton, and belongs to the palæontological collection of the British Museum.

M. D'Orbigny has given the name of *Montlivaltia regularis*[1] to a coral found in the Kelloway Rocks, in France, which very much resembles this fossil by its general form, but has its epitheca more strongly wrinkled: none of the known specimens show the calice, and we are, therefore, unable to decide whether it be or not a distinct species. This form is not seen in any other species of the genus *Montlivaltia*, but we must remark that if *M. Labechii*[2] became taller it would become, in that respect, very like *M. Waterhousei*, and that some ill-preserved specimens of *M. regularis*, belonging to the collection of M. Hebert, of Paris, are almost as short as the former. *M. Labechii* differs from *M. Waterhousei* by its septa being more numerous and terminated by a straight edge inwardly.

Genus CALAMOPHYLLIA, (p. xxxiii.)

CALAMOPHYLLIA RADIATA.   Tab. XXII, figs. 1, 1a, 1b, 1c.

> TUBIPORA, *W. Smith*, strata identified by Org. Foss., p. 30, Upper Oolite, figs. 1 and 2, 1816 (good figures).
>
> EUNOMIA RADIATA, *Lamouroux*, Expos. Method., p. 83, tab. lxxxi, figs. 10, 11 (unsatisfactory figure, representing a very bad specimen), 1821.
>
> —   —   *Lamouroux*, Encyclopèdie, Zooph., p. 382, 1824.
>
> —   —   *Bronn*, Syst. der Urwelt., tab. iv, fig. 13, 1824.
>
> —   —   *Defrance*, Dict. de Sc. Nat. vol. xlii, p. 393, 1826.
>
> TUBIPORA or EUNOMIA, *Phillips*, Geol. of Yorkshire, vol. i, p. 147, 1827.
>
> FAVOSITES RADIATA, *Blainville*, Dict. des Sc. Nat., vol. lx, p. 367, tab. xlii, fig. 4, 1830; and Manuel d'Actinologie, p. 403.
>
> EUNOMIA RADIATA, *Holl*, Handbuch der Petref., p. 414, 1830.
>
> —   —   *Bronn*, Lethea Geogn., tab. xvi, fig. 23, 1836-7.
>
> —   —   *Morris*, Catal. of Brit. Fossils, p. 36, 1843.
>
> LITHODENDRON EUNOMIA, *Michelin*, Iconogr. Zooph., p. 223, tab. xxxiv, fig. 6, 1849.
>
> EUNOMIA RADIATA, *Milne Edwards* and *J. Haime*, Monogr. des Astreides, Ann. des Sc. Nat., s. 3, vol. xi, p. 260, 1849.
>
> CALAMOPHYLLIA RADIATA, *Milne Edwards* and *J. Haime*, Monogr. des Polyp. Foss. des Terr. Palæoz., etc., p. 81, 1851.

---

[1] Prod. de Paléont., vol. i, p. 346.

[2] Tab. xxvi, fig. 3.

Corallum fasciculate, forming a globose tuft, which in some specimens appears to be almost a foot high, but was only four or five inches in the largest specimens seen by us. *Corallites* very tall, straight or slightly bent, almost cylindrical or somewhat prismatical, dichotomising at long intervals, very closely set and spreading out like a sheaf laterally and upwards. *Walls* presenting circular tumefactions more or less characterised, and annular expansions, which sometimes extend from one corallite to another by means of small conical processes, somewhat in the same manner as in *Syringopora* (figs. 1 and 1*a*); but these lateral buttresses are compact and not tubular, as in the latter corals. Costal striæ delicate, of equal size, rather closely set, and but little prominent. Calices of various forms, sometimes almost circular, sometimes oval, almost triangular or subpolygonal (fig. 1*c*). *Columella* quite rudimentary. From sixteen to twenty septa, large or small alternately, and closely set; the smaller ones bending towards the principal ones in a somewhat irregular manner; septal systems not distinctly recognisable in the adult specimens. Diameter of the corallites and of their calice about two thirds of a line.

This fossil is found near Bath. The specimens here described were communicated to us by Mr. Bowerbank and Mr. Pratt. William Smith, who first discovered the species, mentions its existence at Comb-Down, Broadfield Farm, and Westwood; Professor J. Phillips appears to have found it at Terrington, and Mr. Morris says that it has been met with at Farley Downs, Hampton Cliff, and Murrel, near Bradford. It is found also in France, at Langrune, Luc, and Ranville, near Caen, and according to M. Michelin, at Billy, near Chanceaux, Departement de la Côte d'Or.

In general this coral is met with in a state so modified by the process of fossilisation, that it is very difficult to recognise its real zoological affinities. Lamouroux, who described it as the type of his genus *Eunomia*, was only acquainted with specimens in which extraneous matter had been first deposited around and between the corallites, so as to form a cast, and in which the corallites themselves had been afterwards completely destroyed and replaced by a distinct stony deposit, so that the original structure had completely disappeared. The appearance thus produced naturally induced Lamouroux to suppose that this coral was nearly allied to *Tubipora*, and Blainville considered it as being the cast of a Favosites. But M. Michelin, having found some specimens in which the septa had been partially preserved, recognised their affinity to Schweigger's *Lithodendron*. In our Monograph of the Astreidæ, and in the introduction to this work, the genus *Eunomia* was, however, still admitted on account of a peculiar disposition of the epitheca, which appeared in some casts to distinguish it from *Calamophyllia*. But the British specimens communicated to us by Mr. Pratt, and some finely preserved specimens from Normandy, which we have seen in M. D'Orbigny's collection, prove that no sufficient grounds for a distinction of that kind do in reality exist, and that the *Eunomia* of Lamouroux must no longer be separated from *Calamophyllia*.

C. radiata is the smallest species known, and it differs also from the other species of the same genus by the low number of its septa.

Genus CLADOPHYLLIA.[1]

CLADOPHYLLIA BABEANA. Tab. XXII, figs. 2, 2*a*, 2*b*.

MADREPORA FLEXUOSA, *Smith*, Strata Identified by Org. Foss., p. 30, Upper Oolite, fig. 5,
1816 (not *M. flexuosa*, Linnæus).
EUNOMIA BABEANA and CALAMOPHYLLIA PRIMA, *D'Orbigny*, Prod. de Palæont., t. i,
p. 292, 1850.
CLADOPHYLLIA BABEANA, *Milne Edwards* and *J. Haime*, Polyp. Palæoz., etc., p. 81, 1851.

Corallum fasciculate. Corallites placed at unequal distances, cylindrical, with well
marked accretion swellings, and a thick epitheca, the wrinkles of which are quite horizontal.
Calices circular, or somewhat oval when large. *Septa* thin, straight, of unequal size, and
forming three complete cycla. The tertiary ones are sometimes almost rudimentary, and
in some corallites two of the primary ones, placed opposite each other, are more developed
than the others, so as to divide the visceral chamber into two equal parts, a circumstance
which appears to indicate a commencement of fissiparous multiplication. Diameter of the
corallites one and a half or two lines.

This fossil is found in the Great Oolite at Bradford Hill, near Bath, and the specimens
here described belong to the collections of Mr. Walton and Mr. Pratt. It is mentioned by
Smith as having been met with at Castle Combe. M. Terquem has also found it in the
Inferior Oolite of St. Quentin, near Metz, in Lorraine, and M. D'Orbigny, at Langres,
Departement de la Haute Marne.

Cladophyllia Babeana is very much like *C. Conybearii*, here above described, from the
Coral Rag. It appears, however, to differ from it by the regular horizontal direction of the
wrinkles of the epitheca, and the feeble development of its tertiary septa. The fossil
mentioned by M. D'Orbigny under the name of *Calamophyllia prima* is only a variety of
this species with the corallites smaller than in the preceding specimens, and not larger, as
is stated in the short description given by that palæontologist.

Genus ISASTREA.[2]

1. ISASTREA CONYBEARII. Tab. XXII, fig. 4.

Corallum composite, massive, terminated by an almost flat surface. *Calices* nearly
equal, subtetragonal, and circumscribed by a simple edge common to the two adjoining
corallites, or separated only by a slight furrow. No *Columella*. *Septa* thick, in general
straight and much modified by the process of fossilisation ; the well developed ones not
numerous, and alternating with rudimentary ones. Systems unequally developed ; three
complete cycla, and a fourth cyclum in four of the systems ; the principal septa join in the
centre of the visceral chamber. Long diagonal of the calices six or seven lines.

[1] See p. 91.   [2] See p. 73.

We have seen but one specimen of this species; it was very ill preserved, and had been found at Comb-Down, near Bath, by Mr. Pratt.

This fossil, as far as we are able to judge of its characters, appears to differ from all other Isastrea by the tetragonal form of its calices and the small number of septa, relatively to the size of the corallites.

2. ISASTREA LIMITATA.   Tab. XXIII, figs. 2, 2a, and Tab. XXIV, figs. 4, 4a, 5.

> ASTROITES, etc., *R. Plot*, Nat. Hist. of Oxfordshire, p. 88, tab. xi, fig. 6, 1676 (good figure: we are inclined to think, that fig. 7 represents a specimen in which the centre of the calices had been accidentally filled up, so as to produce the appearance of a styliform columella).
>
> MADREPORA, *J. Walcott*, Description and Figure of Petref., found near Bath, p. 47, fig. 63, 1779.
>
> ASTREA LIMITATA, *Lamouroux*, in Michelin's Iconogr. Zooph., p. 229, tab. xciv, fig. 10, 1849.
>
> —   —   *M'Coy*, Ann. and Mag. of Nat. Hist., s. 2, vol. 2, p. 418, 1848.
>
> PRIONASTREA LIMITATA, *Milne Edwards* and *J. Haime*, Ann. des Sc. Nat., s. 3, vol. xii, p. 137, 1849.
>
> PRIONASTREA LIMITATA, P. ALIMENA, and P. LUCIENSIS, *D'Orbigny*, Prod. de Palæont., t. i, p. 322, 1850.
>
> ISASTREA LIMITATA, *Milne Edwards* and *J. Haime*, Polyp. Palæoz., etc., p. 103, 1851.

Corallum massive, terminated by a flat or somewhat gibbose surface. *Calices* almost equal in some parts, very unequal in others; the small ones usually situated in the depressions, and the larger ones on the gibbose parts of the upper surface. The calices are polygonal, not very deep, and terminated by a thin, straight, mural edge. The small ones contain scarcely twenty septa, but in the larger ones the number of these laminæ amounts to about thirty, so that there appears to be in that case three complete cycla, and an incomplete fourth cyclum, but the whole of the septal apparatus presents very great irregularity; thus we have often seen between two principal septa two smaller ones, one of which, more developed than the other, belonged probably to the second cyclum, and the other must have belonged to the third cyclum, but had no corresponding one in the other half of the system so composed. All the *septa* are thin, straight, or only slightly curved, and sparingly granulated, but presenting well characterised radiate striæ on their lateral surfaces. They are but slightly exsert, and, far from passing from one visceral chamber to another, they in general alternate exteriorly with those of the adjoining corallite. The great diagonal of the large calices is about two and a half lines; and their depth one line; the small calices are little more than one line broad.

This fossil was found by Mr. Pratt in the Great Oolite near Bath. Walcott mentioned it having been met with at Hampton Downs. It exists also in the corresponding deposits near Caen, at Langrune, Luc, and Ranville.

Isastrea limitata differs from *I. helianthoides*,[1] *I. explanata*,[2] *I. Munsterana*,[3] *I. crassa*,[4] *I. lamellosissima*,[5] *I. Bernardana*,[6] *T. tenuistriata*,[7] *I. ornata*,[8] and *I. serialis*,[9] by the small size of its calices, and by its septa being more numerous. It differs from *I. polygonalis*,[10] *I. oblonga*,[11] and *I. Michelini*,[12] by its walls being very thin, although well formed; from *I. explanulata*[13] by the concavity of its calices, and from *I. Richardsoni*[14] by the tenuity of its septa. It bears strong resemblance to *T. Guettardana*,[15] but in the latter the calices are rather deeper, and the septa more delicate, less numerous, and developed in a more equal manner. Having lately been enabled to examine the fossils mentioned by M. D'Orbigny under the names of *Pronastrea alimena* and *P. Luciensis*,[16] we have no longer any doubt as to their being specifically identical with the coral here described. We are also inclined to think that the two fossils for which the same author has established the genus *Dendrastrea*[17] are only subdendroid varieties of *Isastrea limitata;* but they are too ill preserved for us to be able to decide the question.

3. ISASTREA EXPLANULATA. Tab. XXIV, figs. 3, 3*a*.

ASTREA EXPLANULATA, *M'Coy*, Ann. of Nat. Hist., s. 2, vol. ii, p. 400, 1848.

Corallum massive, terminated by an almost flat surface. *Calices* shallow, not very unequal, polygonal. Walls rudimentary, but the corallites distinct, even when deprived of a mural investment, and the costal edge of their septa alternating in general with those of the adjoining corallite. *Columella* rudimentary. *Septa* thin, closely set, straight or slightly bent, striated laterally, delicately and regularly denticulated along their upper edge, and very unequal in size. In general three complete cycla, and a few septa belonging to a fourth cyclum appearing in a very irregular manner; sometimes one half of certain systems being more developed than the other; sometimes the septa of the fourth order are formed without being accompanied by those of the fifth, which equally enter into the composition

---

[1] *Astrea helianthoïdes*, Goldfuss, Petref. Germ., pl. xxii, fig. 4*a*.

[2] Tab. xxiv, figs. 3, 3*a*.

[3] *Prionastrea Munsteriana*, Milne Edwards and J. Haime, Ann. Sc. Nat., t. xii, p. 136.

[4] *Agaricia crassa*, Goldfuss, Petref. t. xii, fig. 13.

[5] *Astrea lamellosissima*, Michelin, Icon., pl. vi, fig. 1.

[6] *Prionastrea Bernardina*, D'Orbigny, Prodr., t. i, p. 293.

[7] Tab. xxx, fig. 1.

[8] *Prionastrea ornata*, D'Orbigny, Prodr., t. i, p. 293.

[9] Tab. xxiv, fig. 2.

[10] *Astrea polygonalis*, Michelin, Icon., pl. iii, fig. 1.

[11] Tab. xii, fig. 1.

[12] *Montastrea Michelini*, De Blainville, Dict. Sc. Nat., t. lx, p. 339.

[13] Tab. xviii, fig. 1.   [14] Tab. xxix, fig. 1.

[15] *Astrea formosissima*, Michelin, Icon., pl. vi, fig. 24. (Non Sowerby.)

[16] Op. cit., vol. i, p. 322.

[17] *Dendrastrea Langrunensis*, D'Orbigny, and *D. dissimilis*, D'Orb., Prodr., vol. i, p. 322.

of the fourth cyclum, and in other instances the irregularity becomes still greater, two small and almost equal septa being situated between two principal ones. The diameter of the calices is not quite two lines.

This fossil was found in the Great Oolite at Comb-Down, near Bath, by Mr. Pratt. Mr. M'Coy has met with specimens of the same species in the Inferior Oolite at Dundry, and near Bath.

Isastrea explanulata is remarkable for the rudimentary state of the walls and the shallowness of its calices.

4. ISASTREA SERIALIS. Tab. XXIV, figs. 2, 2a.

Corallum massive, terminated by a flat or slightly convex surface. *Calices* of very unequal size; in general elongated, gemmating near the margin, and forming sometimes short series where two young individuals are thus produced at the same time from two opposite points of the parent calice. Walls thin, compact, irregularly polygonal and not projecting much between the calices. The latter rather shallow. No distinct columella. Septal systems developed in a very irregular manner, and scarcely recognisable. In the large calices which have not yet begun to gemmate, the number of well developed septa amounts often to fifty; they are thin and very unequal in size; most of the small ones become united to a neighbouring large one along the inner edge, and form with it a very acute angle. The principal septa are often bent near the inner part; they appear to have a denticulated edge, and the lateral surfaces somewhat granulated. Between each of these a very small but quite distinct rudimentary septa is always seen.

Long diagonal of the large calices about three lines; depth one line.

The unique specimens of this species that we have seen was found at Comb-Down, near Bath, by Mr. Pratt, and given by that palæontologist to the Museum of the Geological Society. The fossil which Prof. M'Coy[1] mentions as having been found at Minchinhampton, and refers to the *astrea confluens* of Goldfuss, belongs probably to the same species of *Isastrea*.

I. serialis very much resembles *I. Lotharinga*,[2] but differs from it by its flat, low form, the number of its septa, and the delicacy of these laminæ. In *I. Munsterana*,[3] which is also nearly allied to the preceding species, the septa are thicker towards their inner edge.

At first sight the genus *Isastrea* appears to be very different from the genus *Latomeandra*[4]; but these two groups are in reality closely allied, and the passage between them

[1] Ann. of Nat. Hist., s. ii, v. ii, p. 418, 1848.

[2] *Meandrina lotharinga*, Michelin, Icon., pl. xxii, fig. 2.

[3] *Prionastrea Munsteriana*, Milne Edw. and J. Haime, Ann., vol. xii, p. 136.

[4] D'Orbigny, Note Sur des Pol. Foss., p. 8, 1849. (Introd., p. xxxiv.)

is in part established by the species which, like *I. serialis* and *I. Lotharinga*, form sometimes a short series of calices where gemmation is very active, and by those *Latomeandra*
which are of a massive form.   But in all well-preserved fossils belonging to the latter
genus we have found the common basal plate naked, and presenting simple straight costæ,
whereas in *Isastrea* the basal plate is covered with a complete epitheca, and when that
tunic is worn away the costal striæ assume the appearance of radiate fossulæ.   M.
D'Orbigny has recently formed the genus *Meandrophyllia*[1] for the species which are in
some respects intermediate between these two types, having the calices elongated and often
arranged in short series; but it appears useless to separate generically all the degrees by
which one form passes to another, and when the principal characters peculiar to one or the
other of the above-mentioned types become obscure, as in the present case, we prefer
having recourse to the secondary characters just pointed out, in order to determine the
genus to which the doubtful species is to be referred.

Genus CLAUSASTREA.[2]

CLAUSASTREA PRATTI.   Tab. XXII, fig. 5.

Corallum massive, terminated by a slightly convex surface.   *Calices* large, somewhat unequal, and not separated by a distinct wall.   *Columella* spongiose, well developed.
Septa of the adjoining corallites quite confluent; some of them enlarged or much bent at
their point of junction with the corresponding ones from another individual.   They are
about thirty in each calice; they are rather thin, unequal in size, closely set, and present
vertical striæ on their lateral surfaces: the smaller ones are united to the neighbouring
large ones by their inner edge.   The loculi are closed by well-formed and rather numerous
dissepiments.   The common basal plate is covered with thick granulated costal striæ, and
does not appear to have any epitheca.   Diameter of the calices five or six lines.

We have as yet seen but one ill-preserved specimen of this species; it was found at
Comb-Down, near Bath, and belongs to the collection of the Geological Society.   It has not
enabled us to give a complete description of this species, but may easily be distinguished
from the other two species which remain in the genus *Clausastrea* as now circumscribed,
that is to say characterised, by the absence of walls and columella.[3]   In *C. Pratti* the
septa are much thicker than in *C. tessellata*[4] and in *C. consobrina*.[5]

[1] D'Orbigny, Note Sur des Pol. Foss., p. 8.

[2] D'Orbigny, Note Sur des Pol. Fossiles, p. 9, 1849.

[3] See the Introduction to our Memoir on the Palæoz. Fossil Corals, etc., in the 'Archives du Museum,'
vol. v, p. 107.

[4] D'Orbigny, Prod., vol. i, p. 293.

[5] *Synastrea consobrina*, D'Orbigny, loc. cit.   *Clausastrea? consobrina*, Milne Edwards and Haime,
loc. cit.

Genus THAMNASTREA, (p. xliii.)

1. THAMNASTREA LYELLI.   Tab. XXI, figs. 4, 4*a*, 4*b*.

SIDERASTREA LAMOUROUXI, *M'Coy*, Ann. of Nat. Hist., s. ii, vol. ii, p. 419, 1848.   (Not Thamnastrea Lamourouxi, Lesauvage.)

We have seen but a few fragments of this fossil: some were very large, irregularly cylindrical, and somewhat mammillose; others were slender, and these differences must have given to the entire mass a general aspect somewhat different from that of the *T. dendroidea* (or *T. Lamourouxi*) found near Caen, in Normandy.   In the latter the columnar branches, which constitute the compound mass of the coral, appear to vary very small in diameter, however large the size of this mass may be.   The *calices* are very unequally approximated, and where they are the less crowded in one direction most of the septa assume a transverse direction.   The fossula is not surrounded by a circular elevation corresponding to the wall, but when the corallites have been worn down the latter becomes visible, and, although very thin and feebly developed, shows that the radiate laminæ are formed by the costæ as well as by the septa.   The *columella* is small, but in general well characterised and composed of one or two round papillæ.   The fossula is not deep, but never quite superficial.   The *septa* form three cycla, which are often complete, but sometimes those of the last cyclum are deficient in one or two of the systems.   They are thin, denticulated, rather closely set, not very exsert, and somewhat unequal alternately; most of them are flexuous towards the circumference of the corallites.   Those of the second cyclum differ but little from the primary ones, but are not quite so broad; the tertiary ones are much narrower and thinner; they do not appear to incline towards each other, and become united at their inner edge.   In some well-preserved calices very distinct paliform lobules are placed between the columella and the septa of the first two cycla; the primary ones are narrower and more central than those corresponding to the secondary septa; the latter do not occur in the systems where the tertiary cyclum is incomplete.   Diameter of the calices one and a half line.

This fossil is found at Stonesfield, and is in the collection of the Geological Society and of Mr. D. Sharpe.   A specimen belonging to the Cambridge Museum was met with at Minchinhampton, and a cast found near Bath, by Mr. Bowerbank, appears to belong to the same species, although the calices are rather smaller and more crowded than in the above-described specimens.

Thamnastrea Lyelli is very much like *T. affinis*[1] and *T. dendroidea* (or *T. Lamouroux i*),[2]

[1] Milne Edwards and J. Haime, Monogr. des Astreides, Ann. des. Sc. Nat., s. iii, vol. xii, p. 158.

[2] It is the same fossil that Lamouroux described under the name of *Astrea dendroidea*, Expos. Method., pl. lxxviii, fig. 6, and afterwards called by Dr. Lesauvage *Thamnastrea Lamourouxi*, Mem. de la Soc. d'Hist. Nat. de Paris, vol. i, tab. xiv.

to which latter species it has been referred by Prof. M'Coy. It differs, however, from *C. affinis* by its septa, which in the latter are much more unequal in height as well as in thickness, and form generally three complete cycla. In *T. dendroidea* the branches are more cylindrical, and vary much less in diameter; the septa are thicker, strongly denticulated, and do not appear to have any paliform lobules. *T. cadomensis*[1] and *C. Waltoni*[2] are equally of an arborescent form, but in the former the septa are thin, flexuous, and become united by their inner edge; and in the latter the aspect of the calice is quite different, on account of the septa being thick towards their external edge, and on the contrary very thin inwards.

2. THAMNASTREA MAMMOSA, Tab. XXIII, figs. 3, 3 *a*.

Corallum massive, tall, subpyriform, terminated by a mammose surface, and composed of superposed layers, which are intimately united, and most distinct near the basis. *Calices* small, placed at unequal distances, and often disposed in a radiate order towards the summit of the mammillary protuberances. *Fossula* well characterised, but not deep, and containing a small tubercular columella. Sixteen or eighteen *septa*, somewhat unequal and rather thin; some smaller than the rest, are placed irregularly, the others are straight, or bend towards their outer edge, where most of them join the corresponding one from an adjoining corallite. Diameter of the calices, half a line.

This fossil belongs to the collection of Mr. Stokes, and appears to have been found in the Great Oolite at Sapperton, in Gloucestershire.

T. mammosa differs from most species of the same genus by the small size of its calices, its strongly gibbose surface, and its mode of growth by superposed layers. It most resembles *T. scita*,[3] *T. tenuissima*,[4] and *T. concinna*,[5] but its septa are more equally developed than in the latter, and are much less numerous than the first two.

3. THAMNASTREA SCITA, Tab. XXIII, figs. 4, 4 *a*.

Corallum massive, terminated by an almost flat surface, and composed of thin superposed layers, the uppermost of which are often incomplete. *Calices* small, almost equally dispersed, and shallow; fossula small; columella rudimentary. *Septa* delicate, closely set, and differing somewhat in thickness and in breadth alternately; they are very unequally confluent, and some are straight, whereas others are flexuous or strongly geniculated; their lateral surfaces appear to be delicately granulated. Diameter of the corallites about three quarters of a line.

---

[1] *Astrea cadomensis*, Michelin, Iconogr., tab. liv, fig. 4.
[2] Tab. xxix, fig. 4.   [3] Tab. xxiii, fig. 4.
[4] *Synastrea tenuissima*, Milne Edwards and J. Haime, Ann. Sc. Nat., 3d sér., vol. xii, p. 191.
[5] Tab. xviii, fig. 3.

This species was found in the Great Oolite at Hampton cliffs, by M. Walton. It is easily distinguished from the other species belonging to the same genus by the delicacy and the great number of its septa relatively to the small size of its calices.

4. THAMNASTREA WALTONI, Tab. XXIX, figs. 4, 4 *a.*

Corallum arborescent. *Calices* closely set, somewhat unequal. *Walls* subpolygonal, and becoming apparent when the upper surface of the corallites has been worn away. From twenty to twenty-four septa, varying a little in breadth and size alternately, but almost all of the same thickness; in general strongly bent, and very thick where they pass from one corallite to another, but thin towards the fossula; their upper edge appears to be almost entire, and their lateral surfaces but feebly granulated. Fossula well characterised. *Columella* tubercular. Diameter of the calices two-thirds of a line.

We have seen only a small cylindrical fragment of this fossil that was found by Mr. Walton in the Great Oolite, near Bath. It is very nearly allied to *T. Cadomensis,*[1] but appears to differ from that species by the smallness of its calices, and by its septa being thicker and less numerous. The same characters distinguish it from *T. Lyelli,*[2] *T. affinis,*[3] and *T. dendroidea,*[4] to which it resembles by general form of the corallum.

Prof. M'Coy[5] refers to the ASTREA GRACILIS of Goldfuss,[6] a cast of which was found at Minchinhampton, and belongs to the Cambridge Museum. There is also in Mr. Bowerbank's collection a specimen of the same kind, from the environs of Bath. These corals all belong to the genus *Thamnastrea,* but the specimens that we have seen are not in a sufficiently good state of preservation to enable us to characterise them specifically.

Family FUNGIDÆ, (p. xlv.)

Genus ANABACIA, (p. xlvii.)

ANABACIA ORBULITES, Tab. XXIX, figs. 3, 3 *a,* 3 *b,* 3 *c,* 3 *d,* 3 *e.*

BUTTON STONE, *R. Plot,* Nat. Hist. of Oxfordshire, p. 139, tab. viii, fig. 9, 1676.
PORPITE, *Knorr* and *Walch,* Rec. des Monum. des Catastr., v. ii, p. 23, tab. F 3, figs. 6, 7, 1775.
MADREPORA PORPITES, *W. Smith,* Strata identif. by Org. Fossils, p. 30, Upper Oolite, fig. 4, 1816.

---

[1] Michelin, Icon. Zooph., tab. liv, fig. 14.
[2] Tab. xxi, fig. 4.
[3] Milne Edwards and Jules Haime, Ann. Sc. Nat., s. iii, vol. xii, p. 198.
[4] *Astrea dendroidea,* Lesauvage, loc. cit., pl. lxxviii, fig. 6.
[5] Ann. of Nat. Hist., s. ii, vol. ii, p. 418.
[6] Petref. Germ., v. i, tab. xxxviii, fig. 13.

FUNGIA ORBULITES, *Lamouroux*, Expos. Method., p. 86, tab. lxxxiii, figs. 1, 2, 3, 1821.

FUNGIA LÆVIS, *Goldfuss*, Petref. Germ., v. i, p. 47, tab. xiv, fig. 2, 1826.

CYCLOLITES LÆVIS, *Blainville*, Dict. des Sc. Nat., v. lx, p. 301, 1830.

ANABACIA ORBULITES and ANABACIA BAJOCIANA, *D'Orbigny*, Prod. de Paléont., v. i, pp. 321-2, 1850.

—   —   *Milne Edwards* and *J. Haime*, Polyp. Foss. des Terr. Palæoz., etc., p. 122, 1851.

Corallum simple, circular, and affecting the form of a plano-convex lens, with a thick, rounded edge, and a small, shallow, circular fossula. *Septa* very numerous (140 or 150), extremely closely set, of equal thickness and of equal height, but varying in breadth (from the centre to the circumference of the corallum), and very delicately and regularly denticulated; the smaller ones joining the neighbouring large one at their inner edge, so as to make the latter appear to bifurcate. In general, forty-eight principal septa reach to the edge of the fossula. The under surface is often concave. Breadth, six or seven lines. Height, three lines.

The genus *Anabacia* has been very judiciously established by M. D'Orbigny[1] for those simple, lenticular corals which had been usually placed among the Fungia or the Cyclolites, but differ from them by the absence of the mural disc. *Anabacia orbulites* was the first species known to naturalists, and good figures of this fossil were given in the works of Plot and of Knorr. Three other species present the same generical characters, but are easily distinguished from it, specifically, by their general form; one of these, *A. Normaniana*,[2] being very flat; the second, *A. hemispherica*,[3] being on the contrary much taller, and more convex; and the third, *A. Bouchardi*,[4] being almost conical.

Anabacia orbulites appears to be a common species. Specimens found in the Bradford clay at Bradford, and in the Great Oolite at Comb-Down, are in the collection of the Geological Society. Mr. Walton and Mr. Pratt have kindly submitted to our investigation other specimens found at Hampton, near Bath. Mr. Lonsdale mentions its occurrence in the Cornbrash at Atford, and W. Smith met with it at Broadfield Farm five miles from Bath, near Phillips Norton, Somersetshire. It has also been found in the Inferior Oolite; Mr. Bowerbank and Prof. Phillips have communicated to us specimens from Dundry, and Mr. Walton a specimen from Charlcomb.

The same fossil is found in France, in the Great Oolite near Caen, Departement du Calvados, and in the Inferior Oolite at Conlie, Departement de la Sarthe. Goldfuss mentions also its occurrence in the Swiss Jura.

---

[1] Note sur des Polypiers Fossiles, p. 11, 1849.

[2] D'Orbigny, Prod., vol. i, p. 241.

[3] Tab. xxix, fig. 2.

[4] *Fungia orbulites* (pars); Michelin, Icon., tab. liv, fig. 1; *Anabacia Bouchardi*, Milne Edwards and J. Haime, Polyp. Palæoz., etc., p. 122.

Genus COMOSERIS.[1]

COMOSERIS VERMICULARIS, Tab. XXIV, fig. 1.

MEANDRINA VERMICULARIS, *M'Coy*, Ann. of Nat. Hist., s. ii, v. ii, p. 402, 1848.

Corallum composite, massive, convex; its upper surface overrun with strong, cristate, sharp ridges, which are very flexuous, somewhat ramified, and closely set. The septal laminæ, which are very thin and crowded together, ascend parallel to the top of these ridges, where a delicate mural line is visible: about ten of these laminæ are comprised in the space of one line, and they vary a little in size alternately. The depressions situated between these ridges are rather deep, but not very broad; so that when the structure of the corallum is hidden by incrustations of extraneous matter, the general aspect of the fossil resembles very much that of *Meandrina*. But in well-preserved specimens, it is easy to see that the above-mentioned depressions contain a series of distinct *calices*, with confluent septa, but separate, well-defined fossulæ. Each calice has twelve septa which are closely set, slightly denticulated along their edge, and somewhat thickened towards the middle. In the corallites that are situated at the bottom of the depressions, most of the septa follow the general direction of these furrows; but in those situated nearer to the top of the ridges, the septa become almost all perpendicular to the common mural lines; some of them, however, are always more or less curved. Diameter of the calices, about one line; breadth of the depressions, two or three lines.

We have seen only two specimens of this species; one was found in the Great Oolite near Bath, by Mr. Lonsdale, and given by that Palæontologist to the Geological Society's collection; the other belongs to the Cambridge Museum, and was found in the Inferior Oolite at Leckhampton.

C. vermicularis differs from the other two species of the same genus above described,[2] by the form of its sharp edged ridges, and its thin, closely set septa.

Family PORITIDÆ, (p. lv.)

Genus MICROSOLENA, (p. lvi.)

1. MICROSOLENA REGULARIS. Tab. XXV, figs. 6, 6*a*, 6*b*.

ALVEOPORA MICROSOLENA, *M'Coy*, Ann. of Nat. Hist., s. ii, v. ii, p. 419, 1848.

Corallum massive, subturbinate or lobulated, and more or less convex. The English

---

[1] See page 101.

[2] *C. irradians*, tab. xix, fig. 1., and *C. meandrinoides*, D'Orbigny, loc. cit.

specimen here figured is much weather worn, and has lost its epitheca as well as the edge of the trabicular septa, which assume the appearance of moniliform costal striæ. The sclærenchymatous nodules that constitute these styliform septal processes or trabiculæ are placed nearly at equal distances from each other in the same series and in the adjoining series, so that when they occupy a large surface they appear to be arranged in a very regular manner according to three straight lines : one almost vertical, and the two others oblique and crossing each other at right angles, (fig. 6*b*.) *Calices* quite superficial ; the fossula are not deep, but are well defined and placed at a considerable distance from each other. The septo-costal radia are numerous (about thirty or forty), very thin, broad, especially those that are placed perpendicularly to the edge of the corallum, almost equal in size, rather closely set, completely confluent, and formed of a series of nodular styliform processes as already stated. The specimen which is figured in this work, and belongs to Mr. Walton's rich collection, is about one inch high and two broad ; the corallites are about one and a half lines in diameter.

This fossil is found in the Great Oolite at Bradford Hills, and at Dunkerton ; Prof M'Coy states its having been met with at Minchinhampton.

Lamouroux, who established the genus *Microsolena* had very false ideas of its structure and zoological affinities. He supposed that the trabiculæ which constitute the septa were tubes bored in a common mass. M. Michelin recognised the resemblance between *Microsolena* and *Porites*, but placed the former in the genus *Alveopora* of Messrs. Quoy and Gaimard, from which it differs much. An attentive examination of various specimens of the *Microsolena porosa*, of Lamouroux, found near Caen, and of some other species, has enabled us to ascertain that the genus Microsolena must not be discarded but placed in the Family of the *Poritidæ*, near the genus *Coscinaræa*, from which it differs principally by the existence of a common epitheca, and by the septal trabiculæ being placed further apart. We also refer to this generical type some species of a somewhat dendroid form that M. Michelin placed in the genus *Alveopora*, and have been considered by M. D'Orbigny as constituting two new genera : *Dendraræa* and *Dactylaræa*.[1] M. D'Orbigny characterises the first of these divisions as *Dendriform Microsolena*, and the second as *Dendriform Synastræa*, but we have ascertained that the typical species of both present the same structure as *Microsolena*, and the insignificant differences which exist between massive, gibbose, or subdendroid forms, are not in our opinion of sufficient zoological value to be employed as characteristic of generical divisions.

Microsolena regularis resembles very much, by its general form, *Microsolena porosa*,[2] but differs from it essentially by its septa being much more numerous and closer set.

---

[1] Note sur des Pol. Foss., p. 11.
[2] Lamouroux, Exp. meth., tab. lxxiv., figs. 24, 25, 26.

Microsolena tuberosa,[1] *M. racemosa*,[2] *M. excelsa*,[3] and *M. incrustata*,[4] differ from it by the septa being thicker and the general form of the corallum being subdendroid.   As to *Microsolena irregularis*,[5] it appears to be an undeterminable specimen of some *Thamnastrea;* nor does the *Dactylastrea subramosa*[6] of M. D'Orbigny belong to this genus, being identical with our *Thamnastrea affinis*.[7]

2. MICROSOLENA EXCELSA.   Tab. XXV, fig. 5.

SIDERASTRÆA INCRUSTATA, *M'Coy*, Ann. and Mag. of Nat. Hist., s. ii, v. ii, p. 419, 1848.
(Not *Siderastrea incrustata*, Michelin, Icon., 1845.)

Corallum subdendroid, composed of erect cylindrical digitiform ramified branches. Basis covered with a thick, wrinkled, common epitheca, which forms also a few small zones at various heights up the branches.   The rest of the surface covered with *calices*, the centre of which is occupied by a well-defined but shallow fossula.   The corallites are crowded together, almost equally developed, and their *calices* are somewhat polygonal.   The *columella* appears to be papillose, but rudimentary.   In general, there are about twenty-four septa, and consequently three cycla, but sometimes a certain number of the tertiary ones are wanting. The septa are confluent, almost equally developed, rather closely set, thin, and bent or flexuous outwards.   They are composed of distinct *trabiculæ*, arranged much in the same manner as in the preceding species.

This fine fossil coral forms probably long tufts, but we have seen but fragments of about three inches long; the branches are six or seven lines in diameter, and the calices about half a line.   The specimen here described belongs to Mr. Walton's collection, and was found in the Great Oolite, near Bath.   Prof. M'Coy mentions its having been met with in the Great Oolite at Minchinhampton.

M. excelsa is very much like *M. incrustata*,[8] to which Prof. M'Coy referred it; but in the latter the epitheca is much more abundant and the calices are shallower.   *M. tuberosa*[9] is distinguished by its general form being massive and mammose, but not dendroid, and *M. ramosa*[10] by the septa being much thicker and less numerous.

The fossil described by Prof. M'Coy under the name of GONIOPORA RACEMOSA[10] appears to differ very little from *Microsolena excelsa;* it was found in the Great Oolite at Minchinhampton.

[1] *Alveopora tuberosa*, Michelin, Icon., tab. xxix, fig. 7.

[2] *Alveopora racemosa*, ibid., tab. xxix, fig. 6.

[3] Tab. xxix, fig. 5.

[4] *Alveopora incrustata*, Michelin, Icon., tab. xxix, fig. 8.

[5] D'Orbigny, Prodr., tab. i, p. 222.

[6] D'Orbigny, Prodr., tab. ii, p. 97.

[7] Milne Edwards and J. Haime, Ann. Sc. Nat., tab. xii, p. 158.

[8] Michelin, tab. xxix, fig. 8.   [9] Ib., tab. xxix, fig. 7.   [10] Ib., tab. xxix, fig. 6.

CHAPTER XI.

CORALS FROM THE INFERIOR OOLITE.

The corals found in the Inferior Oolite of England belong to twenty-seven species, seventeen of which have not as yet been met with on the continent. Most of these fossils (twenty-one species) belong to the family of *Astreidæ*; two species belong to the family of *Turbinolidæ*, and two to the family of *Fungidæ*; one appears to belong to the family of *Cyathophyllidæ*; we refer it, with some hesitation, to the genus *Zaphrentis*, and must particularly point out its existence here as being the last representative of that important family, which was so abundant in the older geological periods, and is almost exclusively characteristic of the Palæozoic Formations. Most of the corals here described have been seen only in the Inferior Oolite; but three species (*Stylina solida, Anabacia orbulites,* and *Comoseris vermicularis*) exist also in the Great Oolite. The principal localities from which these fossils were obtained, are Dundry, Bath, and Castle Cary in Somersetshire, Burton Bradstock in Dorsetshire, Wotton-under-edge and Crickley in Gloucestershire.

Family TURBINOLIDÆ, (p. xi.)

Genus DISCOCYATHUS, (p. xiii.)

DISCOCYATHUS EUDESI, Tab. XXIX, figs. 1, 1*a*, 1*b*.

> CYCLOLITES EUDESII, *Michelin*, Icon. Zooph., p. 8, tab. ii, fig. 2, 1840; (bad figure.)
> —   TRUNCATA, *Defrance*, MS. collection.
> DISCOCYATHUS EUDESII, *Milne Edwards* and *J. Haime*, Ann. des Sc. Nat., s. 3, vol. ix,
>   p. 297, tab. ix, fig. 7, 1848.
> —   —   *D'Orbigny*, Prod. de Paléont., vol. i, p. 291, 1850.

Corallum simple, discoid; its under surface flat or slightly concave, presenting a small central dimple, and a thick epitheca with circular wrinkles. *Calice* shallow, slightly depressed towards the centre. *Columella* lamellar, rather thin, free to a considerable distance from its upper end, and terminated by an entire edge. *Septa* straight, rather thin, very exsert exteriorly as well as upwards, and terminated by an arched, delicately-crenulated edge; they form four complete cycla, and an incomplete fifth cyclum in two or four of the systems, very rarely in all; those of the second cyclum almost as large as the primary ones; the others developed proportionally to the age of the cyclum to which they belong, except in the systems where a fifth cyclum exists, for there the septa of the fourth cyclum are nearly as broad as the tertiary ones. *Pali* twelve in number, well developed, equal in size, and always corresponding to the tertiary septa, the existence of the incomplete fifth cyclum not appearing to have any influence on their position. Diameter of the corallum eight or nine lines; height three lines.

We have not as yet been able to examine any well-preserved specimens of this fossil, and although we have ascertained the existence of characters that separate it generically from all other corals, we are not quite satisfied respecting its real zoological affinities. We have been induced to consider it as belonging to the family of *Turbinolidæ*, and indeed it closely resembles *Cyathininæ* by the great development of the pali; but in one specimen we have perceived some appearance of dissepiments, and consequently, when better known, the genus *Discocyathus* may prove to be an *Astreida*.

D. Eudesi is the only specimen belonging to this genus. It was figured for the first time by M. Michelin, but from specimens so deeply imbedded in the matrix, that no satisfactory idea of their form could be obtained. In our Monograph of the *Turbinolidæ* we have given a new figure of the same species taken from a fossil belonging to the collection of M. Defrance at Sceaux, but it must be noted that this species which shows the calice very well, presents a complete fifth cyclum, a circumstance which does not usually occur.

Discocyathus Eudesi is found on the Inferior Oolite in Dorsetshire, at Burton Bradstock and Greenland, and, in France, at Bayeux and Port en Bessin. The British specimens here described belong to the collections of Sir H. De la Beche and Mr. Walton.

Genus TROCHOCYATHUS, (p. xiv.)

TROCHOCYATHUS MAGNEVILLIANUS. Tab. XXVI, figs. 1, 1*a*, 1*b*.

TURBINOLIA MAGNEVILLIANA, *Michelin*, Icon. Zooph., p. 8, tab. ii, fig. 2, 1840.
— (TROCHOCYATHUS?) MAGNEVILLIANA, *Milne Edwards* and *J. Haime*, Monogr. des Turbinolides, Ann. des Sc. Nat. s. 3, vol. ix, p. 335, 1848.
APLOCYATHUS MAGNEVILLIANA, *D'Orbigny*, Prod. de Paléont., vol. i, p. 291, 1850.
TROCHOCYATHUS MAGNEVILLIANA, *Milne Edwards* and *J. Haime*, Polyp. Palæoz., &c., p. 23, 1851.

Corallum simple, free, hemispherical; its basal surface presenting a central dimple surrounded with a small elevated edge, and a smooth circle. *Costæ* straight, granulose, striated transversely, almost flat, but somewhat thicker and slightly prominent near the calice, and rather unequal in size alternately, or from four to six. *Septa* exsert, somewhat thickened in the middle, and forming four complete cycla; the secondary ones almost as large as those of the first cyclum. *Columella* papillose and well characterised. *Pali* not very thick ; the primary and secondary ones broader than those of the third order. Height two lines. Diameter three or four lines.

The calice was imbedded in extraneous matter in all the specimens of this species which we have examined, and it is only of late that we have been able to make transverse sections of one of these fossils, and thus to ascertain the existence of the characteristic features of the genus *Trochocyathus*. The figure 1*b*, represents one of these sections restored, and shows that the columella and the pali are disposed in the same way as in the tertiary species belonging to the same generical group; but this oolitic fossil differs from all these by the appearance of its basis, and may for that reason be placed in a peculiar section of the genus which we have recently established under the name of *Trochocyathi liberrimi*.[1] Independently of that character, *Trochocyathus Magnevillianus* resembles much by its general form *T. Harveyanus*[2] and *T. obesus*,[3] but differs from them by the structure of the costæ.

This species was first described by M. Michelin, but very incompletely. M. D'Orbigny has placed it in his genus *Aplocyathus*, which, as we have already stated, is composed of our *Trochocyathus* with a circular calice, and is not, in our opinion, admissible.[4]

Trochocyathus Magnevillianus belongs to the jurassic period, and it must be remarked that there is only one more well characterised representation of the same genus in the oolitic formations, whereas they are common in more recent strata. *T. Michelini*, which is found in deposits belonging to the same great geological period, is of a subdiscoidal form, and its basis is not cicatrized, as in *T. Magnevillianus ;* it has as yet been found only in the Great Oolite, and the latter in the Inferior Oolite, but we are inclined to consider it as being referable to the same genus, some ill-preserved corals that are of a more ancient date, and have been found in the lowest formations of the jurassic group and in the Lias.

This species has been found at Burton Bradstock in Dorsetshire, by Mr. Walton, and at Bridport in the same county, by the Members of the Geological Survey. In France it has also been found in the Inferior Oolite, near Bayeux, in Normandy.

[1] See our Mémoire sur les Polyp. Palæoz., &c.

[2] Tab. xi, fig. 4.

[3] Milne Edwards and J. Haime, Monogr. des Turbinolides, Ann. des Sc. Nat., s. 3, vol. ix, tab. x, fig. 2.

[4] See Part I, page 65.

Family ASTREIDÆ, (p. xxiii.)

Genus AXOSMILIA, (p. xxvi.)

AXOSMILIA WRIGHTI.   Tab. XXVII, fig. 6.

Corallum simple, having the form of a very elongated cone, very narrow at its under end, straight or but slightly curved, presenting circular accretion swellings, and covered with an epitheca which appears to extend to the calicular edge. *Calice* circular. *Septa* forming four complete cycla; straight, thin towards the centre of the calice, appearing to be delicately granulose laterally, unequal in size according to their relative age, and not closely set.   Height of the corallum about one inch.   Diameter of the calice four or five lines.

Found at Dundry and at Cheltenham, in the Trigonia beds, by Dr. Wright.

It is not without much uncertainty that we refer this oolitic coral to our genus *Axosmilia*, for in all the specimens which we have seen, the calice was so imbedded in the stone, that we have not been able to observe its most essential characters, such as the styliform columella; by the form of the septa we may infer that their edge was entire, and the calice deep, as in *Axosmilia*, which this fossil resembles also by its general aspect, more than it does *Montlivaltia*; but if the presumed characters do not in reality exist, it may belong to the latter genus.   At all events *A. Wrighti* differs from *Axosmilia extinctorum*,[1] and from *A. multiradiata*[2] by the number of the septa, and the equal development of all the septal systems, for in *A. multiradiata* there are five cycla, and in *A. extinctorum* only three complete ones, and the septa of the fourth cyclum exist only in one half of each system.

Genus STYLINA, (p. xxix.)

STYLINA SOLIDA.   (See page 105, and Tab. XXII, fig. 3.)

This fossil, which is met with in the Inferior Oolite near Bath, is also found in the Great Oolite, and has consequently been described in a preceding chapter of this Monograph.

---

[1] *Caryophyllia extinctorum*, Michelin. Iconogr., tab. ii, fig. 3*a*.

[2] Milne Edwards and J. Haime, Monogr. des Astreides, Ann. des Sc. Nat., s. 3, vol. x, p. 362.

Genus MONTLIVALTIA, (p. xxv.)

1. MONTLIVALTIA TROCHOIDES.   Tab. XXVI, figs. 2, 2*a*, 3, 3*a*, 10; and Tab. XXVII, figs. 2, 2*a*, 4.

MONTLIVALTIA CARYOPHYLLATA, *Bronn*, Leth. Geogn., tab. xvi, fig. 17, 1836.
—   TROCHOIDES, *Milne Edwards* and *J. Haime*, Ann. des Sc. Nat., s. 3, vol. x, p. 299, 1848.

Corallum simple, turbinate, rather tall, and in general straight, but varying much as to proportions.   Basis obtuse in some specimens; subpedicellate in others.   *Epitheca* thick, wrinkled, and extending to a short distance from the calicular edge.   *Calice* circular, or sometimes oval and shallow; the fossula small and circular, or somewhat oval.   No *columella*.   *Septa* forming in adult specimens five complete cycla, and often an incomplete sixth cyclum in one half of some of the systems; those of the second cyclum as large as the primary ones, and differing but little from the tertiary ones; those of the last cyclum very small.   All these septa are thin, closely set, straight, or nearly so, somewhat granulose laterally, and terminated by a delicately crenulated edge.   Height of the corallum in general about one inch and a half; and diameter one inch two or three lines.

Found in the Inferior Oolite at Charlcomb, by Mr. Walton.   A specimen of the same species, belonging to the collection of the Museum in Paris, is catalogued as having been found in Germany.

Montlivaltia trochoides much resembles the species for which this genus was established, the *M. caryophyllata;*[1] but it differs from it by the septa being thinner, and in general more numerous, and more especially by its epitheca extending almost to the edge of the calice; whereas in *M. caryophyllata* this mural tunic ends at a considerable distance below that margin.   It may be easily distinguished from *M. lens,*[2] *M. Delabechii,*[3] and *M. depressa,*[4] which are all of a discoidal form, by its being much taller than broad; from *M. Waterhousei*[5] and *M. regularis,*[6] by its basis not being regularly convex; and from *M. Smithi*[7] and *M. cupuliformis*[8] by not being fixed by a broad basal surface, and having thinner septa; and from *M. deltoides,*[9] *M. rudis,*[10] *M. cornucopiæ,*[11] *M. bilobata,*[12] and *M. irregularis,*[13] by the circular form of the calice, which in all the latter is more or less

[1] Lamouroux, Exposit. Method. des Polyp., tab. lxxix, figs. 8, 9, 10.
[2] Tab. xxvi, fig. 7.   [3] Tab. xxvi, fig. 5.   [4] Tab. xxix, fig. 5.   [5] Tab. xxvii, fig. 7.
[6] D'Orbigny, Prod., vol. i, p. 346.   [7] Tab. xxi, fig. 1.   [8] Tab. xxvii, fig. 1.
[9] Milne Edwards and Haime, Ann. des Sc. Nat., s. 3, vol. x, tab. vi, fig. 3.
[10] *Cyathophyllum rude*, Sowerby, Geol. Trans., s. 2, vol. iii, tab. xxxvii, fig. 2.
[11] Milne Edwards and J. Haime, loc. cit., p. 298.
[12] *Turbinolia bilobata*, Michelin, Iconogr., tab. lxii, fig. 1, (not tab. lxi, fig. 7.)
[13] Milne Edwards and J. Haime, loc. cit., p. 298.

compressed and oval. *M. trochoides* has more general resemblance to a certain number of species of a somewhat conical form, which have five or six cycla of septa, such as *M. dispar*,[1] *M. tenuilamellosa*, &c. The first of these fossils, however, differs from it by having a less regular form, a shallower fossula, and fewer septa; *M. tenuilamellosa*[2] is shorter, its epitheca does not extend so high, and its septa are thinner and curved inwardly; *M. Wrighti*[3] has thicker and more equally developed septa; *M. Lotharinga*[4] has its basis arched and often inflated; *M. Goldfussiana*[5] is always adherent; *M. Guerangeri*[6] has a thicker but very incomplete epitheca; and in *M. Beaumonti*[7] the septa are more numerous and fluted laterally.

2. MONTLIVALTIA TENUILAMELLOSA. Tab. XXVI, figs. 11, 11*a*.

Corallum subturbinate, short, broader than high, somewhat inflated, straight, or very slightly subpedicellate. *Epitheca* very thick, extending over only two thirds of the height of the corallum, and presenting strong circular wrinkles. *Calice* circular or nearly so, and slightly convex; fossula oval and very deep. *Septa* very thin, terminated by a very delicately denticulated edge, almost smooth laterally, and forming six complete cycla; those of the first three cycla almost equally developed; many of them somewhat bent towards the centre of the visceral chamber. Height one inch. Diameter two inches.

This fossil was found in a bed of Fullers Earth at Dunkerton, and at English Batch, by Mr. Walton.

The general form of this coral renders it easy to be distinguished from most of the species of the genus *Montlivaltia*; those which it resembles most, are *M. ponderosa*,[8] *M. brevissima*,[9] *M. Waterhousei*,[10] and *M. regularis*,[11] but the first of these fossils is much more oblong, has a thin and almost smooth epitheca, reaching almost to the calicular margin, and straight septa; in the second the epitheca is also almost complete and smooth, and there are only five cycla of septa; in the last two the corallum has much the same in form towards its basis, but becomes cylindrical higher up, and the septa are stronger and more numerous.

---

[1] Tab. xiv, fig. 2.   [2] Tab. xxvi, fig. 11.   [3] Tab. xxvi, fig. 12.
[4] Milne Edwards and J. Haime, loc. cit., p. 294.   [5] Ib., loc. cit.
[6] Ib., op. cit., p. 293.   [7] Ib., Ann. des Sc. Nat., s. 3, vol. xi, p. 243.
[8] *Thecophyllia ponderosa*, Milne Edwards and J. Haime, Ann. des Sc. Nat., s. 3, vol. xi, p. 242.
[9] Milne Edwards and J. Haime, op. cit., vol. x, p. 293.   [10] Tab. xxvii, fig. 7.
[11] D'Orbigny, Prod. de Paléont., vol. i, p. 349.

3. MONTLIVALTIA STUTCHBURYI.   Tab. XXVII, figs. 3, 3a, and 5.

Corallum turbinate, rather tall, subpedicellate, straight, or slightly bent. *Epitheca* very thick, reaching almost to the calicular margin, and presenting very strong transverse wrinkles or folds. *Calice* circular. *Columella* rudimentary. *Septa* rather thick, straight, of unequal size, and forming four complete cycla. Height of one of the specimens here described, one inch; the other, although broken at both ends, is larger, and was probably nearly two inches long. Diameter of its calice, twelve lines.

These fossils belong to the Bristol Museum, and are entered in the catalogue of that establishment as having been found at Nunney, near Frome.

In most species of this genus the septa are much more numerous than in this species. In *M. detrita*[1] and *M. inæqualis*,[2] which have also only four cycla, the basis of the corallum is broadly adherent, and the septa are very thick: in *M. striatulata*[3] the basis is also widely adherent, and the septa are very thin : *M. sycodes*[4] resembles it most, but its epitheca does not extend so high up, and the septal systems are unequally developed. In two other species *M. lens*[5] and *M. depressa*,[6] where the fifth cyclum exists, the septa belonging to it are sometimes rudimentary; but the discoidal form of these corals distinguishes them at first sight from *M. Stutchburyi*.

4. MONTLIVALTIA WRIGHTI.   Tab. XXVI, figs. 12, 12a.

The unique specimen of this species is in a very bad state of preservation, and has completely lost its epitheca, but is remarkable by its general form; it is regularly turbinate, and almost twice as broad as it is high. *Calice* almost circular; fossula appearing to be somewhat oblong. *Septa* about seventy in number, well developed, but unequal in size alternately; straight and thick, especially towards the wall; no appearance of any rudimentary septa between the large preceding ones. Height of the corallum one inch; diameter of the calice two inches.

We found this fossil in the Inferior Oolite at Crickley, near Cheltenham; and have placed it in the collection of the Paris Museum. We dedicate the species to Dr. Wright, of Cheltenham, who has kindly communicated to us some interesting fossils from that locality.

M. Wrighti differs from most *Montlivaltiæ* by its general form ; it much resembles

---

[1] *Anthophyllum detritum*, Michelin, Iconogr., tab. x, fig. 1.
[2] *Anthophyllum inæquale*, Michelin, Iconogr., tab. l, fig. 4.
[3] *Caryophyllia striatulata*, Michelin, Iconogr., tab. l, fig. 9.
[4] Milne Edwards and J. Haime, op. cit., vol. x, p. 299.
[5] Tab. xxvi, fig. 7.   [6] Tab. xxix, fig. 9.

M. Smithi, but in this latter species the septa are more unequal, and the basis is broadly adherent; in all others the septa are either more numerous and thinner, or, on the contrary, less numerous and thicker, and in no other have we met with any large septa belonging to the fifth cyclum.

5. MONTLIVALTIA CUPULIFORMIS. Tab. XXVII, figs. 1, 1*a*.

Corallum tall, straight, adherent by a very large basis, above which it is slightly constricted, but soon becomes almost cylindrical. *Epitheca* extending high up towards the calicular margin, but almost entirely worn away in the specimen here described. *Calice* circular; central fossula very small, somewhat oblong. *Septa* rather thin, straight, and forming four well-developed cycla, and a rudimentary fifth cyclum; those of the second cyclum as large as the primary ones. Height of the corallum one inch and a half. Diameter of the calice one inch.

This fossil, of which we have seen but one specimen, was found in the Inferior Oolite at Dundry, by Mr. Pratt.

Most of the various species of *Montlivaltia*, which, like *M. cupuliformis*, are adherent by a broad basis, differ from it by their general form as well as by characters derived from the septa. Thus *M. Smithi*[1] and *M. pateriformis*[2] are broader than high, and have thicker septa. *M. detrita*[3] and *M. inæqualis*[4] are entirely cylindrical; without any constriction near the basis, and have very thick septa. *M. striatulata*[5] is a very small coral with very delicate septa; *M. subtruncata*[6] has six cycla of septa; and *M. Lesueuri*[7] seven cycla. *M. Goldfussiana*[8] resembles most the above-described fossil by its general form, but is more turbinate, and presents a greater number of septa.

6. MONTLIVALTIA DELABECHII. Tab. XXVI, figs. 5, 5*a*, 5*b*.

MONTLIVALTIA DECIPIENS, *M'Coy*, Ann. and Mag. of Nat. Hist., s. 2, vol. ii, p. 419, 1848; (not *Anthophyllum decipiens*, Goldfuss.)

Corallum free, very short, almost discoidal, circular; its inferior surface slightly concave, and presenting a small central cicatrix, indicating its original point of adhesion. *Epitheca*

---

[1] Tab. xxi, fig. 1.

[2] *Anthophyllum patiriforme*, Michelin, Iconogr. Zooph., tab. l, fig. 3.

[3] *Anthophyllum detritum*, Michelin, Iconogr., tab. x, fig. 1.

[4] *Anthophyllum inequale*, Michelin, op. cit., tab. l, fig. 4.

[5] *Caryophyllia striatulata*, Michelin, op. cit., tab. l, fig. 9.

[6] *Lasmophyllia subtruncata*, D'Orbigny, Prod. de Paléont., vol. i, p. 321.

[7] Milne Edwards and J. Haime, Ann. des Sc. Nat., s. 3, vol. x, p. 257.

[8] Milne Edwards and J. Haime, loc. cit., p. 294.

strong, presenting concentric folds or wrinkles, and ceasing at a considerable distance from the calicular margin. *Calice* somewhat convex, with a small shallow central fossula. *Columella* rudimentary, and formed by a certain number of denticules arising from the inner edge of the septa. Five cycla of septa; but the last cyclum not developed on one side of some of the systems. The *Septa* straight, closely set, exsert, terminated by a well-denticulated edge, and slightly striate on their lateral surfaces, but very feebly granulated; the primary ones rather thick, especially towards the middle; the secondary ones almost as large as those of the first cyclum; the others unequally developed according to the orders to which they belong.

The largest specimens which we have seen were almost one inch in diameter, and about four lines high.

This fossil is found in the Inferior Oolite in France, as well as in England. Specimens from Castle Cary, Somersetshire, exist in the collections of the Museum of Practical Geology and of the Paris Museum; specimens found at Dundry have been communicated to us by Mr. Bowerbank and Mr. Pratt; others, in Mr. Walton's collection, were found at Silcombe, Hawkesbury, and Camdown, in Somersetshire, at West Swillets in Dorsetshire, and at Sudbury in Gloucestershire. M. Terquem, of Metz, has met with the same species in the environs of that city.

Montlivaltia Delabechii has a very peculiar form, and may be considered as intermediate between the species which are cylindrical with a convex basis, such as *M. Waterhousei*[1] and *M. regularis*;[2] and those which are quite discoidal, as *M. depressa*[3] and *M. lens*.[4] In these last two the wall is entirely horizontal, as in a Cyclolite; and the epitheca does not extend over any part of the exterior edge of the septa; whereas in *M. Delabechii* the epitheca ascends from the basis some way up the sides of the corallum; in that respect it resembles the inferior portion of a *Montlivaltia Waterhousei* or a *M. regularis*, and it might be considered as a variety of one of those specimens, were it not for its septa being more numerous and straight. It also resembles very much *M. cyclolitoides*,[5] but is taller, and is also characterised by the strong striæ which exist on the lateral surfaces of the septa. *M. decipiens*,[6] to which Prof. M'Coy refers this fossil, differs from it by its form, which is much more conical, by its being broadly adherent, and not showing any vestiges of a columella.

7. Montlivaltia lens. Tab. XXVI, figs. 7, 7*a*, 7*b*, 7*c*; fig. 8.

Corallum discoidal, very short, and much resembling a *Cyclolite* by its form. Wall quite horizontal, or somewhat concave towards the centre of the basis, and covered with a

---

[1] Tab. xxvii, fig. 7.   [2] D'Orbigny, Prod. de Paléont., vol. i, p. 316.
[3] Tab. xxix, fig. 5.   [4] Tab. xxvi, fig. 7.
[5] *Thecophyllia cyclolitoides*, Milne Edwards and J. Haime, Ann. des Sc. Nat., s. 3, vol. xi, p. 242.
[6] *Anthophyllum decipiens*, Goldfuss, Petref. Germ., vol. i, tab. lxv, fig. 3.

thick epitheca, presenting concentric folds or wrinkles, and a very small central cicatrix. *Calice* circular, and somewhat convex; fossula shallow and oblong. *Septa* straight, exsert externally, unequally developed, terminated by a strongly denticulated edge, and forming five complete cycla, four of which are well developed; those of the second cyclum differ but little from the primary ones, and are thicker than the others; those of the fifth cyclum rudimentary. Some denticulations larger than the others, and situated at the inner angle of the principal *septa* produce the appearance of small pali. Diameter seven lines; height three lines.

Found by M. Walton in the Inferior Oolite at Charlcomb and English Batch.

This species, together with *M. depressa*,[1] and, in all probability, *M. numismalis*,[2] constitute in the genus *Montlivaltia* a small section, characterised by the discoidal form of the *corallum*, and the completely horizontal position of the wall. At first sight they may easily be considered as referable to another family, and placed in the genus *Cyclolites;* but, on a closer examination, it will be found that they are not provided with synapticulæ, as is the case with all Fungidæ, and do not differ in structure from true *Astreidæ*. This discoidal form alone distinguishes these fossils from the other species of *Montlivaltia;* but that peculiarity does not appear to us of sufficient value to authorise the establishment of a separate genus.

8. MONTLIVALTIA DEPRESSA. Tab. XXIX, figs. 5, 5*a*.

Corallum discoid, much resembling a *Cyclolite* by its general form; its under surface slightly concave, and covered with a thick epitheca; its upper surface convex, fossula circular, and very superficial. *Septa* straight, thin, unequal, and forming four cycla; those of the first cyclum extending almost to the centre of the calice. Diameter one inch; height three lines.

Found at Wotton-under-Edge, by M. Walton.

We have seen but one very ill-preserved specimen of this fossil, and have not been able to ascertain all its characters in a satisfactory manner. It appears, however, to belong to the section of the discoidal *Montlivaltia*, and may easily be distinguished from the two other species of the same lenticular form by the disposition of its *septa*, which are much less numerous than in *M. numismalis*,[3] and much thinner and less strongly denticulated than in *M. lens*.[4]

---

[1] Tab. xxix, fig. 5.

[2] *Thecophyllia numismalis*, D'Orbigny, Prod. de Paléont., vol. i, p. 321.

[3] *Thecophyllia numismalis*, D'Orbigny, Prod., vol. i, p. 321.

[4] Tab. xxvi, fig. 7.

Genus THECOSMILIA, (p. xvi.)

THECOSMILIA GREGARIA. Tab. XXVIII, figs. 1, 1*a*.

MONTLIVALTIA GREGARIA, *M'Coy*, Ann. and Mag. of Nat. Hist., s. 2, vol. ii, p. 419, 1848.

Corallum composite, not very tall, lobate, and formed of a thick common trunk, from which ascend (diverging in different directions) a certain number of large fasciculi of corallites, enveloped in a common strong epitheca. The *calices*, when free all round, are circular; but those belonging to the same group are in general closely united along the line of contact, and then become more or less polygonal : their diameter varies much in the different corallites belonging to the same compound corallum, as well as in different specimens. In some we have been able to distinguish four complete cycla of *septa* and regularly-developed systems; but in most cases, as is usual with fissiparous Corals, it is difficult to recognise the divers orders to which these radiate laminæ are referable. The *septa* are very exsert, not closely set, thicker externally than towards the centre of the calice, in general straight, and terminated by a regularly denticulated edge. The *dissepiments* are numerous. The large specimen figured in this Monograph is about four inches high and six inches broad; the calices are in general about one inch in diameter; but we have seen some that were one and a half inch in diameter.

This species has been found at Dundry, Leckhampton, and Crickley, near Cheltenham; specimens are in the collections of Mr. Walton, Dr. Wright, Mr. Pratt, and the Museums of Cambridge and Paris.

Thecosmilia gregaria is remarkable for the manner in which the corallites, arising from a common parent, remain for a long time united together after they have become completely constituted as individuals. This character is alone sufficient to distinguish it from *T. trichotoma*,[1] *T. annularis*,[2] *T. lobata*,[3] and *T. ramosa*.[4] *T. Konincki*[5] differs from it by having an additional cyclum of septa; and in *T. Terquemi*[6] the septa are thicker, and the corallites become circumscribed much more tardily. As to the other species, which appear to be referable to the same genus,[7] they have not been characterised with sufficient minuteness to enable us to point out the structural peculiarities which may distinguish them

---

[1] *Lithodendron trichotomum*, Goldfuss, Petref. Germ., vol. i, tab. xiii, fig. 6.

[2] Tab. xiii, fig. 1.

[3] *Lobophyllia lobata*, Michelin, Icon., tab. lxvii, fig. 3.

[4] D'Orbigny, Prod. de Paléont., tab. i, p. 292.

[5] Milne Edwards and J. Haime, Ann. des Sc. Nat., s. 3, vol. 10, p. 272.

[6] *Lobophyllia Terquemi*, Michelin, Icon., tab. iv, fig. 6.

[7] See the list of species given in the first part of our ' Mémoire sur les Polyp. des Terr. Palæoz.,' &c.

from the above-described fossil; and we must only add, that most of them have been mentioned by M. D'Orbigny, under the generic denominations of *Amblophyllia* and *Lasmophyllia*.[1]

Genus LATOMEANDRA, (p. xxxiv.)

1. LATOMEANDRA FLEMINGI.   Tab. XXVII, figs. 9, 9a.

Corallum composite, massive, rather tall; its upper surface horizontal, or slightly convex, and presenting large deep calices, which are sometimes completely circumscribed, but are in general confluent in one direction, so as to form furrows of unequal length, but rarely long. The mural ridges situated between these furrows are always simple, not much elevated, and terminated by a well-defined edge, which, when slightly worn down, shows very distinctly the walls themselves. The *calices* are rather deep, especially in the adult corallites, and the young individuals are formed at a considerable distance from the centre of the visceral chamber of the parent. The *septa* are very thin, closely set, narrow towards the top, delicately crenulated along their edge, straight or slightly curved, and unequal in size alternately; in the larger calices there are about sixty of these radiate laminæ; they become almost parallel in the furrows; but even there the different corallites are in general distinct, and have each a separate fossula. Diameter of the calices, or breadth of the furrows in general, 4 lines; depth, 2 lines.

Found in the Inferior Oolite at Crickley, near Cheltenham, by Dr. Wright.

We are inclined to think that the fossil coral mentioned by Mr. Conybeare and Mr. W. Phillips,[2] as being intermediate between *Astrea* and *Meandrina*, and as having been found in the Inferior Oolite, may belong to this species.

The genus *Latomeandra* was established a few years ago by M. D'Orbigny,[3] but that palæontologist restricted the group to those species which assume a somewhat dendroid form, and constitute series which remain free laterally; those which have the same structure, but are of a massive form, are referred to our genus *Oulophyllia*, or to three new generical divisions which he designates under the names of *Axophyllia*, *Microphyllia*, and *Comophyllia*, but which do not appear to be founded on characters of sufficient value. The specimen from which the definition of the genus *Axophyllia* was taken presents, in some calices, the appearance of a styliform columella, but that is due to an accident of fossilization, and is produced by the presence of a small calcareous concretion in the fossula. In the genus *Comophyllia* the calices appear to be very shallow, a peculiarity of small importance, and in the genus

---

[1] D'Orbigny, Prod. de Paléont., vols. i, ii.

[2] Outlines of the Geol. of England, p. 245, 1822.

[3] Note sur des Polypiers Fossiles, p. 8, 1849.

Microphyllia the furrows are deeper, and the costæ, according to M. D'Orbigny, are dichotomous.[1] As to the differences in the general form of the compound corallum, we have found every intermediate degree between the massive astreiform species, and the sub-dendroid species, without seeing any difference in the structure of the corallites, and all these modifications exist sometimes in different parts of the same specimen. We must therefore conclude that, contrary to what is the case in most of the Astreidæ, the greater or lesser degree of approximation of the corallites or their mode of cementation, is here a circumstance of no zoological value, and must not be employed as a basis for generical divisions. We consequently do not deem it advisable to adopt the three genera above mentioned, and prefer placing all these species in the genus *Latomeandra*. The group thus formed is remarkable for the manner in which the submarginal calicular gemmation takes place, and by its costulated, naked walls; this last-mentioned character distinguishes it from the genus *Isastrea*, which resembles it much by the structure of the corallites, but in which the calices are circumscribed.

Latomeandra Flemingi differs from most species of the same genus by its septa being very thin and very numerous. In *L. corrugata* they are, however, even more numerous, and become frequently adherent together, a disposition which is not met with in the above-described fossil.

2. LATOMEANDRA DAVIDSONI. Tab. XXVII, figs. 10, 10*a*.

Corallum composite, massive, and very convex. Intercalicular ridges simple and not much elevated. Furrows short, shallow, and containing a few very distinct calicular centres. *Septa* rather closely set, slightly thick outwards, irregularly unequal in size, generally curved and delicately crenulated at their edge. Breadth of the calices about two lines.

This fossil belongs to Dr. Wright's collection, and was found in the Inferior Oolite at Crickley, near Cheltenham.

L. Davidsoni may be easily distinguished from the other species of the same genus by the small number and the thickness of its septa. The species which it resembles most are the one described above[2] and *L. Meriani*,[3] but in these the septa are at least as thick towards the centre of the calice as outwards, and the calices are shallower.

[1] Op. cit., p. 8.

[2] *Latomeandra Flemingi*, tab. xxvii, fig. 9.

[3] *Comophyllia elegans*, D'Orbigny, Prod. de Paléont., vol. ii, p. 40; *Latomeandra Meriani*, Milne Edwards and J. Haime, Polyp. des Terr. Palæoz., &c., p. 86.

Genus ISASTREA.

1. ISASTREA RICHARDSONI. Tab. XXIX, figs. 1, 1 *a*.

Corallum massive, flat, or slightly gibbose. *Calices* polygonal, very unequal in size, shallow, and separated by a strong single wall. Fossula distinct; no appearance of a *columella*. *Septa* rather thin, often somewhat curved, unequal in size, and forming three cycla; in one or two of the systems, those of the last cyclum are sometimes deficient, and in other cases a few septa belonging to a fourth cyclum are seen; in general, the six primary ones are much larger than the others, and become thicker near their inner edge, but sometimes the secondary ones are almost as much developed in one of the systems. Diameter of the calices in general one line and a half, the large ones two lines.

This fossil was found in the Inferior Oolite at Dundry, by the Rev. B. Richardson, and presented by that gentleman to the cabinet of the Geological Society of London. We are inclined to refer to the same species a coral found at Beachencliff by Mr. Walton.

By its general aspect *Isastrea Richardsoni* resembles *I. limitata*[1] and *I. explanulata*,[2] but it may be easily distinguished from them as well as from most species of the same genus, by the thickness of the principal septa near their inner edge; this character is also met with in *I. Munsterana*,[3] but in the latter the calices are much larger, and the septa are more numerous and closer set.

2. ISASTREA TENUISTRIATA. Tab. XXX, figs. 1, 1*a*.

ASTREA TENUISTRIATA, *M'Coy*, Ann. and Mag. of Nat. Hist., s. 2, vol. ii, p. 400, 1848.

Corallum massive, terminated by an almost flat surface. *Calices* not very unequal in size, and shallow. Walls not well developed. *Septa*, about seventy-two in number, thin, closely set, straight, or slightly curved, and rather unequal in size from four to four, or especially alternately; dissepiments rather closely set. Breadth of the calice half an inch or more; depth one line.

The specimen here described was kindly communicated to us by Dr. Wright, of Cheltenham, and was found by that naturalist in the Inferior Oolite at Crickley; there is another specimen from Dundry in the Cambridge Museum.

I. tenuistriata differs from all the other species of the same genus by the great number of its septa.

---

[1] Tab. xxiv, figs. 4, 9.   [2] Tab. xxiv, fig. 3.
[3] *Prionastrea Munsterana*, Milne Edwards and J. Haime, Ann. des Sc. Nat., s. 3, vol. xii, p. 136.

3. ISASTREA LONSDALII.

All the specimens of this species which we have seen were in so bad a state of preservation, that we did not consider them worthy of being figured in this Monograph. The *calices* are not very unequal in size, and rather shallow; the *septa* are thin, straight, and form four cycla; those of the first two cycla differ but little in size; the last cyclum is more or less incomplete; great diagonal of the calices two lines and a half.

This fossil was found in the Inferior Oolite at the foot of Lansdown, near Bath, and was given to the Geological Society of London by Mr. Pratt.

Some casts found in the Inferior Oolite at Charlcomb and at Dundry, appear to belong to corals of this genus; they show very unequally developed prismatic calices, but cannot be characterised specifically.

Genus THAMNASTREA, (p. xliii.)

1. THAMNASTREA DEFRANCIANA. Tab. XXIX, figs. 3, 3*a*, 3*b*; and 4, 4*a*, 4*b*.

ASTREA DEFRANCIANA, *Michelin,* Icon. Zooph., p. 9, tab. ii, fig. 1, 1840.
SYNASTREA DEFRANCIANA, *Milne Edwards* and *J. Haime,* Ann. des Sc. Nat., s. 3, vol. xii, p. 153, 1849.
— — *D'Orbigny,* Prod. de Paléont., tab. i, p. 292, 1850.
THAMNASTREA DEFRANCIANA, *Milne Edwards* and *J. Haime,* Polyp. des Terr. Palæoz., &c., p. 110, 1851.

Corallum in general thin, discoid or somewhat turbinate; its common basal plate covered with a complete epitheca strongly wrinkled; its upper surface almost flat, in some slightly convex, in others concave. *Calices* superficial, arranged in concentric lines, rather irregular, and having a very small central fossula. *Columella* quite rudimentary. *Septa* very closely set, very thin, rather flexuous, and in general more developed in the direction of the radii of the compound corallum, than in the opposite direction, completely confluent, and presenting very closely set, regular crenulations at their upper edge. In general they form three complete cycla; but in some individuals the third cyclum is rather incomplete, and in others rudiments of a fourth cyclum appear in some of the systems. Breadth of the calices one line and a half.

When the calices are worn down, the polygonal *walls* become very distinct (fig. 3*b*), and this coral then assumes an appearance very similar to that of a fossil, much weatherworn, that was found at Dundry by our friend Mr. Bowerbank, and is figured in our XXIXth Plate under the number 4; the size of the calices is not quite the same, and we at first sight considered them as belonging to two distinct species, but on closer examination that no longer appeared to be the case.

We are also inclined to think that the *Astrea helianthoïdes* of Prof. M'Coy[1] is a worn specimen of the same species, and that the *Agaricia elegans*[2] of that author may be a young individual of this *Thamnastrea*.

The above-described fossil is found in the Inferior Oolite at Dundry, and in France near Bayeux. Specimens are in the collections of the Geological Society, Mr. Bowerbank, Mr. Walton, and Mr. Pratt.

Thamnastrea Defranciana is remarkable for its general form, which resembles that of certain sponges, and for the tenuity and close approximation of its septa. It is very nearly allied to a fossil of the cretaceous formations, *Thamnastrea conferta*,[3] but in the latter the columella is more developed, and the septa more flexuous. Some other species, *T. tenuissima*[4] and *T. velamentosa*,[5] for example, also bear great resemblance to *T. Defranciana*, but differ from it by having much smaller calices, thicker septa, and a more developed columella.

2. THAMNASTREA TERQUEMI. Tab. XXX, figs. 2, 2a, 2b.

Corallum in general thin, circular, and adherent at the centre of its basis, but sometimes appearing to have been quite free; its common basal plate covered with a thick epitheca, presenting concentric folds or wrinkles; its upper surface flat or slightly convex. *Calices* quite superficial and rather closely set. *Columella* small. *Septa* from twelve to sixteen in number, unequal alternately, irregularly confluent, often geniculate and grossly granulated laterally. Breadth of the calices two lines. In one specimen the corallum was made up of superposed layers arranged obliquely.

This species was found by Dr. Wright in the Inferior Oolite near Cheltenham, and by M. Terquem at St. Quentin, near Metz.

T. Terquemi differs from most species of the same genus, by the irregular and rude appearance of the septa; which are thick, but not as much so as in *T. Lennisi*[6] and *T. Belgica*.[7] The same character distinguishes it also from *T. Mettensis*,[8] which, in other respects, it resembles very much.

---

[1] Ann. and Mag. of Nat. Hist., s. 2, vol. ii, p. 401.

[2] Op. cit., p. 418.

[3] *Synastrea conferta*, Milne Edwards and J. Haime, Ann. des Sc. Nat., s. 3, vol. xii. p. 190.

[4] *Synastrea tenuissima*, Milne Edwards and J. Haime, loc. cit., p. 191.

[5] *Astrea velamentosa*, Goldfuss, Petref. Germ., vol. i, tab. xxiii, fig. 4.

[6] *Astrea Lennisii*, Roemer, Verst. des Norddeut. Kreid., tab. xvi, fig. xxvi.

[7] Milne Edwards and J. Haime, Polyp. Foss. des Terr. Palæoz., &c., p. 100.

[8] See tab. xxx, fig. 3.

3. THAMNASTREA METTENSIS. Tab. XXX, figs. 3, 3a.

Corallum explanate; its upper surface flat or slightly convex; and its common basal plate covered with an epitheca presenting concentric wrinkles. *Calices* superficial, rather closely set, and forming, near the edge of the compound corallum, concentric series in each of which the corallites are more approximated than in the contrary direction. *Columella* rudimentary. *Septa* closely set, varying in thickness alternately, and not much geniculated; between sixteen and twenty-four round each fossula; the tertiary ones often become united to the secondary ones along their inner edge. Breadth of the calices two lines.

Found in the Inferior Oolite at Crickley, near Cheltenham, by Dr. Wright; and at St. Quentin, near Metz, by M. Terquem.

In this fossil the columella is quite rudimentary, and appears even to be quite deficient in many of the corallites. *T. Mettensis* differs in that respect from most species of *Thamnastrea*, in which a papillose columella is usually visible in the centre of the calice. In *T. Defranciana*[1] the columella is also very obscurely defined, but the septa are thinner than in the above-described species.

4. THAMNASTREA FUNGIFORMIS. Tab. XXX, figs. 4, 4a.

Corallum fungiform, pedunculate, and terminated by a convex surface. *Calices* somewhat unequal in size; the largest rather prominent. *Septa* about fifty in number, somewhat unequally developed alternately, thin, closely set, and irregularly denticulated at their upper edge. Breadth of the calices from two to three lines.

Found in the Inferior Oolite at Charlcomb, by Mr. Walton.

We have seen but a few ill-preserved specimens of this species, which appears to be very nearly allied to *T. arachnoides*,[3] but differs from it by its septa being more numerous and more deeply denticulated.

5. THAMNASTREA M'COYI. Tab. XXIX, figs. 2, 2a.

We have not been able to refer to any known species this fossil, of which we have seen but a single specimen in a very bad state of preservation, and much worn; its walls are polygonal, and circumscribed; calices of unequal size, in the centre of which a columella appears to have existed. The *septa*, from twenty to twenty-four in number, are unequally

---

[1] Tab. xxix, fig. 3.   [2] Tab. xxx, fig. 2.   [3] Tab. xvii, fig. 1.

developed alternately; somewhat flexuous and rather thickened externally. Great diagonal of the calices, about a line.

This fossil was found in the Inferior Oolite at Combdown, by Mr. Pratt, and given to the Geological Society by that gentleman.

Family FUNGIDÆ, (p. xlv.)

Genus ANABACIA, (p. xlvii.)

ANABACIA HEMISPHERICA.   Tab. XXIX, figs. 2, 2*a*.

PORPITA, BUTTON STONE, *J. Walcot*, Descriptions and Figures of Petrifications found near Bath, p. 47, fig. 62, var. E, 1779.

Corallum simple, almost hemispherical; its under surface slightly concave towards the centre; its upper surface very convex, and presenting in the centre a well-defined but rather shallow, circular, or elliptic fossula. *Septa* very thin, very closely set, and appearing to dichotomise at the under surface of the corallum; their upper edge is feebly denticulated, and their tissue appears to be much more complete and less trabicular than in the other species belonging to the same genus; their number amounts to about 160. Diameter, four or five lines; height, four lines.

This fossil is found at Dundry. Specimens are in the Collections of the Geological Society, and of Mr. Pratt.

The elevated form of *A. hemispherica* distinguishes this species from all others; these are all shorter and broader, thus *A. Normaniana*[1] is quite flat and discoidal; *A. orbulites*[2] has the form of a plano-convex lens; and *A. Bouchardi*[3] is subconical.

ΧΥΛ

2. ANABACIA ORBULITES.   Tab. XXIX, fig. 3.

This fossil, as we have already stated, is found in the Inferior Oolite at Dundry, as well as in the more recent oolitic formations; it has consequently been described in the preceding chapter. (See page 120.)

---

[1] D'Orbigny, Prod. de Paléont., tab. i, p. 241.

[2] See tab. xxix, fig. 3.

[3] Milne Edwards and J. Haime, Polyp. Foss. des Terr. Palæoz., &c., p. 122; *Fungia orbulites* (*pars*), Michelin, Iconogr., tab. liv, fig. 1.

Genus COMOSERIS.   (See page 102.)

COMOSERIS VERMICULARIS.   Tab. XIV, fig. 1.

This fossil is found at Leckhampton, and is also met with in the Great Oolite. (See page 122.)

Family CYATHOPHYLLIDÆ, (p. lxv.)

Genus ZAPHRENTIS, (p. lxv.)

ZAPHRENTIS? WALTONI.   Tab. XXVII, figs. 8, 8*a*.

It is not without much uncertainty that we refer to the genus *Zaphrentis* the coral here alluded to, for its calice is so deeply imbedded in the surrounding stone, that we have not been able to study it in a satisfactory manner, but we think we have detected some indication of a septal fossula (fig. 8*a*). It is of a conical, elongated form, slightly curved, very narrow at its basis, and surrounded with a thin epitheca presenting some circular wrinkles or dilatations. The calice is almost circular, with a thin margin and rather deep. There appears to be about forty septa; they are very narrow, thin, unequal in size alternately, and closely set. Height, one inch and a half. Diameter of the calice, about six lines.

This interesting fossil belongs to Mr. Walton's valuable collection, and is catalogued as having been found in the Inferior Oolite at Dundry; but we are inclined to think that there may be some mistake about its origin, and that, in reality, it may appertain to some strata of the Carboniferous Formation.

Some other fossils, that we have not had an opportunity of seeing, and of which we are therefore unable to give a description here, have been mentioned by different authors as existing in the Inferior Oolite of England. Such are :

1. CARYOPHYLLIA CONVEXA, *Phillips*, Geol. of Yorkshire, p. 155, tab. xi, fig. 1. Found in the Inferior Oolite at Coldmoor. This fossil is probably a young Montlivaltia.

2. A MEANDRINA, *Phillips*, op. cit., p. 155. From the Inferior Oolite at Blue Wick. (No description of this fossil has been given.)

3. A fossil coral found in the Inferior Oolite by MM. Conybeare and Phillips, ('Outlines of the Geol. of England, p. 249,) and referred to the CYCLOLITES ELLIPTICA of Lamarck.

4. SIDERASTREA CADOMENSIS, *M'Coy*, Ann. of Nat. Hist., s. 2, vol. ii, p. 419. The fossil so named was found in the Inferior Oolite at Leckhampton, and considered by Prof. M'Coy as identical with the *Astrea cadomensis*, described by M. Michelin ('Icon.' tab. 94, fig. 4), or *Thamnastrea cadomensis*, Milne Edwards and J. Haime, (Polyp. des Terr. Palæoz., &c., p. 111.)

5. LITHODENDRON ASTREATUM, *M'Coy*, loc. cit., from Dundry. (No description given.)

CHAPTER XII.

CORALS FROM THE LIAS.

Very few Corals have as yet been found in the Lias. Two species belonging to the family of the *Turbinolidæ* have lately been discovered in a stratum of that formation at Ilminster, by Mr. C. Moore; and there is in the collection of the Geological Society of London a third species, which is labelled as having been found in the Lias, but without any indication of locality: it appears to belong to the family of the *Cyathophyllidæ*, and may, more probably, have been met with in some older deposit, for as yet all the well-characterised *Cyathophyllidæ* are peculiar to the Palæozoic formations. We have also remarked in Mr. Walton's collection a cast that appears to belong to a *Montlivaltia*, and was found by that Palæontologist in the Lias at Wiston; and we must add, that the occurrence of a Coral at Fenny Compton Tunnel, on the Oxford Canal, was pointed out by Messrs. Conybeare and W. Phillips,[1] who considered that fossil as being referable to *Turbinolia* or *Madrepora turbinata* of former zoologists.

Family—TURBINOLIDÆ, (p. xi.)

Genus THECOCYATHUS, (p. xiv.)

THECOCYATHUS MOORII. Tab. XXX, figs. 6, 6*a*.

Corallum simple, turbinate, short and thick, straight and adherent, and provided with a thin epitheca, through which straight, and almost equal costal striæ are visible. *Calice* circular, not very deep. *Columella* well-developed, trabicular. *Septa* rather thin, granulated laterally, and forming four complete cycla; those of the last cyclum converging towards the tertiary ones, and joining them at their inner edge. *Pali* corresponding to all the septa of the first three cycla; those of the first two cycla small, and differing but little; those that correspond to the third cyclum of septa greatly developed, and distinctly bilobate; their inner lobe thin, and much resembling the neighbouring pali; the outer lobe very thick, and granulated. Height, 3 lines; diameter of the calice almost as much.

This interesting fossil was communicated to us by Mr. C. Moore, of Ilminster, who found it on the Upper Lias near that town.

[1] Outlines of the Geol. of England, p. 270.

The genus *Thecocyathus* as yet comprises only two other species, which equally belong to the Upper Lias: *T. tintinnabulum*[1] and *T. mactra;*[2] *T. Moorii* differs from both, by the direction of the septa and by the form of the pali.

Genus TROCHOCYATHUS (p. xiv).

TROCHOCYATHUS? PRIMUS.   Tab. XXX, fig. 8.

Corallum simple, subpedicellate, conical at its basis, cylindrical higher up, and sometimes slightly curved; twelve sub-equal costæ.   Height nearly three lines; diameter of the calice about half a line.

Found in the superior Lias at Ilminster, by Mr. Ch. Moore.

We are very uncertain as to the zoological affinities of this small Coral; it is in a very imperfect state of preservation, and, as far as its characters are known, appears to belong to the genus *Trochocyathus;* it differs from the other species of that group by its slender form and very small number of septa.

Family—CYATHOPHYLLIDÆ (p. lxv).

Genus CYATHOPHYLLUM (p. lxviii).

CYATHOPHYLLUM? NOVUM.   Tab. XXX, fig. 7.

Corallum simple, ceratiform, presenting but very slight circular constrictions, and covered with a thin epitheca.   *Calice* almost circular, deep, and having a thin margin. *Septa* narrow, very thin, unequal alternately, and about 120 in number.   Height, $2\frac{1}{2}$ inches; diameter of the calice, 1 inch 2 lines.

The fossil Coral here described belongs to the collection of the Geological Society of London, and is entered in the catalogue of that establishment has having been found in the Lias, but without any other indication.   It is in such a bad state of preservation that we have not been able to ascertain, with any degree of certainty, its generic characters; but it much resembles, by its general form, some species of the carboniferous periods belonging to the genus *Cyathophyllum*, in which we have placed it provisionally.

The Coral, of which a cast was found in the Lias at Wiston, by Mr. Walton, as above-mentioned, had very delicate and numerous straight septa; its calice was oval, and about 1 inch 3 lines broad in one direction, 1 inch in the other; it appears to be referable to the genus *Montlivaltia*.

---

[1] *Cyathophyllum tintinnabulum*, Goldfuss, Petref. Germ., tab. xvi, fig. 6.
[2] *Cyathophyllum mactra*, Goldfuss, op. cit., tab. xvi, fig. 7.

THE

PALÆONTOGRAPHICAL SOCIETY.

INSTITUTED MDCCCXLVII.

LONDON:

MDCCCLII.

A MONOGRAPH

OF THE

BRITISH FOSSIL CORALS.

BY

H. MILNE EDWARDS,

DEAN OF THE FACULTY OF SCIENCES OF PARIS; PROFESSOR AT THE MUSEUM OF NATURAL HISTORY;
MEMBER OF THE INSTITUTE OF FRANCE;
FOREIGN MEMBER OF THE ROYAL SOCIETY OF LONDON, OF THE ACADEMIES OF BERLIN, STOCKHOLM, ST. PETERSBURG,
VIENNA, KONIGSBERG, MOSCOW, BRUXELLES, HAARLEM, BOSTON, PHILADELPHIA, ETC.,

AND

JULES HAIME.

THIRD PART.

CORALS FROM THE PERMIAN FORMATION AND THE MOUNTAIN LIMESTONE.

LONDON:

PRINTED FOR THE PALÆONTOGRAPHICAL SOCIETY.

1852.

C. AND J. ADLARD, PRINTERS, BARTHOLOMEW CLOSE.

DESCRIPTION

OF

THE BRITISH FOSSIL CORALS.

CHAPTER XIII.

CORALS FROM THE PERMIAN FORMATION.

VERY few Fossils belonging to the great Zoological group of Polypi have as yet been discovered in the Permian Formation. Only five species have been met with in England; and we are entirely indebted to Professor William King for the knowledge of these Corals. We have not been enabled to study them ourselves, and we must therefore beg leave to lay before our readers the descriptions given by that distinguished Palæontologist; but in so doing, we deem it necessary to differ somewhat from the author of the valuable '*Monograph of the Permian Fossils of England*,' respecting the natural affinities of these Zoophytes. The species referred by Professor King to the genera *Calamopora, Stenopora,* and *Alveolites,* appear to have all the exterior characters of *Chætetes;* and we are, therefore, inclined to class them in that generical division : the two other species probably belong to the family of the *Stauridæ,* and, in our opinion, form part of a small genus that Professor King had, in 1849, very properly proposed establishing under the name of *Polycælia,* but has abandoned since that time.

Family FAVOSITIDÆ, (*Introduction,* p. lx.)

Genus CHÆTETES, (*Introduction,* p. lxi.)

1. CHÆTETES? MACKROTHI.

CALAMOPORA MACKROTHII, *Geinitz,* Grund., p. 582, 1846.
STENOPORA INDEPENDENS, *King,* Catal. of the Organic Remains of the Permian Rocks, p. 6, 1848.
— CRASSA, *Howse,* Trans. of the Tyneside Nat. Field Club, vol. i, p. 260, 1848, (not Lonsdale?).

20

STENOPORA MACKROTHII, *Geinitz*, Verst. der Deutsch. Zechst., p. 17, tab. viii, fig. 10, 1848.
CALAMOPORA MACKROTHII, *King*, Permian Fossils of England, p. 26, tab. iii, figs. 3—6, 1850.
CHÆTETES? MACKROTHI, *Milne Edwards* and *Jules Haime*, Pol. Foss. des Terr. Palæoz.,
p. 274, 1851.

" A branching Calamopora : with numerous slender, round or polygonal, transversely-wrinkled tubes, rising perpendicularly in the centre of the branches, and afterwards suddenly curving out to the surface. Interpolated or new tubes numerous, originating on the outside of the old ones. Margin of the apertures with from five to eight spine-like tubercles.

"It is rather a common Coral, being found at Tunstall Hill, Humbleton Quarry, Dalton-le-Dale, Ryhope-Field House Farm, and Whitley, in the Shelly Limestone. The German localities, according to Schlotheim and Geinitz, are Milbitz and Corbusen, in the Lower Zechstein; and Glücksbrunn and Liebenstein, in the Zechstein Dolomite." (*King*, loc. cit.)

2. CHÆTETES? COLUMNARIS.

CORALLIOLITES COLUMNARIS, *Schlotheim*, Taschenb. für die Ges. Miner., p. 59, 1813.
—   —   „   Akad. Münch., vol. vi, p. 23, pl. iii, fig. 10, 1820.
STENOPORA INCRUSTANS, *King*, Catal. of the Org. Rem. of the Permian Rocks, p. 6, 1848.
—   COLUMNARIS, *King*, Perm. Foss. of England, p. 28, pl. iii, figs. 7, 8, and 9, 1850.
CHÆTETES COLUMNARIS, *Milne Edwards* and *Jules Haime*, Pol. Foss. des Terr. Palæoz.,
p. 274, 1851.

"An incrusting Stenopora. Polypidoms tubular, cylindrical, slightly wrinkled more or less transversely, and in close contact, except towards their orifice, where they are a little reduced in diameter, leaving rather wide interspaces, which are often perforated with interpolated tubes. Apertures circular or slightly polygonal, with a tuberculated margin.

"It occurs at Humbleton, Tunstall Hill, and Whitley; but is nowhere a common species. Geinitz's *Alveolites producti*, which may be the same Coral, is found at Corbusen, in Saxony." (*King*, op. cit.)

3. CHÆTETES? BUCHANA.

ALVEOLITES BUCHIANA, *King*, Perm. Foss. of England, p. 30, pl. iii, figs. 10, 11, & 12, 1850.
CHÆTETES BUCHANA, *Milne Edwards* and *Jules Haime*, Pol. Foss. des Terr. Palæoz.,
p. 274, 1851.

"Tubes or cells adjoining, cylindrical, leaning, concavely arcuate ascendingly, alternately overlying each other, and slightly wrinkled more or less transversely. Apertures regularly arranged, circular, occasionally polygonal, margined by a circle of from twelve to fourteen small, closely-packed tubercles, which generally fill up the interspaces.

"It is a scarce fossil, having only occurred to me once in the Shell Limestone at Humbleton Hill Quarry." (*King*, op. cit.)

Family STAURIDÆ, (p. lxiv.)

Genus POLYCŒLIA.

This genus, the establishment of which was proposed by Professor King some years ago, was characterised by that author in the following terms:

"A simple Cyathophyllida. Form conical. Walls solid. Primary vertical plates converging to within a short distance of the centre. Secondary vertical plates reaching about half way to the centre. Transverse plates horizontal, at irregular distances from each other, and extending quite across the cavity. Chambers or lamellar interspaces capacious compared with those of other Cyathophyllidæ. Reproduction within the polypiferous cup." (*King*, 'Ann. and Mag. of Nat. Hist.,' Second Series, vol. iii, April, 1849.)

As we have stated here above, the genus Polycœlia has of late been abandoned by Professor King; but it appears to us that it ought to be adopted, and in the systematic arrangement of the polypi, published by ourselves in 1851, it has been included in the family of the Stauridæ. (See 'Monographie des Polypiers Fossiles des Terrains Palæozoiques, Précedée d'un Tableau Général de la Classification des Polypes,' p. 162.)

1. POLYCŒLIA DONATIANA.

> TURBINOLIA DONATIANA, *King*, Cat. of the Org. Rem. of the Perm. Rocks of Northumberland and Durham, p. 6, 1848.
>
> CALOPHYLLUM DONATIANUM, *King*, Perm. Foss. of England, p. 23, pl. iii, fig. 1, 1850.
>
> POLYCŒLIA DONATIANA, *Milne Edwards* and *Jules Haime*, Pol. Foss. des Terr. Palæoz., p. 317, 1851.

"Calophyllum vermiform; gradually enlarging from the base upwards; transversely wrinkled and longitudinally striated on the outside. Vertical plates dense and somewhat apart from each other; primaries four; secondaries sixteen. Transverse plates horizontal, rather thick, and at irregular distances from each other. Polypiferous cell shallow.

"The specimen figured, which is the only one of the species known to me, was procured at Humbleton Hill, in the upper bed of shell-limestone." (*King*, 'Perm. Foss. of England,' loc. cit.)

2. POLYCŒLIA PROFUNDA.

> CYATHOPHYLLUM PROFUNDUM, *Germer*, Verst. des Mansf. Kupfer Schiefers, p. 37, 1840.
>
> — — *Geinitz*, Neues Jahrb. für Miner. Geol., 1842, p. 579, tab. x, fig. 14.
>
> PETRAIA DENTALIS, *King*, Cat. of the Org. Rem. of Perm. Rocks of Northumberland and Durham, p. 5, 1848.
>
> CYATHOPHYLLUM PROFUNDUM, *Geinitz*, Verst. des Deutsch. Zechst., p. 17, tab. vii, fig. 17, 1848.

CARYOPHYLLIA QUADRIFIDA, *Howse*, Trans. of the Tyneside Nat. F. C., vol. i, p. 260, 1848.
PETRAIA PROFUNDA, *King*, Perm. Foss. of England, p. 23, tab. iii, fig. 2, 1850.
POLYCŒLIA PROFUNDA, *Milne Edwards* and *Jules Haime*, Pol. Foss. des Terr. Palæoz.,
p. 317, 1851.

"Form conical, and slightly curved. Cavity deep and longitudinally furrowed. Plates of two lengths, the longest five or more in number, plain edged (?), and reaching half way to the centre; the shortest from one to four in number. Lamellar interspaces with two very finely denticulated, slightly prominent ridges.

"I have only succeeded in procuring two or three specimens from the shelly magnesian limestone at Humbleton Quarry. Geinitz states its having been found in the lower Zechstein at Eisleben, Ilmenau, Gerbstedt, and between Hettstädt and Leimbach." (*King,* op. cit.)

———————

Aulopora Voigtiana, King, op. cit., p. 31, pl. iii, fig. 13, appears to be a BRYOZOUM.

———————

CHAPTER XIV.

CORALS FROM THE MOUNTAIN LIMESTONE.

THE Fauna of the Mountain Limestone Period is one of the richest in true Polypi; seventy-six species have already been found in the deposits appertaining to this geological division, and the presence of none of these Corals has, as yet, been satisfactorily proved in beds belonging to any other period. Forty-three of these species are British, and they are referable to six families:—Milleporidæ, Favositidæ, Seriatoporidæ, Auloporidæ, Cyathaxoniidæ, and Cyathophyllidæ; but the Favositidæ and the Cyathophyllidæ are the forms which have the most numerous representatives among these Fossils.

The principal localities from which they have been obtained, are Castleton, Bakewell, Oswestry, Derbyshire, Bolland in Yorkshire, Masbury, near Mendip, in Somersetshire, the environs of Bristol, Kendal in Westmoreland, Wellington in Shropshire, Mold, Lilleshall, Frome, Clifton, &c., in England; the Isle of Man; Armagh, Enniskillen, Kulkeag (Fermanagh), Wexford, and Easky (Sligo), in Ireland.

Most of the Carboniferous Fossils that we have represented in the plates joined to this Monograph, belong to the Collections of the Geological Society of London, the Museum of Practical Geology, under the direction of Sir Henry De la Beche, the Museum of Bristol, and the rich Cabinet of our esteemed friend J. S. Bowerbank, Esq. We much regret not having been able to obtain the same liberal aid from the Museum of the University of

Cambridge, and to have been therefore obliged to omit representing in this work a certain number of species that we have not seen in any of the numerous collections so generously placed at our disposal by the great majority of the English Geologists. But the omission that we here allude to is now of less importance than it appeared to us, when our application to the Cambridge Museum was rejected, for, since that time, a young Palæontologist belonging to that scientific establishment, Professor M'Coy, has published very good figures of almost all the Corals that we were desirous of obtaining communication of from the above-mentioned Museum. His recent work[1] will enable us, at least, to complete our Catalogue of the Corals found in the Carboniferous Formation of Great Britain; and having gone to Cambridge in order to see the fossils described by that gentleman, we have easily recognised those species which we had already met with elsewhere, and can without hesitation refer most of the others to generical divisions here adopted.

Family MILLEPORIDÆ, (p. lviii.)

1. *Genus* FISTULIPORA, (p. lix.)

1. FISTULIPORA MINOR.

> FISTULIPORA MINOR, *M'Coy*, Ann. and Mag. of Nat. Hist., 2d Series, vol. iii, p. 130, figs. *a, b*, 1849.
> — — *Milne Edwards* and *Jules Haime*, Pol. Foss. des Terr. Palæoz., p. 220, 1851.
> — — *M'Coy*, Brit. Palæoz. Foss., p. 79, pl. iii B, fig. 12, 1851.

"Cell-tubes with slightly prominent margins at the surface, about four in the space of one line, rather less than their own diameter apart, the intervening space composed of from one to three rows of the minute vesicular cells. The diaphragms in the main tubes slightly irregular, about half their diameter apart; the tubes are from half a line to nearly an inch in length, according to the age of the example, but not altering, materially, their diameter or relative distances.

"Not uncommon in the Carboniferous Limestone of Derbyshire." (*M'Coy*, Brit. Palæoz. Foss., loc. cit.)

---

[1] Description of the British Palæozoic Fossils in the Geological Museum of the University of Cambridge, by F. M'Coy. This work was published in May, 1851, some months after the first part of our '*Monographie des Polypiers des Terrains Palæozoiques*,' and at least a year after the distribution of the first part of our 'Description of the British Fossil Corals' to all the members of the Palæontographical Society. In the beginning of his book (p. 17), Professor M'Coy expresses his regret at not having become acquainted with the latter publication early enough to be able to refer to it; and we feel much gratified in seeing, that the results which Professor M'Coy appears therefore to have obtained solely from his own observation, are often so very similar to those published by ourselves a year before; even by a singular coincidence, he often makes use of the same names for the divisions previously established in the first part of this Monograph.

We have not seen any specimens of this fossil Coral, which constitutes, together with the following species, a small genus nearly allied to *Propora*[1] and *Lyellia*.[2]   In a memoir published in the 'Annals of Natural History' (1849), Professor M'Coy pointed out the existence of infundibuliform tabulæ as being one of its characters, and figured them very distinctly in a woodcut; but in his latter work, the same author represents the tabulæ as being horizontal, without explaining in the text the reason of this change.

2. FISTULIPORA MAJOR.

> FISTULIPORA MAJOR, *M'Coy*, Ann. and Mag. of Nat. Hist., 2d Series, vol. iii, p. 131, 1849.
> —   —   *Milne Edwards* and *Jules Haime*, Pol. Foss. des Terr. Palæoz., p. 220, 1851.

"Cell-tubes two thirds of a line in diameter, and about their own diameter apart; their walls thick, of concentric layers, with closely placed funnel-shaped internal diaphragms; interstices minutely vesicular; four to six rows of vesicular cells between each pair of tubes.

"Rare in the Carboniferous Limestone of Derbyshire." (*M'Coy*, loc. cit.)

2. *Genus* PROPORA, (p. lix.)

1. PROPORA ? CYCLOSTOMA.

> HYDNOPHORA? CYCLOSTOMA, *Phillips*, Geol. of York., vol. ii, p. 202, pl. ii, figs. 9 & 10, 1836.
> ASTREOPORA ANTIQUA, *M'Coy*, Syn. of Carb. Foss. of Ireland, p. 191, pl. xxvi, fig. 9, 1844.
> —   —   *M'Coy*, Ann. and Mag. of Nat. Hist., 2d Series, vol. iii, p. 133, 1849.
> PROPORA? CYCLOSTOMA, *Milne Edwards* and *Jules Haime*, Pol. Foss. des Terr. Palæoz., p. 225, 1851.

"Discoid, convex; surface with large, circular cells, in quincunx, about one third their diameter apart; sides of the cells radiatingly striated; intervening flat spaces with minute, irregular, curving ridges." (*M'Coy*, 'Carb. Foss. of Ireland,' loc. cit.)

The fossil is known to us only by the figures and very brief descriptions given of it by Professor Phillips, and more recently, by Professor M'Coy.   It appears to be very nearly allied to the Corals that form our genus *Propora*, and provisionally, at least, must be referred to that group.   It is the only species of Propora that has, as yet, been found in the Carboniferous Deposits; all the others belong to the Silurian Formation.

Professor Phillips discovered this Coral in Northumberland, and Professor M'Coy mentions its existence in Ireland, at Hook Point, Wexford.

---

[1] Introduction, p. lix.
[2] Milne Edwards and Jules Haime, 'Polypiers Foss. des Terrains Palæozoiques,' p. 226.

Family FAVOSITIDÆ, (p. lx.)

Sub-Family FAVOSITINÆ, (p. lx.)

1. *Genus* FAVOSITES, (p. lx.)

1. FAVOSITES PARASITICA.   Tab. XLV, fig. 2.

> CALAMOPORA PARASITICA, *John Phillips*, Geol. of York., vol. ii, p. 201, pl. i, figs. 61 and
> 62, 1836.
> FAVOSITES PARASITICA, *M'Coy*, Syn. Carb. Foss. of Ireland, p. 192, 1844.
> —   —   *D'Orbigny*, Prod. de Palæont., vol. i, p. 160, 1850.
> —   —   *Milne Edwards* and *J. Haime*, Pol. Foss. des Terr. Pal., p. 244, 1851.

Corallum forming small globular masses, and usually adhering to the stem of an Encrinite. *Walls* very thin. *Calices* unequal in size ; some very small ones near the angles of the larger ones ; the latter 1 or 1½ line in diameter.

Fossil from the Carboniferous Limestone at Bolland, in Yorkshire ; and according to Professor M'Coy, in Ireland.

The fossil designated by Colonel Portlock[1] and by Professor M'Coy,[2] under the name of *Favosites Gothlandica*, appears to belong to this species. The specimens mentioned by the first of these geologists were found in Tyrone and Derryloran ; those described by the latter were met with in the Isle of Man, and in Derbyshire. A collector of the Museum of Paris, M. Marcou, found at Button Mould Knobs, near Louisville, in North America, a Coral, which we equally refer to the above-described species, although its calices are somewhat smaller.

The genus *Favosites*, which is so abundant in the fauna of the Silurian and Devonian Periods, appears to be represented only by the *F. parasitica* in the Carboniferous Formation ; and the other fossils that various authors have described under this generic name, or as *Calamopora*, are now referred to different genera. We must, however, not omit mentioning here two Corals that are not sufficiently well known to be classed zoologically, although they probably are not true Favosites.

One of these fossils is the *Calamopora incrustans* of Professor Phillips.[3] It was found in the Carboniferous Limestone at Bolland, and is known to us only by a very rough figure, given by that distinguished geologist.

The other is the *Calamopora dentifera* of the same author ;[4] it was met with at Bolland, but in the present state of Palæontological science cannot be characterised.

---

[1] Report on the Geology of Londonderry, &c., p. 326.   [2] Syn. Carb. Foss. of Ireland, p. 192.

[3] Geology of Yorkshire, vol. ii, p. 200, tab. i, figs. 63, 64 ; *Favosites incrustans*, D'Orbigny, Prodr., vol. i, p. 160 ; Milne Edwards and Jules Haime, Polyp. Palæoz., p. 246.

[4] Geology of Yorkshire, vol. ii, p. 201, tab. i, figs. 58, 60 ; *Favosites dentifera*, D'Orbigny, op. cit., p. 160 ; Milne Edwards and Jules Haime, loc. cit.

2. *Genus* MICHELINIA, (p. lx.)

1. MICHELINIA FAVOSA. Tab. XLIV, figs. 2, 2*a*, 2*b*, 2*c*.

> HONEY COMB, *Parkinson*, Org. Rem. of a Former World, vol. ii, p. 39, pl. v, fig. 9, 1808.
> MANON FAVOSUM, *Goldfuss*, Petref. Germ., vol. i, p. 4, tab. i, fig. 11, 1826.
> PORITES CELLULOSA, *Fleming*, Brit. Anim., p. 511, 1828.
> FAVASTREA MINON, *Blainville*, Dict. Sc. Nat., vol. lx, p. 340, 1830; Man., p. 375.
> PORITES CELLULOSA, *S. Woodward*, Syn. Table of Brit. Org. Remains, p. 6, 1830.
> MICHELINIA FAVOSA, *De Koninck*, An. Foss. des Terr. Carb. de Belg., p. 30, pl. c, fig. 2, 1842.
> COLUMNARIA SENILIS, *ib.*, p. 25, pl. B, fig. 9. Specimen in a bad state of preservation.
> FAVOSITES ALVEOLATA, *Geinitz*, Grund. der Verst., p. 572, 1845-46.
> MICHELINIA FAVOSA, *Michelin*, Icon. Zooph., p. 254, pl. lix, fig. 2, 1846.
> MICHELINIA FAVOSA and FAVASTREA SENILIS, *D'Orbigny*, Prod., vol. i, p. 160, 1850.
> MICHELINIA FAVOSA, *Milne Edwards* and *Jules Haime*, Pol. Foss. des Terr. Palæoz., p. 249, 1851.

Corallum massive, generally circular; upper surface slightly convex; common basal plate covered with a thick epitheca, that sends off numerous and well-developed radiciform processes. *Calices* somewhat unequal, shallow, and presenting, in well-preserved specimens, margins thickened by small endothecal vesicles. When these vesicles are destroyed near the upper edge of the wall, thirty or forty somewhat unequal small septal striæ become visible, and the wall shows small horizontal series of pores. Diagonal of the calices three or four lines.

Found at Masbury, near Mendip, Somersetshire, and in Derbyshire; at Hook Point, Wexford, and in Enniskillen. The same species has been found at Tournay and Visé in Belgium, and at Ratingen, in Prussia; but it is erroneously that Goldfuss states that it is also met with in the Eifel. Specimens of this Coral are in the Collections of the Geological Society of London, of the Bristol Museum, of J. S. Bowerbank, Esq., &c.

Michelinia favosa differs from *M. antiqua* (see p. 156) and *M. concinna*[1] by the irregular form and vesicular structure of its endotheca. The aspect of its upper surface, due to the unequal development of the calices, distinguishes it from *M. geometrica;*[2] and the radiciform processes of its under surface distinguishes it from *M. convexa,*[3] *M. tenuisepta,*[4] and *M. megastoma.*[5]

---

[1] Lonsdale, in Geol. of Russia, by Murchison, Verneuil, and Keyserling, vol. ii, p. 611, pl. A, fig. 3.
[2] Milne Edwards and Jules Haime, Polyp. Palæoz., tab. xvii, fig. 15.
[3] Op. cit., tab. xvi, fig. 1.   [4] See tab. xliv, fig. 1.   [5] See tab. xliv, fig. 3.

2. Michelinia tenuisepta. Tab. XLIV, figs. 1, 1*a*, 1*b*.

> Calamopora tenuisepta, *John Phillips*, Illust. of Geol. of York., vol. ii, p. 201, pl. ii,
> fig. 30, 1836.
> Michelinia tenuisepta, *De Koninck*, An. Foss. des Terr. Carb. de Belg., p. 31, pl. c,
> fig. 3, 1842.
> —   —   *Michelin*, Icon. Zooph., pp. 83 and 254, pl. xvi, fig. 3, 1843.
> Favosites (Michelinea) tenuisepta, *M'Coy*, Syn. Carb. Foss. of Ireland, p. 193, 1844.
> Michelinea glomerata? *M'Coy*, Ann. and Mag. of Nat. Hist., 2d Series, vol. iii, p. 122,
> 1849.
> Favosites tenuisepta and Michelinia tenuisepta, *D'Orbigny*, Prod. de Palæont., vol. i,
> p. 160, 1850.
> Michelinia tenuisepta, *Milne Edwards* and *Jules Haime*, Pol. Foss. des Terr. Palæoz.,
> p. 250, 1851.
> Michelinea glomerata? *M'Coy*, Brit. Palæoz. Foss., p. 80, pl. iii B, fig. 14, 1851.

Corallum tall; common basal plate with a strong epitheca, striate transversely, but not bearing any radiciform processes. *Calices* polygonal, unequal in size, and containing thirty or forty equally developed septal striæ. *Tubulæ* very delicate, closely set, much blended together, and minutely granulated.

Height of the corallum 4 inches; diagonal of the calices 3 or 4 lines.

M. tenuisepta is found in the environs of Bristol, at Masbury, near Mendip, and at Bolland; Professor M'Coy has also met with it in Ireland; and it exists also on the Continent, at Sablé and Juigné, in France, and at Tournay, in Belgium. Specimens are in the collections of the Museums of Bristol, of Cambridge, of Paris, &c.

This species much resembles *M. favosa*, by the structure of the visceral chambers, but differs from it by the corallites being more elongate, and by the common basal plate not bearing any radiciform appendices. *M. tenuisepta* is also very closely allied to *M. convexa*[1] and *M. megastoma*,[2] but its endothecal vesicles are less convex, and it never attains the size to which this species usually come. The obliquity or irregular arrangement of its tabulæ distinguish it from *M. antiqua*[3] and *M. concinna*,[4] in which the tabulæ are almost horizontal and distinct. *M. geometrica*[5] differs from the above-described species by the great regularity of its polygonal calices.

We refer to this species, but with some doubt, the fossil designated by Professor M'Coy, under the name of *Michelinea glomerata*; in the specimen figured by that geologist, the common basal plate is worn away, so that it is not possible to ascertain

---

[1] D'Orbigny, Prod., vol. i, p. 107; Milne Edwards and J. Haime, Polyp. Palæoz., tab. xvi, fig. 1.

[2] See tab. xliv, fig. 3.

[3] *Dictyophyllia antiqua*, M'Coy, Syn. of Carb. Foss. of Ireland, tab. xxvi, fig. 10.

[4] Lonsdale in Geol. of Russia, by Murch., Verneuil, and Keyserling, vol. i, p. 611, tab. A, fig. 3.

[5] Milne Edwards and J. Haime, Polyp. Palæoz., tab. xvii, fig. 3.

whether there are or not any radiciform appendices; the tabulæ appear, it is true, to be rather more convex than in the above-described species, but the data obtained, as yet, are not sufficient to enable us to characterise this Coral as forming a distinct species.

3. MICHELINIA MEGASTOMA. Tab. XLIV, figs. 3, 3*a*, 3*b*.

> CALAMOPORA MEGASTOMA, *John Phillips*, Illust. of Geol. of Yorkshire, vol. ii, p. 201, pl. ii, fig. 29, 1836.
> FAVOSITES MEGASTOMA, *M'Coy*, Syn. Carb. Foss. of Ireland, p. 192, 1844.
> MICHELINEA GRANDIS. *M'Coy*, Ann. and Mag. of Nat. Hist., 2d series, vol. iii, p. 123, 1840.
> FAVOSITES MEGASTOMA, *D'Orbigny*, Prod. de Paléont., vol. i, p. 160, 1850.
> MICHELINIA MEGASTOMA, *Milne Edwards* and *Jules Haime*, Polyp. Foss. des Terr. Palæoz., p. 251, 1851.
> MICHELINEA GRANDIS, *M'Coy*. Brit. Palæoz. Foss., p. 81, pl. iii *c*, fig. 1, 1851. Good figure.

Corallum subturbinate, convex, pediculate; common basal plate covered with a thick wrinkled epitheca, and not bearing any radiciform appendices. *Calices* very large, rather deep, and somewhat unequal in size. *Septal striæ* very delicate and numerous. *Tabulæ* entirely composed of vesicles, which are very convex, but always broader than high. Diagonal of the *calices* 8 or 9 lines.

This fossil has been found at Kendal and at Bolland, in England; in the Isle of Man; and at Attre, near Mons, in Belgium. Specimens are in the collections of the Geological Society of London, of the Bristol and Cambridge Museums, &c.

M. geometrica[1] is easily distinguished from this species by the regular hexagonal form of its calices; *M. antiqua*[2] and *M. concinna*[3] differ from it by their large tabulæ, being almost horizontal; and *M. favosa*[4] by the presence of radiciform processes. The above-described species differs from *M. convexa*[5] and from *M. tenuisepta*[6] by the large size of its corallites; and it appears to be intermediate between these two species by the form of the endothecal vesicles, these being less inflated than in *M. convexa*, and more convex than in *M. tenuisepta*.

4. MICHELINIA ANTIQUA.

> DICTYOPHYLLIA ANTIQUA, *M'Coy*, Syn. Carb. Foss. of Ireland, p. 191, pl. xxvi, fig. 10, 1844.
> MICHELINIA COMPRESSA, *Michelin*, Icon. Zooph., p. 254, pl. lix, fig. 3, 1846.
> —   ANTIQUA, *D'Orbigny*, Prod. de Palæont., vol. i, p. 160, 1850.
> —   — *Milne Edwards* and *Jules Haime*, Pol. Foss. des Terr. Palæoz., p. 252, 1851.

---

[1] Milne Edwards and Jules Haime, Pol. Foss. des Terr. Palæoz., pl. xvii, fig. 3.
[2] *Dictyophyllia antiqua*, M'Coy, Syn. Carb. Foss. of Ireland, pl. xxvi, fig. 10.
[3] Lonsdale in Murch., Vern., Keys., Russ. and Ural, vol. i, p. 611, pl. A, fig. 3.
[4] See tab. xliv, fig. 2.
[5] D'Orbigny, Prodr., t. i, p. 107; Milne Edwards and Jules Haime, loc. cit., pl. xvi, fig. 1.
[6] Tab. xliv, fig. 1.

Corallum forming a thin incrustating expansion, its upper surface almost horizontal. *Calices* polygonal, very deep, rather unequal in size, and separated by slightly exsert ridges. *Septal striæ* 40 or 50 in number, almost equally developed, subvermiculate, and extending on the tabulæ to a small distance from the walls. *Tabulæ* closely set, almost horizontal towards the centre of the corallites, but very irregular towards their circumference. Diagonal of the calices from 5 to 8 lines.

Found at Hook-point, in Ireland, by Professor M'Coy; and at Tournay, in Belgium.

We have not been enabled to study any British specimens of this species; those that we have examined belonged to the carboniferous formation of Belgium; but the figure published by Professor M'Coy is sufficient to establish the specific identity between the Continental and the Irish fossil here alluded to; and we feel, consequently, no hesitation in inscribing *M. antiqua* in the list of British Corals.

In this species, as well as in *M. geometrica*[1] and in *M. concinna*[2], there are very numerous, almost horizontal, tabulæ; a structural peculiarity, which sufficiently distinguishes them from *M. favosa*,[3] *M. tenuisepta*,[4] *M. megastoma*,[5] and *M. convexa*,[6] in which the endotheca is entirely vesicular. *M. antiqua* differs from *M. geometrica* by the irregularity of its calices, the prolongation of the septal striæ on the upper surface of the tabulæ; the latter character, and its large size, also distinguishes it from *M. concinna*.

3. *Genus* ALVEOLITES, (p. lx.)

1. ALVEOLITES SEPTOSA.   Tab. XLV, figs. 5, 5*a*, 5*b*.

FAVOSITES SEPTOSUS, *Fleming*, Brit. Anim., p. 529, 1828.
  — — *S. Woodward*, Syn. Table of Brit. Org. Rem., p. 5, 1830.
  — — *Phillips*, Geol. of Yorkshire, 2d part, p. 200, pl. ii, figs. 6, 7, 8, 1836.
  — — *M'Coy*, Syn. Carb. Foss. of Ireland, p. 192, 1844.
CHÆTETES SEPTOSUS, *Keyserling*, Reise in Petschora, p. 183, 1846.
ALVEOLITES SEPTOSA, *Milne Edwards* and *Jules Haime*, Pol. Foss. des Terr. Palæoz., p. 259, 1851.
CHÆTETES SEPTOSUS, *M'Coy*, Brit. Palæoz. Foss., p. 82, 1851.

Corallum forming an incrustating, slightly convex, or subgibbose mass, which is in general composed of superposed strata. *Calices* unequally developed, somewhat irregular in form, but in general polygonal and not having a prominent edge. *Walls* rather thin. The solitary septal process well characterised, and sometimes facing two small denticula. Breadth of the calices half a line or a little more.

---

[1] Milne Edwards and Jules Haime, Polyp. Palæoz., tab. xvii, fig. 3.
[2] Lonsdale, in Murch., Vern., and Keyser., Geology of Russia, vol. i, p. 611, pl. A, fig. 3.
[3] See tab. xliv, fig. 2.   [4] See tab. xliv, fig. 1.   [5] See tab. xliv, fig. 3.
[6] D'Orbigny, Prod., vol. i, p. 107; Milne Edwards and J. Haime, op. cit., p. 251, tab. xvi, fig. 1.

This coral is found at Corwen, near Bristol, and has also been met with at Lee, in Northumberland, by Mr. Phillips, in Westmoreland, in Derbyshire, and in Ireland, by Professor M'Coy, and at Novogorod, in Russia, by M. Keyserling. Specimens are in the Museum of Practical Geology, the Bristol Museum, &c.

The fossil mentioned by Colonel Portlock, under the name of *Favosites fibrosa*,[1] and found by that geologist at Armagh, Donaghenry, and Derryloran, in Ireland, appears to belong to this species.

2. ALVEOLITES DEPRESSA. Tab. XLV, figs. 4, 4*a*.

> FAVOSITES DEPRESSUS, *Fleming*, Brit. Anim., p. 529, 1828.
> — — *S. Woodward*, Syn. Table of Brit. Org. Rem., p. 5, 1830.
> — CAPILLARIS, *John Phillips*, Geol. of Yorkshire, 2d part, p. 200, pl. ii, figs. 3, 4, 5, 1836.
> — — *Portlock*, Rep. on Londonderry, p. 327, 1843.
> — — *M'Coy*, Syn. Carb. Foss. of Ireland, p. 191, 1844.
> CHÆTETES CAPILLARIS, *Keyserling*, Reise in Petschora, p. 183, 1846.
> ALVEOLITES DEPRESSA, *Milne Edwards* and *Jules Haime*, Pol. Foss. des Terr. Palæoz., p. 260, 1851.
> CHÆTETES CAPILLARIS, *M'Coy*, Brit. Palæoz. Foss., p. 82, 1851.

Coral much resembling the preceding species, but having the calices much smaller and less irregular. Diameter of the corallites one tenth or one eighth of a line.

The specimens here described were found near Bristol, and in Salop, and were communicated to us by the directors of the Bristol Museum and Museum of Practical Geology. The existence of the same fossil is mentioned by Mr. Phillips, at Gordale and Ribblehead, by Professor M'Coy, at Kendal, Westmoreland, by Col. Portlock, at Armagh, and by M. Keyserling, in Petschora.

Sub-Family CHÆTETINÆ, (p. lxi.)

1. *Genus* CHÆTETES, (p. lxi.)

1. CHÆTETES RADIANS.

> CHÆTETES RADIANS, *Fischer*, Oryct. de Moscov., p. 160, pl. xxxvi, fig. 3, 1830.
> — DILATATUS, CYLINDRICUS, and JUBATUS, *Fischer*, pp. 160, 161, pl. xxxvi, figs. 1, 2, 4, 1830.
> FAVOSITES EXCENTRICA, *Fischer*, pl. xxxv, figs. 5, 6.
> CHÆTETES EXCENTRICUS, *Fischer*, Oryct. de Moscov., 2d edit. p. 159, pl. xxxv, figs. 5, 6, 1837.
> — RADIANS, *Lonsdale*, in Russ. and Ural, vol. i, p. 595, pl. A, fig. 9, 1845.
> — DILATATUS, *Lonsdale*, p. 596.
> — RADIANS, *Milne Edwards* and *Jules Haime*, Pol. Foss. des Terr. Palæoz., p. 263, pl. xx, fig. 4, 1851.

---

[1] Report on Londonderry, p. 327.

Corallum constituting a tall pyriform mass, the upper surface of which is very convex, with polygonal calices and simple well-developed walls. The *calices* are somewhat unequal in size and in form, being sometimes rather triangular, tetragonal, or hexagonal. The *corallites* are very long, and radiate from the basis of the corallum to the top. The *walls* are not perforated. *Tabulæ* horizontal, and placed at about one sixth of a line distance.

This fossil has been met with at Hilsington Barrow, near Kendal, and in various localities in Russia. The only British specimen which we have seen belongs to the Museum of the Geological Society of London, and is not in a state of preservation sufficiently good to render it worth being figured in this Monograph.

C. radians differs from most of the other species belonging to the same genus by its massive, convex form, and apparently also by the absence of mamillæ on its surface. The same characters are met with only in *C. crinitus*[1] and *C. Trigeri*,[2] but the first of these corals differs from the species above described by the existence of superposed layers, and the second by its large calices and slender walls.

2. CHÆTETES TUMIDUS. Tab. XLV, figs. 3, 3*a*, 3*b*.

CALAMOPORA TUMIDA, *Phillips*, Geol. of Yorkshire, 2d part, p. 200, pl. i, figs. 49—57, 1836.
FAVOSITES SCABRA, or CALAMOPORA FIBROSA, *De Koninck*, An. Foss. des Terr. Carb. de
Belgique, p. 9, pl. B, figs. 1, 5, 1842.
Worn specimen.
CALAMOPORA INFLATA, *ibid.*, p. 10, pl. A, fig. 8.
ALVEOLITES IRREGULARIS, *ibid.*, p. 11, pl. B, fig. 2.
FAVOSITES TUMIDA, *Portlock*, Rep. Geol. on Londonderry, p. 326, pl. xxii, fig. 4, 1843.
— *M'Coy*, Syn. Carb. Foss. of Ireland, p. 193, 1844.
ALVEOLITES TUMIDA, SCABRA and IRREGULARIS, *Michelin*, Icon. Zooph., p. 259, 260, pl. lx,
figs. 2, 3, 4, 1846.
FAVOSITES INFLATA, *M'Coy*, Ann. and Mag. of Hist., 2d Ser., vol. iii, p. 134, 1849.
FAVOSITES TUMIDA, CHÆTETES KONINCKII, CERIOPORA IRREGULARIS, TUMIDA and INFLATA,
D'Orbigny, Prodr. de Pal., Vol. i, p. 160, 161, 1850.
CHÆTETES TUMIDUS, *Milne Edwards* and *Jules Haime*, Pol. Foss. des Terr. Palæoz., p. 270,
1851.
STENOPORA INFLATA and TUMIDA, *M'Coy*, Brit. Palæoz. Fos., p. 82, 1851.

Corallum forming cylindrical branches of various sizes. *Calices* unequally developed, with rather thick margins. In general eight calices occupy the space of about one line, but on the slightly projecting gibbosities of the surface of the corallum the calices are somewhat larger, and almost circular.

This fossil has been found at Harrowgate, Greenhow Hill, Brough, Kirby Lonsdale;

---

[1] *Stenopora crinita*, Lonsdale, in Strzelecki, New South Wales and Van Dieman's Island, tab. viii, fig. 5.
[2] Milne Edwards and Jules Haime, Polyp. Palæoz., p. 269, tab. xvii, fig. 6.

Middleham, Florence Court, and Arran, by Professor Phillips; in Derbyshire, at Kendal, in the Isle of Man, at Kulkrag in Fermanagh, at Clogher and Benburn, (Tyrone,) by Colonel Portlock, and in Belgium. A specimen, in a very bad state of preservation, that belongs to the collection of the Geological Society, was met with in the Llandeilo Flags in Marloes Bay, and appears to be specifically identical with the above-described carboniferous fossils.

C. tumidus can easily be distinguished from the other species of the same genus which have a similar form, by the very small size of their calices, their thick margin, and almost circular form.

We are inclined to think that the fossil described by Professor M'Coy, under the name of *Verticillopora dubia*,[1] may belong to this species. The *Ramose milleporite* of Parkinson,[2] appears also to be referable to it. This last-mentioned fossil was found in Wiltshire.

2. *Genus* BEAUMONTIA.[3]

1. BEAUMONTIA EGERTONI. Tab. XLV, fig. 1.

BEAUMONTIA EGERTONI, *Milne Edwards* and *Jules Haime*, Pol. Foss. des Terr. Palæoz., p. 276, 1851.

Corallum forming a tall, lobate mass. *Corallites* basaltiform, somewhat flexuous, and showing distinctly costal striæ under the epitheca. *Calices* very variable in size; the largest about three lines in diameter. *Tabulæ* closely set, mostly horizontal, and very slightly convex, some incomplete and subvesicular.

The specimen here described belongs to the collection of the Geological Society of London, and was found by Sir P. Egerton, at Sracrapagh, Fermanagh (Ireland).

The genus Beaumontia has been established by us since the publication of the first part of this Monograph; it is, therefore, necessary to mention here that it comprises the Favositidæ with non-perforated walls and a more or less vesicular endotheca. This division has the same relation to Chætetes as Michelinia has to Favosites.

Beaumontia Egertoni differs from *B. venelorum*[4] and *B. laxa*, by its tabulæ being mostly horizontal, and but very slightly convex. In a fourth species, *B. Guerangeri*,[5] the calices are much more irregular, and smaller.

---

[1] Synop. of Carboniferous Fossils of Ireland, p. 194, tab. xxvii, fig. 12; *Ceriopora dubia*, D'Orbigny, Prod., vol. i, p. 161.

[2] Organ. Rem., vol. ii, tab. viii, fig. 3; *Millepora ramosa*, Woodward, Synop. Table of Brit. Org. Rem., p. 5.

[3] Milne Edwards and Jules Haime, Pol. Foss. des Terr., Palæoz., p. 276, 1851.

[4] Op. cit., p. 276, tab. xvi, fig. 6.

[5] Op. cit., tab. xvii, fig. 1.

2. BEAUMONTIA LAXA.

> COLUMNARIA LAXA, *M'Coy*, Ann. and Mag. of Nat. Hist., 2d Series, vol. iii, p. 122, 1849.
>
> BEAUMONTIA LAXA, *Milne Edwards* and *J. Haime*, Pol. Foss. des Terr. Palæoz., p. 277, 1851.
>
> COLUMNARIA LAXA, *M'Coy*, Brit. Palæoz. Foss., p. 92, pl. iii c, fig. 11, 1851.

Corallites very long, sometimes free laterally, and cylindrical; at other times, aggregate and prismatical, or presenting some intermediate state between these two forms. *Epitheca* strong, complete, and not showing any traces of the costal striæ in the free state. In the aggregate prismatic corallites, the transverse wrinkles of the epitheca become thinner, and the straight, rather closely set costal striæ are then very visible. The visceral cavity or interior of the corallites is completely filled up with very large, irregular vesicles, that are convex upwards, decline outwardly, and never form complete tabulæ.

Height, nearly 8 inches; diameter of the corallites 4 lines.

Found at Wellington, and in Derbyshire. Specimens are in the collections of the Cambridge Museum, and of M. E. de Verneuil, at Paris.

This Coral differs from the other species of *Beaumontia*, by the tendency of the corallites to remain separate, or to coalesce but incompletely. By the entirely vesicular structure of its endotheca, it approximates to *B. venelorum*,[1] in which species the corallites are always basaltiform, and vary much in breadth.

Sub-Family HALYSITINÆ, (p. lxi.)

Genus SYRINGOPORA, (p. lxii.)

1. SYRINGOPORA RAMULOSA. Tab. XLVI, figs. 3, 3*a*, 3*b*, 3*c*.

> TUBIPORA, *Knorr* and *Walch*, Rec. des Mon. des Catastr., vol. iii, p. 168, Supp., pl. 6 F, fig. 1, 1775.
>
> SYRINGOPORA RAMULOSA, *Goldfuss*, Petref. Germ., vol. i, p. 76, pl. xxv, fig. 7, 1826.
>
> — — *Morren*, Descr. Corall. in Belg. repert., p. 69, 1832.
>
> — — *John Phillips*, Geol. of York, vol. ii, p. 201, pl. ii, fig. 2, 1836.
>
> — — *Milne Edwards*, Ann. de la 2e Edit. de Lamarck, vol. ii, p. 327, 1836.
>
> — — *Portlock*, Rep. on Londonderry, p. 337, 1843.
>
> — — *M'Coy*, Syn. Carb. Foss. of Ireland, p. 190, 1844.
>
> HARMODITES RAMULOSUS, *Keyserling*, Reise in Petschora, p. 174, 1846.
>
> — — *D'Orbigny*, Prod. de Palæont., vol. i, p. 162, 1850.
>
> SYRINGOPORA RAMULOSA, *Milne Edwards* and *J. Haime*, Pol. Foss. des Terr. Palæoz., p. 289, 1851.
>
> — — *M'Coy*, Brit. Palæoz. Foss., p. 83, 1851.

---

[1] Milne Edwards and Jules Haime, op. cit., tab. xvi, fig. 6.

Corallites elongated, flexuous, rather widely separated from each other, and subgeniculated at the origin of the connecting tubes. *Epitheca* delicately wrinkled transversely. Connecting tubes placed at the distance of about 3 lines; diameter of the corallites $1\frac{1}{4}$ or $1\frac{1}{2}$ lines.

The British specimens which we have had an opportunity of examining were found by the officers of the Geological Survey at Oswestry, Mold, and Bradwell. Professor Phillips has found the same species at Bolland, Kirby Lonsdale, Ash-Fell, and Mendip; Professor M'Coy mentions its existence in the Carboniferous Formation of the Isle of Man; and Col. Portlock has met with it at Kilcronaghan, Derry, and at Clogher, Tyrone. This fossil is also found on the Continent, at Visé and Tournay, in Belgium; Olne, in Limburg; Ratingen, in Prussia, and Utkinsk, in Russia.

Syringopora ramulosa is remarkable on account of the distance between the corallites and their geniculate forms, peculiarities which distinguish it from *S. geniculata*,[1] a species much less deserving the specific appellation given to it.

2. SYRINGOPORA RETICULATA. Tab. XLVI, figs. 1, 1*a*.

> TUBIPORA STRUES, AFFINIS, &c., *Parkinson,* Org. Rem., vol. ii, pl. ii, fig. 1, 1808.
> ERISMATOLITHUS TUBIPORITES (CATENATUS) (pars), *William Martin,* Petref. Derb., pl. xlii, fig. 2, 1809, but not fig. 1.
> SYRINGOPORA RETICULATA, *Goldfuss,* Petref., vol. i, p. 76, tab. xxv, fig. 8, 1826.
> TUBIPORA STRUES, *Fleming,* Brit. An., p. 529, 1828.
> — — *S. Woodward,* Syn. Table of Brit. Org. Rem., p. 5, 1830.
> HARMODITES RADIANS, *Bronn,* Leth. Geogn., vol. i, p. 51, tab. v, fig. 7, 1835.
> SYRINGOPORA RETICULATA, *John Phillips,* Geol. of York, vol. ii, p. 201, 1836.
> — — *Milne Edwards,* Ann. de la 2e Edit. de Lamarck, vol. ii, p. 328, 1836.
> — — *Portlock,* Rep. on Lond., p. 337, pl. xxii, fig. 7, 1843.
> — CATENATA, *M'Coy,* Syn. Carb. Foss. of Ireland, p. 189, 1844.
> HARMODITES STRUES, *D'Orbigny,* Prod. de Palæont., vol. i, p. 162, 1850.
> SYRINGOPORA RETICULATA, *Milne Edwards* and *J. Haime,* Pol. Foss. des Terr. Palæoz., p. 290, 1851.
> — — *M'Coy,* Brit. Palæoz. Foss., p. 84, 1851.

Corallites very long, diverging slightly, separated by a space equal to once or twice their diameter, and straight or slightly flexuous. Connecting tubes thick, somewhat irregularly arranged, and placed $1\frac{1}{2}$ or 2 lines apart.

Diameter of the corallites 1 line or less.

This fossil has been found in the Carboniferous Deposits of Bristol, Lilleshall; Winster, Buxton (Martin); Ashfell, Derbyshire (Professor Phillips); Kendal, Westmoreland,

---

[1] Milne Edwards and Jules Haime, Polyp. Foss. des Terr. Palæoz., tab. xx, fig. 1; *Harmodites catenatus,* De Koninck, Anim. Foss. des Terr. Carbonif. de la Belgique, tab. B, fig. 4.

the Isle of Man (Professor M'Coy); and Clogher and West Longfield, Tyrone (Colonel Portlock). It is also met with at Olne in the province of Limbourg.

Syringopora reticulata is remarkable for the small number of its connecting tubes, and for the existence of a delicate transverse lamina which passes through the concentric infundibula, and is shown by a transverse section of the corallum, (tab. xlvi, fig. 1*a*.)

3. SYRINGOPORA GENICULATA.   Tab. XLVI, figs. 2, 2*a*, and fig. 4.

> TUBIPORA MUSICA, AFFINIS, *Parkinson*, Org. Rem., vol. ii, pl. i, figs. 1, 2, 1808.
> — CATENATA, *Fleming*, Brit. Anim., p. 529, 1828. (Not Martin.)
> — RAMULOSA, *S. Woodward*, Syn. Table of Brit. Org. Rem., p. 5, 1830. (Not *Syring. ramulosa*, Goldfuss.)
> SYRINGOPORA GENICULATA, *John Phillips*, Geol. of Yorkshire, vol. ii, p. 201, pl. ii, fig. 1, 1836.
> —   —   *Portlock*, Rep. on Londonderry, p. 337, pl. xxii, fig. 6, 1843.
> —   —   *M'Coy*, Syn. Carb. Foss. of Ireland, p. 190, 1844.
> —   —   *Milne Edwards* and *Jules Haime*, Pol. Foss. des Terr. Palæoz., p. 291, 1851.
> —   —   *M'Coy*, Brit. Palæoz. Foss., p. 83, 1851.
> HARMODITES GENICULATA, *D'Orbigny*, Prodr. de Pal., vol. i, p. 162, 1850.

Corallites very long, diverging slightly towards their upper extremity, cylindrical, very closely set, and surrounded with a thick, wrinkled epitheca. Connecting tubes numerous, not appearing to have any regular mode of arrangement, placed at one line or one line and a half apart, and in general very short, in consequence of an agglomeration of the corallites. A horizontal section of a specimen transformed into a mass of marble, shows very distinctly that the connecting tubes are hollow, and establish a free communication between the visceral chambers of the connected corallites. *Walls* rather thick. *Septa* in general fourteen in number, thin, equally developed, straight, not extending much towards the centre of the visceral chamber, and not closely set. Length of the corallites in general from 5 to 8 inches. Diameter about 1 line or somewhat less; distance between these usually half a line.

This fossil has been found by the collectors of the geological survey at Kendal, Westmoreland, and Professor Phillips mentions its having been met with at Ashfell and at Mendip. Some young corals found at Oswestry, (Tab. XLVI, fig. 4), and differing from the above-described specimens by the diameter of the corallites being smaller, appear to belong to this species, which, according to Colonel Portlock, is also met with in Ireland, at Derryloran, Erigal-Keerogue (Tyrone), and Crevinish, near Kesh (Fermanagh). Specimens are in the collections of Messrs. Phillips, Stokes, Bowerbank, &c.

Syringopora geniculata is easily recognisable by the very slight geniculation of its very closely set corallites and its numerous connecting tubes.

4. Syringopora catenata.

Erismatolithus (Tubiporites) catenatus, *Martin*, Petref. Derb., pl. xlii, fig. 1, 1809.
(Not the fig. 2 which belongs to the
S. reticulata.)
Syringopora catenata, *M'Coy*, Brit. Palæoz. Foss., p. 83, 1851.

" *Corallum* forming large masses of nearly equal, sub-parallel, very slightly diverging tubes, averaging half a line in diameter, and about their diameter apart, connected by nearly equal, transverse tubuli, slightly more than the diameter of the tubes apart, the origin of each producing a slight angular flexuosity in the main tubes; tubular central opening rather large. Found in the carboniferous limestone of Derbyshire." (*M'Coy*, op. cit.)

We have not had an opportunity of examining this fossil, but it appears to be very closely allied to a species found in the Silurian formation, the *Syringopora fascicularis*, or *Tubipora fascicularis* of Linnæus,[1] and seems to differ from it only by the corallites being rather more regular and more closely set.

Syringopora laxa, of Professor Phillips,[2] is as yet but very imperfectly known; it is described as having its corallites irregularly coalescent and distant, with very few connecting tubes. It was found by that geologist at Ash Fell, Derbyshire, and is mentioned by Colonel Portlock[3] as having been met with at Enniskillen.

The fossils described by Professor M'Coy, under the names of *Aulopora campanulata*,[4] *Aulopora gigas*,[5] *Jania bacillaria*,[6] and *Cladochonus brevicollis*,[7] are evidently young syringoporæ. We are inclined to think that *Cladochonus tenuicollis* of the same author[8] may equally belong to this genus, but we entertain more doubt respecting the natural affinities of the fossils which that geologist first described under the names of *Jania antiqua*[9] and *Jania crassa*,[10] and has more recently referred to the genus Cladochonus.[11]

[1] See our Monographie des Polyp. des Terrains Palæozoiques, p. 293.
[2] Geol. of Yorkshire, p. 201.
[3] Report on Londonderry, &c., p. 338.
[4] Synop. Carb. Foss. of Ireland, p. 190, tab. xxvi, fig. 15.
[5] Op. cit., tab. xxvii, fig. 14.
[6] Op. cit., p. 197, tab. xxvi, fig. 11.
[7] Ann. and Mag. of Nat. Hist., 2d Series, vol. iii, p. 128.
[8] Ibid., vol. xx, p. 227, tab. xi, fig. 8.
[9] Carbonif. Foss. of Ireland, p. 197, tab. xxvi, fig. 12.
[10] Op. cit., tab. xxvii, fig. 4.
[11] Ann. and Mag. of Nat. Hist., 2d Series, vol. iii, p. 134.

Family SERIATOPORIDÆ, (p. lxiii.)

Genus RHABDOPORA, (p. lxiii.)

RHABDOPORA MEGASTOMA.

> DENDROPORA MEGASTOMA, *M'Coy*, Ann. and Mag. of Nat. Hist., 2d Series, vol. iii, p. 129.
> 1849; Brit. Palæoz. Foss., p. 79, pl. iii B, fig. 11, 1851.
> RHABDOPORA MEGASTOMA, *Milne Edwards* and *Jules Haime*, Brit. Foss. Corals, Introd.,
> p. lxiii, 1850; Pol. Foss. des Terr. Palæoz., p. 305, 1851.

Corallum subarborescent; its branches coming off at an angle of about 70°; sub-quadrangular, and differing but little in size. Surface of the cœnenchyma granulated, or subechinulated, and obscurely striated. *Calices* arranged in a single row on each surface of the branches, distant from each other, somewhat oval longitudinally, and having slightly prominent edges. Twelve septal tubercles, somewhat unequal in size, and rather thick. Diameter of the branches a little more than half a line; long diameter of the calices about the same.

Found in the carboniferous limestone in Derbyshire, (Cambridge Museum.)

This coral is the only species belonging to the family of *Seriatoporidæ* that has as yet been discovered in the carboniferous formation. It was referred, by Professor M'Coy, to the genus *Dendropora* of M. Michelin, but we have considered it as constituting the type of a peculiar generical division that differs from the former by the septa being more developed and slightly exsert, by the tetragonal form of its branches, the mode of arrangement of its calices, and the structure of the cœnenchyma, which is echinulate, slightly striated, and not very compact, whereas in *Dendropora* it is quite compact, and its surface completely smooth. Professor M'Coy, who appears to have taken only this last-mentioned character into consideration, does not adopt a generical distinction between *Dendropora* and *Rhabdopora*, because he argues that M. Michelin having overlooked the existence of septa in *Dendropora*, may also not have noticed the granulations of the cœnenchyma. But we must beg leave to remark that the observations of M. Michelin are quite foreign to the motives which induced us to establish our genus *Rhabdopora*; it is never from a description, or a simple inspection of a drawing, that we feel authorised to propose new divisions of that value, but it is from an attentive examination of the fossils themselves that we have formed our opinion, and we are fully persuaded that if Professor M'Coy had been enabled to study, as we have done, both the corals described by himself and that figured by M. Michelin, he would have adopted the conclusions we have ourselves come to, and have considered them as appertaining to two perfectly distinct genera.

Family AULOPORIDÆ.[1]

Genus PYRGIA.[2]

PYRGIA LABECHII.   Tab. XLVI, figs. 5, 5*a*.

> PYRGIA LABECHII, *Milne Edwards* and *Jules Haime*, Pol. Foss. des Terr. Pal., p. 311, 1851.

Corallum simple, subturbinate, scarcely bent, and sub-pedicellate. *Epitheca* thick, and wrinkled transversely. *Calice* circular and very deep; 30 or 40 septal striæ. Height 5 lines ; diameter of the calice 2½ lines.

Found at Frome.   Specimens are in the Museum of Practical Geology, &c.

The genus *Pyrgia* is composed of corals which may be considered as being simple and free Aulopora.

It comprises two species; the above-described fossil and *P. Michelini*,[3] which differs from the first by the existence of a long horizontal peduncle and one or two small spur-like radiciform processes.

Family CYATHAXONIDÆ, (p. lxv.)

Genus CYATHAXONIA, (p. lxv.)

CYATHAXONIA CORNU.

> STYLINA SIMPLE, *Parkinson*, Introd. to the Study of Foss. Org. Rem., pl. x, fig. 4, 1822.
> Good figure.
> CYATHOPHYLLUM MITRATUM (pars), *De Koninck*, Anim. Foss. des Terr. Carb. de Belg.,
> p. 22, pl. c, figs. 5*e*, 5*f*, 1842. (Cæt. excl.)
> Not Goldfuss.
> CYATHAXONIA CORNU, *Michelin*, Icon. Zooph., p. 258, pl. lix, fig. 9, 1846.
> —   —   *Milne Edwards* and *Jules Haime*, Pol. Foss. des Terr. Palæoz., p. 320,
> pl. i, fig. 3, 1851.
> —   —   *M'Coy*, Brit. Palæoz. Foss., p. 109, 1851.
> —   MITRATA, *D'Orbigny*, Prod. de Pal., vol. i, p. 158, 1850.

Corallum cylindro-conical, bent in the form of a horn, pointed at the basis, and surrounded with a thin epitheca which has some slight circular wrinkles, but is never echinulated. *Calice* circular, rather deep, and with thin margins. *Columella* cylindrico-conical, very prominent, slightly compressed, and compact, but having a small central canal. *Septal fossula* narrow, but well defined. *Septa* very thin, narrow at their upper

---

[1] Milne Edwards and Jules Haime, Pol. Foss. des Terr. Pal., p. 310, 1851.   [2] Op. cit., p. 310.
[3] Ibid., Polyp. Palæoz., p. 310, tab. xvii, fig. 8.

end and forming four cycla; those of the first three cycla nearly equal, alternating with an equal number of smaller ones, and extending in general to the columella, where they present a small obtuse lobe; the tertiary ones are inclined towards those of the second cyclum, and become united to them near the centre of the visceral chamber. Height of the coral from 5 to 8 lines; diameter of the calice 2 lines. A vertical section shows that the interseptal loculi are quite open.

This fossil has been found at Kendal and at Tournay. A very ill-preserved coral, met with in some part of Yorkshire, also appears to belong to this species. Professor M'Coy mentions its existence in Derbyshire. The only well-preserved British specimens that we have seen belong to the collections of the Cambridge Museum; specimens from Belgium are common in the palæontological collections in Paris.

In our Monograph of the Corals from the Palæozoic Formations we have described five other species of *Cyathaxonia*, which can all be easily distinguished from *C. cornu*: *C. Konincki*[1] by being fixed; *C. cynodon*[2] by its walls being armed with rows of spines; *C. tortuosa*[3] and *C. profunda*[4] by their septa being more numerous, and by their greater size; and *C. Dalmani*[5] by its thick form and its strongly compressed subcristiform columella.

Family CYATHOPHYLLIDÆ, (p. lxv.)

Sub-Family ZAPHRENTINÆ, (p. lxv.)

1. *Genus* ZAPHRENTIS, (p. lxv.)

1. ZAPHRENTIS CORNUCOPIÆ.

CANINIA CORNUCOPIÆ, *Michelin*, Icon. Zooph., p. 256, pl. lix, fig. 5, 1846. Very bad figure.
ZAPHRENTIS CORNUCOPIÆ, *Milne Edwards* and *Jules Haime*, Pol. Foss. des Terr. Palæoz., p. 331. pl. v, fig. 4, 1851.
CYATHOPSIS CORNUCOPIÆ ? *M'Coy*, Brit. Palæoz. Foss., p. 90, 1851.

Corallum conical, elongated, curved, delicately pedunculated and bearing very slight circular wrinkles. *Calice* oval and deep. *Septal fossula* centro-dorsal, elongated. *Septa* numerous; thirty-two large ones alternating with an equal number of thinner but well-developed ones; the former are rather thick at their upper end, but very narrow, and extend to the edge of the septal fossula, on the side of which they are slightly curved, and become united together. Height of the coral one inch, or somewhat more; great diameter of the calice at least 5 lines; its depth 4 or 5 lines.

---

[1] Milne Edwards and Jules Haime, Polyp. Palæoz., p. 321.   [2] Op. cit., tab. i, fig. 4.
[3] Michelin, Iconogr., tab. lix, fig. 8.   [4] Milne Edwards and Jules Haime, op. cit., p. 323.
[5] Milne Edwards and Jules Haime, op. cit., tab. i, fig. 6.

Professor M'Coy mentioned the existence of this species at Red Castle, Maset Rath, Glasgow, the Isle of Man, and Kendal. Specimens from Tournay are in the Paris Museum, and in the collections of M. de Verneuil and M. Michelin. It is from the latter that we have described this fossil, and it is only on the authority of Professor M'Coy that we enter it here in the list of the British Corals.

Z. cornucopiæ is easily distinguished from most species belonging to the same genus by the position and the form of the septal fossula, which extends from the centre of the calice to a small distance from the mural margin towards the large curve or dorsal side of the corallum. This species is, however, very nearly allied to *Z. Konincki*,[1] from which it differs principally by its calice being circular and its septa thicker and broader. In *Z. centralis*[2] the septal fossula is also placed in the centre of the calice, but does not extend outwards, and the septa are strong, and seem inclined to form four groups. *Z. Griffithi*[3] is much stouter, and its calice presents two small lateral septal fossulæ. In *Z. Enniskilleni*[4] the septal fossula extends from the centre of the visceral chamber towards the concave side or small curve of the wall. In *Z. Bowerbanki*[5] the fossula remains limited to the centre of the calice, and in *Z. Phillipsi*[6] it is almost central and circular.

2. ZAPHRENTIS PHILLIPSI.   Tab. XXXIV, figs. 2, 2a, 2b.

ZAPHRENTIS PHILLIPSI, *Milne Edwards* and *Jules Haime*, Pol. Foss. des Terr. Palæoz.,
p. 332, pl. v, fig. 1, 1851.

Corallum slightly curved, somewhat elongate, and encircled with a few well-marked constrictions, sometimes presenting even a series of solutions of continuity in its wall. *Epitheca* strong. *Calice* circular, very deep, and having a thin margin. *Septal fossula* large, situated towards the dorsal side or large curve of the corallum, but near the centre of the calice, deep, enlarged outwardly, and presenting in its middle a septum that is very distinct from the other ones. In the adult specimens thirty-two principal *septa* thin, very narrow, extending to the edge of the fossula, alternating with an equal number of small ones, and forming four groups in consequence of the three primary ones being slightly prominent, and representing, with the fossula, a four-branched cross; in each of the two of these divisions situated on the dorsal side of the calice there are seven principal septa, and in the two others eight; the first of these septa somewhat deviating from the regular radial arrangement. Height of the corallum about 14 lines; diameter and depth of the calice 7 or 8 lines.

Found at Frome, and Slab-house, in England; at Tournay, in Belgium; and at Sablé, in France.

---

[1] Milne Edwards and Jules Haime, Polyp. Palæoz., tab. v, fig. 5.   [2] Ibid., tab. iii, fig. 6.
[3] See tab. xxxiv, fig. 3.   [4] See tab. xxxiv, fig. 1.   [5] See tab. xxxiv, fig. 4.
[6] See tab. xxxiv, fig. 2.

The British specimens of this species that we have seen in the Museum of Bristol and of Practical Geology, were all younger than some of our Belgian specimens, and that circumstance accounts for their not having so many septa, (twenty-six instead of thirty-two,) their calice was also more or less broken down, the upper tabula appeared also more extensive than in the well-preserved adult individuals; but not having discovered any important difference between all these fossils we are confident in their specific identity.

As the position and the form of the septal fossula appear to furnish very good characters for the different species of this genus, *Z. Phillipsi* may at first sight be distinguished from all the species in which that fossula is placed on the ventral or inverted side of the corallum, and from those in which the fossula, although placed, as in this, on the dorsal side, is quite near to the wall of the calice, and extends but little towards the centre of the visceral chamber. The species in which the fossula so far resembles that of *Z. Phillipsi* differ from it by the following peculiarities: in *Z. cornucopiæ*[1] the fossula is long and narrow; in *Z. Konincki*[2] the septa are thicker towards their upper end and have a prominent lobe at their inner edge; in *Z. Griffithi*[3] there are two small septal fossulæ, and in *Z. Michelini*[4] the general form of the corallum is less elongate and less regular, and the septa are stronger and more equal in size.

3. ZAPHRENTIS GRIFFITHI. Tab. XXXIV, figs. 3, 3*a*.

ZAPHRENTIS GRIFFITHI, *Milne Edwards* and *Jules Haime*, Pol. Foss. des Terr. Palæoz., p. 333, 1851.

Corallum short, turbinate, and slightly curved. *Epitheca* thin, and forming small circular ridges. *Calice* circular, not very deep, and having a thin edge. *Septal fossula* large, deep, extending to the centre of the calice, and placed on the dorsal side of the corallum, (that is to say, towards the convex curve.) Some appearance of two other small septal fossulæ placed at right angles with the former one. Thirty-six principal septa, somewhat unequally developed alternately, not closely set, and uniting two by two at their inner edge, where they are slightly bent; those situated near the fossula are somewhat deviated from the normal radiate direction, and unite at their inner edge so as to constitute the lateral margins of the fossula; an equal number of small septa alternating with those above described. Tabulæ well developed. Height of the corallum 12 or 13 lines; diameter of the calice somewhat more.

The only specimen that we have seen belongs to the collection of Mr. Stokes, and was found at Clifton.

This species differs from all the other known Zaphrentes in having its septal fossula centro-dorsal, two small lateral fossulæ, and a short and broad form.

---

[1] *Caninia cornucopiæ*, Michelin, Icon., tab. lix, fig. 5; Milne Edwards and Jules Haime, Polyp. Palæoz., tab. v, fig. 4.   [2] Milne Edwards and Jules Haime, op. cit., tab. v, fig. 5.

[3] See tab. xxxiv, fig. 3.   [4] Ibid., op. cit., tab. iii, fig. 8.

4. ZAPHRENTIS ENNISKILLENI. Tab. XXXIV, fig. 1.

ZAPHRENTIS ENNISKILLENI, *Milne Edwards* and *Jules Haime*, Pol. Foss. des Terr. Palæoz.
p. 334, 1851.

Corallum conical, slightly curved, pointed at its under end, covered with a thin epitheca, and not showing any circular accretion swellings. *Calice* circular, very deep, and terminated by a thin margin. *Septal fossula* well marked, situated towards the concave or ventral side of the corallum, and not reaching quite to the centre of the visceral chamber. Principal *septa* numerous (about forty), very thin, extremely narrow upwards, and straight or but very slightly curved inwards; two of them somewhat larger than the others, and forming an angle at the end of the septal fossula. An equal number of small septa alternating with the principal ones. Height of the corallum 3 inches; depth of the calice more than half that length; diameter of the calice $1\frac{1}{2}$ inch.

The only specimen that we have seen was presented to the Geological Society by Lord Enniskillen, and had been found by that Palæontologist at Loughgill, in the county of Sligo.

This species may easily be distinguished from all the other known Zaphrentis by the great depth of its calice and the position of the septal fossula.

5. ZAPHRENTIS BOWERBANKI. Tab. XXXIV, figs. 4, 4a.

ZAPHRENTIS BOWERBANKI, *Milne Edwards* and *Jules Haime*, Polyp. Foss. des Terr. Palæoz.,
p. 338, 1851.

Corallum very long, almost cylindrical, strongly curved, terminated by a narrow peduncle, covered with a strong *epitheca*, and presenting well-marked circular constrictions, and accretion swellings. *Calice* circular. *Septal fossula* very small, almost central, situated towards the ventral or concave side of the corallum, and divided at its basis by the principal *septum*, which extends to some distance in its cavity. Principal *septa* not numerous (24), very thin, somewhat unequal, and extending almost to the centre of the calice; rudimentary *septa* alternating with the principal ones. Height of the *corallum* 2 or 3 inches; diameter of the *calice*, 6 lines.

Found at Oswestry, at Frome, and in Ireland. Specimens are in the Collections of the Museum of Practical Geology, of the Geological Society, of Mr. Bowerbank, and of the Paris Museum.

This species is remarkable for the smallness of its well circumscribed, sub-central fossula, and by the way in which one of the primary *septa* extends into its cavity. By the great development of this septum *Z. Bowerbanki* approximates somewhat to the genus *Hallia*;[1] but, in the latter, the *septal fossula* does not exist.

[1] See Introduction, page lxvii.

6. ZAPHRENTIS PATULA.

> CANINIA PATULA, *Michelin*, Icon. Zooph., p. 255, pl. lix, fig. 4, 1846.
> ZAPHRENTIS PATULA, *Milne Edwards* and *Jules Haime*, Polyp. Foss. des Terr. Palæoz., p. 338, 1851.
> CYATHOPSIS FUNGITES, *M'Coy*, Brit. Palæoz. Foss., p. 91, 1851. (Not *Turbinolia fungites*, Fleming.)

Corallum conical, somewhat elongate, strongly curved, delicately pedunculated, and showing well-marked circular accretion swellings. *Calice* large and deep. *Septal fossula* deep, broad, not extending to the centre of the calice, and situated towards the dorsal or convex side of the corallum. Principal *septa* numerous (about 40), equally developed, very thin, and extending on the tabula, in the form of slightly curved ridges. Height of the *corallum* $2\frac{1}{2}$ or 3 inches, diameter of the *calice* almost 2, depth 8 lines.

It is on the authority of Prof. M'Coy that we have described this species as being a British Coral; the numerous specimens which we have seen were all from Boulogne or Tournay; Prof. M'Coy mentions its existence at Hook, Wexford; at Craigie, near Kilmarnock; at Ronald's-way (Isle of Man); at Kendal, Westmoreland; also near Glasgow; and at Blyth, Ayrshire.

Z. patula belongs to the section of the genus *Zaphrentis*, in which the tabulæ are very largely developed, and the septal fossula well constituted. It much resembles *Z. Rœmeri*,[1] but differs from this species in being less curved, in having thinner and straighter septa, and by its tabula being not so large. *Z. cylindrica*[2] and *Z. gigantea*,[3] that also belong to the same subdivision, differ from it by their large size and numerous septa, and *Z. Halli*[4] is much more elongate, and has its septal fossula but little developed.

7. ZAPHRENTIS CYLINDRICA. Tab. XXXV, figs. 1, 1a, 1b.

> CYATHOPHYLLUM FUNGITES, *Portlock*, Rep. on the Geol. of Londonderry, p. 332, 1843. (Not *Turbinolia fungites*, Fleming.)
> CANINIA GIGANTEA, *Michelin*, Icon. Zooph., p. 81, pl. xvi, fig. 1, 1843.
> SIPHONOPHYLLIA CYLINDRICA, *Scouler* in *M'Coy*, Syn. of the Carb. Foss. of Ireland, p. 187, pl. xxvii, fig. 5, 1844.
> CANINIA GIGANTEA and SIPHONOPHYLLIA CYLINDRICA, *D'Orbigny*, Prodr. de Pal. Stratigr., vol. i, p, 158, 1850.
> ZAPHRENTIS CYLINDRICA, *Milne Edwards* and *Jules Haime*, Pol. Foss. des Terr. Palæoz., p. 339, 1851.
> CANINIA GIGANTEA, *M'Coy*, Brit. Palæoz. Foss., p. 89, 1851.

---

[1] Milne Edwards and J. Haime, Polyp. Palæoz., p. 341.   [2] See Tab. xxxv, fig. 1.

[3] Ibid., tab. iv, fig. 1; *Caryophyllia gigantea*, Lesueur, Mémoires du Muséum, vol. 6, p. 296.

[4] Ibid., p. 341.

Corallum very long, almost cylindrical, more or less curved, and having large circular accretion swellings. *Septal fossula* rather small in proportion to the size of the visceral chamber, and varying much in its position relative to the bending of the corallum, but always excentrical, and placed at a small distance from the wall.

Principal *septa* numerous (at least 60), thin, closely set, almost equal, alternating with an equal number of rudimentary ones, and extending in the form of striæ almost to the centre of the calice. *Tabulæ* very large, numerous, and closely set. *Interseptal loculi* filled up with vesicular dissepiments, which appear to be independent of the tabulæ.

Height of the corallum, 1 foot or more; diameter from 2 inches to $3\frac{1}{2}$; depth of the calice, 1 inch.

The specimens of this gigantic Coral that we have seen, have been found at Swansea; at Easkey, Sligo, at Kulkeag, Fermanagh; at Tournay, in Belgium; and at Sablé, in France. Col. Portlock mentions its existence at Carnteel, Tyrone; and at Clonoë, Donaghmore; and Professor M'Coy has found it at Castleton Bay, Isle of Man. Specimens are in the Collections of the Geological Society, the Museum of Practical Geology, the Cambridge Museum, the Bristol Museum, Mr. Stokes's, &c.

Z. cylindrica belongs to the same section as the preceding species, and approximates to the genus *Amplexus*. It differs from *Z. patula*[1] and *Z. Halli*[2] by its large size and its numerous septa. By its general form it much resembles *Z. gigantea*,[3] but it differs from it by the structure of the interseptal loculi, which are filled with small vesicles; whereas, in the last-named species, they are occupied only by the exterior portion of the tabulæ.

8. Zaphrentis (?) subibicina.

> Caninia subibicina, *M'Coy*, Ann. Nat. Hist., 2d ser., vol. vii, p. 167, 1851.
> — — *M'Coy*, Brit. Palæoz. Foss., p. 89, 1851.

" *Corallum* much curved, increasing, when young, at the rate of six lines in one inch, to a diameter of one inch three lines; after which, it remains nearly cylindrical for two or three inches more; surface with a thin, nearly smooth, epitheca, marked with obsolete transverse undulations of growth; when the epitheca is removed, the very fine, equal, costal striæ are brought into view, five in two lines at a diameter of one inch two lines; the outer, small, vesicular area, is rather more than a line wide, within which the sixty-five thick primary radiating lamellæ extend, about four lines towards the centre, leaving the broad, flat, smooth, slightly undulated central portion of the diaphragms about six lines in diameter in parts of the circumference; short secondary lamellæ appear one between each of the primary; lateral siphonal depressions strongly marked; *vertical section* shows the

---

[1] *Caninia patula*, Michelin, Icon., tab. lix, fig. 4.

[2] Milne Edwards and J. Haime, Polyp. des Terr. Palæoz., p. 341.

[3] Ibid., tab. iv, fig. 1.

outer vesicular area (at about the above diameter) one and a half line wide, composed of about four very oblique rows of small rounded cells, extending upwards and outwards, from the broad deflected edges of the diaphragms, which latter are thick, tolerably regular, nearly horizontal in the middle, about three interdiaphragmatal spaces in two lines.

"Not uncommon in the carboniferous limestone of Kendal." (*M'Coy*, op. cit.)

This Coral appears to be specifically identical with the fossil which the same author had previously found at Kendal, and had referred to *Cyathophyllum flexuosum* of Goldfuss, under the name of *Caninia flexuosa*;[1] for in speaking of *C. subibicina* he says: "I suspect that this may be the Coral quoted occasionally by authors under the name of the Devonian *Cyathophyllum flexuosum.*" We are also inclined to think that these fossils belong to a species which is found at Tournay, and was described by ourselves under the name of *Zaphrentis tortuosa*.[2] The description given by Professor M'Coy agrees in most respects with the characteristics of this fossil; but, as no figure of *Z. subibicina* has yet been published, and as some of the peculiarities pointed out by that author (the thickness of the septa, and the great size of the fossula, for example,) do not coincide with what we have observed in *Z. tortuosa*, we have considered it advisable, provisionally at least, to retain here the new specific name given to the British specimens.

Genus AMPLEXUS, (p. lxvi.)

1. AMPLEXUS CORALLOIDES. Tab. XXXVI, figs. 1, 1*a*, 1*b*, 1*c*, 1*d*, 1*e*.

AMPLEXUS CORALLOIDES, *Sowerby*, Min. Conchol., vol. i, p. 165, pl. lxxii, 1814.
— — *Bronn*, Syst. der Urw. Konchylien, p. 49, tab. i, fig. 13, 1824.
— SOWERBYI, *Phillips*, Geol. of York., vol. ii, p. 203, pl. ii, fig. 24, 1836.
— CORALLOIDES, *De Koninck*, An. Foss. des Terr. Carb. de Belg., p. 27, pl. B, fig. 6, 1842.
— SOWERBYI, *M'Coy*, Syn. Carb. Foss. of Ireland, p. 185, 1844.
— CORALLOIDES, *Michelin*, Icon., p. 256, pl. lix, fig. 6, 1846.
— — *Milne Edwards* and *Jules Haime*, Pol. Foss. des Terr. Palæoz., p. 342, 1851.
— — *M'Coy*, Brit. Palæoz. Foss., p. 92, 1851.

No complete specimens of this species have, to our knowledge, been met with; only fragments, varying in length from 3 lines, to 4 or 5 inches, have been found; but by their general form it is evident, that this corallum is very long, cylindrical, and irregularly bent; it presents, as usual, some circular accretion swellings; its epitheca is in many places worn away, so as to leave uncovered the outer edge of the septa, which form equidistant vertical lines. We have seen no specimens in which the calice was preserved. The *septa* are

---

[1] Ann. and Mag. of Nat. Hist., 2d series, vol. iii, p. 133.
[2] Milne Edwards and Jules Haime, Pol. Foss. des Terr. Palæoz., p. 335.

equally developed, thin, set wide apart, and quite marginal; they vary in number from 28 to 58, according to the age and the size of the specimens. The *tabulæ* are very large, very closely set, and the greatest part of their surface is smooth. A small depression, corresponding to the septal fossula, is visible near the wall, and is always more distinct on the last tabula than on the others.

This species is found in Ireland, near Dublin; at Kildare; at Carlingford, Lauth, in the county of Clare; in the Valley of Maine, Kerry; at Killarney, and at Cork. According to Prof. Phillips, it is also met with at Bolland, Kettlewell, Menai Bridge, and in the Isle of Man. We have also seen specimens from Tournay and Visé, in Belgium, Casatchi Datchi in the Oural Mountains, and Varsaw in Illinois, United States.

Specimens are in the Collections of the Geological Society, the Bristol Museum, Mr. Bowerbank, the Paris Museum, M. de Verneuil, &c.

A. coralloides is very remarkable by its elongate cylindroid form. These characters distinguish it at first sight from *A. Henslowi*[1]. It never presents any acute transverse ridges, as those seen in *A. nodulosus*,[2] and *A. annulatus*;[3] nor any spines, as in *A. spinosus*.[4] It much resembles *A. cornubovis*[5] and *A. Yandelli*,[6] but differs from them by having the septa less developed, and the septal fossula shallower.

2. AMPLEXUS CORNU-BOVIS.

> CYATHOPHYLLUM MITRATUM, (pars,) *De Koninck*, Anim. Foss. des Terr. Carb. de Belg., p. 22, pl. c, fig. 5*d*, (cæter. excl.) 1842. A young specimen. (Not *Hippurites mitratus*, Schlotheim.)
>
> — PLICATUM, (pars,) *Ibid.*, op. cit., pl. c, figs. 4*c*, 4*d*, 4*e*, (cæt. excl.)
>
> CANINIA CORNU-BOVIS, *Michelin*, Icon., p. 185, pl. xlvii, fig. 8, 1845.
>
> CYATHOPSIS CORNU-BOVIS, *D'Orbigny*, Prod. de Pal. Univ., vol. i, p. 105, 1850.
>
> — — *M'Coy*, Brit. Palæoz. Foss., p. 90, 1851.
>
> AMPLEXUS CORNU-BOVIS, *Milne Edwards* and *Jules Haime*, Pol. Foss. des Terr. Palæoz., p. 343, 1851.

Corallum cylindro-conical, very elongate, strongly curved, often somewhat twisted, and presenting well-marked, circular accretion swellings. *Epitheca* much wrinkled. *Calice* rather deep. *Septal fossula* almost round, and placed very near the wall towards the dorsal or convex side of the corallum. Principal *septa* numerous, (about thirty,) very thin,

---

[1] See tab. xxxiv, fig. 5.

[2] Phillips, Palæoz. Foss., p. 8; *Amplexus serpuloides*, De Koninck, A. Carb. de Belgique, tab. B, fig. 18.

[3] Verneuil and Jules Haime, Bulletin de la Soc. Géol. de France, 2ᵈᵉ sér., vol. vii, p. 151; Milne Edwards and Jules Haime, Polyp. Palæoz., p. 345.

[4] De Koninck, op. cit., tab. c, fig. 1.

[5] Milne Edwards and Jules Haime, op. cit., tab. ii, fig. 1.　　　　　[6] Ibid., tab. iii, fig. 2.

narrow, equally developed, and alternating with an equal number of smaller ones. *Tabulæ* very large, and to a great extent smooth. Height of the corallum about 3 inches or more; diameter of the tabulæ ¾ of an inch.

Professor M'Coy mentions having met with this species at Corwen.[1] All the specimens that we have examined were from Tournay, in Belgium.

A. cornu-bovis differs from a *A. Henslowi*[2] by its elongate cylindrical form; from *A. spinosus*[3] by the absence of mural spines, from *A. nodulosus*[4] and *A. annulosus*[5] by the form of the circular accretion swellings of the wall, which do not constitute acute ridges; and from *A. Yandelli,*[6] which it resembles the most, by its septa being more numerous and its general form less irregular.

3. AMPLEXUS NODULOSUS.

AMPLEXUS NODULOSUS, *Phillips,* Palæoz. Foss., p. 8, 1841.
— SERPULOIDES, *De Koninck,* Anim. Foss. des Terr. Carb. de Belg., p. 28, pl. B, figs. 7, 8, 1842.
— NODULOSUS, *M'Coy,* Syn. Carb. Foss. of Ireland, p. 185, 1844.
— SERPULOIDES, *Michelin,* Icon., p. 257, pl. lix, fig. 7, 1846.
— NODULOSUS, *Milne Edwards* and *Jules Haime,* Pol. Foss. des Terr. Palæoz., p. 345, 1851.

Corallum very long, sub-cylindrical, slightly flexuous, covered with a well characterised epitheca, and presenting, at the distance of about one line and a half apart, a series of circular prominent sharp ridges. *Septa* quite marginal; about thirty. Height of the largest fragments about 3 inches; diameter 3 lines.

Professor Phillips discovered this fossil in England, but does not mention the locality in which it was found. Professor M'Coy mentions its existence in Ireland. The specimens which we have seen are from Visé, in Belgium.

This species is remarkable for the prominent circular ridges of its wall; the same character is observable in *A. annulatus,*[7] but in this latter Coral the mural ridges are not so closely set and a constriction exists above each of them.

[1] Ann. and Mag. of Nat. Hist., 2d series, vol. iii, p. 133.
[2] See tab. xxxiv, fig. 5.
[3] De Koninck, Ann. Foss. des Terr. Carb. de Belgique, tab. c, fig. 1.
[4] Phillips, Palæoz. Fossils, p. 8; *Amplexus serpuloides,* De Koninck, op. cit., tab. B, figs. 7, 8.
[5] De Verneuil and J. Haime, in Bull. de la Soc. Géol. de France, s. 2, vol. vii, p. 151; Milne Edwards and Jules Haime, Pol. Foss. des Terr. Palæoz., p. 345.
[6] Milne Edwards and Jules Haime, op. cit., tab. iii, fig. 2.
[7] De Verneuil and J. Haime, in Bullet. de la Soc. Géol. de France, s. 2, vol. vii, p. 151; Milne Edwards and Jules Haime, Polyp. des Terr. Palæoz., p. 345.

4. AMPLEXUS SPINOSUS.

> AMPLEXUS SPINOSUS, *De Koninck,* Ann. Foss. des Terr. Carb. de Belg., p. 28, pl. c,
> fig. 1, 1842.
> CYATHAXONIA SPINOSA, *Michelin,* Icon., p. 257, pl. lix, fig. 10, 1846.
> — — *D'Orbigny,* Prod. de Pal., vol. i, p. 158, 1850.
> CALOPHYLLUM SPINOSUM, *M'Coy,* Brit. Palæoz. Foss., p. 91, 1851.
> AMPLEXUS SPINOSUS, *Milne Edwards* and *Jules Haime,* Pol. Foss. des Terr. Palæoz., p. 346,
> 1851.

Corallum elongate, cylindro-turbinate, somewhat twisted, acute at its basis, with a rudimentary epitheca and but slightly developed circular accretion swellings. *Costal ridges* flat, sub-equal, closely set, smooth towards the upper part of the corallum, but in the basal half of this armed with a series of small ascendant spines. *Calice* rather deep. *Septal fossula* small. Upper tabula not very broad, especially in young specimens, and sometimes protruding a little in the cavity of the calice. Principal *septa* not numerous, (sixteen,) very thin, very narrow upwards, presenting a concave denticulated edge inwards, slightly bent towards the centre of the corallum, and alternating with an equal number of very small ones. Height of the corallum about 2 inches; diameter of the calice about 5 lines.

Found in the black carboniferous shale at Poolwart, Isle of Man, and at Tournay, in Belgium. Specimens are in the collections of the Cambridge Museum, the Paris Museum, the Ecole des Mines, M. de Verneuil, &c.

This Coral is distinguished from all the other species of Amplexus by the spines which are developed on the surface of the lower part of its wall.

5. AMPLEXUS HENSLOWI. Tab. XXXIV, figs. 5, 5*a.*

> CYATHOPHYLLUM CERATITES, *Michelin,* Icon. Zooph., p. 181, pl. xlvii, fig. 3, 1845. (Not
> Goldfuss.)
> AMPLEXUS HENSLOWI, *Milne Edwards* and *Jules Haime,* Pol. Foss. des Terr. Palæoz., p. 346,
> pl. x, fig. 3, 1851.

Corallum turbinate, not very elongate, not strongly curved, and having but slight circular accretion swellings. *Epitheca* probably delicate, and when worn off leaving uncovered numerous flat equally developed costæ. *Calice* filled up with extraneous matter in all the specimens examined, but appearing to be deep. *Tabulæ* irregularly developed, large, sloping downwards towards the ventral or concave side of the corallum, and reaching almost from wall to wall. *Septa* appearing to be numerous, narrow, and unequal alternately. *Septal fossula* not observable, on account of the filling up of the calice. Height of the corallum in the large specimens 3 inches; diameter of the calice 2 inches or more.

A specimen of this Amplexus was found by Professor Henslow in the Isle of Man, and placed by that geologist in the collection of the Geological Society. The same species is found at Visé, in Belgium, and near Boulogne, in France.

The AMPLEXUS TORTUOSUS of Phillips,[1] which is a fossil of the Devonian formation, is mentioned by Professor M'Coy as existing also in the carboniferous deposits of Ireland.[2] We have not had an opportunity of examining these corals.

3. *Genus* LOPHOPHYLLUM, (p. lxvi.)

LOPHOPHYLLUM (?) ERUCA.

> CYATHOPSIS (?) ERUCA, *M'Coy*, Ann. Nat. Hist., 2d Ser., vol. vii, p. 167, 1851; Brit.
> Palæoz. Foss., p. 90, 1851.

"*Corallum* very small, sub-cylindrical, after a diameter of three lines, diameter three lines and a half; surface marked with coarse, longitudinal, obtuse lamellar striæ, three in the space of one line; radiating lamellæ strong, slightly irregular, connected by several curved, thick, transverse, vesicular plates in the horizontal section, one of the lamellæ stronger than the rest, and extending through the centre, where it is either thickened or confounded with a slight mesial boss of one of the transverse septa, vertical section, middle third traversed by thick, sub-regular, transverse diaphragms, convex upwardly, three inter-diaphragmatal spaces in one line; outer third on each side formed of one or two rows of irregular large cells, formed by the junction and occasional duplicature of the deflected edges of the diaphragms.

" Very common in the black carboniferous limestone and shale of Beith, Ayrshire." (*M'Coy*, op. cit.)

It appears evident, by this description, that the Coral here mentioned must belong to the genus Lophophyllum, and is, probably, specifically different from the Belgian fossils, which were previously known as appertaining to the same division, for Professor M'Coy says that by its external character it bears the most exact resemblance to *Cyathaxonia cornu*, whereas *Lophophyllum Konincki*[3] and *L. Dumonti*[4] are much more conical and less curved. Professor M'Coy refers this fossil to M. D'Orbigny's genus *Cyathopsis*, which is defined by that geologist as being formed of corals resembling *Amplexus*, but with a septal fossula,

---

[1] Palæoz. Fossils, p. 8, tab. iii, fig. 8.
[2] Syn. Carb. Foss. of Ireland, p. 185.
[3] Milne Edwards and Jules Haime, Polyp. des Terr. Palæoz., p. 349, tab. iii, fig. 4.
[4] Ibid., p. 350, tab. iii, fig. 3.

but we have been enabled to ascertain that this last-mentioned character exists in all true Amplexus, and that no other organic peculiarity distinguish these from the typical form of Cyathopsis; we have, therefore, not adopted the new generical name proposed by M. D'Orbigny and employed by Professor M'Coy. We must also remark that the latter author places in the genus Cyathopsis, together with this Lophophyllum, two species of true Zaphrentis, and we do not well understand on what grounds he has proceeded in so doing, or how to interpret the apparent contradictory statements relative to the characters of Cyathopsis, when Professor M'Coy, after having said " These corals differ from Caninia (or Zaphrentis) in wanting the outer perithecal small vesicular area or lining of the walls," adds that they differ " from Calophyllum, (which I only know by name,) by the vesicular edge of the transverse plates between the lamellæ at the walls," &c.

Sub-Family CYATHOPHYLLINÆ, (p. lxvii.)

1. *Genus* CYATHOPHYLLUM, (p. lxviii.)

1. CYATHOPHYLLUM MURCHISONI.   Tab. XXXIII, figs. 3, 3*a*, 3*b*.

PALÆOSMILIA MURCHISONI, *Milne Edwards* and *Jules Haime*, Ann. Sc. Nat., 3ᵐᵉ serie, vol. x, p. 261, 1848.

STREPHODES MULTILAMELLATUM, *M'Coy*, Ann. and Mag. of Nat. Hist., 2d series, vol. iii, p. 5, 1849.

CYATHOPHYLLUM MURCHISONI, *Milne Edwards* and *Jules Haime*, Pol. Foss. des Terr. Palæoz., p. 369, 1851.

STREPHODES MULTILAMELLATUM, *M'Coy*, Brit. Palæoz. Foss., p. 93, pl. iii c, fig. 3, 1851.

Corallum very long, sub-cylindrical, curved, very slightly compressed, and bearing strong circular swellings placed at about 2 or 3 lines apart. *Calice* somewhat oval; its two diameters as 100 : 130, and its long diameter corresponding to the curve of the corallum. *Septa* very thin, very closely set, almost equal, numerous (about 150), straight or slightly bent, and reaching to the centre of the calice. A vertical section shows that the *tabulæ* are very small and distant; the vesicular dissepiments very small and almost vertical, and the septa well developed. Height of the *corallum*, 7 inches; great diameter of the *calice* 2 inches, small diameter 1½ inch.

Found at Frome, Somersetshire; Tyn-y-castle, Clifton, and Mold. Professor M'Coy mentions its existence at Arnside, Kendal, and Lisardrea, Boyle, Roscommon.

Specimens are in the Collections of the Bristol Museum, the Museum of Practical Geology, the Geological Society, the Cambridge Museum, Mr. Bowerbank, Mr. Stokes, and the Museum of Paris.

Before we were enabled to ascertain the internal structure of this *Corallum* by means of a vertical section, we had been misled by its external characters, and had placed it in the vicinity of the genus *Montlivaultia,* under the generic name of *Palæosmilia.*[1] A vertical section shows what are its real zoological affinities, and we do not think that this fossil ought to be distinguished from the true *Cyathophylla,* which correspond nearly to the *Strephodes* of Professor M'Coy.

C. Murchisoni differs from most of the simple species of the same generical division by the smallness of its tabulæ and the thinness of its septa. It much resembles *C. Wrighti,*[2] but this latter Coral is more compressed, harder, and has stronger septa. It is also closely alluded to *C. Stutchburyi,*[3] which differs from it by the septa being stronger and less numerous, and by having a rudimentary septal fossula.

2. CYATHOPHYLLUM WRIGHTI. Tab. XXXIV, figs. 6, 6*a*.

CYATHOPHYLLUM WRIGHTI, *Milne Edwards* and *Jules Haime,* Pol. Foss. des Terr. Palæoz., p. 370, 1851.

This species much resembles the preceding one, but the corallum is shorter, very much compressed towards the calice, and bent near its basis only. *Calice* elongate in the direction of the curve, flat near its edge, and with a narrow somewhat shallow central depression; proportion of the short and long diameter of the calice 100 : 200, or even 220. *Septa* numerous, (at least 130,) somewhat unequal alternately, thin and straight, or slightly curved inwardly. Height of the corallum $3\frac{1}{2}$ inches; calice: great diameter $2\frac{1}{2}$ inches; short diameter 1 inch.

Found at Frome, Somersetshire. Specimens are in the collections of Mr. Bowerbank and of Dr. Wright (Cheltenham).

The compressed form which exists in this species is very rarely met with in Cyathophyllum. *C. angustum,*[4] the transverse section of which is also oblong, is not bent at its basis as *C. Wrighti* is, and its septa are not so closely set.

3. CYATHOPHYLLUM STUTCHBURYI. Tab. XXXI, figs. 1, 1*a*, 2, 2*a*, and Tab. XXXIII, fig. 4.

TURBINOLIA FUNGITES, *Phillips,* Geol. of Yorkshire, 2d part, p. 203, pl. ii, fig. 23, 1836. (Not *T. fungites,* Fleming.)

— EXPANSA, *M'Coy,* Syn. Carb. Foss. of Ireland, p. 186, pl. xxviii, fig. 7, 1844.

CYATHOPHYLLUM EXPANSUM, *D'Orbigny,* Prod. de Palæont., vol. i, p. 159, 1850. (Not Fischer.)

— STUTCHBURYI, *Milne Edwards* and *Jules Haime,* Pol. Foss. des Terr. Palæoz., p. 373, 1851.

---

[1] Comptes-rendus de l'Académie des Sciences, 1848, vol. xxvii, p. 467.
[2] See tab. xxxiv, fig. 6.   [3] See tab. xxxi, figs. 1, 2 ; tab. xxxiii, fig. 4.
[4] Lonsdale, in Murchison's Silurian Syst., tab. xvi, fig. 9.

Corallum straight, or but very slightly curved; sometimes as broad as high, in other specimens very elongate, and having well characterised circular accretion swellings at unequal distances. *Calice* almost circular, with a small, shallow, central cavity, near which some appearance of a small septal fossula is sometimes visible; a broad convex elevation surrounding this central depression, and the exterior portion of the calice forming a flat or somewhat concave zone. *Septa* numerous, (120 to 140), well developed, somewhat unequal alternately, thin, closely set, and for the most part quite straight; the principal ones reach to the centre of the calice, the others almost as far. Specimens 8 or 10 inches long are not uncommon; but others, in which the calice is equally broad, are not more than two inches high.

Found at Bristol, Lilleshall, Clifton, and, according to Professor Phillips, also at Bolland, Ribblehead, Penyghent, Bowes, Hawes, Coverdale, Brough, Ashfell, Orton, in Northumberland, Durham, Derbyshire, Florence Court, Stradone, and Ireland. Specimens are in the collections of the Geological Society, of the British Museum, of Professor Phillips, at York, of the Paris Museum, and of M. de Verneuil.

This coral remains always simple, but bears great affinity to *C. helianthoides*[1] and *C. regium*;[2] but its tabulæ are larger than in either of these species, and its septa are also more numerous than in the first. It differs also from *C. Murchisoni*[3] by the great development of the tabulæ and the thickness of its septa.

4. CYATHOPHYLLUM REGIUM. Tab. XXXII, figs. 1, 1*a*, 2, 3, 4, 4*a*.

> CYATHOPHYLLUM REGIUM, *Phillips*, Geol. of Yorkshire, 2d part, p. 201, pl. ii, figs. 25, 26, 1836.
> ASTREA CARBONARIA, *M'Coy*, Ann. and Mag. of Nat. Hist., 2d Ser., vol. iii, p. 125, 1849.
> FAVASTREA REGIA, *D'Orbigny*, Prodr. de Palæont., vol. i, p. 160, 1850.
> CYATHOPHYLLUM REGIUM, *Milne Edwards* and *Jules Haime*, Pol. Foss. des Terr. Pal., p. 376, 1851.
> ASTREA (PALASTRÆA) CARBONARIA, *M'Coy*, Brit. Pal. Foss., p. 111, pl. 3 A, figs. 7 and 3 B, fig. 1, 1851.

Corallum compound, massive and astreiform. *Calices* polygonal, very unequal in size, and separated by simple linear ridges; their central depression large but not deep, and surrounded by a circular tumefaction; their exterior portion flat or somewhat concave. *Septa* numerous (120 to 130), very thin, closely set, sub-geminate, almost equal exteriorly, but alternatively extending more or less internally; some not reaching quite to the centre of the calice, the others uniting and becoming slightly flexuous, and exsert there, so as to constitute a kind of false columella of an oblong form, that bears a small longitudinal sulcus resembling a rudimentary septal fossula. Diagonal of the calices varying from 1 to 3 inches.

[1] Goldfuss, Petref. Germ., vol. i, tab. xx, fig. 2; tab. xxi, fig. 1.
[2] See tab. xxxii, figs. 1, 2, 3, 4.   [3] See tab. xxxiii, fig. 3.

Found at Bristol, Corwen, Lofthouse in Nidderdale ; its existence in Pembrokeshire and Wrekin is mentioned by Professor Phillips, and at Bakewell, Derbyshire, by Professor M'Coy.   Specimens are in the collections of the Museum of Practical Geology, of Bristol, of Cambridge, of Professor Phillips, at York, of the Paris Museum, &c.

This coral is liable to some variations in form, which are shown in the figures given in this Monograph.   The circular elevation which usually circumscribes the central calicinal fossula, and which is shown in fig. 1, does not exist in the specimen represented in fig. 3, and in the specimens represented in figs. 2 and 4a, the bottom of the fossula is become prominent.   In the specimen, fig. 3, the corallites are pressed very closely together, and the intercalicular mural ridges are very thin and sharp, whereas in figs. 1 and 2 the approximation of the corallites not being carried so far, the mural ridges are thick and blunt.   We may also remark, that in the specimen fig. 3 the septa are thicker than usual, but that peculiarity appears to be dependent on the process of fossilisation only.

C. regium much resembles *C. helianthoides ;*[1] but in the specimens where the corallites remain free laterally, these are of an almost regular turbinate form, and their calice is not inverted exteriorly, so as to assume the form of a mushroom, as is always the case in *C. helianthoides;* the septa are also thinner and more numerous in the above described species than in the latter-mentioned one.

CYATHOPHYLLUM CRENULARE, of Phillips,[2] appears not to differ specifically from *C. regium,* and to be only a variety with smaller calices.   According to Professor Phillips this fossil is found at Clithero, Mendip, Bristol, and in Derbyshire.

5. CYATHOPHYLLUM PARRICIDA.   Tab. XXXVII, figs. 1, 1a, 1b.

CYATHOPHYLLUM PARACIDA, *M'Coy,* Ann. and Mag. of Nat. Hist., 2d ser., vol. iii, p. 7, 1849; Brit. Palæoz. Foss., p. 86, pl. iii c, fig. 9, 1851.

— — *Milne Edwards* and *Jules Haime,* Pol. Foss. des Terr. Palæoz., p. 385, 1851.

Corallum fasciculate and increasing by calicinal gemmation ; the large calices bearing three or four young corallites, which smother by their growth their parent.   The corallites free laterally, conical or cylindroid, and not bearing circular accretion swellings.   *Calices* circular.   *Septa* not numerous (32), almost equal, thin, and united exteriorly by vesicular dissepiments.   *Tabulæ* large and horizontal.   Diameter of the corallum from 3 to 5 lines.

From Mold, Derbyshire.   Specimens are in the collection of the Museum of Practical Geology, of Cambridge, and of Paris.

[1] Goldfuss, Petref. Germ., vol. i, tab. xx, fig. 2, and tab. xxi, fig. 1.
[2] Geol. of Yorkshire, 2d part, pl. ii, figs. 27, 28 ; *Astrea crenularis,* M'Coy, Syn. of Carb. Foss. of Ireland, p. 187 ; *Actinocyathus crenularis,* D'Orbigny, Prod., vol. i, p. 160.

This species differs from all the other corals belonging to the same generical division by the great extent of the tabulæ, and in that respect much resembles *Amplexus* and *Campophyllum;* it may possibly in reality appertain to this last-mentioned genus, which bears to *Cyathophyllum* similar relationship as *Amplexus* does to *Zaphrentis*, but the specimens which we have had an opportunity of examining were not in a state of preservation sufficiently perfect to enable us to ascertain whether the smooth appearance of the tabulæ was due to the absence of septal prolongations or the accidental destruction of their radii.

6. Cyathophyllum? pseudo-vermiculare.

> Cyathophyllum pseudo-vermiculare, *M'Coy*, Ann. and Mag. of Nat. Hist., 2d Series, vol. iii, p. 8, 1849.
> —   —   —   *Milne Edwards* and *Jules Haime*, Pol. Foss. des Terr. Palæoz., p. 388, 1851.
> —   —   —   *M'Coy*, Brit. Palæoz. Foss., p. 86, pl. iii *c*, fig. 8, 1851.

"Elongate, cylindrical, flexuous; surface very irregularly annulated or transversely nodular, coarsely striated longitudinally (about six striæ in one fourth of an inch); branches averaging from half to three fourths of an inch in diameter; small cylindrical branches project at distinct irregular intervals from the sides; *internal structure;*—central area, rather more than half the diameter of the tube, defined, composed of flat, slightly undulated transverse *septa*, bent downwards at the end, bearing at their circumference a series of from 24 to 27 very short, equal, rather distant, radiating lamellæ, *not* reaching half way to the centre; interval between this inner area and the walls filled with loose cellular structure, formed of little more than a single row of large vesicular curved plates, highly inclined upwards and outwards. Not uncommon in the lower carboniferous limestone of Kendal, Westmoreland; (a variety also occurs in the lower carboniferous limestone of Kiltullagh, Roscommon, Ireland)." (M'Coy, *op. cit.*)

7. Cyathophyllum dianthoides.

> Cyathophyllum dianthoides, *M'Coy*, Ann. and Mag. of Nat. Hist., 2d Series, vol. iii, p. 7, 1849.
> —   —   *Milne Edwards* and *Jules Haime*, Polyp. Foss. des Terr. Palæoz., p. 390, 1851.
> —   —   *M'Coy*, Brit. Palæoz. Foss., p. 85, pl. iii c, fig. 7, 1851.

Corallum compound, forming wide conical masses, increasing by calicular gemmation, and very proliferous (from 8 to 16 young corallites rising sometimes from one parent calice). *Corallites* conico-cylindrical. *Septa* numerous, (96 or 100,) thin, straight, crenulate, and somewhat unequal in extent alternately. *Tabulæ* large, nearly horizontal, somewhat vesicular at certain parts; interseptal vesicular dissepiments abundant and pretty regular. Diameter of the *calices* from 6 lines to 1 inch or more. Found at Arnside and Kendal, Westmoreland.

Specimens are in the Cambridge Museum. This species, by its general aspect, much resembles *Cyathophyllum truncatum* [1] of the Wenlock rocks, but its septa are much thinner, and its corallites more cylindrical.

CYATHOPHYLLUM ARCHIACIS. Tab. XXXIV, fig. 7.

Corallum simple, conical; somewhat elongate, curved, very slightly compressed, and presenting a few slight, broad, circular accretion swellings. Epitheca thin. *Calice* oval, with a lamellate edge, a rather deep cavity, and a rudimentary, elongate, septal fossula. Septa very numerous, very thin, closely set, and appearing to be somewhat unequal alternately; towards the centre of the calice they project a little, so as to constitute paliform lobes, which, by their agglomeration, form an oblong ridge. Height of the corallum about 6 inches; long diameter of the calice about $3\frac{1}{2}$ inches; depth of the calice $1\frac{1}{2}$ or 2 inches.

Found in the carboniferous limestone, at Llanymynch, by Sir Roderick I. Murchison. The specimen here figured belongs to the Collection of the Geological Society.

This species differs from all the other simple *Cyathophylla*, by the oval form of its calice, its paliform lobes, and its rudimentary septal fossula. The fossil which Professor M'Coy [2] has referred to the *Clisiophyllum multiplex* of Keyserling [3] appears to belong to this species; it was found at Kendal, Westmoreland.

Professor M'Coy [4] states that TURBINOLOPSIS BINA, T. CELTICA, T. PAUCIRADIALIS, and T. PLURIRADIALIS of Professor Phillips, which appertain to the genus *Cyathophyllum*, and belong to the Devonian formation, are also met with in the carboniferous deposits in Ireland; but as he has given neither description nor figures of the Corals alluded to, we entertain great doubts relative to the exactness of these determinations.

[1] Polyp. Foss. des Terr. Palæoz., p. 379.
[2] Brit. Palæoz. Foss., p. 95.
[3] Petschora, tab. ii, fig. 1.
[4] Syn. of Carb. Foss. of Ireland, p. 186.

2. *Genus* CAMPOPHYLLUM, (p. lxviii.)

CAMPOPHYLLUM MURCHISONI.   Tab. XXXVI, figs. 2, 2*a*, 3.

CAMPOPHYLLUM MURCHISONI, *Milne Edwards* and *Jules Haime*, Pol. Foss. des Terr. Palæoz., p. 396, 1851.

Corallum somewhat elongate, curved, but not twisted, and bearing but slight circular accretion ridges.   Principal *septa* rather numerous (66), not very thin, rather unequal alternately; an equal number of rudimentary ones.   *Tabulæ* very broad; lateral vesicules small, not numerous, and forming only two or three vertical rows.

Height of the corallum 3 or 4 inches; diameter of the calice about 2 inches.

Specimens of this Coral are in the Collections of the Geological Society and of the Bristol Museum, but we do not know in what part of England they were found.

This species differs from *C. flexuosum*,[1] by its general form, which is not remarkably elongate nor flexuous; and from *C. Duchateli*,[2] by its septa being more numerous, and its interseptal vesicules less abundant.

3. *Genus* CLISIOPHYLLUM, (p. lxx.)

1. CLISIOPHYLLUM TURBINATUM.   Tab. XXXIII, figs. 1, 1*a*, 2.

TURBINOLIA FUNGITES (PARS)? *Fleming*, Brit. Anim., p. 510, 1828.
CYATHOPHYLLUM FUNGITES, *De Koninck*, An. Foss. des Terr. Carb. de Belg., p. 24, pl. D, fig. 2, 1842.
CLISIOPHYLLUM TURBINATUM, *M'Coy*, Ann. Nat. Hist., s. 2, vol. vii, p. 169, 1851.
—   KONINCKI, *Milne Edwards* and *Jules Haime*, Pol. Foss. des Terr. Palæoz., p. 410, 1851.
—   TURBINATUM, *M'Coy*, Brit. Palæoz. Fossils, p. 88 and 96, figs. *a*, *b*, *c*, 1851.

Corallum conical, curved, sometimes rather short and stout, in other specimens long and slender; circular accretion ridges thick and irregular; epitheca strong.   Calice circular, rather deep, with a thin, everted edge.   Forty-four thin, principal *septa*, half of which project towards the centre, and bend slightly on the sides of a well-developed lamellar *columella*.   Rudimentary *septa* alternating with the principal ones.   A vertical section shows, that the exterior area of the visceral chamber is occupied by long, oblique vesicules;

[1] Milne Edwards and Jules Haime, Pol. Foss des Terr. Palæoz., pl. viii, fig. 4.—*Cyathophyllum flexuosum*, Goldfuss, Petref. Germ., vol. i, tab. XVII, fig. 3.

[2] Milne Edwards and Jules Haime, op. cit., p. 396.

that the central area is distinct from the preceding one; and that, in the central area, the oblique lines, resembling a small tent, indicate the position of the small tabulæ which are crossed by the principal septa.

Height of the corallum (in the large specimens) 2½ inches; diameter of the calice very variable.

Found at Oswestry; Nunney, near Frome; Castleton, Derbyshire; Wellington, in Shropshire; and, according to Professor M'Coy, at Beith, Ayrshire.

Specimens are in the Collections of the Museum of Practical Geology, of Bristol, of Mr. Bowerbank, &c.

This species is characterised by its well-developed *columella*, and the very regular arrangement of its *septa*.

The Fossil mentioned by Col. Portlock under the name of *Turbinolia mitrata*,[1] and found by that Geologist in the carboniferous formation at Benburb, appears to belong to this species.

2. CLISIOPHYLLUM CONISEPTUM.   Tab. XXXVII, figs. 5, 5*a*.

> CYATHOPHYLLUM CONISEPTUM, *Keyserling*, Reise in Petschora, p. 164, pl. ii, fig. 2, 1846.
> CYATHAXONIA CONISEPTA, *D'Orbigny*, Prodr. de Pal., t. i, p. 158, 1850.
> CLISIOPHYLLUM CONISEPTUM, *Milne Edwards* and *Jules Haime*, Pol. Foss. des Terr. Palæoz.,
>   p. 411, 1851.

Corallum cylindro-conical, very tall, curved, and presenting well-marked, but small accretion ridges. *Calice* circular. *Columellarian* protuberance conical, prominent, and bearing at its summit a small columellarian lamella. *Septa* not very numerous (60 or 70), thin, unequally developed alternately, some of the largest advancing quite to the centre of the calice, and ascending the columellarian protuberance, under the form of flexuous ridges.

Height of the corallum, in general, 3 or 4 inches, and diameter of the calice about 1½ inch; sometimes much larger.

Found at Ticknell, Mold, and Corwen, in England; and, according to Count Keyserling, at Ylytsch in Russia. The large specimen figured in this Monograph belongs to the Collection of the Bristol Museum.

This species is remarkable for its elongate, cylindro-conical form, the smallness of the central lamina placed at the top of the columellarian protuberance, and the great development of this conical protuberance itself. *C. coniseptum* differs also from *C. Hisingeri*[2] by

---

[1] Report on the Geology of Londonderry, p. 331.

[2] Milne Edwards and Jules Haime, Monogr. des Polyp. des Terr. Palæoz., tab. vii, fig. 5.

its *septa* being more numerous and flexuous inwardly, and from *C. Bowerbanki*,[1] by the irregular arrangement of the *septa* near the columellarian protuberance. *C. Keyserlingi*[2] is more bent, has larger accretion ridges, the interseptal loculi more vesicular, and the septa being less numerous.

3. Clisiophyllum Bowerbanki. Tab. XXXVII, figs. 4, 4a.

Clisiophyllum Bowerbanki, Milne Edwards and Jules Haime, Pol. Foss. des Terr.
Palæoz., p. 411, 1851.

Corallum conical, elongate, curved, very narrow at its basis, and presenting but very slight circular accretion swellings. *Calice* circular. *Septa* 70, or more ; the principal ones rising up towards the centre of the corallum, where they become flexuous ; eight of them larger than the rest, and reaching to the top of the columellarian protuberance. Height of the corallum about $2\frac{1}{2}$ inches; diameter of the calice about 12 or 14 lines.

The specimen here described was found in the Carboniferous Deposits of Ireland, and belongs to the collection of our friend, Mr. J. S. Bowerbank.

This species is characterised principally by the unequal development of its principal septa, eight of which only extend to the top of the columellarian protuberance. It most resembles *C. coniseptum*,[3] but independently of its being much shorter, it differs from this Coral by its septa being much more numerous in proportion to the size of the visceral chamber.

Clisiophyllum bipartitum of Professor M'Coy,[4] much resembles this species, but appears to differ somewhat from it by the mode of arrangement of the principal septa. It was found in Derbyshire.

4. Clisiophyllum Keyserlingi.

Clisiophyllum Keyserlingii, M'Coy, Ann. and Mag. of Nat. Hist., 2d series, vol. iii, p. 2,
1849.
— — Milne Edwards and Jules Haime, Pol. Foss. des Terr.
Palæoz., p. 412, 1851.
— — M'Coy, Brit. Palæoz. Foss., p. 94, pl. iii c, fig. 4, 1851.

Corallum conical, and very elongate; curved, and presenting rather strong circular swellings. *Calice* circular. *Columellarian protuberance* conical, and formed by the prolongation of the principal septal radii, twisted round the axis of the corallum. Principal

----

[1] See tab. xxxviii, fig. 4.   [2] Professor M'Coy, Ann. and Mag. of Nat. Hist., s. 2, vol. iii, p. 2.

[3] See tab. xxxvii, fig. 5.   [4] Ann. and Mag. of Nat. Hist., s. 2, vol. iii, p. 2.

septa forty or fifty in number, rather thick in their outer half, and alternating with an equal number of very small ones. The exterior portion of the corallum very vesicular; the columellarian area very distinct. Height of the corallum from 3 to 5 inches. Diameter of the calice about $1\frac{1}{2}$ inch or more.

Found at Oswestry, Derbyshire, and at Visé, in Belgium. Specimens in the Collections of the Museum of Practical Geology, of Bristol, and of Paris. The British specimens communicated to us for the preparation of this Monograph, were not in a sufficiently good state of preservation to be figured.

C. Keyserlingi differs from all the other species of the same genus, except *C. Danaanum*,[1] by its septa being rather thick, and its walls tumefied; it differs also from the latter, in having the septa more numerous and more unequal.

5. Clisiophyllum? costatum.

> Cyathaxonia costata, *M'Coy*, Ann. and Mag. of Nat. Hist., 2d series, vol. iii, p. 6, 1849.
> Clisiophyllum costatum, *Milne Edwards* and *Jules Haime*, Pol. Foss. des Terr. Palæoz., p. 412, 1851.
> Cyathaxonia costata, *M'Coy*, Brit. Palæoz. Foss., p. 109, pl. iii c, fig. 2, 1851.

This species has been established for a small coral, which appears to be a young Clisiophyllum, and belongs probably to one of the preceding ones, although we are not able to determine its precise specific character. It is conical, with a circular calice containing twenty-six septa. It is about 1 inch high, and 3 lines in diameter at the calice.

Found in Derbyshire, and belonging to the Cambridge Museum.

6. Clisiophyllum bipartitum.

> Clisiophyllum bipartitum, *M'Coy*, Ann. Nat. Hist., 2d series, vol. iii, p. 2, 1849.
> —   —   *M'Coy*, Brit. Palæoz. Foss., p. 93, pl. iii c, fig. 6, 1851.

" Very elongate, conic, nearly cylindrical, with a diameter of one and a quarter inch, for the greater part of its length; strongly and regularly striated externally (about five striæ in one fourth of an inch); external striæ corresponding in number to the radiating lamellæ; in the transverse rough section, the central area is rather more than one third the whole diameter, composed of the edges of confusedly-blended vesicular plates, crossed by a few faint extensions of the radiating lamellæ, and divided into two symmetrical portions by a strong median fissure; the space between this inner area and the outer wall is narrow and regularly radiated with about fifty-eight equal, thin, rather distant

---

[1] Milne Edwards and Jules Haime, Polyp. des Terr. Palæoz., p. 412.

lamellæ, connected by numerous delicate transverse vesicular plates; between each pair at the circumference, a shorter radiating lamella occurs, which only reaches half way to the axis, and where they occur, the connecting vesicular plates are smaller and more numerous than from thence to the axis, the intermediate open cellular space less than the outer one in width; *vertical section* indistinctly triareal; outer area defined, about one sixth of the width on each side, composed of small, much curved, vesicular plates, forming minute semicircular cells, arranged in very oblique rows upwards and outwards, about seven in a row; middle zone rather less than the outer one in width, passing gradually into the central structure, formed of few larger and less-curved vesicular plates than the outer zone, and having a nearly horizontal direction, one or one and a half reaching across the space; central area composed of large, thin, close, little curved vesicular plates, forming a strongly arched series of narrow, elongate cells, the convexity of the arch upwards, conforming to the shape of the central boss in the cup.   If the vertical section be at right angles to the medial fissure, or crest of the central boss, there is a line visible down the middle of the section; *terminal cup* deep, lined by the vertical lamellæ, and having a large oval prominent boss in the centre, traversed by a sharp mesial crest; about one half or one third of the radiating lamellæ ascend the central boss, always in a direct line, those at the sides of the mesial crest being at right angles to it, the others joining at a more acute angle at the approach of the extremity; and, opposite one end of the crest, we generally observe one or two of the radiating lamellæ shorter than the rest, producing a sort of siphon-like irregularity, such as we see in *Caninia* (Zaphrentis).

"In the Carboniferous Limestone of Derbyshire; Shale of Beith, Ayrshire." (*M'Coy*, op. cit.)

4. *Genus* AULOPHYLLUM, (p. lxx.)

1. AULOPHYLLUM FUNGITES.   Tab. XXXVII, fig. 3.

> FUNGITES, *David Ure*, History of Rutherglen and East Kilbride, p. 327, pl. xx, fig. 6, 1793.
> TURBINOLIA FUNGITES, *Fleming*, Brit. Anim. p. 510, 1828.
> —   —   *S. Woodward*, Syn. Table of Brit. Org. Rem., p. 7, 1830.
> CYATHOPHYLLUM FUNGITES, *Geinitz*, Grund. der Verst., p. 571, 1845-6.
> CLISIOPHYLLUM PROLAPSUM, *M'Coy*, Ann. and Mag. of Nat. Hist., 2d series, vol. iii, p. 3, 1849.
> AULOPHYLLUM PROLAPSUM, *Milne Edwards* and *Jules Haime*, Brit. Foss. Corals, Introd., p. lxx, 1850.
> AULOPHYLLUM FUNGITES, *Milne Edwards* and *Jules Haime*, Pol. Foss. des Terr. Palæoz., p. 413, 1851.
> CLISIOPHYLLUM PROLAPSUM, *M'Coy*, Brit. Palæoz. Foss., p. 95, pl. iii c, fig. 5, 1851.

Corallum elongate, cylindro-conical, subpedicellate, curved, presenting small circular accretion ridges, and covered with a well-developed epitheca.   Calice not known; upper end of the corallum almost circular. The circle formed by the inner wall, only half the size of that formed by the outer wall. Septo-costal radii numerous, about 180, thin, almost straight, and unequal in size, alternatively; half of them only pass through the inner wall, and extend to the centre of the visceral chamber; the others occupy only the external zone. Height of the *corallum* about 4 inches; diameter of the exterior wall about 13 lines, that of the inner wall being 4 lines. Found at Kildare and in Derbyshire. Specimens are in the Collections of the Museum of Bristol, Cambridge, and Paris.

Professor M'Coy, in his recently published work on 'Palæozoic Fossils,' rejects the genus *Aulophyllum*, that we had previously proposed the existence of; the inner wall being, as he remarks, "merely a question of degree." That is very true, but we considered such a difference in the degree of development of the constituent part of the corallum as being of sufficient value to authorise generic distinction, because we do not find any gradual passage between the organic form belonging to *Cyathophyllum*, in which the inner wall is rudimentary, or does not exist at all, and that peculiar to *Aulophyllum*, where the inner wall is greatly developed, and almost central. As to the genus *Clisiophyllum*, to which Professor M'Coy refers the above-described corals, it differs from our genus *Aulophyllum*, not only by the characters here alluded to, but also by the central elevation of the tabulæ, and the existence of a true *sublamellar columella*.

Aulophyllum fungites differs from *A. Bowerbanki*[1] by its septa being more numerous, and its inner wall wider in proportion to the diameter of the corallum.

Professor M'Coy[2] mentions a small variety of this species, found in the carboniferous limestone of Lowick, Northumberland; and at Beith, Ayrshire.

2. AULOPHYLLUM BOWERBANKI. Tab. XXXVIII, fig. 1.

AULOPHYLLUM BOWERBANKI, *Milne Edwards* and *Jules Haime*, Pol. Foss. des Terr. Palæoz., p. 414, 1851.

Corallum very elongate, subcylindrical, curved, and presenting laterally a prominent line that appears to correspond to a series of rudimentary septal fossulæ. Diameter of the inner wall about half that of the corallum. *Septo-costal lamellæ* about 120 in number, unequal in size alternatively; the large ones rather thick.

The specimen here described was broken at both extremities, but it may easily be seen that its height must have been at least 10 inches. It was found in the Carboniferous Limestone in Ireland, and belongs to Mr. Bowerbank's collection.

[1] See tab. xxxviii, fig. 1.
[2] Op. cit., p. 96.

5. *Genus* LITHOSTROTION.[1]

1. LITHOSTROTION BASALTIFORME.   Tab. XXXVIII, figs. 3, 3*a*, 3*b*.

LITHOSTROTION, *Luid,* Lithophyllacii Britannici Ichnographia, epistola 5, tab. xxiii, 1760.

—   *Parkinson,* Org. Rem., vol. ii, pl. ix, figs. 3 and 6, 1808.

ASTREA BASALTIFORMIS, *Conybeare* and *William Phillips,* Outlines of Geol. of Engl. and Wales, p. 359, 1822.

ASTREA ARACHNOIDES, *Defrance,* Dict. Sc. Nat., vol. xlii, p. 383, 1826.

LITHOSTROTION STRIATUM, *Fleming,* Brit. Anim., p. 508, 1828.

COLUMNARIA STRIATA, *De Blainville,* Dict. Sc. Nat., vol. lx, p. 316, 1830.—Man. d'Actin., p. 360, pl. lii, fig. 3.

LITHOSTROTION STRIATUM, *S. Woodward,* Syn. Table of Brit. Org. Rem., p. 5, 1830.

CYATHOPHYLLUM BASALTIFORME, *Phillips,* Geology of York, vol. ii, p. 202, pl. ii, figs. 21, 22, 1836.

COLUMNARIA STRIATA, *Milne Edwards,* Ann. de la 2de edit. de Lamarck, vol. ii, p. 343, 1836.

ASTREA HEXAGONA, *Portlock,* Rep. on the Geol. of Londonderry, &c., pp. 332, pl. xxiii, fig. i, 1843.

ASTREA BASALTIFORMIS, Ibid., p. 333.

LITHOSTROTION STRIATUM, *M'Coy,* Syn. Carb. Foss. of Irel., p. 188, 1844.

—   MICROPHYLLUM? *Keyserling,* Reise in Petschora, p. 156, tab. i, fig. 2, 1846.

NEMAPHYLLUM MINUS, *M'Coy,* Ann. and Mag. of Nat. Hist., 2d series, vol. iii, p. 17, 1849.

LITHOSTROTION BASALTIFORME and MICROPHYLLUM, *D'Orbigny,* Prodr. de Pal., vol. i, p. 159, 1850.

—   —   *Milne Edwards* and *Jules Haime,* Pol. Foss. des Terr. Pal., p. 441, 1851.

NEMATOPHYLLUM MINUS, *M'Coy,* Brit. Palæoz. Foss., p. 99, pl. iii B, fig. 3, 1851.

STYLASTREA BASALTIFORMIS, *M'Coy,* ibid., p. 107.

Corallum composite, astreiform. Corallites prismatic, and completely united by their walls. *Calices* very unequal in size. A horizontal section shows that the outer walls are very thin and distinct, and that the existence of the inner walls is indicated only by the limit of the vesicular dissepiments which occupy the exterior zone of the interseptal loculi. Columella small and compressed, but slightly inflated in the middle. *Septa* rather closely set (40 or 50), very thin, delicately flexuous, and varying somewhat in size alternately; the largest only extend near to the columella. Great diagonal of the calices 6 or 8 lines; diameter of the zone occupied by the inner wall, $2\frac{1}{2}$ or 3 lines.

The British specimens here described were found at Bristol, Norfolk, and Kendal. Professor Phillips mentions the existence of the same species at Ribble Head, Moughton Scar, Hesket, Newmarket, and Wrekin; and Colonel Portlock, at Desertmartin, Derry, and

---

[1] See 'Pol. Foss. des Terr. Palæoz.,' p. 432.

Derryloran, in Ireland. According to M. Keyserling, it is also met with in Petschora. Specimens are in the collections of the Bristol, Cambridge, and Paris Museums, of Professor Phillips of York, &c.

The name of *Lithostrotion* was introduced almost a century ago by Luid (1760), and applied to a fossil Coral, which must be either the above-described species, or a species very nearly allied to it, and presenting the same generical characters.

Luid's designation was more recently extended by Fleming to a generical division characterised by that Zoologist, in the following terms, "Corals of aggregate prismatical parallel tubes, with simple stellular discs," ('British Animals,' p. 508.) The genus *Lithostrotion*, thus established in 1828, contained four species, the first of which was Luid's original Lithostrotion, the species No. 4 (*L. marginatum*, Flem.), although too imperfectly characterised to be determinable, evidently belongs to the same generical division, but the species No. 3 (*L. oblongum*), differs from the two preceding ones, and belongs to our genus *Isastrea*, and the species No. 2 (*L. floriforme*), is referable to neither of these forms, and must be placed in a distinct generical division. It is to this last-mentioned genus, (designated recently by Professor M'Coy, under the name of *Lonsdaleia*,) that Mr. Lonsdale applied the generical name of *Lithostrotion*, which, according to the rules generally followed in zoological nomenclature, evidently belongs to the first, that is to say to the group formed by Fleming with Luid's *Lithostrotion* and the allied species.

Goldfuss was not acquainted with any well-characterised *Lithostrotion*, and referred to his genus *Columnaria*, (the typical form of which is *C. alveolata*,) an almost undeterminable fossil, which he called *C. lævis*,[1] and which resembles Luid's *Lithostrotion* by its generical features. M. Dana, in his elaborate work on Zoophytes, published in 1846, very judiciously separates these last-mentioned corals from those which are in reality the typical *Columnaria* of Goldfuss, and which he refers to a new genus, proposed by Mr. Hall, under the name of *Favistella;* he was thus led to apply the name of *Columnaria* to Luid's *Lithostrotion* and to the allied species, that is to say to the genus *Lithostrotion* of Fleming, which must, however, remain distinct from the genus *Columnaria*, of Goldfuss, established essentially for the well-characterised fossil described by the German Palæontologist under the name of *C. alveolata*.

Professor M'Coy had adopted the natural group designated by Fleming under the name of *Lithostrotion*, and by M. Dana under that of *Columnaria*, but has given to it the new name *Nemaphyllum*.

In our opinion the limits of the natural group, so well represented by Luid's *Lithostrotion*, ought not to be restricted to the corals which constitute compact masses, in consequence of the complete lateral coalescence of the corallites, but should also comprise those which, having the same structure and the same mode of multiplication, are not so closely set and form fasciculate aggregations. Sometimes the two forms are met with not

[1] Petref. Germ., tab. xxiv, fig. 8.

only in different specimens of the same species, but even in different parts of the same specimen. The genus *Axinura*, established in 1843 by Count Castelnau for these fasciculate *Lithostrotions*, or the division to which Professor Phillips had previously applied Schweigger's generical name *Lithodendron*, and Professor M'Coy has more recently called *Siphonodendron*, must consequently be abandoned. The genus *Acrocyathus* of M. D'Orbigny is identical with M. Castelnau's genus *Axinura*, and therefore is our system of classification united to Fleming's *Lithostrotion*.

In most species of this group the multiplication of corallites evidently takes place by gemmation, but the young individual which thus shoots from the side of the parent corallites is sometimes produced very near the calicular margin; and, in some of these cases, rising up almost perpendicularly, makes the parent corallite deviate slightly from its primitive direction, and may at first sight be mistaken for an instance of fissiparous reproduction; but the calice never showing signs of incipient division attendant on fissiparity, the appearance of a young corallite thus placed at the side of an adult one, and compressing its calice, is not sufficient to authorise us to admit the existence of that mode of multiplication. Mr. Lonsdale admits that some corals, otherwise resembling Lithostrotion, are in reality fissiparous, and it is on that ground that he has established the genera *Stylastræa* and *Diphyphyllum*[1] which differ only from each other in being aggregate, and consequently astreiform, or free laterally and fasciculate; but the arguments which that distinguished Palæontologist makes use of in favour of this opinion, do not appear conclusive, and we therefore do not see sufficient reason for separating these genera from the ordinary Lithostrotion. It is also on the presumed fissiparous mode of reproduction that Professor M'Coy has separated from the latter (which, as has already been stated, he calls *Nemaphyllum*,) the fossils that constitute his new genus *Stylaxis*,[2] and differ from the *Stylastræa* of Mr. Lonsdale, by the existence of a *Columella*, whereas that organ is not seen in the latter; but its absence is probably only accidental, and due to the process of fossilisation, as is often evidently the case in common Lithostrotions. Till the alleged difference in the mode of multiplication be more satisfactorily demonstrated, we therefore deem it advisable not to separate *Stylaxis* from the old genus *Lithostrotion*, in which we also leave, as above stated, *Axinura*, *Stylastræa*, and *Diphyphyllum*.

L. basaltiforme differs from the other astreiform Lithostrotions by its numerous and thin septa. It is distinguished from *L. Portlocki*[3] and *L. M'Coyanum*[4] by the large size of its calices. In *L. ensifer*[5] the columella is more prominent, the septa thicker, and the walls very slightly developed. *L. aranea*[6] is closer allied to the above-described species, but differs from it by its septa, which are not so closely set and more flexuous, and by its columella being stouter.

---

[1] In Murchison, Verneuil, and Keyserling, Russia and Ural, vol. i, p. 621, 1845.

[2] Ann. of Nat. Hist., vol. iii, p. 119, 1849.

[3] See tab. xlii, fig. 1.   [4] Ib., fig. 2.

[5] See tab. xxxviii, fig. 2.   [6] See tab. xxxix, fig. 1.

2. LITHOSTROTION ENSIFER. Tab. XXXVIII, figs. 2, 2a.

> LITHOSTROTION ENSIFER, *Milne Edwards* and *Jules Haime*, Pol. Foss. des Terr. Palæoz., p. 442, 1851.

Corallum massive, with a flat or sub-convex surface. *Corallites* separated only by a very thin epithecal wall, which in some places is scarcely visible. *Calices* polygonal, often ill-circumscribed, almost flat towards their circumference, and presenting, in their centre, a shallow fossula. *Columella* stout, compressed, and very prominent. Principal *septa* about 30 in number, thin, nearly straight, somewhat unequal alternately; some rudimentary ones. Diameter of the corallites 4 or 5 lines.

From Clifton, (Bristol Museum.) In this fossil the columella is more prominent than in any other species of the same genus, and the walls much thinner; by the mode of coalescence of the corallites it bears some resemblance to the genus *Phillipsastræa*, in which the walls disappear completely.

3. LITHOSTROTION ARANEA. Tab. XXXIX, figs. 1, 1a.

> ASTREA HEXAGONA, var. MINOR? *Portlock*, Rep. on Londonderry, p. 332, pl. xxiii, fig. 2, 1843.
> ASTREA ARANEA, *M'Coy*, Syn. Carb. Foss. of Irel., p. 187, 1844.
> NEMAPHYLLUM ARANEA, *M'Coy*, Ann. and Mag. of Nat. Hist., 2d series, vol. iii, p. 135, 1849.
> LASMOCYATHUS ARANEA, *D'Orbigny*, Prod. de Palæont., vol. i, p. 160, 1850.
> LITHOSTROTION ARANEA, *Milne Edwards* and *Jules Haime*, Pol. Foss. des Terr. Palæoz., p. 443, 1851.

Corallum massive. Corallites irregularly polygonal, some of their sides being sometimes curved, whilst most of them are straight. Inner wall rather distinct. *Columella* compressed; its transverse section fusiform. *Septal radii* very thin, but well developed, and slightly flexuous; 22 or 24 principal ones, extending almost to the columella; an equal number of smaller ones that do not reach to the inner wall. Great diameter of the *Calices* about 6 lines; that of the inner wall about half. Dissepiments of the exterior zone very numerous, and forming small, closely set vesicles. *Tabulæ* appearing to be numerous, and much raised towards the centre by the Columella.

Found at Armagh, Ireland; the specimen represented in this Monograph belongs to the collection of M. de Verneuil.

The fossil which Col. Portlock considers as a small variety of the *Astrea hexagona* belongs probably to this species; it was found in Ireland.

L. aranea much resembles *L. basaltiforme;*[1] it differs from it by its columella being stouter, its septa less numerous and straighter, and its endotheca less condensed.

4. LITHOSTROTION PORTLOCKI.   Tab. XLII, figs. 1, 1*a*, 1*b*, 1*c*, 1*d*, 1*e*, 1*f*, 1*g*.

> ASTREA IRREGULARIS, *Portlock,* Rep. on the Geol. of Londonderry, p. 333, pl. xxiii, figs. 3, 4, 1843. (Not *Defrance.*)
> —   —   *M'Coy,* Syn. of the Carb. Foss. of Ireland, p. 187, 1844.
> ASTREA PORTLOCKI, *Bronn,* Index Palæont., p. 128, 1848.
> NEMAPHYLLUM CLISIOIDES, *M'Coy,* Ann. and Mag. of Nat. Hist., 2d series, vol. iii, p. 18, 1849.
> LITHOSTROTION PORTLOCKI, *Milne Edwards* and *Jules Haime,* Polyp. Foss. des Terr. Palæoz., p. 443, 1851.
> NEMATOPHYLLUM CLISIOIDES, *M'Coy,* Brit. Palæoz. Foss., p. 98, pl. iii B, fig. 2, 1851.

Corallum astreiform.   Corallites somewhat unequal in size, prismatic, and completely united by their exterior walls, which are thin, but very distinct.   Inner wall scarcely visible in some of the calices.   *Septa* (22 to 36), very unequally developed alternately, not closely set, extremely thin, and slightly flexuous; the principal ones extending almost to the *Columella,* which is large, slightly compressed, and prominent.   A vertical section shows, that in the exterior zone of the corallites the vesicular dissepiments form 2 or 3 longitudinal series, and are much inclined inwardly, and that the *tabulæ* are well developed, much raised in the centre, and somewhat divided exteriorly.   Great diagonal of the corallites 3 or 4 lines.

Found at Bristol, Graigbenayth, Wellington, Corwen, in Derbyshire; and, according to Col. Portlock, at Kildress and at Kesh, in Ireland.

Specimens are in the collections of the Museum of Practical Geology, of Cambridge, of Paris, and of M. de Verneuil.

This species differs from *L. aranea,*[2] and from *L. basaltiforme,*[3] by the smaller size of the corallites, by its septa being stouter and less numerous, and its columella being more developed.   It differs from *L. ensifer*[4] by the greater development of its walls and from *L. M'Coyanum*[5] by its calices being smaller, and more unequal in size, and its septa less numerous.

---

[1] See tab. xxxviii, fig. 3.   [2] See tab. xxxix, fig. 1.   [3] See tab. xxxviii, fig. 3.
[4] See tab. xxxviii, fig. 2.   [5] See tab. xlii, fig. 2.

5. LITHOSTROTION M'COYANUM. Tab. XLII, figs. 2, 2a, 2b.

> LITHOSTROTION M'COYANUM, *Milne Edwards* and *Jules Haime*, Pol. Foss. des Terr. Palæoz., p. 444, 1851.

This fossil much resembles *L. Portlocki*, but the corallites are smaller and much more unequally developed; its inner walls are more distinct, and the septa less numerous, (20 or 24,) somewhat thicker, and less unequal in size alternately; they form a prominent circle round the columella, which is also prominent. Diagonal of the large individuals 1½ line, rarely 2 lines.

Found at Oswestry and Matlock, Derbyshire. Specimens are in the collection of the Museum of Practical Geology, of Mr. Bowerbank, and of the Paris Museum.

6. LITHOSTROTION (?) CONCINNUM.

> DIPHYPHYLLUM CONCINNUM, *Lonsdale*, in Murch., Vern., Keys., Russ. and Ural, vol. i, p. 624, pl. A, fig. 4, 1845.
> — LATISEPTATUM, *M'Coy*, Ann. and Mag. of Nat. Hist., 2d ser., vol. iii, p. 8, 1849.
> LITHOSTROTION (?) CONCINNUM, *Milne Edwards* and *Jules Haime*, Pol. Foss. des Terr. Palæoz., p. 446, 1851.
> DIPHYPHYLLUM LATESEPTATUM, *M'Coy*, Brit. Palæoz. Foss., p. 88, pl. iii, fig. 10, 1851.

Corallites elongate, cylindrical, presenting slight circular growth swellings, and surrounded with a thin epitheca. Inner *walls* rather distinct. Principal *septa* 32 in number, very thin, and alternating with an equal number of small ones. *Tabulæ* well developed, smooth towards the centre, the exterior zone occupied by oblique, slightly vesicular dissepiments. Diameter from 3 to 5 lines.

Found at Corwen and in the Oural Mountains. Specimens are in the collection of the Cambridge Museum, and of M. de Verneuil.

All the fossils of this species that we have examined were in a bad state of preservation, and the genus *Diphyphyllum*, established for them by Mr. Lonsdale, does not appear to us sufficiently characterised, for it differs from *Lithostrotion* only by the absence of the columella, and we have much reason to think that the non-existence of that organ is here merely accidental, and due to the process of fossilisation. The considerations which induced Mr. Lonsdale to form this new generic division were founded on the supposed fissiparous mode of multiplication of these corals; but after close examination of their structure we are fully convinced that they are not in reality fissiparous, and that the appearance, which at first sight may be taken for a fissiparous division of the calice, is due to the rapid lateral coalescence of the young individual produced by gemmiparity and the parent corallite.

26

7. Lithostrotion (?) septosum.

Nemaphyllum septosum, *M'Coy,* Ann. and Mag. of Nat. Hist., 2d ser., vol. iii, p. 19, 1849.
Lithostrotion septosum, *Milne Edwards* and *Jules Haime,* Pol. Foss. des Terr. Palæoz.,
p. 444, 1851.

" *Corallum* of long, inseparable, slightly diverging, five or six-angled tubes, with an average diameter of five lines. *Vertical section :* axis straight, thin, flat, three fourths of a line wide ; inner area composed of large, rather distant, slightly arched plates, each of which generally extends across the entire area, so that one lengthened cell, (rarely more,) reaches from one side to the other of this area, having the axis in the middle ; outer area broad, of numerous, minute, much arched, vesicular plates, inclining obliquely upwards and outwards, about four of the little cells in the oblique line from the inner area to the outer wall. *Transverse rough fracture* showing the inner area to be composed of slightly conical or cup-shaped plates, their diameter equal to that of the area, and pierced in the centre by the flat persistent axis. *Polished transverse section,* radiating lamellæ forty-eight, thin, twenty-four of which reach the centre, while the intervening ones are nearly marginal, not reaching half way to the inner zone ; interlamellar vesicular plates very numerous and delicate in the outer zone, apparently absent in the inner zone.

" Very common in the carboniferous limestone of Tullyard, Armagh, Ireland." (*M'Coy,* loc. cit.)

8. Lithostrotion decipiens.

Nemaphyllum decipiens, *M'Coy,* Ann. and Mag. of Nat. Hist., 2d ser., vol. iii, p. 18, 1849.
Lithostrotion decipiens, *Milne Edwards* and *Jules Haime,* Pol. Foss. des Terr. Pal.,
p. 441, 1851.
Nemaphyllum decipiens, *Ibid.,* Brit. Palæoz. Foss., p. 99, 1851.

According to Professor M'Coy, this coral is of the same size as *L. irregulare,* from which it differs by its septa being straighter and its exterior vesicles much more oblique. Found in the carboniferous limestone of Derbyshire.

9. Lithostrotion junceum.   Tab. XL, figs. 1, 1*a,* 1*b.*

Juncei lapidei, *David Ure,* Hist. of Rutherglen, p. 337, tab. xix, fig. 12, 1793.
Caryophyllia juncea, *Fleming,* Brit. Anim., p. 509, 1828.
—   —   *S. Woodward,* Syn. Table of Brit. Org. Rem., p. 6, 1830.
Lithodendron junceum, *Keferstein,* Nat. der Erdkorp., vol. ii, p. 785, 1834.
—   sexdecimale, *Phillips,* Geol. of Yorkshire, vol. ii, p. 202, pl. ii, figs. 11, 12,
13, 1836.
Caryophyllia sexdecimalis, *De Koninck,* Foss. des Terr. Carb. de Belg., p. 17, pl. D,
fig. 4, 1842.
Cladocora sexdecimalis, *Morris,* Cat. of Brit. Foss., p. 33, 1843.
Lithodendron coarctatum, *Portlock,* Rep. on Londonderry, p. 336, pl. xxii, fig. 5, 1843.

LITHODENDRON SEXDECIMALE and COARCTATUM, *M'Coy*, Syn. Carb. Foss. of Ireland, pp. 188-89, 1844.

CLADOCORA SEXDECIMALIS, *Geinitz*, Grund. der Vert., p. 570, 1845-46.

DIPHYPHYLLUM SEXDECIMALE *D'Orbigny*, Prod. de Pal., vol. i, p. 159, 1850.

LITHOSTROTION JUNCEUM, *Milne Edwards* and *Jules Haime*, Pol. Foss. des Terr. Palæoz., p. 435, 1851.

SIPHONODENDRON SEXDECIMALE, *M'Coy*, Brit. Palæoz. Foss., p. 109, 1851.

Corallum composite, fasciculate. *Corallites* very elongate, cylindrical, straight, or but slightly irregular, strongly bent upwards near their basis, and placed at various distances, but rarely coalescent. *Epitheca* delicately wrinkled transversely. *Columella* rather large, slightly compressed. Principal *septa* 16 or 18, alternating with an equal number of smaller ones, and extending very near to the columella. Diameter of the corallites about $1\frac{1}{2}$ line.

Found at Mold, Wellington, Oswestry, Allendale; according to Professor Phillips, at Kirby Lonsdale, Kettlewell, Penyghent, Aldstone Moor, Veynal; according to D. Ure, at Rutherglen, Lanarck; according to Professor M'Coy, at Kendal, Lowick, and Burdiehouse; and according to Colonel Portlock, in Ireland, at Derryloran and Corkstown. It is met with also at Visé, in Belgium, and in the Oural Mountains.

Specimens are in the collections of the Geological Society, of the Museum of Practical Geology, of Mr. Bowerbank, &c.

10. LITHOSTROTION MARTINI. Tab. XL, figs. 2, 2*a*, 2*b*, 2*c*, 2*d*, 2*e*, 2*f*, 2*g*.

ERISMATOLITHUS, &c., *William Martin*, Petref. Derb., pl. xvii, 1809.

CARYOPHYLLIA FASCICULATA, *Fleming*, Brit. Anim., p. 509, 1828. (Not *Lamarck.*)

LITHODENDRON FASCICULATUM, *Phillips*, Geol. of Yorkshire, vol. ii, p. 202, pl. ii, figs. 16, 17, 1836.

CARYOPHYLLIA FASCICULATA, *De Koninck*, Anim. Foss. des Terr. Carb. de Belg., p. 17, pl. D, fig. 5, and pl. G, fig. 9, 1842.

LITHODENDRON CÆSPITOSUM, *M'Coy*, Syn. Carb. Foss. of Ireland, p. 188, 1844. (Not *Goldfuss.*)

LITHODENDRON FASCICULATUM, *Lonsdale*, in Murch., Vern., et Keys., Russ. and Ur., vol. i, p. 600, 1845.

CLADOCORA FASCICULATA, *Geinitz*, Grundr. der Verst., p. 570, 1845-46.

DIPHYPHYLLUM FASCICULATUM, *D'Orbigny*, Prod. de Palæont., vol. i, p. 159, 1850.

LITHOSTROTION MARTINI, *Milne Edwards* and *Jules Haime*, Pol. Foss. des Terr. Palæoz., p. 436, 1851.

SIPHONODENDRON FASCICULATUM, *M'Coy*, Brit. Palæoz. Foss., p. 108, 1851.

Corallum fasciculate. Corallites very tall, cylindrical, slightly flexuous, and often coalescent. *Epitheca* thin, and not hiding the delicate, straight, flat striæ, formed by the closely set costæ. *Calices* circular. *Columella* rather thin, and very much compressed. *Septa* extremely thin, straight, or but slightly curved, and rather closely set; 26 principal ones, reaching almost to the outer of the tabulæ, and 26 small ones that do not extend far from the wall. *Tabulæ* placed at about half a line apart, almost horizontal in the centre, but becoming almost erect near the edge. Diameter of the *Calices* 4 or 5 lines; diameter of the smooth portion of the *tabulæ* 1 line.

Found at Rugley, Oswestry, Corwen; according to Prof. Phillips, at Ribblesdale, Teesdale, Ashfell, Bristol, and in Northumberland; according to Martin, at Bakewell, Winster, and Castleton; and, according to Prof. M'Coy, in Ireland. It is also met with at Visé in Belgium. Specimens are in the collections of the Museum of Practical Geology, of Prof. Phillips, Mr. Bowerbank, Mr. Stokes, the Paris Museum, &c.

L. Martini differs from *L. junceum,*[1] *L. irregulare,*[2] *L. antiquum,*[3] *L. harmodites,*[4] and *L. Stokesi,*[5] by the large size of its corallites, and their numerous septa. It is on the contrary smaller than *L. affine,*[6] which is multiradiate, and has numerous small and regular vesicles in the interseptal loculi. *L. Phillipsi*[7] is the species nearest allied to the above-described fossil, but differs from it by the columella being less compressed, and by the frequent coalescence of the corallites. The other species of this genus, such as *L. aranea,*[8] *L. ensifer,*[9] *L. basaltiforme,*[10] *L. Portlocki,*[11] and *L. M'Coyanum,*[12] are easily distinguished from *L. Martini* by the massive, astreiform structure of their compound corallum.

11. LITHOSTROTION IRREGULARE. Tab. XLI, figs. 1, 1*a*, 1*b*, 1*c*, 1*d*, 1*e*.

SCREW STONE, *Robert Plot*, Nat. Hist. of Staffordshire, p. 195, tab. xii, fig. 5, 1686.

MADREPORA, &c., *Parkinson*, Org. Rem., vol. ii, pl. vi, fig. 8, 1808.

CARYOPHYLLÆA, *Conybeare* and *W. Phillips*, Geol. of England and Wales, p. 359, 1822.

CARYOPHYLLIA FASCICULATA, *De Blainville*, Dict. Sc. Nat., vol. lx, p. 311, 1830. Man. d'Actinol, p. 345. (Not *Lamarck*.)

—   —   *Woodward*, Syn. Table of Brit. Org. Rem., p. 6, 1830.

LITHODENDRON IRREGULARE, *John Phillips*, Geol. of Yorkshire, vol. ii, p. 202, pl. ii, figs. 14, 15, 1836.

CLADOCORA IRREGULARIS, *Morris*, Cat. of Brit. Foss., p. 33, 1843.

LITHODENDRON FASCICULATUM, *Portlock*, Rep. on Londonderry, p. 335, 1849. (Not *Phillips*.)

LITHODENDRON IRREGULARE, *Portlock*, Rep. on Londonderry, p. 336, 1849.

LITHODENDRON PAUCIRADIALE, *M'Coy*, Syn. Carb. Foss. of Ireland, p. 189, 1844.

SIPHONODENDRON PAUCIRADIALE, *M'Coy*, Ann. of Nat. Hist., 2d series, vol. iii, p. 139, 1849.

---

[1] See tab. xl, fig. 1.   [2] See tab. xli, fig. 1.

[3] *Lithodendron cespitosum*, Goldfuss, tab. iii, fig. 4; *Lithostrotion antiquum*, Milne Edwards and Jules Haime, Polyp. Foss. des Terr. Palæoz., p. 439.

[4] Milne Edwards and J. Haime, op. cit., tab. xv, fig. 1.

[5] Ibid., tab. xx, fig. 2.   [6] See tab. xxxix, fig. 2.   [7] See tab. xxxix, fig. 3.

[8] See tab. xxxix, fig. 1.   [9] See tab. xxxviii, fig. 2.   [10] See tab. xxxviii, fig. 3.

[11] See tab. xlii, fig. 1.   [12] See tab. xlii, fig. 2.

DIPHYPHYLLUM PAUCIRADIALE, *D'Orbigny*, Prod. de Palæont., vol. i, p. 159, 1850.

DIPHYPHYLLUM IRREGULARE, *D'Orbigny*, Prod. de Paléont., vol. i, p. 159, 1850.

LITHOSTROTION IRREGULARE, *Milne Edwards* and *Jules Haime*, Polyp. Foss. des Terr. Palæoz., p. 436, 1851.

LITHOSTROTION PAUCIRADIALE, *Milne Edwards* and *Jules Haime*, ibid., p. 439.

SIPHONODENDRON AGGREGATUM, *M'Coy*, Brit. Palæoz. Foss., p. 108, 1851.

Corallum fasciculate. Corallites very tall, cylindrical, and flexuous, especially towards their basis, where they are proliferous, and many of their young branches seem to have avorted, and to have become cemented to the neighbouring individuals. *Columella* but slightly compressed. *Septa* extremely thin, and rather widely set; the principal ones (18 in the young corallites, and 24 in the adult,) extending almost to the centre of the Corallite; the intermediate ones coming very near to the principal ones near the *Columella*, and the small ones almost rudimentary. *Tabulæ* placed at about a quarter of a line apart, and presenting only a very small smooth portion. Diameter of the calices about $2\frac{1}{2}$ lines; that of the smooth part of the tabulæ not quite 1 line.

Found at Castleton, Corwen, and Oswestry; according to Professor Phillips, at Bristol, Ashfell, and in Northumberland; according to Col. Portlock, at Martindesert, Desertcreat; and, according to Professor M'Coy, Magheramore and Tobercury. It is also met with in the Oural mountains.

Specimens are in the collections of the Museum of Practical Geology, Professor Phillips, Mr. Bowerbank, the Paris Museum, &c.

In this species the corallites are broader and more lamelliferous than in *L. junceum*,[1] and on the contrary, smaller and not provided with as many septa as in *L. affine*,[2] *L. Martini*,[3] and *L. Phillipsi*.[4] It is more difficult to distinguish *L. irregulare* from *L. antiquum*,[5] *L. harmodites*,[6] and *L. Stokesi*.[7] The latter differs from it by the existence of mural expansions, which are never met with in the above-described species, and *L. harmodites* differs from it by the existence of connecting tubes, which resemble those of *Syringopora*. *L. antiquum* is most closely allied to it, but its septa are less numerous, although the diameter of the calice be somewhat larger, and its columella is stouter and more compressed.

We are inclined to think that the *Diphyphyllum gracile* of Professor M'Coy[8] is a specimen of *Lithostrotion irregulare*, in which the Columella has been accidentally destroyed by the process of fossilisation. This Coral was found at Lowick in Northumberland.

---

[1] See tab. xli, fig. 1.   [2] See tab. xxxix, fig. 2.   [3] See tab. xl, fig. 2.

[4] See tab. xxxix, fig. 3.

[5] Milne Edwards and J. Haime, Polyp. Palæoz., p. 439; *Lithodendron cespitosum*, Goldfuss, Petref. Germ., tab. xiii, fig. 4.

[6] Milne Edwards and J. Haime, op. cit., tab. xx, fig. 2.

[7] Milne Edwards and Jules Haime, op. cit., tab. xv, fig. 1.

[8] Ann. and Mag. of Nat. Hist., vol. iii, p. 2; vol. vii, p. 168; and Brit. Palæoz. Foss., p. 88, figs. *d, e, f.*

12. LITHOSTROTION AFFINE. Tab. XXXIX, figs. 2, 2a, 2b.

MADREPORA, *Knorr* and *Walch*, Rec. des Mon. des Catastr., pl. lxi*, fig. 2, 1775.
— PECTINATA, &c., *Parkinson*, Org. Rem., vol. ii, pl. vi, fig. 5, and perhaps fig. 9, 1808.
ERISMATOLITHUS MADREPORITES (AFFINIS), *William Martin*, Petref. Derb., pl. xxxi, 1809.
CARYOPHYLLIA AFFINIS, *Fleming*, Brit. Anim., p. 509, 1828.
— — *De Blainville*, Dict. Sc. Nat., vol. lx, p. 311, 1830; Man. d'Actin. p. 346.
— — *Woodward*, Syn. Table of Brit. Org. Rem., p. 6, 1830.
LITHODENDRON AFFINE, *Keferstein*, Nat. der Erdk., vol. ii, p. 785, 1834.
— LONGICONICUM and SOCIALE, *Phillips*, Geol. of Yorkshire, vol. ii, p. 203, pl. ii, figs. 18, 19, 1836.
— SOCIALE and LONGICONICUM, *Portlock*, Rep. on Lond., pp. 335-36, 1843.
— AFFINE and SOCIALE, *M'Coy*, Carb. Foss. of Ireland, pp. 188-89, 1844.
DIPHYPHYLLUM LONGICONICUM and SOCIALE, *D'Orbigny*, Prod., vol. i, p. 159, 1850.
LITHOSTROTION AFFINE, *Milne Edwards* and *Jules Haime*, Pol. Foss. des Terr. Palæoz., p. 437, 1851.

Corallum fasciculate, dendroid. *Corallites* erect, cylindro-turbinate, very tall, giving rise to many young individuals, which bend upwards immediately, often coalescent and cemented together, and covered from top to bottom with a thin slightly wrinkled epitheca. In the parts where the epitheca is worn away, the costæ become visible, and are flat, closely set, and equal in size. *Calice* circular, broad and deep. *Columella* compact, compressed, and projecting in the centre of the calicinal tabulæ in the form of a small crista. *Septa* narrow and closely set; the principal ones almost equal (30 or 32), alternating with almost rudimentary ones, which do not articulate on the surface of the tabulæ, are thin, and appear to be denticulate. *Tabulæ* very closely set, convex towards the centre, but bending upwards towards the circumference, simple, regular, and run through by the columella, to which they are intimately united. The smooth part of their upper surface, on which the septa do not extend, occupies somewhat more than one third of the diameter of the corallum, which is about 5 or 6 lines.

The specimens here described were found at Castleton. Professor Phillips mentions the existence of this species at Kulkeagh mountain, Florencecourt, and Settle; according to Martin, it is met with also at Winster, and, according to Colonel Portlock, at Derryloran and Kilcronaghan. Specimens are in the collections of Mr. Bowerbank, of the Paris Museum, and of M. de Verneuil.

This species is the largest of all the fasciculate Lithostrotions except *L. canadense*,[1] in which the corallites are sometimes cylindrical and quite free laterally, and at other places prismatic and completely cemented together, and in which, also, the centre of the calice is

[1] Milne Edwards and Jules Haime, Pol. Palæoz., tab. xiii, fig. 1.

conical and the columella small. *L. affine* is remarkable for the turbinate form of the young corallites, the existence of a well-marked septal fossula, and the abundance of the interseptal vesicles. The coexistence of these characters distinguishes it sufficiently from all the other species.

13. LITHOSTROTION PHILLIPSI. Tab. XXXIX, figs. 3, 3*a*.

> LITHODENDRON FASCICULATUM, *Keyserling*, Reise in Petsch., p. 170, pl. iii, fig. 2, 1846. (Not *Phillips*.)
> LITHOSTROTION PHILLIPSI, *Milne Edwards* and *Jules Haime*, Pol. Foss. des Terr. Palæoz., p. 439, 1851.

Corallum fasciculate, much resembling *L. Martini,*[1] but differing from it by the frequent coalescence of the corallites, which unite laterally, so as to form small rows, somewhat as in the Halysites. There appears to be about 30 septa. *Columella* somewhat compressed. Diameter of the calices about 4 lines.

Found in Ireland and in Russia. The specimen figured in this Monograph belongs to the collection of Mr. Stokes.

The flexuous form of the corallites, and their mode of partial union by means of longitudinal sutures, distinguish this species from all the other fasciculate Lithostrotions.

14. LITHOSTROTION (?) DERBIENSE.

> STYLASTREA IRREGULARIS, *M'Coy*, Ann. and Mag. of Nat. Hist., 2d ser., vol. iii, p. 9, 1849.
> LITHOSTROTION (?) DERBIENSE, *Milne Edwards* and *Jules Haime*, Pol. Foss. des Terr. Palæoz., p. 445, 1851.
> STYLAXIS IRREGULARIS, *M'Coy*, Brit. Palæoz. Foss., p. 101, pl. iii A, fig. 5, 1851.

This fossil is very imperfectly known, having been found only in a very bad state of preservation. The Corallites are irregular and bent in zigzags; the Calices are polygonal; the septæ flexuous, about 30 in number, rather unequal in size alternately. *Tabulæ* broad and almost horizontal; exterior zone accompanied by two or three rows of vesicular dissepiments. Diameter of the corallites 2 lines. Found in the carboniferous limestone of Derbyshire.

15. LITHOSTROTION MAJOR.

> STYLAXIS MAJOR, *M'Coy*, Ann. and Mag. of Nat. Hist., 2d series, vol. iii, p. 120, 1849.
> — *Milne Edwards* and *Jules Haime*, Polyp. Foss. des Terr. Palæoz., p. 454, 1851.
> — *M'Coy*, Brit. Palæoz. Foss., p. 101, pl. iii A, fig. 4, 1851.

---

[1] See tab. xl, fig. 2.

"Tubes averaging 6 lines in diameter, mostly hexagonal; external surface coarsely striated longitudinally, and transversely marked with strong curved irregularities of growth, the convexity of the curves upwards; *horizontal section*, 63 slender radiating lamellæ, converging from the walls towards the flat central style or axis, which is about one line in width; one half of the lamellæ reach the centre, the intervening ones reach rather more than half way; outer area exhibiting numerous transverse vesicular plates between the radiating lamellæ.

"*Vertical section*, axis straight, riband-like; inner area broad, of slightly-curved vesicular plates, forming rows of lengthened irregular cells, extending obliquely downwards and outwards from the axis, about three in a row; *outer area* of rows of small, hemispherically-curved plates, including small, rounded cells, extending very obliquely upwards and outwards, about 5 or 6 in each row.

"From the carboniferous limestone of Derbyshire." (M'Coy, *op. cit.*)

16. Lithostrotion arachnoideum.

Nemaphyllum arachnoideum, *M'Coy*, Ann. and Mag. of Nat. Hist., 2d series, vol. iii, p. 15, figs. A, B, and p. 16, 1849.

Stylaxis arachnoidea, *Milne Edwards* and *Jules Haime*, Pol. Foss. des Terr. Palæoz., p. 454, 1851.

Nematophyllum arachnoideum, *M'Coy*, Brit. Palæoz. Foss., p. 97, pl. iii A, fig. 6, 1851.

"*Stars*, with from four to seven angles, and averaging from 6 to 9 lines in diameter; *axis* very thin, 1 line wide; *vertical section*, inner vesicular area wider than the outer of little arched plates, inclining slightly downwards from the axis; it takes about two (rarely one) of those plates to reach from the axis to the extent of this area, or two irregularly elongate, unequal cells in a slightly oblique line, from the axis to the wall of the inner area; outer area separated from the inner by a sharp, distinct line on each side, and composed of much smaller and more highly curved vesicular plates, so that there are from 5 to 7 small, nearly equal, rounded cells, extending in a line obliquely upwards and outwards, from the inner area to the outer walls of the tube; *horizontal section*, boundary or divisional walls thin, stars radiated with from 50 to 55 very thin lamellæ, of equal thickness, but alternately long and short, the long reaching to the centre, the short barely entering the edge of the inner area; *weathered surface*, stars flattened, separated by a depressed line; *inner area* forming a gentle convex, oval, or circular boss, with the axis forming a short impressed line in the middle; the radiating lamellæ exhibit numerous delicate, curved, interstitial plates in the outer area, but none in the inner area.

"Forms large masses in the carboniferous limestone of Derbyshire." (M'Coy, *op. cit.*)

17. Lithostrotion Flemingi.

STYLAXIS FLEMINGII, *M'Coy*, Ann. and Mag. of Nat. Hist., 2d series, vol. iii, p. 121, 1849.
— — *Milne Edwards* and *Jules Haime*, Pol. Foss. des Terr. Palæoz., p. 494, 1851.
— — *M'Coy*, Brit. Palæoz. Foss., p. 160, pl. iii A, fig. 3, 1851.

" *Corallum* of very long, prismatic, generally hexagonal, easily separable tubes, averaging 3 lines in diameter; outer surface strongly striated longitudinally, and marked with direct transverse rugosities of growth; bipartite division of the columns frequent;[1] *vertical section,* exhibiting the thin flat axis, surrounded by an inner zone of small, slightly curved, interstitial plates, inclining downwards and outwards from the axis, forming on each side a row of nearly simple oblique cells; *outer zone* of small, vesicular, much-curved plates, inclined in an opposite direction, or upwards and outwards, four or five in a row; *horizontal section,* axis thin, half a line wide, surrounded by about 43 thin, radiating lamellæ, from the walls, half of which only reach half way; numerous small, thin, transverse, connecting plates between the lamellæ in the outer zone.

"Common in the carboniferous limestone of Derbyshire." (M'Coy, *op. cit.*)

6. *Genus* PHILLIPSASTRÆA, (p. lxx.)

1. PHILLIPSASTRÆA RADIATA.   Tab. XXXVII, figs. 2, 2*a.*

ERISMATOLITHUS TUBIPORITES (radiatus), *W. Martin*, Petrif. Derb., pl. xviii, 1809.
TUBIPORA RADIATA, *S. Woodward*, Syn. Table of Brit. Org. Rem., p. 5, 1830.
ASTREA HENNAHII (pars), *Phillips*, Palæoz. Foss., pl. vii, fig. 15 D, (Cæt. excl.), 1841.
     (Not *Lonsdale*).
SARCINULA PLACENTA and PHILLIPSII, *M'Coy*, Ann. and Mag. of Nat. Hist., 2d series, vol. iii, p. 124, 125, 1849.
PHILLIPSASTREA HENNAHII (pars), *D'Orbigny*, Prod. de Paléont., vol. i, p. 107, 1850.
PHILLIPSASTREA RADIATA, *Milne Edwards* and *Jules Haime*, Pol. Foss. des Terr. Palæoz., p. 448, 1851.
SARCINULA PHILLIPSII and PLACENTA, *M'Coy*, Brit. Palæoz. Foss., p. 110, pl. iii B, fig. 9, 1851.

Corallum massive, astreiform, with a flat surface. *Calices* irregularly placed; their edges very slightly prominent, and their central fossula rather deep. *Columella* slender, compressed, and in general not very distinct. Septo-costal radii almost completely confluent exteriorly; 24 or 30 in number, very thin, and becoming alternately unequal in size near the walls, where some of them terminate. Diameter of the Calices about

---

[1] We have here above explained how this appearance may be produced by submarginal gemmation. (See p. 192.)

1½ line; depth one half the diameter. A vertical section shows that the walls are somewhat indistinct; the visceral chambers are closed by large, concave, somewhat irregular *tabulæ*, and that the vesicles occupying the intercostal loculi are small, irregular, twice as high as they are broad, and arranged pretty regularly in horizontal rows.

Found in Derbyshire, and at Corwen; and, according to Martin, at Winster. Specimens are in the Collections of the Museums of Practical Geology, of Cambridge and of Paris.

This species differs from *P. Verneuili*,[1] by its septo-costal radii being very slender, and somewhat unequal, and it appears to differ from *P. tuberosa*[2] by its calices not having prominent edges, and by the horizontal arrangement of the series of interseptal vesicles, which, in this last-mentioned fossil, appear to form concave or flexuous rows. We are, however, not quite sure that these differences may not be accidental.

2. PHILLIPSASTRÆA TUBEROSA.

SARCINULA TUBEROSA, *M'Coy*, Ann. and Mag. of Nat. Hist., 2d series, vol. iii, p. 124, 1849.
PHILLIPSASTREA TUBEROSA, *Milne Edwards* and *Jules Haime*, Pol. Foss. des Terr. Palæoz., p. 449, 1851.
SARCINULA TUBEROSA, *M'Coy*, Brit. Palæoz. Foss., p. 110, pl. iii B, fig. 8, 1851.

Calices prominent, mammiliferous, and in general placed very distant from each other, but irregularly so. Septo-costal radii (32) extremely thin, confluent, and flexuous externally, but not strongly geniculated. Intercostal dissepiments subpolygonal, twice as broad as they are high, somewhat unequal, and appearing to form concave or flexuous rows. Diameter of the Calices, scarcely 2 lines.

From the carboniferous limestone of Derbyshire. Specimen in the Cambridge Museum.

Sub-Family AXOPHYLLINÆ,[3]

1. *Genus* PETALAXIS, (p. lxxi.[4])

PETALAXIS PORTLOCKI. Tab. XXXVIII, figs. 4, 4*a*.

STYLAXIS PORTLOCKI, *Milne Edwards* and *Jules Haime*, Pol. Foss. des Terr. Palæoz., p. 453. 1851.

---

[1] Milne Edwards and Jules Haime, Polyp. Palæoz., p. 447, tab. 10, fig. 5.
[2] *Sarcinula tuberosa*, M'Coy, Ann. and Mag. of Nat. Hist., 2d series, vol. iii, p. 124.
[3] Milne Edwards and Jules Haime, Pol. Foss. des Terr. Palæoz., p. 452.
[4] Under the name of Nematophyllum.

Corallum astreiform; Corallites prismatic; principal septa extremely slender, broader, and less numerous than in *P. M‘Coyana*.[1]

In a former work we referred this species to the *Genus* STYLAXIS of Professor M‘Coy, because, from the inspection of the woodcuts given by that author in the ' Annals of Natural History,' we had been led to suppose that *Stylaxis* differed from *Lithostrotion*, by the existence of a well-developed septal apparatus, and a lamellar columella; but the figures recently published by the same geologist prove that we were mistaken, and that the *Genus* STYLAXIS of Professor M‘Coy does not in reality differ from *Lithostrotion*; we have, therefore, been obliged to abandon that name, and to establish here a new generical division (*Petalaxis*), for the fossils which we formerly named *Stylaxis M‘Coyana*[2] and *S. Portlocki.*

2. *Genus* AXOPHYLLUM, (p. lxxii.)

AXOPHYLLUM RADICATUM, *Milne Edwards* and *Jules Haime*, Pol. des Terr. Palæoz., p. 456; tab. xii, fig. 1, 1851.

We are inclined to think that the fossil coral, found by Col. Portlock in the carboniferous limestone at Stewartston, Tyrone,[3] and referred by that Geologist to the *Turbinolia verrucosa* of Hisinger (which is an Omphyma of the Silurian formation), belongs to this species, or is closely allied to it.

3. *Genus* LONSDALEIA, (p. lxxii.[4])

1. LONSDALEIA FLORIFORMIS. Tab. XLIII, figs. 1, 1*a*, 1*b*, 1*c*, 1*d*, 1*e*, 2, 2*a*.

Stone found in Wales (?), *Llwyd*, Phil. Trans., vol. xxi, p. 187, No. 252, figs. 3, 4, 1700. Very rough figures.

ERISMATOLITHUS MADREPORITES (FLORIFORMIS), *William Martin*, Petref. Derb., tab. xliii, figs. 3, 4, and tab. xliv, fig. 5, 1809.

STYLINA COMPOUND, *Parkinson*, Introd. to the Study of Foss. Org. Rem., pl. x, fig. 5, 1822. Good.

ASTREA FLORIDA, *Defrance*, Dict. Sc. Nat., vol. xlii, p. 383, 1826.

LITHOSTROTION FLORIFORME, *Fleming*, Brit. Anim., p. 508, 1828.

—   —   *S. Woodward*, Table of Brit. Org. Rem., p. 5, 1830.

COLUMNARIA FLORIFORMIS, *De Blainville*, Dict. Sc. Nat., vol. lx, p. 316, 1830; Man., p. 350.

CYATHOPHYLLUM FLORIFORME, *Phillips*, Geol. of York, vol. ii, p. 202, 1836.

---

[1] Milne Edwards and Jules Haime, Pol. Foss. des Terr. Palæoz., p. 453, pl. xii, fig. 5.

[2] Polyp. Palæoz., p. 453, tab. xii, fig. 5.   [3] Report on Londonderry, p. 331.

[4] Under the name of *Lithostrotion.*

ASTREA EMARCIDA, *Fischer*, Oryct. de Moscou, p. 154, pl. xxxi, fig. 5, 1837.
— PENTAGONA (?), *Ibid.*, p. 154.
— MAMMILLARIS, *Fischer*, ibid., p. 154, pl. xxxi, figs. 2, 3.
CYATHOPHYLLUM EXPANSUM, *Ibid.*, p. 155, pl. xxxi, fig. 1, 1837. (Named *Astrea expansa*, in the first edition, 1830.)
LITHOSTROTION MAMILLARE, and ASTROIDES, *Lonsdale*, in Murch., Vern., Keys., Russ. and Ur., vol. i, pp. 606, 607, figs. *a, b, c*, 1845.
CYATHOPHYLLUM ASTREA, *Bronn*, Ind. Palæont., p. 367, 1848.
STROMBODES CONAXIS, *M'Coy*, Ann. and Mag. of Nat. Hist., 2d ser., vol. iii, p. 10, 1849.
LITHOSTROTION MAMILLARE, *D'Orbigny*, Prodr. de Pal., vol. i, p. 159, 1850.
LONSDALIA FLORIFORMIS, *Milne Edwards* and *Jules Haime*, Pol. Foss. des Terr. Palæoz., p. 458, 1851.
STROMBODES CONAXIS, *M'Coy*, Brit. Palæoz. Foss., p. 102, pl. 3 B, fig. 4, 1851.
— FLORIFORME, *Ibid.*, p. 103.

Corallum astreiform. *Corallites* prismatic, very unequal in size, and separated by well-developed exothecal walls. *Calices* rather deep. *Columella* stout, very prominent, compressed at its end, which assumes the form of a small crest, and presents, on its lateral sides, ascendant curved ridges. Twenty-four principal *septa*, which are thin, narrow; form in general, a slight annular protuberance round the central fossula, and alternate with an equal number of smaller septa. The costal prolongation of the septal radii pretty well marked on the exterior zone. Diagonal of the large corallites 8 or 10 lines, sometimes half as much more; diameter of the inner walls from 3 to 5 lines.

A vertical section shows that the interseptal dissepiments are very closely set, (about half a line apart,) and almost horizontal, or sloping slightly upwards towards the columella; the inner walls but little developed, and the vesicles of the exterior zone very unequal in size and very oblique inwards. A horizontal section shows that four or five of these vesicles are placed between the outer and inner walls, and that the regular radiate laminæ pass through the concentric laminæ of the columella, which is dense in the axis of the corallite.

Found at Bristol, Lilleshall, Mold, Oswestry, Whitehaven, Maryport, (Cumberland,) according to Professor Phillips, at Bolland, and according to Professor M'Coy, at Bakewell, in Derbyshire. It is also met with in the carboniferous formation in Russia. Specimens are in the collections of the Geological Society, of the Museum of Practical Geology, of the Bristol and Cambridge Museums, of Mr. Bowerbank, of Professor Phillips, of the Paris Museum, &c.

The generic division, designated here under the name of *Lonsdaleia*, has been considered by Mr. Lonsdale and by many of our contemporaries as being the genus *Lithostrotion*, of Fleming; but the figure in Llwyd's work, quoted by that geologist, can admit of no uncertainty as to the real signification of the latter group, which, as we have already said, was evidently intended to receive the coral described above under that name. Professor M'Coy, who does not adopt Fleming's genus, *Lithostrotion*, applies to the *Lithostrotion* of

Mr. Lonsdale the name of *Strombodes*; the zoological value of which had been previously fixed in a manner quite different by Schweigger and by Goldfuss, and he has at the same time given the name of *Lonsdaleia* to a new division comprising the corals which do not differ from his ill-denominated *Strombodes* by their structure, but are only fasciculate and not coalescent. This latter distinction does not appear to us advisable, and in order not to augment without necessity the number of generic names applied to the same objects, we have employed the name of *Lonsdaleia* in a wider acceptation, and made it equivalent to the *Lithostrotion* of Mr. Lonsdale, and to *Lonsdaleia* and *Strombodes* of Professor M'Coy.

Lonsdaleia floriformis is easily distinguished from *L. rugosa*[1] and *L. duplicata*[2] by the disposition of its corallites to coalesce completely. It differs from *L. Bronni*[3] by the unequal development of its septa, and from *L. papillata*[4] by its columella being larger and more prominent, and by the part of the corallites situated near the exterior wall not being entirely vesicular.

2. Lonsdaleia papillata.

> Cyathophyllum papillatum, *Fischer*, Oryct. de Moscou, p. 155, pl. xxxi, fig. 4, 1837.
> Columnaria Troosti, *Castelnau*, Ter. Sil. de l'Amer. du Nord., pl. xix, fig. 2, 1843.
> Lithostrotion floriforme, *Lonsdale*, in Murch. Vern. Keys. Russ. and Ur., vol. i, p. 609,
>   figs. *a, b, c*, 1845. (Not *Fleming*.)
> —   emarciatum, *Ibid.*, p. 603, figs. *a, f*.
> —   floriforme, *Keyserling*, Reise in Petschora, p. 154, tab. i, fig. 1, 1846.
>   (Synonymis exclusis.)
> Strombodes emarciatum, *M'Coy*, Ann. and Mag. of Nat. Hist., 2d ser., vol. iii, p. 136,
>   1849.
> Lithostrotion floriforme, *D'Orbigny*, Prod. de Pal., vol. i, p. 159, 1850.
> Lonsdalia papillata, *Milne Edwards* and *Jules Haime*, Pol. Foss. des Terr. Palæoz.,
>   p. 460, pl. xi, fig. 2, 1851.
> Strombodes emarciatum, *M'Coy*, Brit. Palæoz. Foss., p. 103, 1851.

Corallum massive; upper end of the corallites polygonal, often tetragonal, with simple, thin edges; exterior zone almost flat; calicinal fossula rather large and deep. *Columella* not very prominent, somewhat attenuated upwards, and bearing laterally subvertical, oblique, slightly curved ridges. Principal *septa* 22 or 24 in number, thin, not extending quite to the columella, and alternating with an equal number of small ones. Great diagonal of the corallites, in general, about 8 lines. Diameter of the walls $3\frac{1}{2}$ or 4 lines.

A vertical section shows that the inner walls are slender, but very distinct, and formed by the inner edge of the vesicles of the exterior zone, which are somewhat unequal, strongly

---

[1] See tab. xxxviii, fig. 5.

[2] *Lonsdalia crassiconus*, M'Coy, Ann. and Mag. of Nat. Hist., s. 2, vol. iii, p. 12.

[3] Milne Edwards and Jules Haime, Polyp. Foss. des Terr. Palæoz., p. 459, tab. ii, fig. 1.

[4] Ibid., p. 460, tab. ii, fig. 2.

arched, slightly inclined inwards, broader than high, and bear on their surface small rudiments of costal prolongations. A horizontal section usually passes through three of these vesicles in the space comprised between the two walls, and shows that the dissepiments in the inner zone are simple, almost horizontal, or slightly arched, and placed at about half a line apart.

Found in Derbyshire, in Russia, and in North America. Specimens are in the collection of M. de Verneuil.

This species much resembles *L. floriformis*,[1] but differs from it by its columella being less stout and less prominent, and by its rudimentary costal radii. It differs from *L. Bronni*[2] by the unequal development of its septa, and from *L. rugosa*[3] and *L. duplicata*[4] by the fasciculate forms of these two latter species.

3. LONSDALEIA RUGOSA. Tab. XXXVIII, fig. 5.

LONSDALIA RUGOSA, *M'Coy*, Ann. and Mag. of Nat. Hist., 2d series, vol. iii, p. 13, 1849.
  —   —   *Milne Edwards* and *Jules Haime*, Pol. Foss. des Terr. Palæoz., p. 461, 1851.
  —   —   *M'Coy*, Brit. Palæoz. Foss., p. 105, pl. iii B, fig. 6, 1851.

Corallites in general free laterally, sub-cylindrical, having very strong circular growth swellings, and covered with a very thick epitheca. *Columella* broad. *Septa* thin, and almost equal, about 10. Diameter of the Corallites very variable;—the largest about 8 lines.

Found at Mold and Corwen. Specimens are in the collections of the Museums of Practical Geology, of Cambridge and of Paris.

This species differs from *L. Bronni*,[5] *L. floriformis*,[6] and *L. papillata*,[7] by the tendency of the Corallites to remain free laterally; and from *L. duplicata*,[8] which it resembles in that respect, by the irregular tumefactions of the surface of the walls of the corallites.

---

[1] See tab. xliii, figs. 1, 2.

[2] Milne Edwards and Jules Haime, Polyp. des Terr. Palæoz., tab. xi, fig. 1.

[3] M'Coy, Brit. Palæoz. Fossils, p. 105.

[4] *Lonsdalia crassiconus*, M'Coy, Ann. and Mag. of Nat. Hist., 2d ser., vol. iii, p. 12. *Erismatolithus duplicatus*, Martin, Petrif. Derb., pl. xxx.

[5] Milne Edwards and Jules Haime, Polyp. Palæoz., tab. xi, fig. 1.

[6] See tab. xliii, figs. 1, 2.   [7] Milne Edwards and Jules Haime, op. cit., tab. xi, fig. 2.

[8] *Lonsdalia crassiconus*, M'Coy, Ann. of Nat. Hist., 2d series, vol. iii, p. 12.

4. LONSDALEIA DUPLICATA.

ERISMATOLITHUS (MADREPORITES) DUPLICATUS, *W. Martin*, Petrif. Derb., pl. xxx, 1809.
CARYOPHYLLIA DUPLICATA, *Fleming*, Brit. Anim., p. 509, 1828.
—　　　　　　—　　*S. Woodward*, Syn. Table of Brit. Org. Foss., p. 5, 1830.
CLADOCORA DUPLICATA, *Geinitz*, Grund. der Verst., p. 570, 1845-6.
LONSDALEIA? STYLASTRÆAFORMIS, *M'Coy*, Ann. of Nat. Hist., 2d series, vol. iii, p. 461,
　　　　1849.
LONSDALEIA CRASSICONUS, *M'Coy*, Ann. and Mag. of Nat. Hist., 2d series, vol. iii, p. 12,
　　　　1849.
LONSDALIA CRASSICONUS, *Milne Edwards* and *Jules Haime*, Pol. Foss. des Terr. Palæoz.,
　　　　p. 461, 1851.
LONSDALEIA CRASSICONUS, *M'Coy*, Brit. Palæoz. Foss., p. 103, pl. iii B, fig. 5, 1851.
LONSDALEIA DUPLICATA, *M'Coy*, ibid., p. 105.
LONSDALEIA STYLASTREIFORMIS, *M'Coy*, ibid., p. 106, pl. iii B, fig. 7.

Corallites in general free laterally, and having but slight, circular, mural growth swellings. *Septa* rather thin, and sub-equal; 24 or 26; exterior zone filled with very large vesicles. Diameter of the Corallites about 1 inch.

Found at Arnside, Kendal, Bakewell, Derbyshire, (Museum of Cambridge).

This species differs from *L. rugosa*[1] by the small development of the growth swellings of its epitheca; from *L. Bronni*,[2] by its *septa* being less numerous, and its *columella* smaller, and from *L. floriformis*,[3] and *L. papillata*,[4] by its septa being almost equally developed, and by the great size of the vesicles of its exterior zone.

GENERA INCERTÆ SEDIS.

1. *Genus* MORTIERIA, (p. lxxiv.)

MORTIERIA VERTEBRALIS.

MORTIERIA VERTEBRALIS, *De Koninck*, Anim. Foss. des Terr. Carb. de Belg., p. 12, pl. B,
　　　　fig. 3, 1842.
—　　　　—　　*Michelin*, Icon. Zooph., p. 253, pl. lix, fig. 1, 1846.
—　　　　—　　*Milne Edwards* and *Jules Haime*, Pol. Foss. des Terr. Palæoz.,
　　　　p. 467, 1851.

Corallum short, cylindroid, having the form of a biconcave centrum of the vertebra of some fishes. About 100 septal radii. Diameter two or three inches; height, according to M. de Koninck, varying from 3 lines to 2 inches.

---

[1] See tab. xxxviii, fig. 5.　　　　　[2] Milne Edwards and Jules Haime, op. cit., tab. xi, fig. 1.
[3] See tab. xliii, figs. 1, 2.　　　　[4] Milne Edwards and Jules Haime, op. cit., tab. xi, fig. 2.

Till lately this singular fossil had been found only in the carboniferous deposits of Tournay, in Belgium, but a specimen bearing the indication of Derbyshire was sent, together with other fossils, to the Museum of Paris, by Lady Hastings.

2. *Genus* HETEROPHYLLIA, (p. lxxiii.)

1. HETEROPHYLLIA GRANDIS.

> HETEROPHYLLIA GRANDIS, *M'Coy*, Ann. and Mag. of Nat. Hist., 2d ser., vol. iii, p. 126, figs. *a, b,* 1849.
>
> —   —   *Milne Edwards* and *Jules Haime,* Pol. Foss. des Terr. Palæoz., p. 467, 1851.
>
> —   —   *M'Coy,* Brit. Pal. Foss., p. 112, pl. iii A, fig. 1, 1851.

" Stem slightly flexuous, about 5 lines in diameter, scarcely tapering in 3 inches; longitudinally marked with deep unequal grooves, and few, large, polygonal, unequal ridges, giving a very irregularly angulose section to the stem; surface smooth; horizontal section, few, distant lamellæ, destitute of any order of arrangement, but irregularly branching and coalescing in their passage from the solid external walls towards some indefinite point near the centre, where the few main lamellæ irregularly anastomose. *Vertical section* showing about the middle an irregularly flexuous line, (the edge of one or two of the radiating vertical lamellæ,) from which, on each side, a row of thin, distant, sigmoidally curved plates extends obliquely upwards and outwards, forming a row of large rhomboidal cells on each side.

" Rare in the carboniferous limestone of Derbyshire." (*M'Coy*, op. cit.)

2. HETEROPHYLLIA ORNATA.

> HETEROPHYLLIA ORNATA, *M'Coy*, Ann. and Mag. of Nat. Hist., 2d ser., vol. iii, p. 127, 1849.
>
> —   —   *Milne Edwards* and *Jules Haime,* Pol. Foss. des Terr. Palæoz., p. 467, 1851.
>
> —   —   *M'Coy,* Brit. Pal. Foss., p. 112, pl. iii A, fig. 2, 1851.

" Stems sub-cylindrical, long, flexuous, averaging one and a half lines in diameter, with about sixteen narrow, sub-equal, longitudinal ridges, sharply defined, and separated by flat spaces rather wider than the ridges they separate; the ridges are set with small round tubercles more than their own diameter apart; surface very minutely granulose; internal structure as in the preceding species. *Horizontal section,* lamellæ about fourteen at the margin, (one usually coinciding with each external ridge).

" Rather rare in the carboniferous limestone of Derbyshire." (*M'Coy*, op. cit.)

THE

PALÆONTOGRAPHICAL SOCIETY.

INSTITUTED MDCCCXLVII.

LONDON:

MDCCCLIII.

A MONOGRAPH

OF THE

BRITISH FOSSIL CORALS.

BY

H. MILNE EDWARDS,

DEAN OF THE FACULTY OF SCIENCES OF PARIS; PROFESSOR AT THE MUSEUM OF NATURAL HISTORY;
MEMBER OF THE INSTITUTE OF FRANCE;
FOREIGN MEMBER OF THE ROYAL SOCIETY OF LONDON, OF THE ACADEMIES OF BERLIN, STOCKHOLM, ST. PETERSBURG,
VIENNA, KONIGSBERG, MOSCOW, BRUXELLES, HAARLEM, BOSTON, PHILADELPHIA, ETC.,

AND

JULES HAIME.

FOURTH PART.

CORALS FROM THE DEVONIAN FORMATION.

LONDON:
PRINTED FOR THE PALÆONTOGRAPHICAL SOCIETY.
1853.

DESCRIPTION

OF

THE BRITISH FOSSIL CORALS.

CHAPTER XV.

CORALS FROM THE DEVONIAN FORMATION.

THE British Corals appertaining to the Devonian Formation are in general so completely imbedded and filled up with extraneous calcareous matter, that it is difficult to distinguish them otherwise than by the study of polished sections; but these usually show their structural characters in a very satisfactory manner, and enable the Palæontologist to recognise their zoological affinities. In France and in Germany the corals belonging to the same geological period are, on the contrary, often met with in an excellent state of preservation, and show all the details of their exterior surface, as well as the most minute parts of their interior organisation; but in this Monograph we have only figured British specimens. For the more complete representation of some species, we must consequently refer to other works, such as the excellent publication of ' Goldfuss on the German fossils,' and our *Monographie des polypiers des terr. Palæozoiques.* The corals discovered in the Devonian Formation, in the different parts of the world, belong to about 150 well-defined species, 46 of which have been met with in England. To these British corals may be added 3 species that are very imperfectly known, casts only of them having been as yet found, and few specimens that have received names, but are not determinable zoologically. Almost half of the British species have not as yet been found in other countries; 22 have been discovered on the Continent; and we may also remark that most of the American species are not seen here, only 6 of the latter have been met with in England; among these, 5 are at the same time Continental. The corals belonging to the family of the Cyathophyllidæ are very predominant, and form 33 of the 46 above-mentioned species. The family of Favositidæ is represented by 10 species, and the three remaining species belong one to each of the three families Stauridæ, Milleporidæ, and Poritidæ; with the exception of one species of Poritidæ that we have not seen, all these fossils belong, therefore, to the two sub-orders *Zoantharia tabulata* and *Zoantharia rugosa*, one of which has no representative in the actual Fauna, nor in the Tertiary and Secondary Formations. Three of these Devonian fossils exist also in the Silurian rocks, but all the others appear to be peculiar to the Devonian period.

The principal localities from which these corals have been obtained are Torquay, Plymouth, and Newton Bushel. The specimens described in this Monograph belong mostly to the collections of Dr. Battersby, Mr. Pengelly, Mr. Bowerbank, Mr. Phillips, and the Geological Society; some of the latter were more particularly valuable to us, being the type specimens of the species described in 1840 by Mr. Lonsdale in the memoir of Messrs. Sedgwick and Murchison, on the Devonian Formation of England.

Family MILLEPORIDÆ, (*Introd.*, p. lviii.)

1. *Genus* HELIOLITES, (*Introd.*, p. lviii.)

HELIOLITES POROSA.　Tab. XLVII, figs. 1, 1*a*, 1*b*, 1*c*, 1*d*, 1*e*, 1*f*.

HELIOLITHE PYRIFORME, à étoiles d'une demi-ligne de diamètre, &c., *Guettard*, Mém. sur les Sc. et les Arts, vol. iii, p. 454, pl. xxii, figs. 13 and 14, 1770.
ASTREA POROSA, *Goldfuss*, Petref. Germ., vol. i, p. 64, tab. xxi, fig. 7, 1826.
HELIOPORA PYRIFORMIS, *De Blainville*, Dict. Sc. Nat., vol. lx, p. 357, 1830; Manuel d'Actinologie, p. 392.
—　　　—　　　*Steininger*, Mém. Soc. Géol. France, vol. i, p. 346, 1831.
—　　INTERSTINCTA, *Bronn*, Leth. Geogn., vol. i, p. 48, tab. v, fig. 4, 1835.
PORITES PYRIFORMIS, *Lonsdale*, Geol. Trans., 2d series, vol. v, pl. lviii, fig. 4, 1840. (Not Lonsdale in Silur. System.)
—　　　—　　　*Phillips*, Palæoz. Foss., p. 14, pl. vii, fig. 19, 1841.
EXPLANARIA INTERSTINCTA, *Geinitz*, Grundr. der Verst., p. 568, 1845-46.
GEOPORITES POROSA and PHILLIPSII, *D'Orbigny*, Prodr. de Paléont., vol. i, pp. 108, 109, 1850.
HELIOLITES POROSA, *Milne Edwards* and *Jules Haime*, Pol. Foss. des Terr. Palæoz., p. 218, 1851.
PALÆOPORA PYRIFORMIS, *M'Coy*, Brit. Palæoz. Fossils, p. 67, 1851.

Corallum compound, forming generally a globular mass, which in some specimens is subgibbose, in others cylindrical; sometimes composed of very distinct superposed layers. *Calices* somewhat unequal in size, placed rather irregularly at distances equal to about two or three times their diameter, surrounded by a very thin rim, and slightly elevated above the general surface of the corallum. The calicular fossula large and rather deep. *Septa*, twelve in number, somewhat unequal in size alternately, almost straight, thick exteriorly, and extending almost to the centre of the visceral cavity. The pores of the *Cænenchyma* small, almost equal in size and nearly regularly hexagonal; one eighth of a line in diameter. Calices almost half a line in diameter.

A vertical section of this corallum shows that the *Tabulæ* are horizontal, or slightly oblique, and less closely set than in the other species. The laminæ of the tubes of the Cænenchyma are thin, but assume the appearance of vertical lines, much more strongly marked than those formed by the dissepiments. The latter appear to be quite independent of the adjacent tubes, and are not, in general, placed so as to correspond together horizontally.

Found at Torquay, Teignmouth Beach, Walcombe Beach, Woolborough Quarry, Babbacombe, Newton, Plymouth, and Marychurch; and also in Germany in the Eifel Mountains, and on the banks of the river Lahn.

Specimens are in the Museums of the Geological Society of London, of Practical Geology, and in the collections of Messrs. J. S. Bowerbank, Battersby, and Pengelly.

This coral has often been confounded with the *Heliolites interstincta*,[1] but differs from it by the calices being much less closely set, and by the Cænenchyma being more developed. In *H. Murchisoni*[2] the calices are also very distant, but in the above-described species the tabulæ are less numerous, and the dissepiments of the Cænenchyma are thinner than the vertical laminæ of the tubes of the same tissue, and do not correspond among themselves so as to constitute horizontal strata.

We have not adopted the name of *Pyriformis*, which Blainville, Mr. Lonsdale, and Mr. M'Coy apply to this species, as having been given to it by Guettard; because the French epithet *pyriforme* was given by Guettard himself to several other species, but not made use of as a specific name, and because Goldfuss had called it *Astrea porosa* before Blainville proposed employing the former designation.

We also see no reason for giving to the genus, to which this specimen belongs, the name of *Palæopora* in preference to that of *Heliolites*, the latter having been revived from Guettard's writings in 1846 by Mr. Dana, and the former having been introduced only in 1848 by Mr. M'Coy. The name of *Géoporites*, given more recently to the same group of corals by M. D'Orbigny, must also, in consequence of the law of priority, be rejected.

2. Genus BATTERSBYIA.[3]

BATTERSBYIA INÆQUALIS. Tab. XLVII, figs. 2, 2a, 2b.

BATTERSBYIA INÆQUALIS, *Milne Edwards* and *Jules Haime*, Monogr. des Pol. Foss. des Terr. Palæoz., p. 227, 1851.

Corallum composite, massive. *Corallites* very unequal in size, with thick non-costulate walls, united together by a thin, spongiose, irregular Cænenchyma. *Calices* almost circular, never subpolygonal. *Septa* small but well defined, somewhat unequal in size alternately, rather thick towards the wall, but very thin inwardly; 26 of them in the large calices. *Tabulæ* appearing to be vesicular, and filling the visceral chambers. *Cænenchyma*

---

[1] *Astrea porosa*, Hisinger, Leth. Succ., p. 98, tab. xxviii, fig. 2, 1837.

[2] Milne Edwards and Jules Haime, Pol. Foss. des Terr. Palæoz., p. 215, 1851.

[3] Milne Edwards and Jules Haime, Polyp. Foss. des Terr. Palæoz., pp. 151, 227, 1851.

as in the other Madreporites. Diameter of the large individuals rather above one and a half line; that of the young corallites less than half a line, and between these two extremes numerous differences.

This remarkable fossil cannot be placed in any of the generical divisions enumerated in the Introduction to this Monograph, and constitutes the type of a new division, which we have established in our work on the Corals of the Palæozoic formations. It is characterised by its spongy Cænenchyma, and its subvesicular Tabulæ.

Found at Teignmouth by Dr. Battersby.

We have dedicated this genus to Dr. Battersby, of Torquay, in commemoration of the liberality with which that gentleman has communicated to us, for description, a most valuable series of Devonian fossils.

Family FAVOSITIDÆ, (*Introd.*, p. lx.)

Sub-family FAVOSITINÆ, (*Introd.*, p. lx.)

1. *Genus* FAVOSITES, (*Introd.*, p. lx.)

1. FAVOSITES GOLDFUSSI. Tab. XLVII, figs. 3, 3*a*, 3*b*, 3*c*.

CALAMOPORA GOTHLANDICA (pars), *Goldfuss*, Petref., p. 78, pl. xxvi, figs. 3*b*, 3*c*, 1829.
(Cœt. exclusis.)

FAVOSITES — *Phillips*, Palæoz. Foss., p. 16, pl. vii, fig. 21, 1841.

CALAMOPORA — ? *Ad. Rœmer*, Verst. des Harzgeb., p. 6, tab. iii, fig. 2, 1843.

FAVOSITES — ? *Lonsdale*, in Strzelecki's Description of New South Wales and Van Diemen's Land, p. 266, 1845.

— GOTHLANDICUS, *Steininger*, Verst. des Ueberg. geb. der Eifel, p. 9, 1849.

— GOLDFUSSI, *D'Orbigny*, Prodr. de Paléont., vol. i, p. 107, 1850.

— — *De Verneuil* and *J. Haime*, Bull. de la Soc. Géol. de France, 2d ser., vol. vii, p. 162, 1850.

— — *Milne Edwards* and *J. Haime*, Pol. Foss. des Terr. Palæoz., p. 235, pl. xx, fig. 3, 1851.

Corallum composite, forming a convex globular, or pyriforme mass. *Calices* for the most part nearly of the same size, but sometimes intermingled with some very small ones. The inner surface of the *walls* rendered rugose by the presence of small points; their sides unequally developed, and presenting 1, 2, or 3 vertical rows of small pores or holes (almost always 2 rows), which are regularly circular, more closely set than in *F. gothlandica*, and sometimes alternate, but in other parts opposite. Diagonal of the large corallites somewhat more than a line.

Found at Barton, near Torquay, by Dr. Battersby. Mr. Phillips mentions its having been met with also at Shoreham Point, Plymouth, and Babbacombe. The same species is found also at Nehou and Visé, in France; at Millar, in Spain; at Paffrath, in the Eifel and in the Hartz Mountains; in the Oural, in Russia; in the States of Indiana, Ohio, and Kentucky, in America; and (according to Mr. Lonsdale) at Yass plains, in New South Wales.

This fossil was, till of late, confounded with *F. gothlandica*,[1] to which it bears in fact great resemblance exteriorly; but it differs from it by the mural pores being more distant from each other, and arranged in two vertical lines on each side of the wall. In *F. alveolaris*[2] and *F. aspera*[3] these pores are always situated in the angles formed by the prismatic walls of the corallites, and in *F. basaltica*[4] and *F. polymorpha*[5] they only form a single line, placed in the middle of each side of the wall. In *multipora*[6] and *F. Troosti*,[7] on the the contrary, there are always three series of pores on each side of the wall. *F. parasitica*[8] and *F. Forbesi*[9] differ from *F. Goldfussi* by their calices being very unequal in size, and *F. Hisingeri* may be distinguished from the latter by the great development of the dissepiments. As to the ramose species of Favosites, they are sufficiently distinct from the above-described fossil, in consequence of their form.

2. FAVOSITES RETICULATA. Tab. XLVIII, figs. 1, 1*a*, 1*b*.

> CALAMOPORA SPONGITES (var. RAMOSA), *Goldfuss*, Petref. Germ., vol. i, p. 80, tab. xxviii, fig. 2*a*—*g*, 1829. (Cœt. excl.)
>
> ALVEOLITES RETICULATA, *De Blainville*, Dict., vol. lx, p. 369, 1830.—Man., p. 404.
>
> CALAMOPORA SPONGITES, *Geinitz*, Grundr. der Verst., pl. xxiii A, fig. 13, 1845-46.
>
> — — *Keyserling*, Reise in das Petsch., p. 178, 1846.
>
> ALVEOLITES SPONGITES, *D'Orbigny*, Prodr. de Paléont., vol. i, p. 108, 1850. (Not *Steininger*.)
>
> FAVOSITES ORBIGNYANA, *De Verneuil* and *Jules Haime*, Bull. Soc. Géol. de France, 2d ser., vol. vii, p. 162, 1850.
>
> — RETICULATA, *Milne Edwards* and *Jules Haime*, Pol. Foss. des Terr. Palæoz., p. 241, 1851.

Corallum dendroid, composed of thick branches (from half a line to one line in diameter), which intermingle much, and often coalesce. *Calices* almost equal in size, with thick walls, and having somewhat less than half a line in diameter.

Found at Torquay in England, at Nehou and Brest in France, Palapaya and Ferrones in Spain, Eifel in Germany, and (according to M. Keyserling) at Uchta in Russia.

[1] *Calamopora gothlandica* (pars), Goldfuss, Petref. Germ., t. i, p. 78, pl. xxvi, figs. 3*a* and 3*e*, 1829.

[2] *Calamopora alveolaris* (pars), Goldfuss, ibid., p. 77, pl. xxvi, figs. 1*a*, 1*c*.

[3] Id. (pars.), Goldfuss, ibid., p. 77, pl. xxvi, fig. 1*b*.

[4] *Calamopora basaltica*, Goldfuss, ibid., p. 78, pl. xxvi, figs. 4*c*, 4*d*.

[5] *Calamopora polymorpha*, var. *tuberosa*, Goldfuss, ibid., p. 79, pl. xxvii, figs. 2*b*, 2*c*, 2*d*, 3*b*, 3*c*. (Cœt. excl.)

[6] Lonsdale, Silur. Syst., p. 683, pl. xix, bis fig. 5, 1839.

[7] Milne Edwards and Jules Haime, Pol. Foss. des Terr. Palæoz., p. 238, pl. xviii, fig. 1.

[8] Tab. xlix, fig. 2.

[9] *Calamopora basaltica* (pars), Goldfuss, Petref. Germ., t. i, p. 78, tab. xxvi, fig. 4*b*, 1829.

We have, provisionally, admitted, as forming distinct species, various ramose Favosites, which have previously been described as such by Blainville, but which may probably, when better known, be found to be only varieties of the same species. Such are *F. reticulata*, *F. cervicornis*,[1] and *F. dubia*.[2] The latter differs, however, from the first by its branches not being coalescent, nor so closely set, and by the calices being rounded and obliquely placed on the surface of the branches. *F. cervicornis* has its calices more unequal in size, its walls thinner, and its branches larger and more irregular. All these have only a single line of pores on each side of the walls, and these pores are large and placed at a distance from each other.

3. FAVOSITES CERVICORNIS.   Tab. XLVIII, fig. 2.

> CALAMOPORA POLYMORPHA, var. RAMOSA DIVARICATA, *Goldfuss*, Petref., vol. i, p. 79, pl. xxvii, figs. 3a, 4a, 4b, 4c, 1829. (Cœt. exclus.)
>
> ALVEOLITES CERVICORNIS, *De Blainville*, Dict. Sc. Nat., vol. lx, p. 369, 1830.—Man., p. 405.
>
> THAMNOPORA MILLEPORACEA (pars.), *Steininger*, Mém. Soc. Géol. de France, vol. i, p. 338, 1831.
>
> CALAMOPORA POLYMORPHA, *Ad. Rœmer*, Verst. des Harzgeb., p. 6, tab. ii, fig. xvi, 1843.
>
> FAVOSITES CRONIGERA and ALVEOLITES CELLEPORATUS, *D'Orbigny*, Prodr. de Paléont., vol. i, p. 107, 1850.
>
> —   —   *De Verneuil* and *Jules Haime*, Bull. Soc. Géol. de France, 2d ser., vol. vii, p. 102, 1850.
>
> —   CERVICORNIS, *Milne Edwards* and *Jules Haime*, Pol. Foss. des Terr. Palæoz., p. 243, 1851.
>
> —   POLYMORPHA, *M'Coy*, Brit. Palæoz. Foss., p. 68, 1851.

Corallum subdendroid.   *Calices* unequal in size; walls somewhat thick.

Found at Torquay, and, according to Professor M'Coy, at Newton Bushel, Teignmouth, New Quay, Plymouth, and Bedruthen Slope, St. Eual; at Mons in Belgium; near Brest in France; at Ferrones and Consejo de Llaviera in Spain; and in the Eifel, at Heiflerstein, Villmar, and Bensberg, in Germany.

This coral differs from *F. reticulata*[3] and *F. dubia*[4] by its branches being thicker and more irregular, its calices being more unequal in size, and its walls being thinner.

4. FAVOSITES DUBIA.

> CALAMOPORA POLYMORPHA, var. GRACILIS, *Goldfuss*, Petref. Germ., vol. i, p. 79, tab. xxvii, fig. 5, 1829.
>
> ALVEOLITES DUBIA, *Blainville*, Dict., vol. lx, p. 370, 1830.—Man., p. 405.
>
> THAMNOPORA MADREPORACEA, *Steininger*, Mém. de la Soc. Géol., vol. i, p. 338, 1831.

---

[1] Tab. xlviii, fig. 2.

[2] *Calamopora polymorpha*, var. *gracilis*, Goldfuss, Petref. Germ., t. i, p. 19, tab. xxvii, fig. 5, 1829.

[3] Tab. xlviii, fig. 1.

[4] *Calamopora polymorpha*, var. *gracilis*, Goldfuss, Petref. Germ., t. i, p. 19, tab. xxvii, fig. 5, 1829.

FAVOSITES POLYMORPHA, *Phillips*, Palæoz. Foss., p. 15, pl. viii, fig. 20, 1841.
ALVEOLITES CERVICORNIS, *Michelin*, Icon., p. 187, pl. xlviii, fig. 2, and pl. xlix, fig. 3, 1845.
—   —   *D'Orbigny*, Prodr. de Paléont., vol. i, p. 107, 1850.
FAVOSITES DUBIA, *Milne Edwards* and *Jules Haime*, Pol. Foss. des Terr. Pal., p. 243, 1851.

Corallum dendroid; its branches placed wide apart, not coalescent, and about half a line in diameter. *Calices* somewhat oblique, deep, with the exterior part of their edge rounded or subpolygonal; walls thick; some very small calices often situated between the large ones, which are about two thirds of a line in diameter; a single line of large pores on each side of the walls.

Found at Torquay, and, according to Professor Phillips, at Lee Quarry near Combe Martin; West Hagginton; Hillsborough near Ilfracombe; Babbacombe, Hope, Sharkham Point, Mudstone Bay; in France, at Ferques (Pas-de-Calais), Viré, Chassegrain (Sarthe); in Germany at Bensberg; in America at the Falls of Ohio, and in the Clarke county.

This coral differs from *F. reticulata*[1] and *F. cervicornis*[2] by the obliquity of its calices, —a character which gives it some resemblance to Alveolites.

The only British specimen of this species that we have seen belongs to the collection of the Geological Society, but is in too bad a state of preservation to be worth being represented in the plates joined to this Monograph.

5. FAVOSITES FIBROSA.   Tab. XLVIII, figs. 3, 3*a*, 3*b*.

CALAMOPORA FIBROSA, var. TUBEROSA RAMOSA, *Goldfuss*, Petref. Germ., vol. i, p. 82,
tab. xxviii, fig. 3*a*, 3*b*, 1829. (Cœt. excl.)
FAVOSITES MICROPORUS, *Steininger*, Mém. Soc. Géol. de France, vol. i, p. 337, 1831.
ALVEOLITES FIBROSA, *Lonsdale*, Sil. Syst., p. 683, pl. xv, fig. 1, 1839.
FAVOSITES FIBROSA (pars), *Lonsdale*, ibid., p. 683, pl. xv bis., fig. 6, 1839 (but not fig. 7).
—   —   *Phillips*, Palæoz. Fossils, p. 17, pl. ix, fig. 25, 1841.
CALAMOPORA FIBROSA, *Ad. Rœmer*, Verst. des Harzgeb., p. 6, pl. iii, fig. 4, 1843.
—   —   *Keyserling*, Reise in das Petschora land, p. 177, 1846.
ALVEOLITES FIBROSUS, *D'Orbigny*, Prodr. de Paléont., vol. i, p. 108, 1850.
FAVOSITES FIBROSA, *Milne Edwards* and *Jules Haime*, Pol. Foss. des Terr. Pal., p. 244, 1851.

Corallum massive, very convex; sometimes subpyriform or sublobate. *Corallites* prismatical, radiating from the base of the corallum to its surface; straight, or slightly flexuous, and almost equal in size. *Tabulæ* very closely set (12 or 15 in the space of a line). Mural pores large in proportion to the size of the corallites, closely set, alternating with the tabulæ, and arranged in single vertical lines at the angles of the prismatical walls. Diameter of the calices about one tenth of a line.

Found in the Devonian formation at Torquay, and, according to Mr. Phillips, at Darlington near Totness, Sharkham Point, and Babbacombe; in France, at Viré (Sarthe); in Germany in the Eifel and in the Hartz Mountains. Found also in the superior

---

[1] Tab. xlviii, fig. 1.   [2] Ibid., fig. 2.

Silurian deposits at Storderley Edge (Upper Ludlow Rocks), Larden, and, according to Mr. Lonsdale, at Churn-bank, on Palmer's Kairn near Ludlow; in Ireland, according to Mr. M'Coy, in numerous localities of the counties of Galway, Kerry, Wexford, Kildare, Mayo, Tyrone, Waterford, and Wicklow; in Russia at Waschkina (according to Keyserling); in America at Casskill, and at Lexington, Kentucky (Goldfuss). Found in the inferior Silurian deposits at Landovery.

We are inclined to think that it is by mistake that Professor M'Coy mentions this species as having been met with in the carboniferous formation in Ireland, and presume that the fossil so alluded to by that geologist was the *Alveolites septosa*.

The above-described species is easily distinguished from all the other species of Favosites by the smallness of its calices. It resembles *F. alveolaris*[1] and *F. aspera*[2] by the angular position of its mural pores; but these two species differ from it by the calices being much more irregular in size, as well as much larger.

We have not remarked any material difference between the specimens found in the Devonian and the Silurian formations; but all these corals are so ill-preserved, that we are not inclined to attach much importance to that supposed specific identity.

2. *Genus* EMMONSIA.[3]

EMMONSIA HEMISPHERICA.   Tab. XLVIII, figs. 4, 4*a*.

> FAVOSITES ALVEOLARIS, *Hall*, Geol. of New York, p. 157, No. 31, figs. 1, 1*a*, 1843. (No. 8 of Blainville.)
> FAVOSITES HEMISPHERICA, *Yandell* and *Shumard*, Contrib. to Geol. of Kentucky, p. 7, 1847.
> ALVEOLITES HEMISPHERICA, *D'Orbigny*, Prod. de Paléont., vol. i, p. 49, 1850.
> FAVOSITES HEMISPHERICA, *Jules Haime* and *Verneuil*, Bull. Soc. Géol. de France, 2d ser., vol. vii, p. 162, 1850.
> EMMONSIA HEMISPHERICA, *Milne Edwards* and *Jules Haime*, Pol. Foss. des Terr. Palæoz., p. 247, 1851.

Corallum composite, forming a subspherical mass, which becomes sometimes very tall, and composed of superposed layers. *Calices* irregular, polygonal, and varying in size. *Septa* (12) well developed, straight or slightly curved, and extending to the centre of the upper tabulæ. Mural *pores* large, placed at about a quarter of a line apart, and arranged in pairs on some of the sides of the wall, but forming single lines on others. *Tabulæ* very closely set, somewhat irregularly horizontal. In the visceral chambers, where they are broken down, they leave fragments adhering to the walls; and in general, above the space included between two of these fragments, a third fragment exists, so as to constitute an

---

[1] *Calamopora alveolaris* (pars), Goldfuss, Petref. Germ., t. i, pl. xxvi, figs. 1*a*, 1*c*, 1826.

[2] *Calamopora alveolaris* (pars), Goldfuss, ibid., tab. xxvi, fig. 1*b*.

[3] Milne Edwards and Jules Haime, Monographie des Polyp. Foss. des Terr. Palæoz.; Archives du Museum, vol. v, p. 246, 1851.

alternate mode of superposition. Great diagonal of the calices about half a line; the tabulæ are placed at about one tenth of a line apart.

Found in the Devonian formation at Torquay; in Spain at Contejo de Castrillon, near Aviles; in America at Caledonia, New York, at the Falls of Ohio, at Charleston Landing (Indiana), in the Isle of Mackinaw, and, according to Mr. Hall, at Williamsville, Erie county. Found also in the superior Silurian deposits at Springfield, Ohio, and in Perry county, Tennessee.

The genus Emmonsia contains two other species—*E. alternans*[1] and *E. cylindrica*.[2] The first differs from the species here described by the alternate arrangement of the mural edges, by the incomplete tabulæ being more regular, and by its calices being larger. *E. cylindrica* has calices still larger, and often circular; it is also characterised by the existence of four or five lines of pores on each side of the walls.

3. *Genus* ALVEOLITES, (p. lx.)

1. ALVEOLITES SUBORBICULARIS. Tab. XLIX, figs. 1, 1*a*.

ALVEOLITES SUBORBICULARIS, *Lamarck*, Hist. des Anim. sans Vert., vol. ii, p. 186, 1816; 2d ed., p. 286.

— ESCHAROIDES, *Lamarck*, Hist. des An. sans Vert., vol. ii, p. 186, 1816; 2d ed., p. 286.

ESCHARITES SPONGITES, *Schlotheim*, Petrefactenkunde, 1st part, p. 345, 1820.

ALVEOLITES SUBORBICULARIS and ESCHAROIDES, *Lamouroux*, Encycl. (Zooph.) pp. 41, 42, 1824.

CALAMOPORA SPONGITES, var. TUBEROSA, *Goldfuss*, Petref. Germ., vol. i, p. 80, tab. ii, fig. 1*a*—*h*, 1829. (Cæt. excl.)

ALVEOLITES ESCHAROIDEA and SUBORBICULARIS, *De Blainville*, Dict. Sc. Nat., vol. lx, p. 269, 1830.—Manuel, p. 404.

— SPONGITES, *Steininger*, Mém. de la Soc. Géol. de France, vol. i, p. 334, pl. 20, fig. 4, 1831.

CALAMOPORA — *Morren*, Descr. Cor. in Belg. Rep., p. 74, 1832.

FAVOSITES — *Phillips*, Palæoz. Foss., p. 16, pl. viii, fig. 23, 1841.

CALAMOPORA SUBORBICULARIS, *Michelin*, Icon., p. 188, pl. 48, fig. 7, 1845.

— SQUAMOSA or IMBRICATA, ibid., p. 189, pl. xlix, fig. 15, 1845.

FAVOSITES SUBORBICULARIS and ALVEOLITES TUBEROSA, *D'Orbigny*, Prod. de Palæont., vol. i, pp. 107, 108, 1850.

ALVEOLITES — *Milne Edwards* and *Jules Haime*, Pol. Foss. des Terr. Palæoz., p. 255, 1851.

Corallum composite, forming irregular incrustated masses, which are in general fixed on ramose favosites or on a cyathophyllum, and are composed of superposed layers,

---

[1] Milne Edwards and Jules Haime, Pol. Foss. des Terr. Palæoz., p. 248, 1851.

[2] *Favosites cylindrica*, Michelin, Icon., p. 255, pl. lx, fig. 1, 1846.

terminated by an irregular or subgibbose surface. *Calices* very oblique, closely set (but unequally so), elongated transversely, subtriangular, and turned towards the edge of the corallum. The outer or under side of these calices bears interiorly a small elongated ridge, which appears to represent a septum, and is placed opposite to a small notch. Transverse diameter of the calices about two fifths of a line ; small diameter about half that length.

Found in the Devonian deposits of Torquay, Tor Abbey, Babbacombe, Teignmouth, and, according to M. Phillips, at Hope.

Specimens are in the collections of the Geological Society, Messrs. Battersby and Pengelly at Torquay.

Alveolites Labechei[1] is a massive subgibbose species, very nearly allied to the above-described coral, but differs from it by the interior dentation of the calicular edge being but slightly developed, and by the calices being more irregular in size. In *A. Battersbyi*[2] the septum represented by this interior expansion is, on the contrary, formed by very strong spiniform processes, and the mural pores are very large.

In *A. compressa*[3] the calices are much more irregular, and the inner processes are very small.

The other species of this genus are not massive.

2. ALVEOLITES BATTERSBYI. Tab. XLIX, figs. 2, 2*a*.

ALVEOLITES BATTERSBYI, *Milne Edwards* and *Jules Haime*, Pol. Foss. des Terr. Palæoz., p. 257, 1851.

Corallum forming a subspherical mass. *Calices* unequal in size, and somewhat irregular. Vertical and horizontal sections show that the *walls* are thin, and perforated by large circular pores rather closely set, and that in different places they give rise to strong ascending spiniform processes, which, by their superposition, represent unpaired septa. *Tabulæ* very thin and irregular.

Found at Torquay. Specimens are in the collections of Dr. Battersby and of Mr. Pengelly.

This species is remarkable for the slight obliquity of its calices, the large size of its mural pores, and more especially for the great development of its septal processes.

3. ALVEOLITES VERMICULARIS. Tab. XLVIII, figs. 5, 5*a*.

ALVEOLITES VERMICULARIS, *M'Coy*, Ann. and Mag. of Nat. Hist., 2d ser., vol. vi, p. 377, 1850.
— — *M'Coy*, Brit. Palæoz. Foss., p. 69, 1851.

Corallum dendroid, with slender, cylindrical coalescent branches that bifurcate at

---

[1] Milne Edwards and Jules Haime, Pol. Foss. des Terr. Palæoz., p. 257.
[2] Tab. lxix, fig. 2.   [3] Tab. xlix, fig. 3.

almost right angles. A vertical section shows that the corallites diverge from the centre towards the surface of the branches in an oblique ascendant direction, are somewhat flexuous, and terminate by a calicular margin that is prominent at its under part. The *walls* are thick. In some places the *tabulæ* appear closely set, but in most parts of the specimen submitted to our investigation they were completely destroyed. The indications of the mural pores were also obscure. Diameter of the branches about $1\frac{1}{2}$ line, that of the calices about $\frac{1}{5}$th of a line.

This Coral was found at Torquay by Dr. Battersby, and is known to us only by a polished specimen belonging to the collection of that palæontologist. We at first thought that it might be referred to the *Ceriopora Goldfussi* of Michelin, but Professor M'Coy, who appears to have had some better preserved specimens, has since that recognized the existence of triangular calices and a fissiparous mode of multiplication. He therefore places this fossil in the genus Alveolites, and after more ample investigation we have been led to adopt his opinion. Professor M'Coy adds that *Alveolites Vermicularis* is polymorphous and is met with at Teignmouth, at Newquay, and at Bedruthen Steps, St. Eual.

ALVEOLITES COMPRESSA. Tab. XLIX, fig. 3.

Corallum massive. *Calices* arranged in a circular manner round divers places on the surface of the corallum, compressed, elongated, and very unequal in size, the larger ones being about half a line across. *Walls* thick exteriorly, and convex. The three septal processes somewhat unequally developed; short, but quite distinct.

This species resembles *Alveolites orbicularis*[1] by its general appearance, but differs from it by the calices being much more unequal in size, arranged in circular lines, and provided with three septal processes that do not differ in size.

Found at Torquay by Mr. Pengelly.

Family PORITIDÆ, (p. lv.)

The singular fossil coral to which Goldfuss gave the name of *Pleurodictyum problematicum*[2] has been met with in the Meadsfoot Sands near Torquay, by Prof. Phillips;[3] but we have not as yet seen any British specimen of that species, and we must therefore refrain from describing it here. We hope to be able to have it figured in an appendix to our Monograph.

[1] Tab. xlix, fig. 1.

[2] Petref. Germ., vol. i, p. 113, pl. xxxviii, fig. 18. See also our Monographie des Polypiers Fossiles des Terrains Palæozoiques, p. 210, pl. xviii, figs. 3, 4, 5, 6.

[3] Palæozoic Fossils, p. 19, pl. xix, fig. 24.

Family STAURIDÆ, (p. lxiv.)

Genus METRIOPHYLLUM BATTERSBYI.

METRIOPHYLLUM BATTERSBYI. Tab. XLIX, fig. 4.

METRIOPHYLLUM BATTERSBYI, *Milne Edwards* and *Jules Haime*, Pol. Foss. des Terr. Palæoz., p. 318, 1851.

This species has been established by the study of a polished slab belonging to the collection of Dr. Battersby, and showing a transverse section made at a small distance below the calice. The quadrifascicular mode of arrangement of the septa is very distinct. The principal septa are forty-eight in number, somewhat thick, and extending to the centre of the coral; they alternate with an equal number of smaller ones which are also thinner; towards the middle of each group they are slightly flexuous, and towards the outer part they become shorter and somewhat oblique. Some dissepiments are also visible. Diameter about eight lines.

Found at Torquay.

This species differs from *Metriophyllum Bouchardi*[1] by its septæ being twice as numerous and slightly thickened near the centre.

Family CYATHOPHYLLIDÆ, (p. lxv.)

1. *Genus* AMPLEXUS, (p. lxvi.)

AMPLEXUS TORTUOSUS. Tab. XLIX, figs. 5, 5*a*.

AMPLEXUS TORTUOSUS, *Phillips*, Palæoz. Foss., p. 8, pl. iii, fig. 8, 1841.
— — *Milne Edwards* and *Jules Haime*, Pol. Foss. des Terr. Palæoz., p. 347, 1851.
— YANDELLI, (pars), *ibid.*, p. 344.
— TORTUOSUS, *M'Coy*, British Palæoz. Foss., p. 70, 1851.

Corallum elongate, cylindrico-conical, curved, and slightly tortuous; circular wrinkles well developed and irregular. *Epitheca* strong, and wrinkled transversely. *Calice* suboval, with 4 distinct septal fossulæ, (the one placed near the convex side of the corallum larger and deeper than the three others). *Tabulæ* not very closely set, irregular, and presenting in the centre a large smooth space. *Septa* (30 to 50 in the adult

---

[1] Milne Edwards and Jules Haime, Pol. Foss. des Terr. Palæoz., p. 318, pl. vii, figs. 1, 2.

individuals) slender, but little developed, not very unequal in size; some rudiments of smaller septa between the former.

The specimens figured in this Monograph (tab. xlix, fig. 5) about 4 inches long, and about 1½ inch wide. Professor M'Coy justly remarks that the specimens described by Mr. Phillips were young individuals, and mentions a gigantic specimen the diameter of which was 1 inch 9 lines.

The specimens here described were found at Torquay and Plymouth, and belong to the collections of Dr. Battersby and Mr. Pengelly. The same species exists at Barton and South Petherwin, according to Mr. Phillips, and at Newton Bushel according to Professor M'Coy.

This species resembles very much *Amplexus Yandelli*,[1] and *A. cornubovis*,[2] but differs from both by the septa being less numerous and almost equal in size, and by the existence of 4 septal fossulæ. *Amplexus coralloides*[3] differs also from *A. tortuosus* by not having the depressions on the tabulæ, and by the circular wrinkles being larger.

2. *Genus* HALLIA, (p. lxvii.)

HALLIA PENGELLYI.   Tab. XLIX, figs. 6, 6*a*, 6*b*.

HALLIA PENGELLYI, *Milne Edwards* and *Jules Haime*, Pol. Foss. des Terr. Pal., p. 354, 1851.

The *calice* of this coral is almost circular, with 54 principal *septa*, which alternate with an equal number of small and thinner ones; the former are very thick, straight, and disposed in a regular radiate manner towards the circumference of the visceral chamber; towards their inner edge they are provided with a large and thin paliform lobe. The cristiform septum is not as large as in *Hallia insignis*,[4] and it is the lobes belonging to the principal septa situated near this that affect a pinnate mode of arrangement. The dissepiments are very slender and closely set. Diameter 1 inch or more; the area occupied by the paliform lobes forms an ellipse of about 9 lines long and 6 lines broad.

Found at Torquay, (Coll. of the Geological Society of London, and of Mr. Pengelly at Torquay). We are inclined to refer to this species some young corals from Pethervin, which are in a very bad state of preservation, and belong to the collection of the Geological Society. They have a strong epitheca.

In *Hallia Pengellyi* the characters of the generic type are not as distinct as in *H. insignis*,[5] the large septum being less developed, and the adjacent septa not assuming as regular a pinnate mode of arrangement; it is also to be remarked that in *H. insignis* all the septa are provided with a very large paliform lobe.

[1] Milne Edwards and Jules Haime, Pol. Foss. des Terr. Palæoz., p. 344. pl. iii, fig. 2, 1851.
[2] Ibid., p. 343, pl. ii, fig. 1.
[3] Tab. xxxvi, fig. 1.
[4] Milne Edwards and Jules Haime, Pol. Foss. des Terr. Palæoz., p. 353, pl. vi, fig. 3, 1851.   [5] Ibid.

3. *Genus* CYATHOPHYLLUM, (p. lxviii.)

1. CYATHOPHYLLUM CERATITES. Tab. L, fig. 2.

> CYATHOPHYLLUM TURBINATUM, (pars), *Goldfuss*, Petref. Germ., vol. i, p. 50, pl. xvi, figs. 8*c*,
> *d, f, g, h*, 1826. (Not *Madrepora turbinata*, Linné.)
> — CERATITES, (pars), *ibid.*, pl. xvii, figs. 1, 2*f*, and perhaps also fig. 5.
> — TURBINATUM, *Holl*, Hanb. der Petref., p. 416, 1830.
> — CERATITES, *Deshayes*, Coq. cærvet. des Terr., p. 247, pl. xi, fig. 2, 1831.
> — TURBINATUM, *D'Orbigny*, Prodr. de Palæont., t. i, p. 105, 1850.
> — CERATITES, *Milne Edwards* and *Jules Haime*, Pol. Foss. des Terr. Palæoz.,
> p. 361, 1851.
> — — *M'Coy*, Brit. Palæoz. Foss., p. 70, 1851.

Corallum, simple (sometimes two or three individuals are united by their bases, but their union is evidently accidental), turbinate, elongate, slightly curved, and presenting rather well marked growth swellings. *Epitheca* very strong. *Calice* deep and with a thin margin; one or two rudimentary septal fossulæ. *Septa* delicate, alternately larger and smaller but not differing much in size, narrow at their upper end, straight, and not extending quite to the bottom of the central fossula which, as well as the interseptal loculi, is somewhat vesicular. The number of the septa varies, according to the size of the corals, from 60 to 120. The large individuals are sometimes 6 inches wide, with the calice about 3 inches in diameter, and $1\frac{1}{2}$ or two inches deep, but most specimens are not more than two inches in diameter.

Found at Torquay, and according to Prof. M'Coy, at Newton Bushel. In the Eifel Mountains in Germany.

The only British specimen of this species that we have seen is the one figured in this Monograph; it is a young individual in a very indifferent state of preservation.

Cyathophyllum ceratites differs from the other simple species of the same generical group by the depth of its calice, its rudimentary septal fossula, and its septa being almost equally developed.

2. CYATHOPHYLLUM ROEMERI. Tab. L, fig. 3.

> CYATHOPHYLLUM DIANTHUS, (pars), *Goldfuss*, Petref. Germ., vol. i, p. 54, tab. xvi, fig. 1*e*,
> 1826.
> — ROEMERI, *Milne Edwards* and *Jules Haime*, Pol. Foss. des Terr. Palæoz.,
> p. 362, pl. viii, fig. 3, 1851.

Corallum simple, conical, elongated, curved, and free. *Epitheca* presenting some prominent folds, principally on the side of the convex curve. *Calice* almost circular, large and deep, 74 or more. *Septa* alternately somewhat thicker or thinner, very closely set, not exsert, denticulated, narrow, slightly arched at their upper edge, and extending to the centre of the visceral chamber, where they become slightly curved. Height 2 inches; diameter somewhat more than 1 inch; depth about 8 lines.

The specimen found at Torquay belongs to Dr. Battersby. This species is also met with at Bensberg and in the Eifel Mountains in Germany.

The type specimen of this species is from the Eifel Mountains, and has only 74 septa. The Torquay fossil that we consider as belonging to the same species, and have figured here, present more than 100 septa; but that difference evidently depends on an accidental multiplication of these laminæ in one part of the septal system where they are more closely set than elsewhere. The species most nearly allied to *C. Roemeri* is *C. Michelini*,[1] but the latter is adherent, its septal fossula is rudimentary, and its septa less closely set.

3. CYATHOPHYLLUM OBTORTUM. Tab. XLIX, fig. 7.

> STROMBODES VERMICULARIS, *Lonsdale*, Trans. of the Geol. Soc. of London, 2d ser., vol. v,
> pl. lviii, fig. 7, 1840. (Not *Cyathophyllum vermiculare* of
> Goldfuss.)
> — — *Phillips*, Palæoz. Foss., p. 11, pl. vii, fig. 14, 1841.
> CYATHOPHYLLUM OBTORTUM, *Milne Edwards* and *Jules Haime*, Pol. Foss. des Terr. Palæoz.,
> p. 366, 1851.
> STREPHODES VERMICULARIS, *M'Coy*, British Palæoz. Foss., p. 73, 1851.

Corallum simple, elongated, subcylindrical. *Calice* circular. Principal *septa* (32 or 34) very thin towards their inner edge and somewhat thicker exteriorly, much curved and twisted near the centre of the calice, and alternating with an equal number of others that are smaller and still thinner. Vesicular dissepiments well developed on the exterior part of the visceral chamber. Height about $2\frac{1}{2}$ inches; diameter of the calice, 1 inch.

Found at Torquay, Plymouth, and Newton Bushel, by Mr. Lonsdale; and at Darlington by Mr. Phillips.

Collections of the Geological Society, and of Prof. Phillips, at York.

This species is very remarkable on account of the septa being so strongly twisted near the centre of the visceral chamber,—a character which distinguishes it easily from *C. Michelini*[2] and *C. Roemeri*[3] that in other respects resemble it very much.

4. CYATHOPHYLLUM DAMNONIENSE. Tab. L, fig. 1.

> CYSTIPHYLLUM DAMNONIENSE, *Lonsdale*, Geol. Trans., 2d ser., vol. v, p. 703, pl. lviii,
> fig. 11, 1840.
> — — *Phillips*, Palæoz. Foss., p. 9, pl. iv, fig. 11, 1841.

---

[1] Milne Edwards and Jules Haime, Pol. Foss. des Terr. Palæoz., p. 366, 1851.
[2] Ibid.　　　　　　　　　　　　　　　[3] Tab. l, fig. 3.

Cyathophyllum damnoniense, *Milne Edwards* and *Jules Haime*, Pol. Foss. des Terr. Palæoz., p. 371, 1851.
Cystiphyllum — *M'Coy*, Brit. Palæoz. Foss., p. 71, 1851.

Corallum simple, elongate, subturbinate, and almost straight. *Septa* (100 or more) somewhat unequal in size alternately, closely set, very slender exteriorly, thick towards their inner part, and slightly curved. *Dissepiments* very closely set, vesicular, somewhat irregular, smaller and more numerous towards the outer part of the visceral chamber. Some small *tabulæ* somewhat irregular, and very closely set, at the centre of the coral. Height sometimes 3 inches.

Found at Torquay, Newton Bushel, Plymouth, and also, according to Professor Phillips, at Sharkham Point and Babbacombe. Specimens are in the collections of the Geological Society of London, of Dr. Battersby, and Mr. Pengelly.

The fossil Coral designated by the name of Cyathophyllum celticum,[1] and found in the Devonian deposits of Cornwall and Devonshire,[2] is as yet so imperfectly known that we are not able to characterize it in a satisfactory manner. The specimens met with are only natural casts from which the real coral has more or less completely disappeared; they show, however, that the *septa*, (to the number of 36 or 48), must have been alternately of unequal size, and that the principal ones extended to the centre of the visceral chamber, where they became somewhat twisted.

We have given the name of Cyathophyllum Bucklandi to a species which Professor M'Coy described under that of *Petraia gigas*,[3] but which is quite distinct from the *Cyathophyllum gigas*, previously described in MM. Yandell and Shumard's Paper on the Geology of Kentucky, and therefore could not retain the same name. It is a simple coral like the preceding ones, and is known only by the casts it has left in the surrounding rock. Till some better preserved specimens be met with, we therefore do not think it necessary to have this fossil figured in our Monograph.

[1] *Turbinolia celtica*, Lamouroux, Exp. Meth., p. 85, tab. lxxviii, figs. 7, 8, 1821. *Deslongchamps*, Encycl. (Zooph.), p. 761, 1824; Milne Edwards, 2d edit. of Lamarck, vol. ii, p. 362, 1836. *Petraia celtica*, Lonsdale, Geol. Trans., 2d. ser., vol. v, p. 697, pl. lviii, fig. 6, 1840. *Turbinolopsis celtica*, Phillips, Palæoz. Foss., p. 3, pl. i, fig. 1, 1841. *Cyathophyllum celticum*, D'Orbigny, Prod. de Palæont., vol. i, p. 105, 1850; Milne Edwards and Jules Haime, Pol. Foss. des Terr. Palæoz., p. 373, 1851. *Petraia celtica*, M'Coy, Brit. Palæoz. Foss., p. 74, 1851.

[2] Prof. Phillips mentions this fossil as having been found at South Petherwin, Saint Columb, Pobruan and Fowey in Cornwall; and at Combes, Mudstone Bay, Yealm, Torquay, and Brushford in Devonshire.

[3] *Petraia gigas*, M'Coy, Ann. and Mag. of Nat. Hist., 2d ser., vol. iii, p. 1, 1849; (not *Cyathophyllum gigas*, Yandell and Shumard). *Cyathophyllum Bucklandi*, Milne Edwards and Jules Haime, Pol. Foss. des Terr. Palæoz., p. 390, 1851. *Petraia gigas*, M'Coy, British Palæoz. Foss., p. 74, 1851.

Professor M'Coy describes this fossil in the following terms:—"*Corallum* elongate, conic, gradually increasing, (at an angle from the apex of about 30° externally), slightly oblique; section apparently elliptical, the axes in the proportion of 70 to 100; internal cast obtusely conic, expanding at an angle of about 50° in

The fossils described by Mr. Lonsdale under the name of *Turbinolopsis bina*[1] appear to belong also to the genus Cyathophyllum, but have as yet been found only in the form of casts which are scarcely determinable. They show in general 72 septa of unequal size, alternately denticulated, and slightly curved towards the centre of the visceral chamber. Prof. Phillips mentions the existence of this coral in the Devonian deposits of Combe, near Ashburton.[2]

The corals to which the names of *Turbinolopsis pauciradialis*,[3] *T. elongata*,[4] *T. rugosa*,[5] and *T. pluriradialis*[6] have been given by Prof. Phillips, appear to be specifically identical, or very nearly allied to the preceding species; but the specimens as yet known are so imperfect that we cannot lay before the reader any useful information concerning their structure.

8. CYATHOPHYLLUM HELIANTHOIDES. Tab. LI, figs. 1, 1*a*.

> CYATHOPHYLLUM HELIANTHOIDES, *Goldfuss*, Petref. Germ., vol. i, p. 61, tab. xx, fig. 2*a—k* and tab. xxi, fig. 1, 1826.
> FAVASTREA HELIANTHOIDEA, *De Blainville*, Dict. Sc. Nat., vol. lx, p. 341, 1830.—Man., p. 375.
> TURBINOLIA HELIANTHOIDES, and ASTREA HELIANTHOIDEA, *Steininger*, Mém. Soc. Géol. de France, vol. i, pp. 344, 345, 1831.
> MONTICULARIA AREOLATA, *Ibid.*, p. 346, pl. xx, fig. 10.
> CYATHOPHYLLUM HELIANTHOIDES, *Morren*, Descr. Corall. Belg., p. 58, 1832.
> —　　　　—　　*Milne Edwards*, 2d edit. of Lamarck, vol. ii, p. 429, 1836.

compressed specimens, its small end obtuse from the filling up of a considerable length of the base of the coral, by nearly solid sclerenchyme; external walls thick, dense; lamellæ averaging 74 in the adult cups; with the diameter of two and a half inches, the primary ones extending towards the centre, nearly straight for above one third the diameter, then abruptly diminishing in strength, and gradually convoluted spirally towards the broad central area; the secondary lamellæ much finer than the primary, extending about one fifth of the diameter towards the centre; *internal casts* with thirty-three to thirty-seven broad, flattened, smooth ribs, separated by deep smooth-edged sulci (representing the primary lamellæ); these sulci in some specimens, divided by connecting filaments of matrix, produced by perforation in the original plate; each rib divided in the middle by a very fine slit, not reaching quite to the narrow base, (representing the secondary lamellæ,) becoming nearly as strong as the primary towards the broad edge of the cup. No transverse vesicular laminæ. Lengths of imperfect casts about two and a half inches; width of same specimen pressed flat, nearly four and a half inches; width of ribs between the primary sulci at edge of cup varying from two to three lines. Very common in the fine grey Devonian slates of New Quay." (M'Coy, *l. c.*)

[1] *Turbinolopsis bina*, Lonsdale, in Murchison, Silur. Syst., p. 692, pl. xvi bis, fig. 5, 1839. *Turbinolopsis bina*? Ibid., p. 693, pl. xvi bis, fig. 6. *Petraia bina*, M'Coy, Syn. Sil. Foss. of Ireland, p. 60, 1846. *Streptelasmæ bina*, D'Orbigny, Prod. de Pal., vol. i, p. 47, 1850. *Cyathophyllum binum*, Milne Edwards and Jules Haime, Pol. Foss. des Terr. Palæoz., p. 374, 1851.

[2] Palæoz. Foss., p. 4, pl. i, fig. 2.

[3] Phillips, Palæoz. Foss., p. 5, pl. i, fig. 4. From Corffe Quarry, near Tawstock.

[4] Ibid., p. 6, pl. ii, fig. 68. From Horderley, May Hill, and Lickey Hill.

[5] Ibid., p. 7, pl. ii, fig. 7*c*. From Snowdon.

[6] Ibid., pp. 5, 6, pl. ii, figs. 5*a*, 5*β*. From Brushford, Linton, Pilton, and Fowey Harbour.

ASTREA HELIANTHOIDES, *Lonsdale*, Geol. Trans., 2d. ser., vol. v, p. 697, 1840.
DISCOPHYLLUM HELIANTHOIDES, *D'Orbigny*, Prod. de Paléont., vol. i, p. 106, 1850.
CYATHOPHYLLUM HELIANTHOIDES, *Milne Edwards* and *Jules Haime*, Pol. Foss. des Terr. Palæoz., p. 375, pl. viii, fig. 5, 1851.
STREPHODES HELIANTHOIDES, *M'Coy*, Brit. Palæoz. Foss., p. 73, 1851.

Corallum simple or composite.

When simple this species is subturbinate, short, broad, with the edge of the calice reverted, so as to form an obtuse prominent ridge around the central fossula. Sixty or eighty equally-developed *septa*, slightly thickened towards the exterior by the granulations and striæ that arise from their lateral surfaces; almost all of these extend to the centre of the calice, where they become slightly curved, and present, in the well-preserved specimens, small but well-characterised paliform lobes, which, by their agglomeration, form a sort of crown near the centre of the calicular fossula. The edge of the calice is circular and slightly lamellate. The height of the corallum is usually about two inches, and in that case the diameter of the calice is about double, or somewhat more, and that of the paliform circles about four lines.

When composite this corallum assumes an astreiform appearance, and the corallites, which are united together side by side, are circumscribed by polygonal lines, usually not very prominent. The *calices* are in general smaller than in the simple specimens, very unequal in size, and not provided with as many septa. In a variety of this species, the calicular swelling is large and prominent. Vertical sections show that the central part of the visceral chambers is occupied by slightly-developed tabulæ, and the outer parts filled with numerous and somewhat regular vesicles.

Found at Plymouth, Teignmouth Beach, and Mudstone Beach; in France at Viré (Sarthe); in Germany, in the Eifel, Rokeskill, Blankenheim, Steinfeld, Luxembourg, Reinfeld, Sigmaringen; in America, at Harrisville, Ohio, and in the Isle of Mackinaw.

British specimens are in the collections of Mr. J. S. Bowerbank, Dr. Battersby, and Mr. Pengelly.

The species of cyathophyllum that most approximates *C. helianthoides* is *C. Regium*,[1] from the mountain limestone; but in the latter the simple corallites are more regularly turbinate, and the calice is not everted so as to assume the form of a mushroom; the septa are also more numerous and slender.

9. CYATHOPHYLLUM HEXAGONUM. Tab. L, figs. 4, 4*a*.

MADREPORA TRUNCATA? *Esper*, (Pflanz.) Petref., tab. iv; (not Linné).
CYATHOPHYLLUM HEXAGONUM, *Goldfuss*, Petref. Germ., vol. i, p. 61, tab. xx, fig. 1, 1826.
FAVASTREA HEXAGONA, *De Blainville*, Dict. Sc. Nat., vol. lx, p. 340, 1830.—Man., p. 375.
ASTREA HEXAGONA, *Steininger*, Mém. Soc. Géol. de France, vol. i, p. 345, 1831.

---

[1] Tab. xxxii, figs. 1—4.

CYATHOPHYLLUM HEXAGONUM, *Morren*, Descr. Corall. in Belg. Repert., p. 57, 1832.
— — *Milne Edwards*, 2d edit. of Lamarck, vol. ii, p. 429, 1836.
ASTREA ANANAS, *Ad. Rœmer*, Verst. der Harzegeb., p. 5, tab. ii, fig. 11, 1843.
CYATHOPHYLLUM HEXAGONUM, *Milne Edwards* and *Jules Haime*, Pol. Fos. des Terr. Palæoz.,
p. 382, 1851.

Corallum composite, astreiform; gemmation calicular and extracalicular. *Calices* polygonal, varying much in size, rather deep, and circumscribed by *walls* that are not very prominent, but quite distinct; thin, and always simple. Forty-six septa, alternately large and small, the twenty-three latter not extending far from the wall; the large ones thin, terminated by a denticulate edge, which is horizontal near the margin, but very strongly arched upwards and inwards, and presenting near the centre of the visceral chamber a small paliform lobe; these lobes form a very distinct crown round the centre of the calice. Height of the corallum about $2\frac{1}{2}$ inches, depth about 2 lines, and diameter of the circle of paliform lobes somewhat more than 1 line.

Found at Torquay; at Montignies, St. Christophe, and Chimay in Belgium; at Bensberg, Refrath, and Grund in Germany.

British specimen in Dr. Battersby's collection.

This species differs from *C. quadrigeminum*[1] by its calices being larger and shallower, and by the paliform lobes of the septa; it differs from *C. cæspitosum*[2] by these lobes being larger than in the latter and the septa more equally developed. In *C. boloniense*[3] the calices are larger, the septa less prominent and more developed, and the paliform lobes are quite rudimentary. In *C. marmini*[4] the calices are deep and the septa rather unequal. In *C. Sedgwicki*[5] the septa are less numerous than in the above-described species, and become thicker towards three fourths of their breadth.

10. CYATHOPHYLLUM CÆSPITOSUM. Tab. LI, figs. 2, 2*a*, 2*b*.

CYATHOPHYLLUM CÆSPITOSUM, *Golfuss*, Petref. Germ., vol. i, p. 60, tab. xix, fig. 2, 1826.
— HEXAGONUM (pars.), *Ibid.*, tab. xix, fig. 5*a, b, c*. (Cœt. excl.)
CARYOPHYLLIA DUBIA, *De Blainville*, Dict. Sc. Nat., vol. lx, p. 311, 1830.—Man., p. 345.
CYATHOPHYLLUM CÆSPITOSUM, *Milne Edwards*, 2d edit. of Lamarck, vol. ii, p. 428, 1836.
— — *Lonsdale*, Geol. Trans., 2d ser., vol. v, 3d part, pl. lviii,
fig. 8, 1840.
— — *Phillips*, Palæoz. Foss., p. 9, pl. 3, fig. 10, 1841.
CLADOCORA GOLDFUSSI, *Geinitz*, Grundr. der Verst., p. 569, 1845-46.
DIPHYPHYLLUM CÆSPITOSUM, *D'Orbigny*, Prod. de Paléont., vol. i, p. 106, 1850.
CYATHOPHYLLUM — *Milne Edwards* and *Jules Haime*, Pol. Foss. des Terr. Palæoz.,
p. 384, 1851.
— — *M'Coy*, British Palæoz. Foss., p. 69, 1851.

---

[1] Goldfuss, Petref. Germ., t. i, p. 59, tab. xix, figs. 1, 5*f*, and tab. xviii, fig. 6, 1826.
Tab. li, fig. 2.   [2] Tab. lii, fig. 1.   [4] Tab. lii, fig. 4.   [5] Tab. lii, fig. 3.

Corallum composite, fasciculated, or astreiform; tall; gemmations principally calicular. *Corallites* almost cylindrical, and presenting but slight growth swellings. *Calices* in general circular, sometimes agglomerated and polygonal; rather deep. Forty or fifty *septa*, somewhat unequal in size-alternately, thin, narrow at the top, straight, and bearing a small paliform lobe near the centre of the calice. Diameter of the calices about 4 lines. *Tabulæ* well developed; interseptal vesicles small. In horizontal sections of this corallum the spot where the dissepiments cease has the appearance of an inner wall, placed at a small distance from the exterior one.

Found at Teignmouth Beach near Torquay, at Newton, and at Plymouth.

Specimens are in the collections of Mr. Bowerbank and of Dr. Battersby.

In a variety of this species found at Torquay, the calices are not more than 2 or $2\frac{1}{2}$ lines in diameter.

M. D'Orbigny has placed this coral in the genus *Diphyphyllum* of Mr. Lonsdale. As Professor M'Coy very justly remarks, some specimens appear so distinctly dichotomous that they evidently belong to this division, whereas in other specimens the gemmation is quite lateral, as in the common Cyathophylla.[1]

This species resembles *C. marmini*[2] by its general arrangement, being in some specimens fasciculate and in others astreiform; but in the latter fossil the calices are deeper, and the septa are not only less unequal, but also produce at their upper edge the appearance of an inner wall. The same variations in the general form of the composite corallum is sometimes met with also in *C. quadrigeminum*,[3] which, however, differs from *C. cæspitosum* by the septa being still more slender, and not bearing any paliform lobes.

11. CYATHOPHYLLUM BOLONIENSE.   Tab. LII, figs. 1, 1*a*.

MONTASTREA BOLONIENSIS, *Blainville*, Dict. Sc. Nat., vol. lx, p. 339, 1830.—Man., p. 394.
CYATHOPHYLLUM HEXAGONUM, *Michelin*, Icon. Zooph., p. 181, pl. xlvii, fig. 2, 1845. (Not *Goldfuss*.)
LITHOSTROTION ARACHNOIDES, *D'Orbigny*, Prod. de Paléont., t. i, p. 106, 1850. (Not *Astrea arachnoides*, de France.)
CYATHOPHYLLUM BOLONIENSE, *Milne Edwards* and *Jules Haime*, Pol. Foss. des Terr. Palæoz., p. 385, pl. 9, fig. 1, 1851.

Corallum composite, astreiform, forming a subcircular rather flat mass. *Calices* polygonal, very unequal in size, deep, and separated by thin, straight walls. Forty-two or forty-six *septa*, almost equal in size, very slender, striated laterally, delicately denticulated, and straight; half of them do not extend quite to the centre of the calice, the other

---

[1] It is surprising that after having recognized these variations the latter author should not have come to a similar result respecting the Diphyphyllum of the carboniferous formation which bear the same relation to Lithostrotions.

[2] Milne Edwards and Jules Haime, Pol. Foss. des Terr. Palæozoiques, p. 386, pl. ix, figs. 2, 3, 1851.

[3] Goldfuss, Petref. Germ., t. i, p. 59, tab. xix, figs. 1, 5*f*, and tab. xviii, fig. 6, 1826.

advance a little more, and present a very small paliform lobe, which is in general rather indistinct. All the septa are broad, and their upper edge is somewhat oblique till at some distance from the wall, but becomes slightly convex further inwards. Diagonal of the large calices about 8 lines.

Found at Ogwell, Torquay; and at Ferques near Boulogne.

In the collection of Dr. Battersby, &c.

This species is very closely allied to *Cythophyllum hexagonum*,[1] but its septa are more similar in size, less prominent at a small distance from the walls, and have much smaller paliform lobes. It bears also great resemblance to *C. Sedgwicki*,[2] which differs from it principally by the septa becoming thicker at a small distance from the centre of the calice. *C. Davidsoni*[3] presents also the same aspect, but has the calices much smaller, and the septa smaller and less numerous.

12. CYATHOPHYLLUM MARMINI. Tab. LII, figs. 4, 4a.

> CYATHOPHYLLUM PROFUNDUM, *Michelin*, Icon. Zooph., p. 184, pl. xlviii, fig. 1, 1849. (Not Geinitz.)
> — CÆSPITOSUM, *Ibid.*, p. 184, pl. xlvii, fig. 5. (Not Goldfuss.)
> LITHOSTROTION PROFUNDUM, *D'Orbigny*, Prod. de Pal., vol. i, p. 106, 1850.
> CYATHOPHYLLUM MARMINI, *Milne Edwards* and *Jules Haime*, Pol. Foss. des Terr. Palæoz., p. 386, pl. ix, figs. 2, 3, 1851.

Corallum composite, subfasciculate or astreiform; gemmation almost always lateral. *Corallites* very unequal in size, circular when free laterally, polygonal when aggregated. *Calice* deep and broad. About 40 *septa*, more or less similar among themselves, thin, delicately denticulated, and extending for the most part to the bottom of the calicular cavity, where they present only rudimentary paliform lobes; at a short distance from the wall they are somewhat prominent, and assume there the appearance of an interior wall. Dissepiments numerous. Diameter of the calices 4 or 5 lines; depth about $2\frac{1}{2}$.

Found at Teignmouth and Torquay. (Collection of Dr. Battersby.)

This species differs from *C. cæspitosum*[4] by its calice being deeper, its septa less unequal, and more especially by its mode of gemmation, which is almost always lateral; and from *C. quadrigeminum*[5] by the septa being less regular and the existence of paliform lobes, which are absent in the latter species.

10. CYATHOPHYLLUM SEDGWICKI. Tab. LII, figs. 3, 3a.

> CYATHOPHYLLUM SEDGWICKI, *Milne Edwards* and *Jules Haime*, Pol. Foss. des Terr. Palæoz., p. 387, 1851.

---

[1] Tab. l, fig. 4. [2] Tab. lii, fig. 1.
[3] Milne Edwards and Jules Haime, Pol. Foss. des Terr. Palæoz., p. 389. [4] Tab. li, fig. 2.
[5] Goldfuss, Petref. Germ., tab. i, p. 59, tab. xix, figs. 1, 5f, and tab. xviii, fig. 6, 1826.

Corallum composite, astreiform. *Calices* unequal in size, polygonal, and separated by nearly straight walls. Gemmation calicular as well as lateral. *Septa* (32 or 40) well developed, somewhat unequal in size; the smaller ones thin all along, the larger ones thin outwardly, but becoming thicker at about two thirds of their breadth, and again thin towards the centre of the calice, where they are slightly curved, and present a very small paliform lobe. The vesicular *dissepiments* are mostly small, but are rather unequal in size, and do not extend beyond the middle of the thick part of the principal septa. Great diagonal of the calices usually about 6 lines; diameter of the circle of paliform lobes about two thirds of a line.

Found at Babbacombe Beach, Torquay. (Collections of Mr. Bowerbank and Dr. Battersby.)

This species is intermediate between *C. hexagonum*[1] and *C. boloniense*,[2] but approximates most to the latter, from which it differs principally by the thickness of the principal septa at a small distance from the centre of the calice.

---

14. CYATHOPHYLLUM ÆQUISEPTATUM. Tab. LII, fig. 1.

   CYATHOPHYLLUM ÆQUISEPTATUM, *Milne Edwards* and *Jules Haime*, Pol. Foss. des Terr.
                                Palæoz., p. 389, 1851.

Corallum composite, fasciculate. *Corallites* distant from each other, multiplying by lateral gemmation, surrounded by an epitheca, and appearing subcylindrical. *Calices* deep, and with a thin edge. *Septa* (about 36) very narrow upwards, not remarkably thin, and almost equal in size. Diameter of the corallites about 4 lines.

Found at Ilfracombe in Devonshire.

In the Collection of the Geological Society.

This species is remarkable for the equal development of all the septa.

We are inclined to consider the *Strephodes gracilis*[3] of Prof. M'Coy as belonging to the genus Cyathophyllum, but it may be a species of *Ptychophyllum*.

[1] Tab. l, fig. 4.   [2] Tab. lii, fig. 1.

[3] *Strephodes gracilis*, M'Coy, Ann. and Mag. of Nat. Hist., 2d ser., vol. vi, p. 378, 1850; M'Coy, Brit. Palæoz. Foss., p. 72, 1851.

Professor M'Coy describes this fossil in the following terms:—"*Corallum* simple, very gradually tapering, irregularly twisted, averaging three inches long, and eight lines in adult diameter; *horizontal section*, outer wall very thick, solid; radiating lamellæ at the above diameter about 56, very thin, extending in a slightly irregular manner towards a large central space, which the primary ones fill with their irregular complicated extremities; secondary lamellæ as thick as the primary, of irregular lengths, but seldom extending one fourth the distance to the centre; transverse vesicular plates extremely delicate, rather few, irregular; vertical section showing in the middle a few irregularly flexuous delicate longitudinal lines (edges of the complicated ends of the vertical radiating lamellæ); sides occupied by very open vesicular tissue, composed of large, curved, delicate, oblique plates, forming about two rows of great cells on each side; outer wall very thick, forming a nearly smooth surface; when decorticated, the lamellar sulci average 5 in 2 lines; terminal cup deep, strongly radiated to the flattened centre. Locality; Newton Bushel."

4. *Genus* ENDOPHYLLUM.[1]

1. ENDOPHYLLUM BOWERBANKI. Tab. LIII, fig. 1.

> ENDOPHYLLUM BOWERBANKI, *Milne Edwards* and *Jules Haime*, Pol. Foss. des Ter. Palæoz.,
> p. 394, 1851.

Corallum composite, astreiform. *Corallites* more or less intimately united together by rudimentary exterior walls and an irregular vesicular tissue. Inner *walls* well constituted, circular, and often double. Principal *septa* (30 or 32) pretty well developed, rather slender, very flexuous inwardly, extending almost to the centre of the calice, and alternating with an equal number of smaller septa. They do not project much outside the wall, so as to form costal striæ, that soon disappear in the vesicular tissue. *Tabulæ* well developed and somewhat irregular. Diameter of the mural circles about 8 lines, distance between them 5 or 7 lines.

Found at Barton near Torquay.

In the Collections of Mr. Bowerbank, Dr. Battersby, &c.

The Genus ENDOPHYLLUM has been established since the publication of the Introduction to this Monograph, and is intermediate between *Cyathophyllum* and *Acervularia*, having most of the structural characters of the first, but presenting completely vesicular tissue exteriorly to well-defined walls. In Acervularia there is a well-developed epitheca, which does not exist in Endophyllum, and the septal system is much more developed in the space between the two mural investments.

Endophyllum Bowerbanki differs from *E. abditum*[2] by its outer walls being rudimentary, its inner walls being well constituted, and its septa thicker though still rather slender.

2. ENDOPHYLLUM ABDITUM. Tab. LII, fig. 6.

> ENDOPHYLLUM ABDITUM, *Milne Edwards* and *Jules Haime*, Pol. Foss. des Terr. Palæoz.,
> p. 394, 1851.

Corallites more or less closely united by polygonal walls, which are rather strong. Inner walls thin, often double and rather irregularly circumscribed. The space comprised between the two walls is filled with large vesicles, on which some costal striæ may be recognised. Principal *septa* (34 to 40) very slender, especially inwardly, where they become much curved, an equal number of smaller septa alternating with the principal ones. Diagonal of the large calices almost 2 inches; diameter of the mural circle about 12 lines.

---

[1] Milne Edwards and Jules Haime, Pol. Foss. des Terr. Palæoz.; Archives du Muséum, vol. v, p. 393, 1851.

[2] Tab. lii, fig. 6.

Found at Teignmouth Beach by Dr. Battersby.

This fossil differs from *E. Bowerbanki*[1] by its well-developed outer wall and its very slender septa.

5. *Genus* CAMPOPHYLLUM, (p. lxviii.)

The fossil described by Mr. Phillips under the name of *Cyathophyllum turbinatum*,[2] and found by that able geologist at Babbacomb, appears to be specifically identical with Goldfuss's *Cyathophyllum flexuosum*,[3] a coral, appertaining to our genus *Campophyllum*;[4] but we have seen as yet no British specimen of this species.

6. *Genus* PACHYPHYLLUM, (p. lxviii.)

PACHYPHYLLUM DEVONIENSE.   Tab. LII, figs. 5, 5a.

PACHYPHYLLUM DEVONIENSE, *Milne Edwards* and *Jules Haime*, Pol. Foss. des Terr. Palæoz., p. 397, 1851.

This species is known to us only from a polished slab, communicated to us by Dr. Battersby, but appears to be sufficiently characterised to authorise its admission in a methodical arrangement of the Devonian corals.

The *corallites* are not circumscribed, but their radii are not completely confluent. The exterior portion of each individual is principally formed by a vesicular tissue, through which well-defined but very slightly constituted costæ extend. At some distance from the centre of each corallite, a well-marked subcircular or elliptical zone is formed by a slight enlargement of the septa, and appears to represent a rudimentary wall. *Septa* (44 or 48) very slender, unequally developed alternately, the larger ones very slender inwardly, where they become somewhat flexuous and appear to have a paliform lobe, and extending only to a short distance from the centre of the calice. Breadth of the corallites about 8 lines; diameter of the mural zones about 4 lines.

Found at Torquay by Dr. Battersby.

This species differs from *P. Bouchardi*[5] by its septa being more numerous, more slender and unequal, and by the principal ones bearing a small paliform lobe.

[1] Tab. liii, fig. 1.

[2] Palæoz. Fossils, p. 8, pl. vii, fig. 9.

[3] Petref. Germ., vol. i, p. 57, tab. xvii, figs. 3a, 8.

[4] *Campophyllum flexuosum*, Milne Edwards and Jules Haime, Monogr. des Polyp. des Terr. Palæoz., p. 395, pl. viii, fig. 4.

[5] Milne Edwards and Jules Haime, Pol. Foss. des Terr. Palæoz., p. 397, pl. vii, fig. 7, 1851.

7. *Genus* CHONOPHYLLUM, (p. lxix).

CHONOPHYLLUM PERFOLIATUM. Tab. L, fig. 5.

> CYATHOPHYLLUM PLICATUM, *Goldfuss*, Petref. Germ., vol. i, p. 59, tab. xviii, fig. 5, 1826.
> (Not tab. xv, fig. 12.)
> — — *Milne Edwards*, 2d edit. of Lamarck, vol. ii, p. 431, 1836.
> — PERFOLIATUM, *Goldfuss*, MSS. in Bonn Museum.
> CHONOPHYLLUM PERFOLIATUM, *Milne Edwards* and *Jules Haime*, Pol. Foss. des Terr. Palæoz.,
> p. 405, 1851.

Corallum simple, straight, rather elongate. *Calice* not remarkably deep, and of a subconical form. *Septa* (60 to 74) equally developed, straight, and extending almost to the centre of the corallum. Some vestiges of a rudimentary septal fossula are visible. Height about 3 inches, diameter about 2 inches.

Found at Torquay. (Collection of Dr. Battersby.)

A fossil found at Wenlock, and belonging to the collection of M. D'Archiac, appears to belong also to this species.

C. perfoliatum differs from *C. elongatum*[1] by its general form being broader and less elongate, and by its tabulæ being less infundibuliform.

8. *Genus* HELIOPHYLLUM, (p. lxix.)

HELIOPHYLLUM HALLI. Tab. LI, fig. 3.

> STROMBODES HELIANTHOIDES, *Phillips*, Fig. and Descr. of Palæoz. Foss., p. 10, pl. v,
> fig. 13a, 1841. (Not *Cyathophyllum helianthoides*,
> Goldfuss.)
> — — *Hall*, Geol. of New York, 4th part, p. 209, No. 48, fig. 3,
> 1843.
> CYATHOPHYLLUM TURBINATUM, *Ibid.*, No. 49, fig. 1. (Not Goldfuss.)
> — — *Castlenau*, Ter. sil. de l'Amer. du Nord., pl. xvi, fig. 5, 1843.
> HELIOPHYLLUM HALLI, *Milne Edwards* and *Jules Haime*, Brit. Foss. Corals, Introd., p. lxix,
> 1850.
> — — *Milne Edwards* and *Jules Haime*, Pol. Foss. des Terr. Palæoz.,
> p. 408, pl. vii, fig. 6, 1851.

Corallum simple, turbinate, or cylindro-conical, usually elongate, and slightly curved at its base, provided with an epitheca, and presenting slight circular swellings. *Calice*

---

[1] Milne Edwards and Jules Haime, Polyp. Foss. des Terr. Palæoz., p. 406, pl. viii, fig. 1, 1851.

circular, rather deep, with a small septal fossula. *Septa* (80 or even more) very thin, closely set, rather broad at their upper end, where they are arched and denticulate, alternately larger and smaller, slightly twisted near the centre of the visceral chamber. A vertical section shows that the lateral processes of the septa are arched and ascendant; those situated towards the upper end of the corallum terminate at the edge of the septa; those situated lower down unite near the centre of the visceral chamber, so as to constitute irregular *tabulæ*. The interseptal loculi are filled up with these lamellate processes, which are situated at about half a line apart, and united by closely-set simple dissepiments that form right angles with them. Diameter of the calice from 1 to 2 inches.

The specimen submitted to our investigation was found at Torquay. Prof. Phillips has also met with this fossil at Plymouth, Babbacombe, and Sharkham Point. The same species is found in North America.

9. *Genus* ACERVULARIA, (p. lxx).

1. ACERVULARIA GOLDFUSSI. Tab. LIII, figs. 3, 3*a*.

CYATHOPHYLLUM ANANAS, *Goldfuss*, Petref., vol. i, p. 60, pl. xix, fig. 4*a*, 1826. (Not fig. 4*b*.)
— — *Hall*, Handb. der Petref., p. 416, 1830.
— — *Morren*, Descr. Corall. in Belgio Repert., p. 56, 1832.
— — *Milne Edwards*, 2d edit. of Lamarck, vol. ii, p. 429, 1836.
ASTREA BASALTIFORMIS, *Ad. Roemer*, Verst. des Harzgeb., p. 5, tab. ii, fig. 12, 1843.
ACERVULARIA GOLDFUSSI, *De Verneuil* and *Jules Haime*, Bull. Soc. Géol. de France, 2d ser., vol. vii, p. 161, 1850.
LITHOSTROTION ANANAS, (pars), *D'Orbigny*, Prod. de Paléont., vol. i, p. 106, 1850.
ACERVULARIA GOLDFUSSI, *Milne Edwards* and *Jules Haime*, Pol. Foss. des Terr. Palæoz., p. 417, 1851.

Corallum composite, massive, astreiform; the polygonal lines on its surface well marked and rather zigzagged. Great diagonal of the corallites about 3 lines. The inner wall well constituted, rather strong, with the septa somewhat exsert, and being only about 1 line in diameter. *Septa* (24 or 26) almost straight, very slender, and extending alternately more or less towards the centre. *Dissepiments* rather closely set.

Found at Torquay, by Dr. Battersby.

This species much resembles *A. coronata*[1] by the development of its inner and outer walls, but differs from it by the septa being somewhat unequal in size. The costo-septal radii are still more similar and closer set in *A. pentagona*,[2] the corallites of which are also much smaller. In *A. Roemeri*[3] these radii are very slender, and flexuous outwardly, and the outer walls are very vaguely indicated.

---

[1] Tab. liii, fig. 4.   [2] Tab. liii, fig. 5.   [3] Tab. liv, fig. 3.

2. ACERVULARIA CORONATA. Tab. LIII, figs. 4, 4a, 4b.

> ACERVULARIA CORONATA, *Milne Edwards* and *Jules Haime*, Pol. Foss. des Terr. Palæoz., p. 416, 1850.

A polished section of this coral shows that the corallites are united by means of well-defined polygonal epithecal walls. The inner walls are also well constituted, and circumscribe a very small area comparatively to the breadth of the corallites. *Septa* (generally 28) very slender, somewhat thickened by lateral granulations near the outer wall, where most of them become slightly curved. In the space comprised between the two mural investments the septa are equally developed; but only one half of them penetrate into the visceral chamber, and extend almost to the centre of the corallite, where they bear a small but well-defined paliform lobe. *Exothecal dissepiments* very closely set. Diagonal of the corallites 5 or 6 lines.

Found at Barton near Torquay.

In the Collections of Dr. Battersby and Mr. Pengelly.

This species, by the development of its two mural investments, differs from *A. intercellulosa*[1] and *A. limitata*,[2] in which the inner wall is only indicated by a slight thickening of the septa; and from *A. Battersbyi*,[3] in which the exterior wall is rudimentary. In *A. Roemeri*[4] the septa are much more slender, and more curved outwardly. In *A. pentagona*[5] and *A. Goldfussi*,[6] the septa all reach very near to the centre of the visceral chamber, whereas in the above-described coral half of them do not extend beyond the inner walls.

3. ACERVULARIA INTERCELLULOSA. Tab. LIII, figs. 2, 2a.

> ASTREA INTERCELLULOSA, *Phillips*, Palæoz. Foss. of Cornwall, &c., p. 12, pl. vi, fig. 17, 1841.
> FAVASTREA   —   *D'Orbigny*, Prod. de Paléont., vol. i, p. 107, 1850.
> ACERVULARIA —   *Milne Edwards* and *Jules Haime*, Pol. Foss. des Terr. Palæoz., p. 417, 1851.

Corallites polygonal, unequal in size, circumscribed by well-marked zigzagged exterior walls. *Inner walls* rendered distinct by a thickening of the septa, and forming circles which are very large in proportion to the size of the polygones. *Septa* (40 to 44) slightly developed in the exterior parts of the corallites, where they become quite lost in the vesicular tissue; in the part corresponding to the inner wall they are thick, but they become slender again more inwardly, where one half of them reach almost to the centre of the corallite, and are provided with a paliform lobe. Great diagonal of the polygonal corallites about 6 lines; diameter of the calice about 4 lines.

---

[1] Tab. liii, fig. 2.   [2] Tab. liv, fig. 1.   [3] Tab. liv, fig. 2.
[4] Tab. liv, fig. 3.   [5] Tab. liii, fig. 5.   [6] Tab. liii, fig. 3.

Found at Torquay.

In the Collections of Dr. Battersby and Mr. Pengelly.

This species resembles *A. limitata*[1] in having the inner wall rudimentary, and indicated only by a thickening of the septa ; it differs from it, as well as from the other Acervularia, by the great number of the septa.

4. ACERVULARIA PENTAGONA.　　Tab. LIII, figs. 5, 5*a*, 5*b*.

> CYATHOPHYLLUM PENTAGONUM, *Goldfuss*, Petref. Germ., vol. i, p. 60, tab. xix, fig. 3, 1826.
> FAVASTREA PENTAGONA, *De Blainville*, Dict. Sc. Nat., vol. lx, p. 340, 1830.—Man., p. 375.
> CYATHOPHYLLUM PENTAGONUM, *Morren*, Desc. Corall. Belg., p. 56, 1832.
> —　　　　　　—　　　*Milne Edwards*, 2d ed. of Lamarck, vol. ii, p. 430, 1836.
> ASTREA PENTAGONA, *Lonsdale*, Geol. Trans., 2d ser., vol. v, pl. 57, fig. 1, 1840. (Not fig. 1*a*.)
> —　　　—　　*Phillips*, Palæoz. Foss., p. 11, pl. vi, fig. 15, 1841.
> ACERVULARIA PENTAGONA, *Michelin*, Icon., p. 180, pl. xlix, fig. 1, 1845.
> —　　　ANANAS, *Ibid.*, p. 180, pl. xlvii, fig. 1.
> LITHOSTROTION PENTAGONUM, *D'Orbigny*, Prod., vol. i, p. 106, 1850.
> ACERVULARIA PENTAGONA, *Milne Edwards* and *Jules Haime*, Pol. Foss. des Terr. Palæoz.,
> 　　　　p. 418, 1851.
> —　　　—　　*M'Coy*, Brit. Palæoz. Foss., p. 91, 1851.

Corallum forming an astreiform mass. *Corallites* somewhat unequal in size, polygonal, their great diagonal measuring in general about 2 lines, and the diameter of the inner walls about half a line. *Septa* (18 to 24) subequal, slender, and almost straight. The lines of demarcation between the corallites slightly zigzagged.

Found at Torquay and at Ogwell. Prof. Phillips mentions having met with this species at Newton Bushel, Sharkham Point, and Babbacombe. It is found also in the Eifel, and in the province of Limbourg.

In the Collections of Messrs. Bowerbank and Pengelly.

This is the smallest species of Acervularia known, and that in which the septa are the least numerous. *A. coronata*,[2] to which it bears most resemblance, differs from it by the great inequality of the septa.

5. ACERVULARIA LIMITATA.　　Tab. LIV, fig. 1.

> ASTREA PENTAGONA (pars), *Lonsdale*, Geol. Trans., 2d ser., vol. v, pl. lviii, fig. 1*a*, 1840.
> 　　　　(Not fig. 1.)
> ACERVULARIA LIMITATA, *Milne Edwards* and *Jules Haime*, Pol. Foss. des Terr. Palæoz.,
> 　　　　p. 419, 1851.

A polished slab of this coral shows that the corallites are circumscribed by well-marked zigzagged polgyonal lines. The inner walls are, on the contrary, but slightly marked,

---

[1] Tab. liv, fig. 1.　　　　　　　　　　　　　　[2] Tab. liii, fig. 4.

and are principally indicated by a small thickening of the septa. In general 26 *septa*, rather slender, granulated on the sides, and often slightly curved in the space comprised between the two mural investments; half of them do not extend further than the inner wall, and those which penetrate into the central area do not appear to have any paliform lobes. Diagonal of the corallites 3 lines; diameter of the inner walls 1 line.

Found in Newton Quarry near Torquay.

In the Collections of the Geological Society and Dr. Battersby.

In this species the inner wall is rudimentary, as in *A. intercellulosa*,[1] but the septa are less numerous, and do not give rise to paliform processes.

6. ACERVULARIA BATTERSBYI.   Tab. LIV, fig. 2.

ACERVULARIA BATTERSBYI, *Milne Edwards* and *Jules Haime*, Pol. Foss. des Terr. Palæoz., p. 419, 1851.

A horizontal section of this species shows that the Corallites are very closely united together and limited only by a very thin exterior wall, which form zigzags, is slightly marked, and constitutes polygonal divisions. The *inner walls* are, on the contrary, very thick, and circumscribe a central area, which is very small in proportion to the space occupied by the whole of each Corallite; they appear to be composed of a dense exothecal tissue. *Septa* (36), of equal size in the outer area, very slender, for the most part much curved, almost confluent, and slightly thickened, where enclosed in the inner wall; half of them extend almost to the centre of the visceral chamber, where they present a small paliform lobe. *Dissepiments* very abundant and closely set in the exterior area, but almost completely absent in the inner area. Great diagonal of the Corallites 6 or 8 lines; diameter of the calices 2 lines, or somewhat more.

Found at Torquay and at Newton.

In the Collections of the Geological Society and Dr. Battersby.

This species, by the feeble development of the outer walls and its subconfluent septa, leads to the genus *Phillipsastrea*. It differs from *A. Roemeri*, where the exterior wall is also but slightly developed, by the septa being more numerous and provided with a paliform lobe.

7. ACERVULARIA ROEMERI.   Tab. LIV, fig. 3.

ASTREA HENNAHII, *Ad. Roemer*, Verst. des Harzgebirges, p. 5, tab. ii, fig. 13, 1843.   (Not Lonsdale.)
—   PARALLELA ? *Ibid.*, tab. iii, fig. 1.
PHILLIPSASTREA PARALLELA ? *D'Orbigny*, Prod. de Paléont., vol. i, p. 107, 1850.

[1] Tab. liii, fig. 2.

ACERVULARIA ROEMERI, *De Verneuil* and *Jules Haime*, Bull. Soc. Géol. de France, 2d ser.,
vol. vii, p. 162, 1850.
—   —   *Milne Edwards* and *Jules Haime*, Polyp. Foss. des Terr. Palæoz.,
p. 420, 1851.

Corallum massive with an almost flat surface. *Corallites* prismatic, and intimately
united together. Outer walls very slender, and often difficultly recognised in certain states
of fossilization. Great diagonal of the Corallites 2 or 3 lines or more; diameter of the
inner walls less than 1 line. 26 or 28 septa costal laminæ, very slender, and strongly curved
or flexuous towards the centre of the Corallites.

Found at Torquay; at Grund in the Hartz; and at Puerto de las Volcas near Pola de
Gordon, in the province of Leon in Spain.

In the Collection of Dr. Battersby.

Professor M'Coy mentions this species as having been met with also at Barton and
Teignmouth; but he does not distinguish it from *Acervularia intercellulosa*.

This species differs from all the other Acervulariæ by its septa being much curved, and
its outer walls rudimentary. *A. Battersbyi*[1] has the septa more numerous, more slender,
and provided with a paliform lobe.

10. *Genus* SMITHIA.[']

1. SMITHIA HENNAHII.   Tab. LIV, fig. 4.

ASTREA HENNAHII (pars), *Lonsdale*, in Sedgwick and Murchison, Geol. Trans., 3d ser.,
vol. v, p. 697, pl. lviii, fig. 3, 1840.
—   —   *Phillips*, Palæoz. Foss., p. 12, pl. vi, fig. 16, 1841.
CYATHOPHYLLUM HENNAHII, *Bronn*, Index Paléont., vol. i, p. 368, 1848.
LITHOSTROTION HENNAHII, ACTINOCYATHUS HENNAHII, and PHILLIPSASTREA HENNAHII
(pars), *D'Orbigny*, Prod. de Paléont., vol. i, pp. 106, 107, 1850.
SMITHIA HENNAHII, *Milne Edwards* and *Jules Haime*, Polyp. Foss. des Terr. Pal., p. 421, 1851.
ARACHNOPHYLLUM HENNAHII, *M'Coy*, Brit. Palæoz. Foss., p. 72, 1851.

A polished horizontal section of this compound astreiform corallum shows that the *mural
circles,* although slender, are well characterised, and placed at a distance from each other,
equal to 2, 3, or even 4 times their diameter, but varying sometimes very much in the same
specimen. *Costal radii* (24 or 26 in a corallite) slender, appearing to be slightly granulated
on their sides, and in general much more developed, more confluent and straighter
in one direction than in the other, where they become irregular, flexuous, angular or
geniculate; half of the radii do not extend beyond the wall; the others become somewhat
thicker at that part, and pass on towards the centre of the visceral chamber, where some
traces of small paliform lobes are seen. Diameter of the mural circles about $1\frac{1}{3}$ line.

---

[1] Tab. liv, fig. 2.
[2] Milne Edwards and Jules Haime, Pol. Foss. des Terr. Palæoz., p 421, 1851.

A vertical section shows that the intercostal loculi are filled up with vesicles, which are very small and pretty regular. The *dissepiments* of the interseptal loculi are almost horizontal, and unite at the centre of the visceral chamber so as to form a series of small and very closely set *tabulæ*.

Found at Torquay, Plymouth, and Newton.

In the Collections of Mr. Bowerbank, Dr. Battersby, and Mr. Pengelly.

S. Pengelli[1] differs from the above-described species by its septa being more numerous and its walls rudimentary.

In *S. Bowerbanki*[2] the septa are less numerous and more vermiculate.

In *S. Boloniensis*[3] the calices are smaller, and costal radii completely confluent.

2. SMITHIA PENGELLYI.   Tab. LV, fig. 1.

ASTREA HENNAHII, (pars), *Lonsdale*, in Sedgwick and Murchison, Geol. Trans., 2d ser., vol. v, 3d part, p. 697, pl. lviii, fig. 3a, 1840.

SMITHIA PENGELLYI, *Milne Edwards* and *Jules Haime*, Pol. Foss. des Terr. Pal., p. 422, 1851.

Mural circles not very distinct, and indicated principally by a slight thickening of the septa, which are placed at very unequal distances. *Costal radii* (not more than 40) alternately unequal in thickness, granulated laterally; in general larger and more confluent in one direction than in the other, where they are flexuous and even angular; half of them do not extend further than the wall; the others become somewhat thicker on that part, and, passing on, become very slender towards the centre of the visceral chamber, where they bear paliform appendices. Diameter of the mural circle 2 lines, or somewhat more. *Dissepiments* very closely set.

Found at Torquay and Plymouth.

In the Collections of the Geological Society, Mr. Bowerbank, Dr. Battersby, and Mr. Pengelly.

This species differs from the preceding one by its septa being more numerous, thicker, and closer set. It is also well characterised by its walls being rudimentary.

3. SMITHIA BOWERBANKI.   Tab. LV, fig. 2.

SMITHIA BOWERBANKI, *Milne Edwards* and *Jules Haime*, Pol. Foss. des Terr. Palæoz., p. 423, 1851.

Mural circles well developed, and placed very widely apart, but at unequal distances. *Costal radii* (18 or 20) completely confluent, slender, larger, and straighter in one direction than in the other, but in general flexuous and vermiculate (the more so as they

---

[1] Tab. lv, fig. 1.   [2] Tab. lv, fig. 2.

[3] Milne Edwards and Jules Haime, Pol. Foss. des Terr. Palaeoz., p. 423, 1851.

extend farther from the calice), becoming somewhat thicker in the wall, where they also become unequal in size; the larger ones do not appear to extend quite to the centre of the visceral chamber, and show no traces of paliform lobes. *Dissepiments* very small. Diameter of the wall about two thirds of a line.

Found at Torquay by Dr. Battersby.

This corallum differs from all the other species of the same genus by the small dimensions of the calices and the considerable distance between them; it is also remarkable for the complete confluency of its costal radii.

11. *Genus* SPONGOPHYLLUM.[1]

SPONGOPHYLLUM SEDGWICKI.   Tab. LVI, figs. 2, 2a, 2b, 2c, 2d, 2e.

SPONGOPHYLLUM SEDGWICKI, *Milne Edwards* and *Jules Haime*, Pol. Foss. des Terr. Pal., p. 425, 1851.

Corallum composite, massive, astreiform. *Calices* polygonal, very unequal in size, and circumscribed by pretty strong walls. No *columella*. *Septal radii* (14 or 16) extremely slender, extending in general to a short distance from the centre of the visceral chamber, slightly flexuous, and often not very distinct in the midst of the vesicular tissue that fills the cavity of the corallites. These septa alternate with an equal number of rudimentary ones.

A vertical section shows that the outer parts of the visceral chambers are occupied by vesicles unequal in size, in general much elongated and rather irregular, but that in the centre there are small horizontal tabulæ. Diagonal of the large calices $2\frac{1}{2}$ or 3 lines.

Found at Torquay.

In the Collection of Dr. Battersby.

This fossil is the only known species of our genus *Spongophyllum*, which is characterized principally by the rudimentary state of the radial laminæ; these appear to form slight ridges on the surface of the vesicles, but not to pass through them, and resemble the costal system of the corals forming the genus *Lonsdalia*.

12. *Genus* SYRINGOPHYLLUM, (p. lxxii.)

SYRINGOPHYLLUM CANTABRICUM.   Tab. LV, fig. 3.

PHILLIPSASTREA CANTABRICA, *De Verneuil* and *Jules Haime*, Bull. Soc. Géol. de France, 2d ser., vol. vii, p. 162, 1850.
SYRINGOPHYLLUM? CANTABRICUM, *Milne Edwards* and *Jules Haime*, Pol. Foss. des Terr. Palæoz., p. 451, 1851.

---

[1] Milne Edwards and Jules Haime, Pol. Foss des Terr. Palæoz.; Archives du Muséum, vol. v, p. 425, 1851.

Corallum composite, forming almost flat masses. *Calices* slightly prominent and placed at unequal distances (in general about 1⅓ their diameter). *Costæ* irregularly confluent, large, rather thin, equally developed, flexuous or geniculated, delicately crenulated, and closely set (about a quarter of a line apart); 15 or 16 principal ones slightly exsert, terminated by an arched edge, extending almost to the centre of the visceral chamber, where they become very slender, bearing a small paliform lobe and alternating with an equal number of small septa; *wall* well developed, rather thick. *Columella* appearing to be slightly compressed. Diameter of the calices 1½ line; depth almost half a line.

Found at Torquay; and at Valcos in the province of Leon in Spain.

In the Collection of the Geological Society of London, &c.

This species differs from *S. organum*[1] and *S. Torreanum*[2] by the costal radii being more numerous and more confluent.

Family CYSTIPHYLLIDÆ, (p. lxxii.)

Genus CYSTIPHYLLUM, (p. lxxii.)

CYSTIPHYLLUM VESICULOSUM.   Tab. LVI, figs. 1, 1*a*, 1*b*.

> CYATHOPHYLLUM VESICULOSUM, *Goldfuss*, Petref. Germ., p. 58, pl. xvii, fig. 5, and pl. xviii, fig. 1, 1826.
> — SECUNDUM, *Ibid.*, p. 58, pl. xviii, fig. 2.
> — CERATITES (pars), *Ibid.*, pl. xvii, fig. 2*k*.
> — — *Milne Edwards*, 2d edit. of Lamarck, vol. ii, p. 430, 1836.
> CYSTIPHYLLUM VESICULOSUM, *Phillips*, Palæoz. Foss., p. 10, pl. iv, fig. 12, 1841.
> — — *De Verneuil* and *Jules Haime*, Bull. Soc. Géol. de France, 2d ser., vol. vii, p. 162, 1850.
> — SECUNDUM and VESICULOSUM, *D'Orbigny*, Prod. de Pal., vol. i, p. 106, 1850.
> — VESICULOSUM, *Milne Edwards* and *Jules Haime*, Pol. Foss. des Terr. Palæoz., p. 462, 1851.
> — — *M'Coy*, Brit. Palæoz. Foss., p. 71, 1851.

Corallum simple, very long, slightly bent, subcylindrical, provided with a very strong epitheca, and presenting rather strong subhorizontal circular wrinkles. *Calicular* cavity rather deep; the septal striæ, when visible, more distinct towards the outer part of the calice. Vesicules unequal in size; the largest occupying the centre of the visceral cavity and about 1 line in length. Height of the coral in general about 3 or 4 inches. We have seen in the collection of Mr. Pengelly a specimen that measured above 1 foot in length, and 1½ inch in diameter.

The British specimens submitted to our examination were found at Torquay, Plymouth,

---

[1] *Sarcinula organum*, Hisinger, Leth. Succ., p. 97, tab. xxviii, fig. 8, 1837.
[2] Milne Edwards and Jules Haime, Pol. Foss. des Terr. Palæoz., p. 452, 1851.

and Mudstone Bay; Prof. Phillips has met with the same species at Babbacombe, and Prof. M'Coy at Newton Bushel. It exists also in Spain at Millar, in the province of Leon; in Germany in the Eifel Mountains; and in North America at Corn Island, Falls of the Ohio.

This species differs from *C. lamellosum*[1] by its cylindroid form and its large transverse wrinkles. *C. cylindricum*[2] and *C. Siluriense*[3] are both easily distinguished from it by the existence of radiciform processes; and in *C. Grayi*[4] the vesicles are much more oblique towards the surface of the coral, and very irregular towards its centre.

[1] *Cyathophyllum lamellosum* and *placentiforme*, Goldfuss, Petref., tab. xviii, figs. 3, 4.

[2] Lonsdale, Silur. Syst., p. 691, pl. xvi bis, fig. 3, 1839.

[3] (Pars), Lonsdale, Ibid., pl. xvi bis, fig. 1.

[4] Milne Edwards and Jules Haime, Pol. Foss. des Terr. Palæoz., p. 465, 1851.

THE

PALÆONTOGRAPHICAL SOCIETY.

INSTITUTED MDCCCXLVII.

LONDON:

MDCCCLIV.

A MONOGRAPH

OF THE

BRITISH FOSSIL CORALS.

BY

H. MILNE EDWARDS,

DEAN OF THE FACULTY OF SCIENCES OF PARIS; PROFESSOR AT THE MUSEUM OF NATURAL HISTORY;
MEMBER OF THE INSTITUTE OF FRANCE;
FOREIGN MEMBER OF THE ROYAL SOCIETY OF LONDON, OF THE ACADEMIES OF BERLIN, STOCKHOLM, ST. PETERSBURG,
COPENHAGEN, VIENNA, KONIGSBERG, MOSCOW, BRUXELLES, HAARLEM, BOSTON, PHILADELPHIA, ETC.,

AND

JULES HAIME.

FIFTH PART.
CORALS FROM THE SILURIAN FORMATION.

LONDON:

PRINTED FOR THE PALÆONTOGRAPHICAL SOCIETY.

1854.

DESCRIPTION

OF

THE BRITISH FOSSIL CORALS.

CHAPTER XVI.

CORALS FROM THE SILURIAN FORMATION.

At no period of the geological history do Polypi appear to have been more abundant, and to have constituted as important a portion of the marine fauna, as at the time during which the Silurian deposits were formed. The variety of species is here as considerable as in most of the richest coralliferous rocks of a more recent date, and the number of specimens is usually greater. But what contributes still more to the importance of the study of the Silurian corals, is the good state of preservation in which they are generally found. The environs of Dudley were long ago, and still are, celebrated for their numerous fossil corals, as well as for the abundance of their trilobites; of late years many other localities, equally rich in palæontological treasures, have been explored in various parts of Great Britain; and, at the present day, more than half of the species discovered in the Silurian deposits of the new as well as of the old world, have been found in England. This result is principally due to the indefatigable researches of Sir Roderick Murchison and his followers. The British Silurian fossils found by that able and justly-celebrated geologist, were described and figured in his standard work by Mr. Lonsdale, who referred most of them to the species previously described by Goldfuss, and found by that naturalist in the Devonian deposits of the Eifel Mountains in Germany. This supposed identity, however, does not exist in any of the well-characterised species. The specimens figured by Mr. Lonsdale in Sir R. Murchison's work, have been communicated to us for examination by the Council of the Geological Society of London, and on comparing them with the specimens figured by Goldfuss, and placed by that author in the Poppelsdorf Museum at Bonn, we have been able to ascertain that almost all of them are specifically different. M. D'Orbigny, without having had an opportunity of making any such direct comparison, came at the same time to a similar conclusion; and the researches of Professors Sedgwick and M'Coy fully confirm this result.

33

The British Silurian corals differ but little from those of Gothland, and resemble also very much those (in small number) that have been met with in the corresponding rocks of Bohemia, but are generally distinct from those of the Silurian deposits of North America. The total number of species discovered in the various Silurian deposits amounts to 129, and with the exception of 8, they all belong to our divisions of *Zoantharia tabulata* and *Zo. rugosa;* one of which zoological forms is comparatively rare in the present time, and the other belongs exclusively to the palæozoic period. Seventy-six of these corals are found in England, and about half of these have not been met with elsewhere. Most of these British fossils (sixty-eight species) belong to the families of Favositidæ and Cyathophyllidæ; and the only species not appertaining to the above-mentioned higher divisions are four Fungidæ.

The principal palæontological collections, from which we have obtained the specimens studied and described in this chapter, are those of the Geological Society of London, of the Museum of Practical Geology, of Mr. Fletcher (Dudley), of Mr. John Gray (Dudley), of Mr. Bowerbank, Mr. Charles Stokes (London), of M. Bouchard-Chantereaux (Boulogne-sur-Mer), of M. de Verneuil, and of the Museum of Natural History of Paris.

Most of these corals belong to the upper Silurian rocks, and those found in the lower Silurian deposits are in general very ill preserved and unsatisfactorily characterised. We, therefore, have not deemed it necessary to devote a separate chapter to their description.

Family FUNGIDÆ, (*Introd.*, p. xlv.)

Genus PALÆOCYCLUS, (*Introd.*, p. xlvi.)

1. PALÆOCYCLUS PORPITA. Tab. LVII, figs. 1, 1*a*, 1*b*, 1*c*.

FOSSILE QUERFURTENSE, *D. S. Buttners*, Corallogr. Subterr., p. 25, tab. iii, fig. 5, 1714.

FUNGITARUM CAPITULA, PARVA, STRIATA, &c., *Magnus Bromel*, Acta Liter. Suec., vol. ii, p. 446, figs. *a—h*, 1728.

MADREPORA SIMPLEX, ORBICULARIS, &c., *Fougt*, Amæn. Acad., t. i, p. 91, tab. iv, fig. 5, 1749.

MADREPORA PORPITA, *Linné*, Syst. Nat., edit. xii, p. 1272, 1767.

CYCLOLITES NUMISMALIS, *Lamarck*, Syst. des Anim. sans Vert., p. 369, 1801.

PORPITES HEMISPHERICUS, *Schlotheim*, Petrefactenkunde, 1st part, p. 349, 1820.

MADREPORITES PORPITA, *G. Wahlenberg*, Nov. Act. Reg. Societ. Scient. Upsal., vol. viii, p. 95, 1821.

CYCLOLITES NUMISMALIS, *Hisinger*, Leth. Suec., p. 100, tab. xxviii, fig. 5, 1837.

PALÆOCYCLUS PORPITA, *Milne Edwards* and *Jules Haime*, Introd., p. xlvi, 1850.—Polyp. Foss. des Terr. Palæoz. (Arch. du Mus., vol. v.), p. 204, 1851.— Ann. des. Sc. Nat., 3d ser., vol. xv, p. 110, 1851.

Corallum discoidal; its under surface flat, covered with a strong epitheca, presenting concentrical wrinkles, and completely free, or provided with a conical, curved, flat peduncle. The upper surface much depressed towards the centre, and surrounded by a thick rim formed by prominent septa. The *columella* (if existing) very short and imperfectly formed. Large septa (28 or 30), uniformly developed, and alternating with smaller ones; they probably form six systems, four of which are composed of seven elements, and two of eleven derivated septa; or, in other words, they appear to constitute four complete cycla, to which septa of a fifth cyclum are superadded in one half of two or three systems, in which the quaternary septa have also attained a greater size. All these septa are thick, very closely set externally, and quite straight, the upper edge of the principal ones is regularly convex, and armed with closely-set, strong teeth or crenulations; two rows of these small points often exist on the same septum, towards the exterior part of the corallum. Diameter, in general, 5 or 6 lines; height about 1 line.

Dudley. It is also met with in Gothland.

Specimens are in the Collections of Mr. T. W. Fletcher, and of Mr. J. Gray, at Dudley, of the Museum of Practical Geology, London; of M. de Verneuil, at Paris, &c.

This species differs from *P. Fletcheri*,[1] and *P. rugosus*,[2] by its extremely flat form. *P. præacutus*[3] is also discoidal, but all its septa are equally developed, whereas, in *P. porpita*, there are alternately large and small septa.

2. PALÆOCYCLUS PRÆACUTUS. Tab. LVII, figs. 2, 2*a*, 2*b*, 2*c*.

> CYCLOLITES PRÆACUTA, *Lonsdale*, in Murchison, Silur. Syst., p. 693, tab. xix, fig. 4, 1839.
>
> CYCLOLITES LENTICULATA, *Ibid.*, p. 693, tab. xv, fig. 5 (but not *Porpites lenticulatus* of Schlotheim).
>
> CYCLOLITES PRÆACUTUS, *Eichwald*, Silur. Schisten Syst., p. 201, 1840.
>
> DISCOPHYLLUM PRÆACUTUM and LENTICULATUM, *D'Orbigny*, Prodr. de Paleont., vol. i, p. 47, 1850.
>
> PALÆOCYCLUS PRÆACUTUS, *Milne Edwards* and *Jules Haime*, Pol. Foss. des Terr. Palæoz. (Arch. du Mus., vol. v), p. 205, 1851.—Ann. des Sc. Nat., 3d ser., vol. xv, p. 110, 1851.

The only specimens of this species that we have seen, are those described by M. Lonsdale, and belonging to the Collection of the Geological Society of London. They are cyclolitoid coralla, much resembling *P. porpita* by their general form, but thinner. Their under surface is almost flat; sometimes it is slightly prominent in the centre, but never presents any appearance of a peduncle, and is covered with a somewhat thin, delicately-wrinkled epitheca. *Septa* 48 in number, not much elevated, regularly crenulated, and not appearing to alternate with smaller ones. Diameter of a large individual, about 8 lines; height about 1 line.

Found at Marloes Bay, Pembrokeshire.

---

[1] See tab. lvii, fig. 3.   [2] See tab. lvii, fig. 4.   [3] See tab. lvii, fig. 2.

This species is of a more regular discoidal form than any others of the same genus. *P. porpita*,[1] which is also very flat, differs from it by the septa being alternately large and small.

3. PALÆOCYCLUS FLETCHERI. Tab. LVII, figs. 3, 3*a*, 3*b*, 3*c*, 3*d*, 3*e*, 3*f*.

PALÆOCYCLUS FLETCHERI, *Milne Edwards* and *Jules Haime*, Pal. Foss. des Terr. Palæoz. (Arch. du Mus., vol. v), p. 205, 1851.—Ann. des Sc. Nat., 3d ser., vol. xv, p. 111, 1851.

Corallum very short, but of a subturbinate form, with a short, strongly curved peduncle, a thick epitheca, and some well developed accretion wrinkles. *Calice* circular, or almost so, with a somewhat deep cavity, and a lamellated edge. 36 or 38 principal septa, alternating with an equal number of small ones; somewhat thick, closely set, not tall, and scarcely exsert; their edge slightly convex, and provided with rather strong denticulations, which, closely set exteriorly, become more distant towards the centre, and scarcely ever form a double row on the same septum. Diameter of the calice about 8 lines; height of the corallum about 4 lines.

In the young specimens the peduncle is more developed proportionally. In an aged individual, that measured almost 12 lines in diameter, and that does not appear to differ specifically from the former, the marginal denticulations of the principal septa are almost obsolete exteriorly.

Dudley. Collections of Mr. T. W. Fletcher, and of M. Bouchard-Chantereaux (Boulogne).

By its general form, this species is intermediate between *P. porpita*[2] and *P. rugosus*;[3] its septa are also more numerous and more strongly denticulated than in either of these fossils.

4. PALÆOCYCLUS RUGOSUS. Tab. LVII, figs. 4, 4*a*, 4*b*, 4*c*, 4*d*.

PALÆOCYCLUS RUGOSUS, *Milne Edwards* and *Jules Haime*, Pol. Foss. des Terr. Palæoz. (Arch. du Mus., vol. v), p. 206, 1851.—Ann. des Sciences Nat., 3d ser., vol. xv, p. 111, 1851.

Corallum cylindro-turbinate, sometimes rather elongate; its basis subpedunculated, much bent, and compressed; *epitheca* thick, and presenting strongly developed accretion wrinkles, which are very oblique near the basis. *Calice* circular, with a large, and somewhat deep cavity. Principal septa (26 or 28) alternating with an equal number of small ones; somewhat thick, with their edges regularly denticulated, and slightly arched interiorly. The

---

[1] See tab. lvii, fig. 1.   [2] See tab. lvii, fig. 1.   [3] See tab. lvii, fig. 4.

large specimens are about 4 or 5 lines in diameter, and somewhat more in height ; the young ones are almost as broad, but scarcely more than half as high.

Found at Dudley and at Wenlock. Specimens in the Collections of Mr. Fletcher, of the Geological Society of London, of the Parisian Museum, &c.

This coral is much taller than the preceding species of the same genus, and its septa are thinner and less strongly denticulated. By its exterior characters it resembles very much the young specimens of *Cyathophyllum*, but its internal structure renders it quite distinct.

Family MILLEPORIDÆ, (p. lviii.)

1. *Genus* HELIOLITES, (p. lviii.)

1. HELIOLITES INTERSTINCTA.   Tab. LVII, figs. 9, 9*a*, 9*b*, 9*c*, 9*d*.

MILLEPORA SUBROTUNDA, &c., *Fougt*, Amæn. Acad., vol. i, p. 99, tab. iv, fig. 24, 1749.

MADREPORA INTERSTINCTA, *Linné*, Syst. Nat., ed. 12, p. 1276, 1767.

PORPITAL MADREPORITE, *Parkinson*, Org. Rem., vol. ii, pl. vii, fig. 2—5, 1808.

MADREPORITES INTERSTINCTUS,*Wahlenberg*,Nov. Acta Soc. Scient. Upsal., vol. viii,p. 98,1821.

SARCINULA PUNCTATA, *Fleming*, Brit. Anim., p. 508, 1828.

     —   —   *S. Woodward*, Syn. Table of Brit. Org. Rem., p. 5, 1850.

ASTRÆA CORONA, *Ch. Morren*, Descr. Cor. Belg., p. 64, tab. xxi, figs. 1, 2, 1832.

ASTREA POROSA, *Hisinger*, Leth. Suec., p. 98, tab. xxviii, fig. 2, 1837. (Not *Goldfuss*.)

PORITES PYRIFORMIS, *Lonsdale*, Silur. Syst., p. 686, pl. xvi, figs. 2, 2*a*, 2*b*, 2*c* (cæt. excl.), 1839.

HELIOPORA INTERSTINCTA, *Eichwald*, Sil. Schist. Syst. in Esthland, p. 199, 1840.

PORITES PYRIFORMIS, *Lonsdale*, Russ. and Ural, vol. i, p. 625, 1845.

PORITES INTERSTINCTA, *Keyserling*,Wiss. Beob. auf ein Reise in das Petschora Land, p. 175, 1846.

GEOPORITES LONSDALEI, PYRIFORMIS, and INTERSTINCTA, *D'Orbigny*, Prodr. de Paleont., vol. i, pp. 49, 50, 1850.

PALÆOPORA INTERSTINCTA, *M'Coy*, Brit. Palæoz. Foss., p. 15, 1851.

HELIOLITES INTERSTINCTA, *Milne Edwards* and *Jules Haime*, Pol. Foss. des Terr. Palæoz. (Arch. du Mus., vol. v), p. 214, 1851.

HELIOLITES PYRIFORMIS, *Hall*, Paleont. of New York, vol. ii, p. 133, pl. xxxvi A, fig. 1, 1852.

Corallum composite, massive, in general more or less round or gibbose, sometimes subdendroid. *Calices* closely set (at a distance equal only to their diameter or to two thirds of it), and of the same size in the different parts of the same compound specimen ; their edge circular, and slightly prominent ; in well preserved specimens a small columellary elevation is visible on the surface of the uppermost tabula ; 12 septa, somewhat unequal in size alternately. Polygonal divisions of the cœnenchyma regular and of equal size; their diameter about one eighth of a line. Diameter of the calices about two thirds of a line.

This species has been found in the inferior Silurian beds at Applethwaite and Caradoc. Sir R. Murchison has found it also at Marloes Bay, in Pembrokeshire. It exists also in the upper Silurian deposits at Wenlock and Dudley; and in Ireland. The British localities mentioned by Sir R. Murchison as presenting this fossil, are—Aymestry, Rutter Edge, Wenlock Edge, Lincoln Hill, Benthall Edge, Haven near Aymestry, Lindell's Park, the Ledbury, Delves Green, and Walsall. Professor M'Coy points out its existence in the Coniston limestone of Long Steddale, Westmoreland; and in Galway, Kerry, Mayo, and Dublin. It is also met with at Nehou and Viré, in France; in Gothland, in Russia, and in North America. Mr. Hall has recently found it in the Niagara limestone at Lockport, and at Milwaukie, Wisconsin.

H. interstincta much resembles *H. porosa*,[1] with which many authors have confounded it; but it differs from that species by the calices being much more closely set, and by the polygonal divisions of the cœnenchyma being proportionally much smaller.

We are inclined to think that the fossil corals which Professor M'Coy mentions as being extremely abundant in the calcareous schists and limestone of Craig Head, Ayrshire; Girvan; and in the fine Caradoc limestone of Mulock Quarry, Dalquorhan, Ayrshire, and designates under the name of *Palæopora favosa*,[2] are ill-preserved specimens of the above-described species.

2. HELIOLITES MURCHISONI. Tab. LVII, figs. 6, 6a, 6b, 6c.

FUNGITES, *Thomas Pennant*, Philos. Trans., vol. xlix, 2d part, p. 513, tab. xv, fig. 2, 1757.
COMPOUND MADREPORITE, *Parkinson*, Org. Rem. of a Form. World, vol. ii, pl. vii, fig. 10, 1808.
PALÆOPORA INTERSTINCTA, var. SUBTUBULATA, *M'Coy*, Brit. Palæoz. Foss., p. 16, pl. i c, fig. 2, 1851.
HELIOLITES MURCHISONI, *Milne Edwards* and *Jules Haime*, Pol. Foss. des Terr. Palæoz. (Arch. du Mus., vol. v), p. 215, 1851.

Corallum composite, massive, irregularly circular; its upper surface convex, and its under surface in general free, and presenting strongly-marked circular rugose ridges. *Calices* equally developed in the same specimen, and varying but little in size in different specimens; about half a line in diameter; their margin very thin, and scarcely exsert, but quite distinct from the surrounding cœnenchyma; and in some of the well-preserved corallites, 12 small septa, somewhat unequally developed alternately, are visible. The calices are not closely set, and vary in their degree of approximation; the space between

---

[1] See tab. xlvii, fig. 1.

[2] British Palæoz. Fossils, p. 15, tab. i c, fig. 15.—Ann. and Mag. of Nat. Hist., 2d series, vol. vi, p. 285 (1850).

them being, in general, equal to twice, and sometimes to thrice their diameter. The polygonal divisions of the cœnenchyma are very regular in form and in size; their breadth is about half of a line.

A vertical section of this fossil shows that the walls are very distinct, and the tabulæ closely set, and in general placed horizontally, but sometimes slightly oblique. The appearance of the cœnenchyma differs in the upper and lower parts of the corallum; near the upper surface, to the depth of about 1 line, the vertical lines corresponding to the section of the prismatical lamina of the tubes, are much more strongly marked than the horizontal lines that correspond to the sections of the trabiculæ or small intratrabicular diaphragms; but in the lower parts of the corallum a contrary disposition exists, and the transversal lines running nearly parallel to the surface of the compound mass, are the only ones which are distinct to the naked eye.

H. Murchisoni is found in the upper Silurian deposits of Wenlock Edge. Professor M'Coy has met with it at Aymestry, in Herefordshire; at Coniston Waterhead, in Lancashire; and at Llansantfraid, in Denbighshire. It is found also in Gothland.

Specimens are in the Collections of the Parisian Museum, of the Museum of Practical Geology of London, of M. de Verneuil, and M. Michelin.

This fossil coral differs from *H. interstincta*[1] by the greater distance between the calices, and the abundance of the cœnenchyma. It much resembles *H. porosa*[2] by its external characters, but differs from it by the structure of its cœnenchyma; the horizontal laminæ of that tissue being much more developed than the vertical ones, whereas the contrary takes place in *H. porosa*.

3. HELIOLITES MEGASTOMA. Tab. LVIII, figs. 2, 2a, 2b, 2c, 2d.

PORITES PYRIFORMIS (pars), *Lonsdale*, in Murchison, Silur. Syst., p. 686, pl. xvi, figs. 2d, 2e (cæt. excl.), 1839.
PORITES MEGASTOMA, *M'Coy*, Silur. Foss. of Irel., p. 62, pl. iv, fig. 19, 1846.
GEOPORITES INTERMEDIA, *D'Orbigny*, Prodr. de Pal., vol. i, p. 49, 1850.
PALÆOPORA MEGASTOMA, *M'Coy*, Brit. Palæoz. Foss., p. 16, pl. i c, fig. 4, 1851.
HELIOLITES MEGASTOMA, *Milne Edwards* and *Jules Haime*, Pol. Foss. des Terr. Palæoz. (Arch. du Mus., vol. v), p. 216, 1851.
HELIOLITES MACROSTYLUS, *J. Hall*, Paleont. of New York, vol. ii, p. 135, pl. xxxvi A, fig. 2, 1852..

Corallum composite, massive, irregularly hemispherical; sometimes adherent, sometimes free. Epitheca of the under surface not much developed. *Calices* having all nearly the same dimensions in the same specimen, but varying from two thirds of a line to $1\frac{1}{4}$ line in different specimens; very closely set, quite circular, and with a margin not prominent, and scarcely

---

[1] See tab. lvii, fig. 5.　　　　　　[2] See tab. xlvii, fig. 1.

discernible from the surrounding cœnenchyma. Septa 12 in number, but little developed, rather slender, and slightly unequal in size alternately.

A vertical section shows that the visceral chamber of each corallite is limited by a thin, but well-characterised *wall*, and is divided by numerous closely-set, well-developed *tabulæ*, that are almost all quite horizontal. The continuation of the vertical septa is often visible in the space comprised between the tabulæ. The *cœnenchyma* is entirely made up with small square cells, formed by the parieties of the vertical canaliculæ, and the horizontal dissepiments meeting at right angles. In some parts of the corallum all those small horizontal diaphragms comprised between two adjoining corallites are placed on the same level, and correspond exactly; in other parts they alternate more or less completely; but in no instance do these intratrabicular dissepiments correspond with the intramural tabulæ.

Found in the lower Silurian deposits at Coniston, and in the upper Silurian beds at Wenlock Edge. Professor M'Coy mentions its existence at Mathyrafal, Montgomeryshire; High Haume, Dalton in Furness, Lancashire; Blayn y Cwm, West of Nantyre, Glyn Ceiriog; and Egool, Bellaghadereen, Mayo. A variety of the same species has been met with by that geologist in the Bala limestone of Maes Meillion, south of Bala, Merionethshire. Mr. Hall has also found it in the Niagara limestone, at Milwaukie, in Wisconsin. Some corals from the Devonian deposits of Nehou, in Normandy, do not appear to differ specifically from the former.

Specimens are in the Collections of the Museum of Practical Geology, of the Geological Society, of the Bristol Museum, the Parisian Museum, &c.

H. megastoma is easily recognised by its large and closely-set calices, and by the slight development of the cœnenchyma.

4. HELIOLITES GRAYI. Tab. LVIII, figs. 1, 1*a*.

HELIOLITES GRAYI, *Milne Edwards* and *Jules Haime*, Pol. Foss. des Terr. Palæoz. (Arch. du Mus., vol. v), p. 217, 1851.

Corallum composite, dendroidal, forming lamellar sublobated expansions, both surfaces of which bear *calices*. These are placed at various distances from each other (one, two, or three times their diameter), and are limited by a small, well-marked, circular ridge, formed by the exsert edge of 12 subequal thick *septa*. The canaliculæ of the cœnenchyma are somewhat irregular, and their parieties are rather thick. Diameter of the calices about one third of a line.

This fine fossil was found in the upper Silurian beds at Walsall, and belongs to the Collection of Mr. J. Gray, of Dudley.

The remarkable frondescent form of this fossil is met with in no other species of the same genus. In most other respects *H. Grayi* much resembles *H. Murchisoni*[1] and *H. inordinata*,[2] but may be distinguished from both by the circular exsert margin of the calices.

The fossil figured in M. Murchison's justly-celebrated work under the name of BLUMENBACHIUM GLOBOSUM Lonsdale,[3] appears to be a mould of a coral belonging to the above-described species. It was found at Wenlock.

5. HELIOLITES INORDINATA. Tab. LVII, figs. 7, 7a.

> PORITES INORDINATA, *Lonsdale*, in Murchison, Silur. Syst., p. 687, pl. xvi *bis*, fig. 12, 1839.
> LONSDALIA INORDINATA, *D'Orbigny*, Prodr. de Paléont., vol. i, p. 25, 1850.
> PALÆOPORA SUBTILIS? *M'Coy*, Brit. Palæoz. Foss., p. 17, 1851.
> HELIOLITES INORDINATA, *Milne Edwards* and *Jules Haime*, Pol. Foss. des Terr. Palæoz.
> (Arch. du Mus., vol. v), p. 217, 1851.

Corallum arborescent, very ramous; its branches slender, subcylindrical, and about one or one and a half line in diameter. *Calices* not prominent, circular, or somewhat elongated in the longitudinal direction of the branches, and set at various distances from each other in different parts of the corallum. *Septa* 12 in number, nearly similar in size, rather thick, and well developed. Cells of the cœnenchyma polygonal and somewhat irregular. Diameter of the calices about half a line.

Found in the lower Silurian beds at Robeston Walthen, Pembrokeshire. According to Professor M'Coy it exists also in the upper Silurian deposits at Ferriter's Cove, Doonquin, and Dingle, in Kerry. The fossil which Professor M'Coy describes under the name of *Palæopora subtilis*,[4] and which (from the description given by that author) we are inclined to consider as not differing specifically from *H. inordinata*, appears to be common in the fine Caradoc sandstone of Mulock, Dalquorhan, Ayrshire. Specimens are in the Collection of the Geological Society.

The above-described species differs from the other Heliolites by its dendroidal form, and by the development of its septa.

2. *Genus* PLASMOPORA, (p. lix.)

1. PLASMOPORA PETALIFORMIS. Tab. LIX, figs. 1, 1a, 1b, 1c.

> PORITES PETALIFORMIS, *Lonsdale*, in Murchison, Silur. Syst., p. 687, pl. xvi, fig. 4, 1839.
> ASTREOPORA PETALIFORMIS, *D'Orbigny*, Prodr. de Paléont., vol. i, p. 50, 1850.
> PALÆOPORA PETALIFORMIS, *M'Coy*, Brit. Palæoz. Foss., p. 17, 1851.
> PLASMOPORA PETALIFORMIS, *Milne Edwards* and *Jules Haime* (*Introd.*, p. lix, 1850),
> Polyp. Foss. des Terr. Palæoz., p. 221, 1851.

---

[1] See tab. lvii, fig. 6. [2] See tab. lvii, fig. 7. [3] Silurian System, tab. xv, fig. 26.

Corallum massive, hemispherical, free; its edge thin, and its under surface slightly concave, and covered with a strong epitheca wrinkled concentrically. *Calices* circular, almost equal in size, not closely set, and terminated by a very thin, not exsert, margin. *Costæ* very slender, and meeting directly those of the adjoining corallites under an angle, or else bifurcating and joining small transverse laminæ, so as to constitute small polygonal divisions between the calices. These costæ are not closely set; they do not always correspond to the septa, and they often bear some small tubercles laterally. The calicular fossulæ are not very deep, and contain 12 very slender *septa*, which extend almost to the centre, and are somewhat irregular in size. Diameter of the calices almost one line; distance between them somewhat more.

A vertical section shows that the *walls* are slender, but still very distinct, and apparently not perforated. The visceral chambers of the corallites are occupied by large *tabulæ* that are rather closely set, and in general almost horizontal, but somewhat irregular. The space between the corallites is filled up with vertical canaliculæ formed by the costæ that are well developed, and are subdivided by horizontal or slightly convex dissepiments into cells of about one fifth of a line broad.

Upper Silurian deposits at Dudley, Walsall, and Delves Green. Professor M'Coy mentions its existence in the Coniston limestone of Sunny Brow near Coniston, Lancashire; in the impure limestone of Golugoed, Mendinam, Caermarthenshire; and at Egool and Bellaghaderreen, Mayo.

Specimens are in the Collections of Mr. Fletcher and Mr. J. Gray, of Dudley; of the Geological Society, the Parisian Museum, &c.

P. petaliformis much resembles *P. follis*,[1] but in this last-mentioned species the calices are smaller and more closely set, and the general form of the corallum appears to be constantly different. *P. scita*[2] is much smaller than *P. petaliformis*, and its septo-costal radii are much more regular.

2. Plasmopora scita. Tab. LIX, figs. 2, 2a.

PLASMOPORA SCITA, *Milne Edwards* and *Jules Haime*, Polyp. Foss. des Terr Palæoz. (Arch. du Mus., vol. v), p. 222, 1851.

Corallum free; basal surface slightly convex, with a lamellated and somewhat prominent edge, and covered with a strong, wrinkled epitheca. *Calices* equally developed, shallow, quite circular, with a thin, slightly prominent edge, and set at a distance from each other that does not exceed their diameter. *Costæ* slender, smooth laterally, set wide apart, in direct continuation with the septa, and always joining directly the corresponding ones of the neighbouring corallites, but often united together at their outer edge by a small transverse lamina, which closes up exteriorly the intercostal loculi. *Septa* 12, almost

---

[1] Milne Edwards and J. Haime, Polyp. Foss. des Terr. Palæoz., p. 223, pl. xvi, fig. 3.

[2] See tab. lix, fig. 2.

equal in size, rather thick towards the walls, but very slender towards the centre of the corallites. Diameter of the calices about one third of a line.

Upper Silurian beds of Dudley. Collection of Mr. J. Gray.

This species differs from *P. petaliformis*[1] by the calices being much smaller, and more closely set, and by the septo-costal radii being more regular.

3. *Genus* PROPORA, (p. lix.)

PROPORA TUBULATA. Tab. LIX, figs. 3, 3*a*, 3*b*.

PORITES TUBULATA, *Lonsdale*, in Murchison, Silur. Syst., p. 687, pl. xvi, fig. 3, 1839.

ASTREOPORA TUBULATA, LONSDALEI, and GRANDIS, *D'Orbigny*, Prodr. de Paléont., vol. i, p. 50, 1850.

PALÆOPORA TUBULATA, *M'Coy*, Brit. Palæoz. Foss., p. 18, 1851.

PROPORA TUBULATA, *Milne Edwards* and *Jules Haime*, Pol. Foss. des Terr. Palæoz. (Archives du Mus., vol. v), p. 224, 1851.

HELIOLITES ELEGANS and SPINIPORA? *J. Hall*, Paleont. of New York, vol. ii, pp. 130, 131, pl. xxxvi, figs. 1, 2, 1852.

Corallum massive, irregularly convex; common basal plate covered with a epitheca presenting concentric folds; upper surface convex or subgibbose. *Calices* circular, somewhat unequal in size, and surrounded with a slightly prominent edge that is crenulated in consequence of the prolongation of the septa beyond the walls, where they constitute small, thick *costæ*, which are sometimes sufficiently developed to attain those of the adjoining corallites. In general, 12 septa, slightly exsert, rather thick exteriorly, and somewhat unequal in size. Diameter of the calices usually about half a line.

A vertical polished section shows that the *walls* are distinct; the *tabulæ* closely set, concave in the middle, sometimes quite horizontal, and at others irregularly placed; the space situated between the walls of the adjoining corallites occupied by an abundant, irregular exothecal tissue. Some of the dissepiments composing this tissue are horizontal, and assume the appearance of small extramural tabulæ, but others constitute vesicular cells. No traces of the costæ are seen in this exothecal mass.

The specimens here described were from Dudley and Wenlock. The localities mentioned by Sir R. Murchison are Woolhope Valley, Benthall Edge, Ledbury, Woodside near Nashsur, Fownhope, and the west parts of the Malvern Hills, between Asten Ingham and May Hill; by Professor M'Coy, Aymestry, in Herefordshire; Altgoch, Llanfyllin, in Montgomeryshire; Golugoed; Mulock, Dalquorhan, in Ayrshire. The same species is found in Gothland and in Bohemia.

Specimens are in the Collections of the Museum of Practical Geology, of the Geological Society, of the Museum of Paris, &c.

---

[1] See tab. lix, fig. 1.

This species differs from *P. conferta*[1] by its calices being much smaller and more closely set.

Mr. J. Hall has lately described, under the name of Heliolites,[2] some fossils that appear to be referable to the above-described species. They were found in the limestone of Niagara and Lockport.

Family FAVOSITIDÆ, (p. lx.)

1. *Genus* FAVOSITES, (p. lx.)

1. FAVOSITES GOTHLANDICA. Tab. LX, figs. 1, 1*a*.

> TUBER SIVE GLOBUS CORALLINUS? *D. S. Buttners*, Corall. Subterr., p. 17, tab. i, figs. 1, 3, 1714.
>
> LAPIS CALCARIUS, *Magnus Bromel*, Acta liter. Suec., vol. ii, pp. 414-15, 1728.
>
> CORALLIUM GOTHLANDICUM, &c., *Fougt*, Amæn. Acad., vol. i, p. 106, tab. iv, fig. 27, 1749.
>
> FUNGITES? *Thomas Pennant*, Philos. Trans., vol. lix, p. 513, tab. xv, fig. 1, 1757.
>
> TUBIPORA PRISMATICA, *Lamarck*, Syst. des Anim. sans Vert., p. 377, 1801 (absq. descript.)
>
> MADREPORA FASCICULARIS? *Parkinson*, Org. Rem., vol. ii, pl. vi, fig. 11, 1808.
>
> FAVOSITES GOTHLANDICA, *Lamarck*, Hist. des Anim. sans Vert., vol. ii, p. 206, 1816.—2d edit., p. 320.
>
> — — *Defrance*, Dict. Sc. Nat., vol. xvi, p. 298, 1820.
>
> — — *Lamouroux*, Exp. Meth. des G. de Pol., p. 66, 1821.
>
> — — *Lamouroux*, Encycl. (Zooph.), p. 388, 1824.
>
> CALAMOPORA GOTHLANDICA (pars), *Goldfuss*, Petref. Germ., vol. i, p. 78, pl. xxvi, figs. 3*a*, 3*e*, 1829.
>
> FAVOSITES GOTHLANDICUS, *Eichwald*, Zool. Spec., vol. i, p. 194, 1829.
>
> — RETICULUM? *Ibid.*, p. 194, tab. xi, fig. 14.
>
> CALAMOPORA GOTHLANDICA (pars), *Morren*, Descr. Cor. in Belg. repert., p. 72, 1832.
>
> — — *Stephen Kutorga*, Beitr. zur Geogr. und Paleont., Dorpat's, p. 24, tab. v, fig. 2, 1835.
>
> — BASALTICA, *Hisinger*, Leth. Suec., p. 96, pl. xxvii, fig. 5, 1837. (Not *Goldfuss*.)
>
> — GOTHLANDICA, *Eichwald*, Sil. Schist. Syst. in Esthland, p. 198, 1840.
>
> FAVOSITES SUBBASALTICA, *D'Orbigny*, Prodr. de Paléont., vol. i, p. 49, 1850.
>
> — GOTHLANDICA, *M'Coy*, Brit. Palæoz. Foss., p. 20, 1851.
>
> — — *Milne Edwards* and *Jules Haime*, Pol. Foss. des Terr. Palæoz. (Arch. du Mus., vol. v), p. 232, 1851.
>
> — NIAGARENSIS, *J. Hall*, Paleont. of New York, vol. ii, p. 125, pl. xxxiv A (*bis*), fig. 4, and p. 324, pl. lxxiii, fig. 1, 1852.

Corallum massive, convex, sometimes rather tall. *Calices* somewhat unequal in size. 10 or 12 septa. Mural pores surrounded by a small rim, and forming on each

---

[1] Milne Edwards and J. Haime, Pol. Foss. des Terr. Palæoz., p. 225.
[2] Paleontology of New York, vol. ii.

portion of the wall two longitudinal series, which alternate. Breadth of the calices somewhat more than 1 line.

Found in the Caradoc sandstone at Cuttimore's Quarry near Tortworth; and in the upper Silurian deposits at Wenlock, and Dinas Court near Aymestry. Professor M'Coy mentions its existence at Dudley; Golugoed, Mendinam, in Caermarthenshire; Ledbury, in Herefordshire; Old Radnor, Presteigne, in Radnorshire. It has also been met with in Groningen, Gothland, Russia, and North America. Mr. Hall has recently mentioned its existence in the Niagara limestone, at Niagara Falls, Lockport, Rochester and Milwaukie, Wisconsin, and in the coralline limestone at Schoharie.

This fossil has, till of late, been confounded with *Favosites Goldfussi*,[1] in which the mural pores are more closely set, and are sometimes opposite, but at other times alternate, in the adjoining series.

2. FAVOSITES ASPERA. Tab. LX, figs. 3, 3*a*.

> CALAMOPORA ALVEOLARIS (pars), *Goldfuss*, Petref., vol. i, p. 77, tab. xxvi, fig. 1*b* (cæt. excl.), 1829.
> — — *Morren*, Descr. Corall. in Belg. Repert., p. 72, 1832.
> FAVOSITES ALVEOLARIS, *Lonsdale*, in Murchison, Silur. Syst., p. 681, pl. xv *bis*, fig. 2, and perhaps also the fig. 1, 1839.
> CALAMOPORA ALVEOLARIS, *Ed. Eichwald*, Sil. Schist. Syst. in Esthland, p. 198, 1840.
> FAVOSITES ALVEOLARIS, *Lonsdale*, in Murch., Vern. and Keys. Russia and Ural, vol. i, p. 610, 1845.
> CALAMOPORA ALVEOLARIS, *Keyserling*, Reise in Petschora, p. 177, 1846.
> FAVOSITES ASPERA, *D'Orbigny*, Prodr. de Paléont., vol. i, p. 49, 1850.
> — — ? *M'Coy*, Brit. Palæoz. Foss., p. 20, 1851.
> — — *Milne Edwards* and *Jules Haime*, Pol. Foss. des Terr. Palæoz. (Arch. du Mus., vol. v), p. 234, 1851.

Corallum massive, with an almost flat surface; calices unequal in size. Tabulæ presenting six large, and well-marked submarginal fossulæ. Mural pores rather closely set, and placed at the angles of the visceral chamber. Diagonal of the large calices somewhat more than 1 line.

Found in the Caradoc sandstone at Powis Castle, Cefn-y-garrey, Llandovery; in Wenlock limestone, at Leinthall Earls near Ludlow. According to Mr. Lonsdale, in the middle and lower Ludlow rocks, at Mocktree Hill, Aymestry, and Tatton Edge; and in the Wenlock limestone, at Purlieux, Malvern, Haven near Aymestry, Hurst Hill near Sedgley, the western side of the Malvern Hills, Abberley, Little Ridge, Easthop, Winslow Mill, Townhope, Westhope, and Woolhope. Professor M'Coy has found it at Ardaun, and Cappacorcogne, Cong, in the county of Galway; Foylathurrig, Dingle, in the county

---

[1] See tab. xlvii, fig. 3

of Kerry; Portrane and Malahide, in the county of Dublin. It exists also in Sweden, Holland, and Russia.

Specimens are in the Collections of the Geological Society, of the Poppelsdorff Museum at Bonn, of M. de Verneuil, Paris, &c.

This species much resembles *Favosites alveolaris*,[1] with which it is often confounded, but differs from it by the size and number of the marginal fossulæ of the tabulæ, as well as by the calices being more unequal in size.

3. FAVOSITES MULTIPORA. Tab. LX, fig. 4.

FAVOSITES MULTIPORA, *Lonsdale*, in Murchison, Silur. Syst., p. 683, pl. xv *bis*, fig. 5, 1839.
— — *M'Coy*, Synopsis of the Silurian Fossils of Ireland, p. 63, 1846.
— — *D'Orbigny*, Prodr. de Paléont., vol. i, p. 48, 1850.
— ALVEOLARIS? *M'Coy*, Brit. Palæoz. Foss., p. 19, 1851.
— MULTIPORA, *Milne Edwards* and *Jules Haime*, Pol. Foss. des Terr. Palæoz. (Arch. du Mus., vol. v), p. 237, 1851.

Corallum hemispherical; common basal plate covered with a concentrically wrinkled epitheca. *Calices* of equal size, and arranged in very regular series, that form somewhat elongated hexagones, the opposite angles of which are of equal value. Each portion or side of the walls generally present three series of pores, closely set, and arranged somewhat irregularly. Large diagonal of the calices about half a line.

Found in the Caradoc sandstone at Haverfordwest, and in the Wenlock limestone at Marloes Bay, in Pembrokeshire. Professor M'Coy has met with it also in Galway and Kerry.

Specimens are in the Collections of the Geological Society of London, and of the Museum of Practical Geology.

This fossil differs from the other species of Favosites by the regularity of its calices, and the abundance of its mural pores.

4. FAVOSITES FORBESI. Tab. LX, figs. 2, 2*a*, 2*b*, 2*c*, 2*d*, 2*e*, 2*f*, 2*g*.

MADREPORA SUBROTUNDA? &c., *Fougt*, Amæn. Acad., vol. i, p. 100, tab. iv, fig. 17, 1749.
CALAMOPORA BASALTICA (pars), *Goldfuss*, Petref. Germ., vol. i, p. 78, tab. xxvi, fig. 1*b*, 1829.
— — *Morren*, Descr. Cor. Belg., p. 73, 1832.
— GOTHLANDICA, *Hisinger*, Leth. Suec., p. 96, tab. xxvii, fig. 4, 1378. (Not *Goldfuss*.)
FAVOSITES GOTHLANDICA, *Lonsdale*, in Murchison, Sil. Syst., p. 682, pl. xv *bis*, fig. 3, 4, 1839. (Not of *Lamarck*.)
— — *D'Orbigny*, Prodr. de l'Paléont., vol. i, p. 48, 1850.
— FORBESI, *Milne Edwards* and *Jules Haime*, Pol. Foss. des Terr. Palæoz. (Arch. du Mus., vol. v), p. 238, 1851.

---

[1] Goldfuss, Petref. Germ., vol. i, pl. xxvi, fig. 1*a*, 1*c*.

Corallum massive, convex or subgibbose.   *Calices* unequal in size; the large ones usually dispersed amongst the smaller ones, almost circular, and about 1 line in diameter; the smallest about a quarter of a line in diameter; others presenting all the intermediate degrees between these two sizes.   In some specimens the difference between the calices is much less marked.   A vertical section shows that the walls are thin, and the tabulæ horizontal, and, in general, closely set, but very unequally so.

Found in the lower Silurian deposits at Under Daniell's Wood, Tortworth; in the Wenlock limestone at Wren's Nest, Dudley, Benthall Edge, Falfield near Tortworth, Much Wenlock; and, according to Mr. Lonsdale, in the lower Ludlow rocks at Sitch Wood, Ledbury, Wentwoon Common, Wenlock; and in the middle Ludlow rocks, at Aymestry, Tatton Edge, Downton on the Rock.   Professor M'Coy mentions its existence at Ardaun, Cong, and Kilbridge, in Galway; and at Egool and Bellaghaderreen, in the county of Mayo.   It is also met with in Sweden.

Specimens in the Collections of the Geological Society, of the Museum of Practical Geology, of Mr. J. S. Bowerbank, of the Bristol Museum, &c.

Some very small specimens, found at Dudley, and belonging to the Collections of Mr. Fletcher and of Mr. Bowerbank, present the general characters of this species, but have much larger calices; they appear to be young colonies of the above-described species, and we easily understand how the mode of multiplication of Favosites must tend to diminish the dimensions of the large corallites, as the number of individuals so produced augments.

In no other species of this genus is the difference in the size of the calices so great as in *F. Forbesi*.

5. FAVOSITES HISINGERI.   Tab. LXI, figs. 1, 1*a*, 1*b*.

TUBULARIA FOSSILIS, &c., *Magnus Bromel*, Acta Liter. Suec., vol. ii, p. 408, 1728.

MADREPORA PORIS, &c., *Fougt*, Amæn. Acad., vol. i, p. 101, tab. iv, fig. 21, 1749.

FUNGITES, *Th. Pennant*, Philos. Trans., vol. xlix, 2d part, p. 513, tab. xv, fig. 4, 1753.

FAVOSITES ALCYON? *Defrance*, Dict. Sc. Nat., vol. xvi, p. 298, 1820.

    —   GOTHLANDICA? *De Blainville*, Dict. Sc. Nat. (Atlas Zooph.), pl. xl, fig. 4, 1830.
              —Man. d'Actin., pl. lxii, fig. 4.

    —   ALCYON? *De Blainville*, Dict., pl. xlii, fig. 5.—Man., pl. lxiv, fig. 5.

FAVOSITES? *Pander*, Beitr. zur Geogn. des Russ. Reiches, pl. xxix, fig. 9, 1830.

CALAMOPORA MINUTISSIMA? *Castelnau*, Terr. Sil. de l'Amér. du Nord, pl. xviii, fig. 2, 1843.

FAVOSITES HISINGERI, *Milne Edwards* and *Jules Haime*, Pol. Foss. des Terr. Palæoz. (Arch. du Mus., vol. v), p. 240, pl. xvii, figs. 2, 2*a*, 2*b*, 1851.

ASTROCERIUM VENUSTUM, *J. Hall*, Paleont. of New York, vol. ii, p. 120, pl. xxxiv, fig. 1, 1852.

Corallum massive, subgibbose.   Calices almost equal in size, separated by walls that are rather thick, in general pretty regularly hexagonal, and about half a line in diameter.

12 septa, subequal, not very thick, extending to the centre of the corallites, and formed by well-developed, slightly curved processes.   Tabulæ slender, rather closely set, and horizontal, or somewhat flexuous.

Lower Silurian deposits at Cullimore's Quarry, Tortworth; upper Silurian rocks of Benthall Edge, Wenlock Edge; Gothland; and Niagara limestone, at Lockport, and Rochester (J. Hall).

Specimens in the Collections of the Museum of Practical Geology, of the Geological Society, of the Bristol Museum, of Mr. Bowerbank, of the Museum of Paris, &c.

In this species the septal processes are bent upwards, and more developed than in the other Favosites.   Mr. Hall has, for that reason, proposed placing it in a new generical division, under the name of *Astrocerium* :[1] but every intermediate degree between this structure and the most rudimentary state of the septal system in some other Favosites are met with, so that the line of separation would be arbitrary, and it is also to be remembered that in many cases the delicate and brittle processes which constitute the septa, have evidently disappeared during the fossilisation of the coral.   In *Favosites Goldfussi*,[2] for example, we have found distinct remains of the septal processes only in two specimens, although we have searched attentively for them in several hundred specimens found in various localities.

We are inclined to think that the Dudley fossil, mentioned by Parkinson under the name of *Porpital madreporite*,[3] and by Dr. Fleming under that of *Sarcinula angularis*,[4] is referable to the above-described species.

6. FAVOSITES CRISTATA.   Tab. LXI, figs. 3, 3*a*, and 4, 4*a*?

> MADREPORITES CRISTATUS, *Blumenbach*, Comment. Soc. Scient. Gött., vol. xv, p. 154, tab. iii,
>   fig. 12, 1803.
> CALAMOPORA POLYMORPHA, *Hisinger*, Leth. Suec., p. 97, tab. xxvii, fig. 6, 1837. (Not
>   of *Goldfuss*.)
>   —   SPONGITES? *Ibid.*, p. 97, tab. xxvii, fig. 7.   (Not of *Goldfuss*.)
> FAVOSITES POLYMORPHA, *Lonsdale*, in Murchison, Silur. Syst., p. 684, pl. xv, fig. 2, 1839.
> CALAMOPORA POLYMORPHA, *Eichwald*, Silur. Schist. Syst. in Esthland, p. 198, 1840.
> FAVOSITES POLYMORPHA, *Lonsdale*, M. V. K. Russ. and Ural, vol. i, p. 610, 1845.
> ALVEOLITES LONSDALEI, *D'Orbigny*, Prodr. de Paléont., vol. i, p. 49, 1850.
> FAVOSITES CRISTATA, *Milne Edwards* and *Jules* Haime, Pol. Foss. des Perr. Palæoz. (Arch.
>   du Mus., vol. v), p. 242, 1851.

Corallum dendroidal; its branches generally spreading, cylindrical, and submamillose. *Calices* somewhat unequal in size, often almost circular, and with a rather thick margin. The large ones about half a line in diameter.

---

[1] Paleontology of New York, vol. ii, p. 120, 1852.

[2] See tab. xlvii, fig. 3.

[3] Org Remains, vol. ii, p. 69, tab. vii, figs. 3, 7.

[4] British Animals, p. 508, 1828.—Woodward, Syn. Tab. of Brit. Org. Rem., p. 5, 1830.

Upper Silurian beds of Wenlock, Ludlow, Dudley, Ireland, Sweden, and Russia.

Specimens in the Collections of the Museum of Practical Geology, of the Geological Society, of Mr. Bowerbank, of the Museum of Paris, &c.

This coral bears great resemblance to *Favosites cervicornis*,[1] and we are even doubtful as to its being specifically different from it; its calices are, however, less unequal in size, and almost circular.

In a Dudley specimen, which we consider as belonging to the above-described species, the large calices were not more than two fifths of a line in diameter.

According to Mr. Lonsdale, the same species appears to exist in the Devonian formation of the Oural Mountains.

7. FAVOSITES FIBROSA. (See p. 217, and Tab. XLVIII, fig. 3.) Tab. LXI, figs. 5, 5*a*.

> STENOPORA FIBROSA, *M'Coy*, Brit. Palæoz. Foss., p. 24, 1851.
> ASTROCERIUM CONSTRICTUM, *J. Hall*, Paleont. of N. Y., vol. ii, p. 123, pl. xxxiv A, figs. 2 and 3, 1852.

Lower Silurian, Horderley, and Llandovery.

According to Professor M'Coy (op. cit.), this species has been found in the Coniston limestone schists of Llansantfraid, the Caradoc sandstone schists of Bala, the Upper Ludlow rocks, the Wenlock limestone, the limestone of Llandeilo, &c., of a great quantity of British localities.

Professor Hall indicates it in the shale of the Niagara group at Lockport.

8. FAVOSITES CRASSA.

> FAVOSITES CRASSA, *M'Coy*, Ann. and Mag. of Nat. Hist., 2d series, vol. vi, p. 284, 1850.
> — — *M'Coy*, Brit. Palæoz. Foss., p. 20, pl. i c, fig. 9, 1851.

" *Corallum* forming large, subcylindrical, curved branches, composed of long, slightly diverging, remarkably regular and equal prismatic tubes, opening as thin-walled polygonal cells on the surface, with a nearly uniform diameter of half a line; two rows of pores on each face of the prismatic tubes, diaphragms either slightly more or less than the diameter of the tubes apart; interpolated young tubes few.

"In the Coniston limestone, Coniston Water-head, Lancashire."

FAVOSITES? OCULATA, M'Coy, Brit. Palæoz. Foss., p. 21—*Ceriopora oculata*, Goldfuss, Petref. Germ., vol. i, tab. lxiv, fig. 14,—appears to appertain to the class of Bryozoa.

[1] See tab. xlviii, fig. 2.

2. *Genus* ALVEOLITES (p. lx).

1. ALVEOLITES LABECHEI.　Tab. LXI, figs. 6, 6*a*, 6*b*.

FAVOSITES SPONGITES (pars), *Lonsdale*, in Murchison, Silur. Syst., pl. xv *bis*, figs. 8, 8*a*, 8*b*
　　　　　　　(cæt. excl.), 1839.　(Not *Calamopora spongites*, *Goldfuss*.)
CALAMOPORA SPONGITES, *Eichwald*, Sil. Syst. in Esthland, p. 197, 1840.
ALVEOLITES LABECHII, *Milne Edwards* and *Jules Haime*, Pol. Foss. des Terr. Palæoz. (Arch.
　　　　　　　du Mus., vol. v), p. 257, 1851.

Corallum massive, convex or subgibbose, very closely resembling the *Alveolites suborbicularis*[1] of the Devonian formation; but differing from it, by the calices being more irregular, scarcely prominent, with very thin edges, and subtriangular, and by the inner process being very indistinct.　Large diameter somewhat more than one third of a line; small diameter one third less.

Found in the upper Silurian deposits at Wenlock, and at Benthall Edge.　Professor M'Coy[2] mentions its existence at Doonquin, Dingle, and Ferriter's Cove, in the county of Kerry; Egool and Bellaghaderreen, in the county of Mayo; River Chapel and Gorey, in the county of Wexford.　According to Eichwald, it exists also in Russia.

Specimens are in the Collections of the Museum of Practical Geology of London, of the Parisian Museum, of M. Bouchard-Chantereaux at Boulogne.

2. ALVEOLITES GRAYI.　Tab. LXI, figs. 2, 2*a*.

ALVEOLITES GRAYI, *Milne Edwards* and *Jules Haime*, Pol. Foss. des Terr. Palæoz. (Arch.
　　　　　　　du Mus., vol. v), p. 258, 1851.

Corallum presenting a flat or submamillose surface.　*Calices* very irregular, inclined in various directions, in general subtriangular, and having their outer margin somewhat arched.　The elevation situated on the inner surface of this inferior calicinal edge in general distinct, but not very prominent.　Walls rather thick.　Size of the calices: large diameter about half a line, small diameter one third less.

Upper Silurian rocks of Wenlock and Dudley.　Specimens in the Collections of the Museum of Practical Geology, of Mr. J. Gray of Dudley, and of the Museum of Paris.

This species much resembles *Alveolites suborbicularis*[3] and *A. Labechei*,[4] but its calices are always larger, and limited by walls that are thicker in proportion to the size of the corallites.

---

[1] See tab. xlix, fig. 1.　　　[2] Synopsis of the Silurian Fossils of Ireland, p. 64.
　　[3] See tab. xlix, fig. 1.　　　　　　[4] See tab. lxi, fig. 6.

3. ALVEOLITES REPENS. Tab. LXII, figs. 1, 1a.

> MILLEPORA REPENS, *Fougt*, Amæn. Acad., vol. i, p. 99, tab. iv, fig. 25, 1749.
> —   RAMIS, &c., *Ibid.*, p. 98, tab. iv, fig. 14.
> —   CERVICORNIS? *Wahlenberg*, Nov. Acta Soc. Upsal., vol. viii, p. 100, 1820.
> CALAMOPORA FIBROSA, var. RAMIS GRACILIBUS, *Goldfuss*, Petref. Germ., vol. i, p. 82,
>   tab. xxviii, fig. 4, 1826.
> POCILLOPORA APPROXIMATA? *Eichwald*, Zool. Spec., vol. i, p. 182, 1829.
> MILLEPORA BURTINIANA? *Morren*, Descr. Cor. Belg., p. 25, tab. vii, figs. 1—4, 1832.
> —   REPENS, *Hisinger*, Leth. Suec., p. 102, tab. xxix, fig. 5, 1837.
> —   RAMOSA, *Ibid.*, p. 103, tab. xxix, fig. 6.   (It is a worn specimen.)
> —   REPENS? *Lonsdale*, in Murchison, Silur. Syst., p. 680, pl. xv, fig. 30 (excl.,
>   fig. 30a), 1830.
> CHÆTETES REPENS, *D'Orbigny*, Prodr. de Paléont., vol. i, p. 49, 1850.
> ALVEOLITES REPENS, *Milne Edwards* and *Jules Haime*, Pol. Foss. des Terr. Palæoz. (Arch.
>   du Mus., vol. v), p. 258, 1851.
> CLADOPORA SERIATA, *J. Hall*, Paleont. of New York, vol. ii, p, 137, pl. xxxviii, fig. 1, 1852.

Corallum ramose; its branches slender (in general not more than two lines in diameter) and frequently coalescent.   *Calices* rather closely set, somewhat broader than high; their exterior margin presenting a medial fissure, on each side of which is a small denticulation of unequal size.   Breadth of the calice about one fifth of a line.

From the upper Silurian rocks of Dudley and Wenlock.   Sir R. Murchison has also found it at Lincoln Hill, Coalbrook Dale, Benthall Edge, Hurst Hill, and Sedgley.   It is also met with in Gothland; and in North America, where Mr. Hall has recently found it at Lockport.

Specimens in the Collections of Mr. Fletcher, of the Paris Museum, &c.

Alveolites vermicularis[1] much resembles this species, but its branches are much thicker, and the angle formed by the bifurcation of its branches is much more open; the outer margin of the calices is also more prominent than in *A. repens*.

4. ALVEOLITES? SERIATOPOROÏDES. Tab. LXII, figs. 2, 2a.

> MILLEPORA RAMIS, &c., *Fougt*, Amæn. Acad., vol. i, p. 98, tab. iv, fig. 15, 1849.
> MILLEPORITES REPENS, *Wahlenberg*, Nov. Acta Soc. Upsal., vol. viii, p. 100, 1821.   (Not
>   of Fougt.)
> MILLEPORA REPENS (pars), *Lonsdale*, in Murchison, Silur. Syst., pl. xv, fig. 30a, 1839.
> ALVEOLITES? SERIATOPORA, *Milne Edwards* and *Jules Haime*, Pol. Foss. des Terr. Palæoz.
>   (Arch. du Mus., vol. v), p. 260, 1851.
> CLADOPORA MULTIPORA, *J. Hall*, Paleont. of New York, vol. ii, p. 140, pl. xxxix, figs. 1a,
>   1b, 1c, 1d (cæt. excl.?), 1852.

---

[1] See tab. xlviii, fig. 5.

This small ramose coral appears to be intermediate bétween true Alveolites and Cœnites; it resembles the latter by the great development of its cœnenchyma, and Alveolites by the almost circular form of the calices, but differs from both by the direction and the mode of arrangement of the corallites, which, instead of being oblique, are almost perpendicular to the axis of the branches, and form vertical series somewhat as in the genus Seriatopora. When this fossil becomes more completely known in its structural characters, it will probably form the type of a new generical division. Mr. Hall places it in his genus Cladopora, which is partly composed of Bryozoa.

Diameter of the branches about one and a quarter line; diameter of the calices about one fifth of a line.

Dudley. Mr. Hall mentions its existence in the lower part of the Niagara limestone at Lockport.

Collections of Mr. Fletcher and of the Parisian Museum.

3. *Genus* MONTICULIPORA.[1]

1. MONTICULIPORA PETROPOLITANA.

> FAVOSITES PETROPOLITANUS, *Pander*, Russ. Reiche, p. 105, tab. i, figs. 6, 7, 10, 11 (excl., fig. 8), 1830.
> CALAMOPORA FIBROSA (pars), *Goldfuss*, Petref. Germ., vol. i, p. 215, tab. lxiv, fig. 9, 1833.
> FAVOSITES HEMISPHERICUS, *St. Kutorga*, Zweit. Beitr. zur Geogn. und Paleont. Dorp., p. 40, tab. viii, fig. 5, and tab. ix, fig. 3, 1837.
> CALAMOPORA FIBROSA, *Eichwald*, Sil. Syst. in Esthl., p. 197, 1840.
> FAVOSITES LYCOPODITES, *Lardner Vanuxem*, Geol. of New York, 3d part, p. 46, fig. 3, 1842.
> — — *Will. Mather*, Geol. of New York, 1st part, p. 397, fig. 3, 1843.
> CHÆTETES PETROPOLITANUS, *Lonsdale*, in Murch., Vern. and Keys., Russ. and Ural, vol. i, p. 596, tab. A, fig. 10, 1845.
> — — *Keyserling*, Reise in Petsch., p. 180, 1846.
> FAVOSITES PETROPOLITANA, *M'Coy*, Syn. of the Silur. Foss. of Irel., p. 64, pl. iv, fig. 21, 1846.

---

[1] D'Orbigny, Prodr. de Paléont., vol. i, p. 25, 1850; *Nebulipora*, M'Coy, Ann. and Mag. of Nat. Hist., 2d series, vol. vi, p. 283, 1850; and Brit. Palæoz. Fossils, p. 22, 1851.
This generical division was proposed by M. D'Orbigny, since the publication of the Classification of Polypi, given in the introduction to this work, and the great resemblance which exists between the corals included in this new group and the common *Chætetes* induced us to reject it; in our *Monographie des Polypiers Fossiles des Terrains Palæozoiques*, we therefore included all these fossils in the old genus Chætetes of Fischer. But since that we have observed some specimens in which the fissiparous mode of reproduction, attributed by Fischer to his original Chætetes, is quite distinct; whereas, in the species to which M. D'Orbigny gives the name of *Monticulipora*, the gemmiparous reproduction is evident. We therefore now think it advisable to admit the generical distinction established by that palæontologist. Besides the

CHÆTETES LYCOPERDON (pars), *Hall,* Paleont. of New York, vol. i, p. 64, pl. xxiii, fig. 1,
and pl. xxiv, figs. 1 *a—h,* and perhaps also pl. lxxv, fig. 2
(cæt. excl.), 1847.

CHÆTETES RUGOSUS, *Ibid.,* vol. i, p. 67, pl. xxiv, fig. 2.

— PETROPOLITANUS, LYCOPERDON, and SUBFIBROSUS, *D'Orbigny,* Prodr. de Paléont.,
vol. i, pp. 25 and 108, 1850.

— — *Milne Edwards* and *Jules Haime,* Pol. Foss. des Terr. Palæoz.
(Arch. du Mus., vol. v), p. 263, 1851.

— LYCOPERDON (pars), *Hall,* Paleont. of New York, vol. ii, p. 40, pl. xvii, fig. 1 *a—f,*
1852. (Cæt. excl.)

Corallum in general free; its basal plate flat or concave, and completely covered with
a concentrically wrinkled epitheca. Upper surface regularly convex, in general hemispherical, and presenting obtuse tuberosities, about one line broad, and varying very much
in height. In some specimens these tubercles appear to have been worn away, and their
existence is indicated only by the presence of small groups of large calices, with thick
walls; the calices are rather unequal in size, generally polygonal, sometimes almost
circular; the largest are about one fifth of a line in diameter; the walls are not perforated;
the tabulæ are horizontal, complete, and placed at about one twelfth of a line from each
other. Some vestiges of septa are often visible. Young specimens are flat and discoidal.

This species is found in the lower Silurian deposits at Caradoc and Ony River. Profr.
M'Coy mentions its existence at Glounmore and Ventry Harbour, in the county of Kerry;
at Knockmahon, Tramore, and Ballydowan Bay, Waterford county. It is also met with

---

species here mentioned, the genus Monticulipora comprehends the following fossils, which, in our work on
the Palæozoic Corals, were placed in the genus Chætetes.

C. Panderi, Milne Edwards and Jules Haime, op. cit., p. 265.

Monticulipora filiasa, D'Orbigny, Prod. de Paléont., tab. i, p. 25.—*Chætetes filiasa,* Milne
Edwards and J. Haime, op. cit., p. 266.

C. Dalei, M. Edwards and J. Haime, op. cit., p. 266, pl. xix, fig. 6.

Monticulipora ramosa, D'Orbigny, op. cit., p. 25.—*Chætetes ramosus,* M. Edwards and J.
Haime, op. cit., p. 266 pl. xix, fig. 2.

— *mammulata,* D'Orbigny, loc. cit.—*Chætetes mammulatus,* M. Edwards and
J. Haime, op. cit., p. 267, pl. xix, fig. 1.

— *frondosa,* D'Orbigny, loc. cit.—*Chætetes frondosus,* M. Edwards and J. Haime,
loc. cit., pl. xix, fig. 5.

Ptilodictya pavonia, D'Orbigny, op. cit., p. 22.—*Chætetes pavonia,* M. Edwards and
J. Haime, loc. cit., pl. xix, fig. 4.

Chætetes rugosus, M. Edwards and J. Haime, op. cit., p. 268, pl. xx, fig. 6.

— *Torrubiai,* De Verneuil and J. Haime, Bullet. Soc. Géol. de France, s. 2, v. vii, p. 162.
—M. Edwards and J. Haime, loc. cit., pl. xx, fig. 5.

— *Trigeri,* M. Edwards and J. Haime, op. cit., p. 269, pl. xvii, fig. 6.

— *heterosolen,* Keyserling, Reise in Petsch., p. 181, fig. *a, b.*

in North America, and in Russia. In his recent work, Mr. Hall mentions its existence in the lower parts of the Clinton group in Wayne county, in Niagara county, and at Flamborough Head, in Canada West.

Specimens are in the Collections of the Geological Society of London, of the Museum of Paris, of M. de Verneuil, M. D'Orbigny, &c.

M. Panderi[1] very much resembles this species, but differs from it by its turbinate form, and the rudimentary state of the tubercles on its upper surface.

2. MONTICULIPORA PAPILLATA.   Tab. LXII, figs. 4, 4*a*.

NEBULIPORA PAPILLATA, *M'Coy*, Ann. and Mag. of Nat. Hist., 2d series, vol. vi, p. 284, 1850.
—   —   *M'Coy*, Brit. Palæoz. Foss., p. 24, pl. i c, fig. 5, 1851.
CHÆTETES TUBERCULATUS, *Milne Edwards* and *Jules Haime*, Pol. Foss. des Terr. Palæoz.
                (Arch. du Mus., vol. v), p. 268, pl. xix, figs. 3, 3*a*, 1851.
RHINOPORA TUBERCULOSA? *J. Hall*, Paleont. of New York, vol. ii, p. 170, pl. 40 E, fig. 4,
                1852.

Corallum very thin, incrustating. Tubercles much compressed, elongated in the same longitudinal direction, rather prominent, about one line in length, one half more in breadth, and set at a distance from each other equal to about twice their breadth; the top of these tubercles is rather compact. Calices somewhat unequal in size and in form; those that are placed on the tubercles being rather larger than the others, and about one third of a line in breadth.

Found in the upper Silurian rocks of Dudley. Professor M'Coy has met with it in the upper Ludlow rocks of Brigster, Kendal, Westmoreland; at Coniston, Lancashire; and at Firbank, Sedbergh, Kendal. It exists also in the blue limestone of Cincinnati, Springfield, and Lebanon, in Ohio.

Specimens are in the Collections of Mr. Fletcher of Dudley, and of M. de Verneuil at Paris.

This species much resembles *M. mammulata*,[2] in which, however, the tubercles are more prominent, more elongated, and more irregular. *Monticulipora Dalei*[3] is also very closely allied to the above-described species, but differs from it by its dendroidal form, and its small round tubercles.

[1] *Chætetes Panderi*, Milne Edwards and Jules Haime, Pol. Foss. des Terr. Palæoz., p. 265, 1851.

[2] *Chætetes mammulatus*, Milne Edwards and Jules Haime, Pol. Foss. des Terr. Palæoz., p. 267, pl. xix, fig. 1.

[3] *Chætetes Dalei*, ibid., p. 266, p. xix, fig. 6.

3. MONTICULIPORA FLETCHERI.   Tab. LXII, figs. 3, 3*a.*

> CALAMOPORA SPONGITES? var. *Goldfuss,* Petref. Germ., vol. i, p. 216, pl. 64, fig. 10 (in parte), 1833.
>
> FAVOSITES SPONGITES (pars), *Lonsdale,* in Murchison, Silur. Syst., pl. xv *bis,* figs. 9, 9*a,* 9*b* (cæt. excl.), 1839.
>
> CHÆTETES FLETCHERI, *Milne Edwards* and *Jules Haime,* Pol. Foss. des Terr. Palæoz. (Arch. du Mus., vol. v), p. 271, 1851.
>
> — LYCOPERDON (pars), *J. Hall,* Paleont. of New York, vol. ii, p. 40, pl. xvii, figs. 1 *g—i* (cæt. excl.), 1852.

Corallum dendroidal; branches about one and a half or two lines in diameter, and not bearing any well-characterised tubercles.   *Calices* of two kinds; some circular, very closely set, and about one eighth of a line in diameter; others subpolygonal, much smaller, and placed between the former ones.

Dudley.   North America, in the Clinton group (J. Hall).

Specimens are in the Collections of the Geological Society, of Mr. Fletcher, of the Museum of Paris, &c.

By its general form *M. Fletcheri* resembles *M. pulchella,*[1] but its branches are slenderer, and bifurcate under a more obtuse angle.   Both these species are almost deprived of the tubercles which are in general so remarkable in the corals of this genus, but the mode of arrangement of the calices differs: in *M. pulchella,* the large calices are in general collected in groups amongst the smaller ones; whereas, in *M. Fletcheri,* the smaller cells are set irregularly between the large ones, which vary but little.

4. MONTICULIPORA PULCHELLA.   Tab. LXII, figs. 5, 5*a,* 5*b.*

> CHÆTETES PULCHELLUS, *Milne Edwards* and *Jules Haime,* Pol. Foss. des Terr. Palæoz., p. 271, 1851.

Corallum ramose; its branches often somewhat compressed, and from two to four lines in diameter.   Tubercles broad, not very prominent, and somewhat stellated.   *Calices* rather regularly hexagonal, and very unequal in size; those that occupy the centre of the tubercles about one fifth of a line in diameter, and at least twice as large as those placed in the intervals between the groups thus formed.

Wenlock, Dudley, Coalbrook Dale.

Specimens in the Collections of the Geological Society, of Mr. Fletcher, &c.

This species, as we have already mentioned, much resembles *Monticulipora Fletcheri,*[2] but differs from it by the mode of grouping of the cells, and the size and the angle of bifurcation of its branches.

---

[1] See tab. lxii, figs. 5, 5*a.*   [2] See tab. lxii, fig. 3.

5. Monticulipora? Bowerbanki. Tab. LXIII, figs. 1, 1*a*, 1*b*, 1*c*.

Favosites spongites (pars), *Lonsdale*, in Murchison, Silur. Syst., p. 683, pl. xv *bis*, figs. 8*c*, 8*d*, 8*e* (cæt. excl.), 1839. (Not the *Calamopora spongites* of Goldfuss.)

Discopora squamata? *Ibid.*, p. 679, pl. xv, fig. 23.

Chætetes? Bowerbanki, *Milne Edwards* and *Jules Haime*, Pol. Foss. des Terr. Palæoz. (Arch. du Mus., vol. v), p. 272, 1851.

The general form of this coral varies much at different ages. In young specimens it is massive, subspherical, and but slightly gibbose; but in older specimens the tuberosities appear to have risen up, so as to constitute cylindrical flexuous branches; and in a very large specimen these branches are ramose, very tall, numerous, and compose a cæspitose mass. The upper end of each of these branches is dilated so as to form a kind of round head, the surface of which does not present any tubercles, and is occupied by closely set subpolygonal slanting calices that are about one fifth or two fifths of a line in diameter, and much resemble those of Alveolites, but are deprived of septal processes.

From Much Wenlock, Benthall Edge, Dudley, Walsall, and, according to Mr. Lonsdale, Hurst Hill, Sedgley.

Specimens are in the Collections of Mr. Bowerbank, Mr. Fletcher, and M. de Verneuil.

The ill-preserved fossil, from Desertcreat, which Mr. Portlock has represented under the name of *Favosites polymorpha*,[1] appears to be referable to this species, which is remarkable for the slight difference of size between its calices, the lozenge form of these, and its general aspect.

6. Monticulipora explanata.

Nebulipora explanata, *M'Coy*, Ann. and Mag. of Nat. Hist., 2d series, vol. 6, p. 283, 1850.—Brit. Palæoz. Foss., p. 23, pl. i c, fig. 6, 1851.

"*Corallum* forming very thin, irregularly expanded laminæ, upwards of two inches long, covered above with nearly regular, quincuncially arranged, flat or slightly depressed, nebular clusters of larger tubes, about one and a half lines in diameter, and rather less than twice their diameter apart (about twelve or fourteen cells between one centre and the next); smaller intermediate tubes about six in one line.

"Coniston limestone, Coniston, Lancashire; limestone of Applethwaite Common, Westmoreland." M'Coy, op. cit.

[1] Report on the Geol. of Londonderry, &c., p. 326, pl. xxi, fig. 2*a*.

7. MONTICULIPORA LENS.

NEBULIPORA LENS, *M'Coy*, Ann. and Mag. of Nat. Hist., 2d series, vol. 6, p. 283, 1850.
— — *M'Coy*, Brit. Palæoz. Foss., p. 23, pl. i c, fig. 7, 1851.

" *Corallum* forming lenticular masses, averaging 10 lines in diameter, and 1½ lines thick in the middle, gradually thinning to the edge; base slightly concave, with small concentric wrinkles; upper surface evenly convex; clusters of large cells rounded, flat, or slightly concave, about 1 line in diameter, and usually a little more than their diameter apart (averaging from sixteen to twenty cells between one centre and another); smaller tubes averaging eight in one line, larger tubes of the clusters averaging four or five in one line; two inter-diaphragmal spaces equal the diameter of the tubes; apparently two irregular, close rows of connecting pores on each face of each tube (?).

" Caradoc sandstone of Horderly West; schists of Moel Uchlas? Pont y Glyn, Diffwys, near Corwen; Cwm of the Cymmerig, Bala." M'Coy, op. cit.

STENOPORA? GRANULOSA, M'Coy, Brit. Palæoz. Foss., p. 26; *Ceriopora granulosa*, Goldfuss, Petref. Germ., vol. i, tab. lxiv, fig. 13; is a fossil coral from Dudley, which appears to belong to the class of Bryozoa.

4. *Genus* LABECHEIA.[1]

1. LABECHEIA CONFERTA. Tab. LXII, figs. 6, 6a, 6b, 6c.

MONTICULARIA CONFERTA, *Lonsdale*, in Murchison, Silur. Syst., p. 688, pl. xvi, fig. 5, 1839.
LABECHEIA CONFERTA, *Milne Edwards* and *Jules Haime*, Pol. Foss. des Terr. Palæoz. (Arch. du Mus., vol. v), p. 280, 1851.

Corallum massive, or forming lamellar expansions, which are often incrusting. Common basal plate covered with a thick, wrinkled epitheca; upper surface flat or sub-mammillar, and covered with small granular conical tubercles, which appear to rise from the edge of the walls; calices confluent and not distinct; visceral chambers filled up with complete, horizontal, closely set tabulæ, and presenting quite rudimentary septa. Walls thick, and not perforated. Breadth of the visceral chamber about a quarter of a line.

In some specimens the marginal mural tubercles are ranged in regular series, so as to assume the appearance of small ridges.

---

[1] See Milne Edwards and Jules Haime, Pol. Foss. des Terr. Palæoz., p. 279, 1851.

From Wenlock, Benthall Edge, Gleedon Hill; Ardaun and Cong, Galway (M'Coy).

Specimens are in the Collections of the Museum of Paris, of the Museum of Practical Geology, of the Bristol Museum, of the Geological Society, of Mr. Fletcher, of Mr. J. Gray, &c.

This fossil was placed by Mr. Lonsdale in the genus Monticularia (Hydnophora), which it resembles in some of its exterior characters; but it differs most essentially, not only from Hydnophora, but also from all the other corals belonging to the large family of the Astræidæ, by existence of its well-developed tabulæ, and rudimentary state of the septal apparatus. This mode of structure is characteristic of the group of Favositidæ. It has great affinity with Chætetes, but differs from that genus by the mode of termination of its walls, and constitutes, in our system of classification, a new generical division, which we have dedicated to our celebrated friend, Sir Henry de la Beche, to whom geology in general, and more especially the geology of England, is most deeply indebted.

5. *Genus* HALYSITES, (p. lxii.)

1. HALYSITES CATENULARIA. Tab. LXIV, figs. 1, 1*a*, 1*b*, 1*c*.

TUBULARIA FOSSILIS, &c., *Magnus Bromel,* Acta Liter. Suec., vol. ii, pp. 410, 411, and 412, figs. 2 and A, 1728.

MADREPORA TUBIS, &c., *Fougt,* Amæn. Acad., vol. i, p. 103, tab. iv, fig. 20, 1749.

TUBIPORA CATENULARIA, *Linné,* Syst. Nat., ed. 12, p. 1270, 1767.

FUNGITE, *Knorr* and *Walch,* Rec. des Mon. des Cat., vol. ii, pl. F 9*, fig. 4, 1775.

TUBIPORA FLABELLARIS? *Othon Fabricius,* Fauna Groenl., p. 432, 1788.

—　CATENULATA, *Gmelin,* Syst. Nat., ed. 13, p. 3753, 1789.

—　　—　*Parkinson,* Org. Rem., vol. ii, pl. iii, figs. 5, 6, 1808.

TUBIPORITES CATENARIUS, *Schlotheim,* Petref., 1st part, p. 366, 1820.

CATENIPORA TUBULOSA, *Lamouroux,* Exp. Meth. des G. de Pol., p. 65, 1821.

—　　—　*Lamouroux,* Encycl. (Zooph.), p. 177, 1824.

—　LABYRINTHICA, *Goldfuss,* Petref. Germ., vol. i, p. 75, tab. xxv, fig. 5, 1826.

HALYSITES ATTENUATA, *Fischer,* of Waldheim, Not. sur des Tubip. Foss., p. 16, fig. 4, 1828.

—　DICHOTOMA, *Ibid.,* p. 17.

—　MICROSTOMA, *Ibid.,* p. 18.

—　STENOSTOMA, *Ibid.,* p. 18.

CATENIPORA APPROXIMATA, *Eichwald,* Zool. Spec., vol. i, p. 192, tab. ii, fig. 9, 1829.

—　DISTANS, *Ibid.,* p. 192, tab. ii, fig. 10.

—　COMMUNICANS? *Ibid.,* p. 193.

—　or TUBIPORA, *R. C. Taylor,* Mag. of Nat. Hist., vol. iii, p. 271, fig. 2, 1830.

—　ESCHAROÏDES, *De Blainville,* Dict. Sc. Nat. Atlas (Zooph.), pl. xliii, fig. 5, 1830.

—Man., pl. lxv, fig. 5.

HALYSITES DICHOTOMA, ATTENUATA and MACROSTOMA, *Fischer,* Oryct. de Moscou, pl. xxxviii, figs. 1, 2, 4, 1830.

CATENIPORA LABYRINTHICA, *Morren,* Descr. Cor. Belg., p. 68, 1832.
— — *St. Kutorga,* Beitr. zur Geogn. und Paleont., Dorpat's, p. 23, tab. v, fig. 1, 1835.
HALYSITES LABYRINTHICA, *Bronn,* Leth. Geogn., vol. i, p. 52, tab. v, fig. 8, 1835.
CATENIPORA LABYRINTHICA, *Milne Edwards,* in Lamk., 2d edit., vol. ii, p. 322, 1836.
HALYSITES LABYRINTHICA, *Fischer,* Oryct. de Moscou, 2d edit., p. 163, pl. xxxviii, figs. 1, 2, 4, 1837.
CATENIPORA LABYRINTHICA, *Hisinger,* Leth. Suec., p. 95, tab. xxvi, fig. 10, 1837.
— ESCHAROÏDES, *Lonsdale,* in Murchison, Silur. Syst., p. 685, pl. xv *bis,* fig. 14, 1839.
— LABYRINTHICA, *Eichwald,* Sil. Syst. in Esthland, p. 199, 1840.
— ESCHAROIDES, *Hall,* Geol. of New York, 4th part, No. 22, fig. 1, 1843.
— AGGLOMERATA, *Ibid.,* No. 22, fig. 2.
— LABYRINTHICA, *Castelnau,* Terr. Sil. de l'Amer. Sept., p. 45, pl. xvii, fig. 1, 1843.
— MICHELINI, *Ibid.,* p. 45, pl. xvii, fig. 2.
— ESCHAROIDES, *Portlock,* Rep. on the Geol. of Londonderry, &c., p. 325, pl. xx, fig. 9, 1843.
— — *Dale Owen,* Rep. on Geol. of Iowa, &c., p. 33, pl. vii, fig. 2, 1844.
HALYSITES LABYRINTHICA, *Keyserling,* Reise in Petschora, p. 175, 1846.
CATENIPORA GRACILIS, *Milne Edwards* and *Jules Haime,* Atlas du Regne Anim. de Cuvier (Zooph.), pl. lxv *bis,* fig. 2, 1849.
— COMPRESSA, *Ibid.,* pl. lxv *bis,* fig. 3.
HALYSITES LABYRINTHICA, *D'Orbigny,* Prodr. de Paléont., vol. i, p. 50, 1850.
— AGGLOMERATA, *Ibid.,* p. 50.
— CATENULARIA, *Milne Edwards* and *Jules Haime,* Pol. Foss. des Terr. Palæoz. (Arch. du Mus., vol. v), p. 281, 1851.
— CATENULATUS (pars), *M'Coy,* Brit. Palæoz. Foss., p. 26, 1851.
CATENIPORA ESCHAROÏDES, *J. Hall,* Paleont. of New York, vol. ii, pp. 44 and 127, pl. xviii, fig. 2, and pl. xxxv, fig. 1, 1852.
— AGGLOMERATA? *Ibid.,* p. 129, pl. xxxv *bis,* fig. 2.

Corallum very tall, forming a loose convex mass, composed of large vertical plates, that are themselves made up with single series of tubular corallites united laterally. These plates meet so as to form various angles, and to constitute, at the surface of the mass, large irregular reticulations; the divisions of which are much larger in one direction than in the other. The sides of the reticulations are composed of a series of corallites, varying in number from three to eight. *Calices* elliptical, and of the same size in the same specimen, but varying very much in different specimens (their great diameter varying from almost half a line to above 1½ line). Walls strong, and covered with a thick epitheca. Septa trabicular (12), reaching almost to the centre of the visceral chamber. Tabulæ well developed, horizontal, and closely set.

Llandeilo flags: Robeston Walshen, Pembrokeshire. Sholeshook, Pembrokeshire (Murchison).

Caradoc sandstone: Hughley, Salop, Lickey (Murchison).

Wenlock limestone: Much Wenlock, Dudley, Aymestry, Benthall Edge. Lincoln Hill, Townhope, Little Ridge, Easthope, Malvern (Murchison).

Middle Ludlow rocks: Aymestry, Tatter Edge near Downton-on-the-Rock Murchison).

In Ireland, Col. Portlock mentions its existence at Desertcreat, Tyrone, and at Portrane, Dublin; and Professor M'Coy in a great number of localities in the counties of Galway, Kildare, Kerry, Mayo, Tyrone, and Dublin.

This coral is also met with in Bohemia, Sweden, Norway, Russia, and North America.

Specimens are in the Collections of the Geological Society of London, of the Museum of Practical Geology, &c.

Mr. Hall has separated from this species some specimens, under the name of *H. agglomerata*,[1] confined in the upper part of the Niagara limestone, in Sweden and Ogden, Monroe county, and in which the reticulations are proportionally smaller; but the numerous gradations that we observed between these forms, induced us to consider them as being only varieties of the same species. We must even add that the distinction between *H. catenularia* and *H. escharoides* is not as yet clearly established; the principal difference between them consisting only in the form of the reticulations, which are much more regular and of more equal dimensions in *H. escharoides*.

2. HALYSITES ESCHAROÏDES. Tab. LXIV, figs. 2, 2*a*.

> FUNGITE, *G. W. Knorr* and *J. E. E. Walch*, Rec. des Monum. des Catastr., vol. ii, pl. ꜰ ix, figs. 1, 2, 3, 1775.
> CORALLITE, *Ibid.*, vol. iii, Suppl., p. 158, pl. vi A, 1775.
> MADREPORA CATENULARIA, *Esper*, Pflanz. (Petref.), tab. v.
> TUBIPORA CATENULATA? *Parkinson*, Org. Rem., vol. ii, pl. iii, fig. 4, 1808.
> CATENIPORA ESCHAROÏDES, *Lamarck*, Hist. des An. sans Vert., vol. ii, p. 207, 1816; 2d edit., p. 322.
> TUBIPORITES CATENULARIA, *Wahlenberg*, Nov. Act. Soc. Scient. Upsal., vol. viii, p. 99, 1821.
> CATENIPORA ESCHAROÏDES, *Lamouroux*, Exp. Meth., p. 65, 1821.
> — — *Krüger*, Gesch. der Urw., vol. ii, p. 264, 1823.
> — — *Lamouroux*, Encycl. (Zooph.), p. 177, 1824.
> — — *Goldfuss*, Petref. Germ., vol. i, p. 74, pl. xxv, fig. 4, 1826.
> HALYSITES JACOWICKYI, *Fischer*, Not. sur des Tubip. Foss., p. 15, figs. 5 and 6, 1828.
> CATENIPORA ESCHAROÏDES, *Eichwald*, Zool. Spec., vol. i, p. 192, 1829.
> — EXILIS, *Ibid.*, p. 193, tab. ii, fig. 13.
> — RETICULATA, *Ibid.*, p. 192, tab. ii, fig. 11.
> — ESCHAROIDES, *De Blainville*, Dict. Sc. Nat. Atlas (Zooph.), pl. xl, fig. 1, 1830; Man. d'Actin., p. lxii, fig. 1.
> TUBIPORA CATENULATA, *S. Woodward*, Syn. Table of Brit. Org. Rem., p. 5, 1830.
> CATENIPORA ESCHAROIDES, *Holl*, Handb. der Petref., p. 412, 1830.

---

[1] Geology of New York, 4th part, No. 22, fig. 2.

CATENIPORA ESCHAROIDES, *Steininger*, Mém. Soc. Géol. de France, vol. i, p. 341, 1831.
—   —   *Morren*, Descr. Corall. Belg., p. 68, 1832.
HALYSITES ESCHAROÏDES, *Fischer*, Oryct. de Mosc., p. 164, pl. xxxviii, fig. 3, 1837.
CATENIPORA ESCHAROIDES, *Hisinger*, Leth. Suec., p. 94, tab. xxvi, fig. 9, 1837.
—   —   *Eichwald*, Sil. Syst. in Esthl., p. 199, 1840.
—   —   *Castelnau*, Terr. Sil. de l'Amer. du Nord., p. 45, pl. xvii, fig. 3, 1843.
HALYSITES ESCHAROÏDES, *Geinitz*, Grundr. der Verst., p. 581, pl. xxiii A, fig. 11, 1845-46.
—   CATENULATA, *Keyserling*, Reise in Petsch., p. 175, 1846.
—   ESCHAROÏDES, *D'Orbigny*, Prodr. de Pal., vol. i, p. 50, 1850.
—   CATENULATA, *Ibid.*, p. 109.
—   ESCHAROÏDES, *Milne Edwards* and *Jules Haime*, Pol. Foss. des Terr. Palæoz. (Arch. du Mus., vol. v), p. 284, 1851.

Reticulations of the upper surface of the *corallum* small and polygonal; their sides in general formed by only 2 or 3 corallites. *Calices* elliptical, and about half a line broad in the direction of the series of individuals, but much less in the other. 12 *septa*. *Tabulæ* strong and closely set.

Found at Benthall Edge; in Gothland, in Groningue, Russia, and North America.

Specimens are in the Collections of M. Bouchard-Chantereaux, the Museum of Paris, &c.

6. *Genus* SYRINGOPORA, (p. lxii.)[1]

1. SYRINGOPORA BIFURCATA.   Pl. LXIV, figs. 3, 3*a*, 3*b*.

TUBIPORITES FASCICULARIS? *Wahlenberg*, Nov. Act. Soc. Ups., vol. viii, p. 99, 1821. (Not Linné.)
AULOPORA SERPENS? *Ibid.* A young specimen.
SYRINGOPORA RETICULATA, *Hisinger*, Leth. Suec., p. 95, tab. xxvii, fig. 2, 1837. (Not Goldf.)
AULOPORA SERPENS, *Ibid.*, p. 95, tab. xxvii, fig. 1. A young specimen?
SYRINGOPORA RETICULATA, *Lonsdale*, in Murchison, Silur. Syst., p. 684, pl. xv bis, fig. 10, 1839.
—   BIFURCATA, *Ibid.*, p. 685, pl. xv bis, fig. 11.
HARMODITES CATENATUS (pars), *Geinitz*, Grundr. der Verst., p. 565, 1845–46.
—   BIFURCATA, *D'Orbigny*, Prodr. de Paléont., vol. i, p. 50, 1850.
SYRINGOPORA BIFURCATA, *M'Coy*, Brit. Palæoz. Foss., p. 27, 1851.
—   —   *Milne Edwards* and *Jules Haime*, Pol. Foss. des Terr. Palæoz. (Arch. du Mus., vol. v), p. 287, 1851.

Corallites straight, or slightly geniculated at the places where the connecting tubes arise; separated from each other at a distance equal to once or twice their diameter. Connecting tubes large, often somewhat ascendant, and placed at about two lines apart. Diameter of the corallites one third or one line.

[1] Under the name of Harmodites.

Found at Dudley; at Gleedon Hill and Wenlock (Murchison); in Groningue, and in Gothland.

Specimens in the Collections of Mr. Fletcher, of Mr. Bowerbank, of the Geological Society, of the Paris Museum, &c.

2. SYRINGOPORA FASCICULARIS.   Tab. LXV, figs. 1, 1a, 1b, 1c.

> TUBIPORA FASCICULARIS, *Linné*, Syst. Nat., edit. 12, p. 1271, 1767.
> —   —   *Othon Fabricius*, Fauna Groenl., p. 429, 1788.
> SYRINGOPORA FILIFORMIS, *Goldfuss*, Petref. Germ., vol. i, p. 113, tab. xxxviii, fig. 16, 1829.
> —   —   *Morren*, Descr. Cor. Belg., p. 70, 1832.
> AULOPORA SERPENS, *De Blainville*, Man. d'Actin., pl. lxxxi, fig. 1, 1834.
> SYRINGOPORA FILIFORMIS, *Milne Edwards*, in Lamarck, Hist. des Anim. sans Vert., 2d edit. vol. ii, p. 328, 1836.
> —   —   *Lonsdale*, in Murchison, Silur. Syst., p. 685, pl. xv bis, fig. 12, 1839.
> AULOPORA TUBÆFORMIS, *Ibid.*, p. 676, pl. xv, fig. 8 ; and, perhaps, also AULOPORA SERPENS, *Ibid.*, p. 675, pl. xv, fig. 6.
> HARMODITES FILIFORMIS, ANGLICA, and IRREGULARIS, *D'Orbigny*, Prodr. de Paléont., vol. i, pp. 50, 51, 1850.
> SYRINGOPORA FASCICULARIS, *Milne Edwards* and *Jules Haime*, Pol. Foss. des Terr. Palæoz. (Arch. du Mus., vol. v), p. 293, 1851.

Corallum, when very young, rampant, and presenting short prominent calicular tubes, so as to resemble very closely *Aulopora ;* but by the progress of age these calicular tubes become very tall, and multiply by lateral gemmation, so as to form a fasciculated, cespitose mass, in which the corallites are rather closely set, being placed at a distance equal to once or twice their diameter, which is itself about one third of a line. *Walls* very thick, and covered with a strong epitheca. Corallites but slightly geniculated, geminating frequently, and united by only few large connecting tubes.

Found at Dudley, Dorming Wood, Benthall Edge, Gleedon Hill. Sir Roderick Murchison has found it in the Wenlock limestone at Eastnor Park, Ledbury, Prescoed Common, Usk, Aston Ingham near Newent, and in the Ludlow rocks at Ristley Wood, near Newent.

It is also met with in Gothland and Groningue.

This species is remarkable by its corallites being very slender and straight, or slightly geniculated, and by the large diameter of the connecting tubes. It much resembles *S. exilis*,[1] in which the corallites are, however, more flexuous and more closely set.

The resemblance which the young specimens of this species bear to Aulopora is so great, that many authors have referred them to that generical division. But in different

---

[1] Milne Edwards and Jules Haime, Pol. Foss. des Terr. Palæoz., p. 295.

parts of the same specimen, and sometimes also in different specimens, we have observed the intermediate form between this state and the fasciculated structure peculiar to Syringopora.

3. SYRINGOPORA SERPENS. Tab. LXV, figs. 2, 2*a*.

> MADREPORA TUBULIS, &c., *Fougt*, Amæn. Acad., vol. i, p. 105, tab. iv, figs. 22, 26, 1749.
> TUBIPORA SERPENS, *Linné*, Syst. Nat., edit. 12, p. 1271, 1767.
>   —   —   *Othon Fabricius*, Fauna Groenl., p. 428, 1788.
> CATENIPORA AXILLARIS, *Lamarck*, Hist. des Anim. sans Vert., vol. ii, p. 207, 1816; 2d edit., p. 322.
>   —   —   *Lamouroux*, Exp. Meth., p. 66, 1821.
> TUBIPORITES SERPENS, *Krüger*, Gesch. der Urw., p. 263, 1823.
> CATENIPORA AXILLARIS, *Lamouroux*, Encycl. (Zooph.), p. 177, 1824.
> AULOPORA CONGLOMERATA, *Lonsdale*, in Murchison, Silur. Syst., p. 675, pl. xv, fig. 3, 1839. (Not Goldfuss.)
>   —   LONSDALEI, *D'Orbigny*, Prodr. de Paléont., vol. i, p. 51, 1850.
> SYRINGOPORA SERPENS, *Milne Edwards* and *Jules Haime*, Pol. Foss. des Terr. Palæoz., p. 294, 1851.

Young specimens of this coral equally resemble Aulopora; by the progress of age and the multiplication of individuals, the corallites become very tall and closely set. *Epitheca* strong. *Walls* thick. 18 *septal striæ*. Connecting tubes few in number. Diameter of the corallite somewhat more than half a line.

Dudley and Benthall Edge. Gothland.

Specimens are in the Collections of Mr. Fletcher, of the Museum of Paris, &c.

This species differs from the preceding ones by the large size of its corallites, the irregular way in which they are set, and the extreme rarity of the connecting tubes.

Syringopora Lonsdaleana,[1] which Professor M'Coy found in the Silurian beds of Portrane and Mahahide, county of Dublin, is evidently distinct from the three Silurian species here described, but is so imperfectly known that it appears difficult to characterise it. The corallites are large, straight, and irregularly placed. The connecting tubes are large, and appear to be very short.

Syringopora cæspitosa, Lonsdale, in Murchison, Silur. Syst., pl. xv bis, fig. 13 (not Goldfuss); *Harmodites Lonsdalei*, D'Orbigny, Prodr. de Paléont., vol. i, p. 50; is a coral from Wenlock, which is entirely unknown to us, and seems not to belong to the genus Syringopora.

[1] M'Coy, Syn. of Sil. Foss. of Irel., p. 65, pl. iv, fig. 20, 1846.

7. *Genus* Cœnites.[1]

1. Cœnites juniperinus. Tab. LXV, figs. 4, 4*a*.

> Cœnites juniperinus, *Eichwald*, Zool. Spec., vol. i, p. 179, 1829.
> Limaria clathrata, *Lonsdale*, in Murchison, Sil. Syst., p. 692, pl. xvi bis, figs. 7, 7 *a*, 1839. (Not Steininger.)
> — Lonsdalei, *D'Orbigny*, Prodr. de Paléont., vol. i, p. 49, 1850.
> Cœnites juniperinus, *Milne Edwards* and *Jules Haime*, Pol. Foss. des Terr. Palæoz. (Arch. du Mus., vol. v), p. 301, 1851.
> Limaria ramulosa, *J. Hall*, Palæont. of New York, vol. ii, p. 142, pl. xxxix, fig. 4, 1852.

Corallum dendroid; its branches cylindrical, somewhat flexuous, scarcely ever coalescent. *Calices* closely set, not prominent, much elongated transversely, or even almost linear. The upper margin concave, and presenting a small medial process. The under edge deeply emarginated in the middle, and presenting a small obtuse process on each side. Cœnenchyma not very abundant. Diameter of the branches about 2 lines; large diameter of the calices about two fifths of a line; and about four times their diameter in the opposite direction.

Dudley and Russia. Mr. Hall mentions its existence at Lockport, Niagara county.

Specimens are in the Collections of Mr. Fletcher and M. de Verneuil.

It is not without some hesitation that we place the genus Cœnites in the class of Polypi; the form of the calices resembling very much what is met with in certain Bryozoa. In the present state of our knowledge concerning the structure of these fossils, their zoological affinities cannot be determined with certainty, but we are inclined to think that they are allied to the Favositidæ.

C. intertextus[2] resembles *C. juniperinus* in its dendroid form, but differs from it in the form of the calices.

2. Cœnites intertextus. Tab. LXV, figs. 5, 5*a*.

> Cœnites intertextus, *Eichwald*, Zool. Spec., vol. i, p. 179, pl. ii, fig. 16, 1829.
> Limaria fruticosa, *Lonsdale*, in Murchison, Silur. Syst., p. 692, pl. xvi bis, figs. 7*b*, 8, 8*a*, 1839. (Not Steininger.)
> Cœnites intertextus, *M'Coy*, Brit. Palæoz. Foss., p. 22, 1851.
> — — *Milne Edwards* and *Jules Haime*, Pol. Foss. des Terr. Palæoz. (Arch. du Mus., vol. v), p. 302, 1851.

Corallum ramose; branches cylindrical, not coalescent. *Calices* not very closely set, very prominent, their opening narrow, subtriangular, and presenting three small obtuse

---

[1] Eichwald, Zool. Spec., vol. i, p. 179, 1829.—Milne Edwards and Jules Haime, Pol. Foss. des Terr. Palæoz., p. 301, 1851.　　　[2] See tab. lxv, fig. 5.

denticulations. Diameter of the branches two or three lines; large diameter of the calices about one fourth of line; small diameter about one eighth.

Dudley. Sir Roderick Murchison has found it at Wenlock, Ledbury, Lincoln Hill, and Coalbrook Dale, Nasch Scar, Presteign, Abberley Hills. Aymestry, Herefordshire (M'Coy). Russia (Eichwald).

This species is well characterised by its prominent mammilar calices, and by the peculiar form of their terminal aperture. *C. fruticosus*,[1] found by M. Steininger in the Devonian formation of the Eifel Mountains, to which Mr. Lonsdale referred this coral, much resembles it in its general form, but differs from it by the oblique direction of its calicular apertures.

3. CŒNITES LINEARIS. Tab. LXV, fig. 3.

> CŒNITES LINEARIS, *Milne Edwards* and *Jules Haime*, Pol. Foss. des Terr. Palæoz. (Arch. du Mus., vol. v), p. 302, 1851.

Corallum massive, convex or subgibbose, and composed of thin superposed layers. *Calices* closely set, not prominent, or but very slightly so, linear, with their margin very obscurely denticulated, about half a line broad and one twelfth in the contrary direction.

Dudley. Collection of Mr. Fletcher.

This species differs greatly from the preceding ones, by its massive form, and its completely linear calices.

4. CŒNITES LABROSUS. Tab. LXV, figs. 6, 6a.

> CŒNITES LABROSUS, *Milne Edwards* and *Jules Haime*, Pol. Foss. des Terr. Palæoz. (Arch. du Mus. vol. v), p. 302, 1851.

Corallum lamellar, cyathoid, and pedunculated; its under surface granulose, and resembling that of a rasp; each tubercle being terminated on its upper side by an almost semilunar aperture, the under lip of which is prominent, slightly emarginated in the middle, and almost covers the aperture. The medial denticulations of the upper lip not much developed; breadth of the calices about one third of a line. The upper surface is completely covered by extraneous matter in both specimens that we have seen.

Dudley. Collection of Mr. Fletcher.

This coral differs from all the other species of the same genus in general form, and in the peculiar disposition of the under lip which hides the marginal denticulation of its calices.

---

[1] *Limaria fruticosa*, Steininger, Mém. Soc. Géol. de France, vol. i, p. 339, 1831.

5. CŒNITES? STRIGATUS.

CŒNITES STRIGATUS, *M'Coy*, Ann. and Mag. of Nat. Hist., 2d ser., vol. vi, p. 280, 1850.
     —   —   *M'Coy*, Brit. Palæoz. Foss., p. 22, pl. i c, fig. 8, 1851.

" *Corallum* forming cylindrical, dichotomous branches, two to three lines in diameter; surface with small, narrow, triangular cells, the base of the triangle below, and the apex usually more or less prolonged upwards into a vermiform channel, often upwards of half a line long; four to five rows of cells in the space of one line measured transversely, about two in the same space measured longitudinally; compact interstitial space between the rows of cell-openings usually rather exceeding their width.

" In the Wenlock limestone of Dudley, Staffordshire."   M'Coy, loc. cit.

Family THECIDÆ; (p. lxiii.)

Genus THECIA, (p. lxiii.)

1. THECIA SWINDERNANA.   Tab. LXV, figs. 7, 7*a*.

AGARICIA SWINDERNIANA, *Goldfuss*, Petref. Germ., vol. i, p. 109, pl. xxxviii, fig. 3, 1829.
     —   —   *Morren*, Descr. Cor. Belg., p. 46, 1832.
PORITES EXPATIATA, *Lonsdale*, in Murchison, Silur. Syst., p. 687, pl. xv, fig. 3, 1839.
    —   SWINDERNANA, *Bronn*, Ind. Paléont., p. 1031, 1849.
ASTREOPORA EXPATIATA, *D'Orbigny*, Prodr. de Paléont., vol. i, p. 50, 1850.
PALÆOPORA? (THECIA) EXPATIATA, *M'Coy*, Brit. Palæoz. Foss., p. 14, 1851.
THECIA SWINDERNANA (Introd., pl. lxiii, 1850), *Milne Edwards* and *Jules Haime*, Pol. Foss.
                                  des Terr. Palæoz. (Arch. du Mus., vol. v),
                                  p. 306, 1851.

Corallum massive, subgibbose, not very thick; its under surface covered with a thin wrinkled epitheca, and sometimes free, sometimes adherent in the middle; upper surface covered with small superficial calices, which vary in size and form—the larger ones placed on the prominent part of the tuberosities, the other in the depressions; most of them circular, some polygonal; most of them confluent, some separated by a slight furrow. Septa varying in number from 12 to 18, well developed, rather thick, slightly flexuous, closely set, terminated by horizontal upper edge, and not extending quite to the centre of the calice, where a small shallow circular fossula is visible, but where we have not been able to discover any columella.   These septa are somewhat unequal in breadth alternately, but are all of equal thickness, and are prolonged externally under the form of horizontal costal ridges.   A vertical section shows that the visceral chambers are separated by an abundant compact spurious cœnenchyma, resulting from the intimate union of the costæ.

Tabulæ horizontal, strong, and occupying only the central part of the visceral chambers, where the septo-mural tissue does not extend. Breadth of the calices about half a line.

Dudley, Benthall Edge; Lincoln Hill, Coalbrook Dale, Aston Ingham near May Hill, Lindells, and Woolhope (Murchison); Gothland and Groningue.

Specimens in the Collections of the Bristol Museum, of the Geological Society, of the Museum of Practical Geology, of Mr. Fletcher, &c.

2. THECIA GRAYANA. Tab. LXV, fig. 8.

THECIA GRAYANA, *Milne Edwards* and *Jules Haime*, Pol. Foss. des Terr. Palæoz. (Arch. du Mus., vol. v), p. 307, 1851.

Corallum massive, thin, adherent; basal surface covered with a very thick, circularly wrinkled epitheca. Calices subequal, not prominent, and presenting a rather deep, circular central fossula. 12 septa, equally developed, closely set, very thick, subconfluent exteriorly, and bearing on their upper edge a double row of granulations. Diameter of the calices about half a line.

Dudley. Collection of Mr. J. Gray.

This species differs from the preceding one by the lesser number of the septa.

Family CYATHAXONIDÆ, (p. lxv.)

Genus CYATHAXONIA, (p. lxv.)

CYATHAXONIA? SILURIENSIS.

CYATHAXONIA SILURIENSIS, *M'Coy*, Ann. and Mag. of Nat. Hist., 2d ser., vol. vi, p. 281, 1850.
— — *M'Coy*, Brit. Palæoz. Foss., p. 36, pl. i c, fig. 11, 1851.

" *Corallum* elongate-conic, about 5 lines long and 2 lines in diameter at the height from the base; strong central axis, nearly one third of the diameter; 60 or 70 strong radiating lamellæ, each extending from the axis to the outer wall, before reaching which it bifurcates, leaving a triangular interlamellar space, about equal in width to the distance between the adjoining lamellæ; surface coarsely ridged longitudinally, the sulci corresponding to the divided edges of the lamellæ, leaving one of the equal intervening ridges to correspond with each of the spaces between the individual lamellæ and between their divided edges.

" Rare in the upper Ludlow rock of Underbarrow, Kendal, Westmoreland." M'Coy, loc. cit.

Family CYATHOPHYLLIDÆ, (p. lxv.)

1. *Genus* AULACOPHYLLUM, (p. lxvii.)

AULACOPHYLLUM MITRATUM. Tab. LXVI, figs. 1, 1*a*, 1*b*.

HIPPURITES MITRATUS (pars), *Schlotheim*, Petref., 1st part, p. 352, 1820.
TURBINOLIA OBLIQUA, *Hisinger*, Anteckningar, vol. v, p. 128, pl. viii, fig. 7, 1831.
   — FURCATA, *Ibid.*, p. 128, tab. vii, fig. 4.
   — MITRATA, *Hisinger*, Leth. Suec., p. 100, pl. xxviii, fig. 10 (var. *obliqua*), and
         fig. 11 (var. *furcata*), 1837.
CYATHOPHYLLUM MITRATUM, *Geinitz*, Grundr. der Verst., p. 571, pl. xxxiii A, fig. 8, 1845-46.
PETRAIA ÆQUISULCATA, *M'Coy*, Ann. and Mag. of Nat. Hist., 2d ser., vol. vi, p. 279, 1850.
   — — *M'Coy*, Brit. Palæoz. Foss., p. 39, 1851.
AULACOPHYLLUM MITRATUM, *Milne Edwards* and *Jules Haime*, Pol. Foss. des Terr. Palæoz.
         (Arch. du Mus., vol. v), p. 356, pl. ii, fig. 6, 1851.

Corallum simple, turbinated, free, subpedicillated, thin, strongly curved; epitheca thin, and presenting slight circular wrinkles. 34 principal septa, rather thick exteriorly, extending mostly to the centre of the visceral chamber, and alternating with an equal number of rudimentary ones. A furrow occupying the place of the septal fossula, not well defined, and indicated by the junction of the septa at the bottom of the calices. Height of the corallum $1\frac{1}{2}$ or 2 inches; diameter of the calice about 8 lines.

Dudley, Walsall. Professor M'Coy mentions also its existence at Coniston, Lancashire; Applethwaite Common, Westmoreland; Mulock Quarry, Dalquorhan near Girvan, Ayrshire; High Haume, Dalton in Furness, Lancashire; Glyn Ceiriog, Denbighshire.

This coral is also met with in Gothland. It does not present the characteristic peculiarities of Aulacophyllum in as distinct a manner as the other species of the same genus; but in adult specimens the pinnate arrangement of the septa is very evident. In *A. sulcatum*[1] they are much more numerous.

2. *Genus* CYATHOPHYLLUM, (p. lxviii.)

1. CYATHOPHYLLUM? LOVENI. Tab. LXVI, figs. 2, 2*a*.

MADREPORA SIMPLEX, &c., var. δ, *Fougt*, Amæn. Acad., vol. i, p. 90, tab. iv, fig. 4, 1749.
FUNGITES, *Thomas Pennant*, Phil. Trans., vol. xlix, 2d part, p. 515, tab. xv, figs. 8 and 9,
         1757.
CYATHOPHYLLUM FLEXUOSUM? *Hisinger*, Leth. Suec., p. 102, pl. xxix, fig. 3, 1837. (Not
         Goldfuss.)

---

[1] Pol. Foss. des Terr. Palæoz., p. 355, pl. vi, fig. 2.

TRYPLASMA ARTICULATA, *Lonsdale*, in Murchison, Vern., and Keys., Russ. and Ural, vol. i,
pl. A, fig. 8, 1845. (Not *Cyathophyllum articulatum*, Hisinger.)
CYATHOPHYLLUM? LOVENI, *Milne Edwards* and *Jules Haime*, Pol. Foss. des Terr. Palæoz.
(Arch. du Mus., vol. v), p. 364, 1851.

Corallum simple, or accidentally aggregate, very tall, nearly cylindrical, subpedicillated, and slightly curved near its basis. The differences in the degree of activity in the development of the polyp are so very great, that the corallum presents an alternate series of circular constrictions and prominent ridges, and has somewhat the appearance of a pile of cyathiform corallites. Costæ rather thick, equally developed and flat. Calice circular, rather shallow. Septa about 60 in number, alternately large and small, closely set, and bearing numerous strong marginal denticulations. The small septa correspond to the middle of the costæ, and the large ones to the intercostal furrows. Height about 12 or 15 lines; breadth of the calice 4 or 5 lines; depth about 2 lines.

Found at Wren's Nest near Dudley; and in Gothland.

Specimens in the Collections of M. de Verneuil and M. Bouchard-Chantereaux.

This species is easily recognised by its prominent circular accretion ridge, which gives to it a lamellated appearance, much resembling that of Chonophyllum; it differs, however, from this genus by the great development of the septal apparatus.

We are inclined to think that the fossil coral found by Professor M'Coy at Egool and Bellaghaderreen, in the county of Mayo, and referred by that palæontologist to *Cyathophyllum flexuosum* of Goldfuss, belongs to this species.

2. CYATHOPHYLLUM ANGUSTUM. Tab. LXVI, figs. 4, 4*a*.

CYATHOPHYLLUM ANGUSTUM, *Lonsdale*, in Murchison, Silur. Syst., p. 690, pl. xvi, fig. 9, 1839.
— — *D'Orbigny*, Prodr. de Paléont., vol. i, p. 47, 1850.
CYSTIPHYLLUM BREVILAMELLATUM, *M'Coy*, Ann. and Mag. of Nat. Hist., 2d ser. vol. 6,
p. 276, 1850.
— — *M'Coy*, Brit. Palæoz. Foss., p. 32, pl. i B, fig. 19, 1851.
CYATHOPHYLLUM ANGUSTUM, *Milne Edwards* and *Jules Haime*, Pol. Foss. des Terr. Palæoz.
(Arch. du Mus., vol. v), p. 365, 1851.

Corallum simple, straight, tall, somewhat compressed (may be accidentally), and presenting but few feebly developed accretion ridges; septa very thin, placed about half a line apart, and united by strong dissepiments, so that when the epitheca is worn away (as in the specimen here figured) the surface of the corallum appears covered with regular square cells. A vertical section shows that the tabulæ are small, closely set, somewhat irregular, and occupy only one fourth or one fifth of the diameter of the corallum; the rest being filled up with vesicular cells that are placed obliquely, and are about half a line broad. The height of the corallum was probably about four inches; diameter about 1½ inch.

Found in the Wenlock shale at Atwoods Shaft, Lickey.

Mr. Lonsdale mentions its existence in the Caradoc sandstone of Coal Moors, Lickey. A specimen is in the Collection of the Geological Society.

This coral differs from most of the other species of Cyathophyllum by the small extent of the tabulæ. This structural character is also met with in *C. damnoniense*[1] and *C. Stutchburyi*.[2] The above-described fossil differs from the first of these by the vesicles being much more regular; and from the second by the absence of large cells between the space occupied by the tabulæ and that occupied by the vesicular tissue. It differs also from both these species by septa being much less numerous.

3. CYATHOPHYLLUM PSEUDO-CERATITES. Tab. LXVI, figs. 3, 3*a*, 3*b*.

MADREPORA SIMPLEX, &c., var. ε, *Fougt*, Amæn. Acad., vol. i, p. 90, tab. iv, fig. 7, 1749.

FUNGITES, *Thomas Pennant*, Philos. Trans., vol. xlix, 2d part, p. 514, tab. xv, fig. 7, 1757.

STREPHODES PSEUDO-CERATITES, *M'Coy*, Ann. and Mag. of Nat. Hist., 2d ser., vol. vi, p. 275, 1850.

— — *M'Coy*, Brit. Palæoz. Foss., p. 30, pl. i B, fig. 20, 1851.

CYATHOPHYLLUM RECURVUM, *Milne Edwards* and *Jules Haime*, Pol. Foss. des Terr. Palæoz. (Arch. du Mus., vol. v), p. 368, 1851.

Corallum simple, turbinated, becoming subcylindrical by the progress of age, very narrow at its basis, strongly curved, and covered with a rather thick epitheca. Accretion ridges not very large. Calice suboval, deep, and presenting rudimentary septal fossula, situated on the side corresponding to the convex part of the corallum. 38 well-developed thin septa, alternating with an equal number of smaller ones. Height of the corallum about 2 inches; breadth of the calice about 8 lines; depth 5 or 6 lines.

Dudley, Wenlock. Old Radnor, Presteign, Radnorshire; Sedgley (M'Coy). It is also met with in Gothland.

C. ceratites,[3] to which Professor M'Coy compares this species, differs from it by its general form, its numerous septa, and its calice, which is almost polygonal.

The coral found by Col. Portlock at Desertcreat, in the county of Tyrone, and referred by that geologist to the *Turbinolopsis elongata* of Phillips, appears to belong to this species.

4. CYATHOPHYLLUM ARTICULATUM. Tab. LXVII, figs. 1, 1*a*.

MADREPORA TURBINATA (pars), *Esper*, Pflanz. (Petref.), tab. iii, figs. 3 and 4.

MADREPORITES ARTICULATUS, *Wahlenberg*, Nov. Act. Soc. Upsal., vol. viii, p. 87, 1821.

CYATHOPHYLLUM VERMICULARE, *Hisinger*, Anteckn., vol. v, p. 130, tab. viii, fig. 8, 1831. (Not Goldfuss.)

---

[1] See tab. iv, fig. 1.   [2] See tab. xxxiii, fig. 4.   [3] See tab. iv, fig. 2.

Lɪᴛʜᴏᴅᴇɴᴅʀᴏɴ ᴄᴀꜱᴘɪᴛᴏꜱᴜᴍ, *Ch. Morren,* Descr. Cor. Belg., p. 47, 1832. (Not Goldfuss.)
Cʏᴀᴛʜᴏᴘʜʏʟʟᴜᴍ ᴠᴇʀᴍɪᴄᴜʟᴀʀᴇ, *Hisinger,* Leth. Suec., p. 102, pl. xxix, fig. 2, 1837.
— ᴀʀᴛɪᴄᴜʟᴀᴛᴜᴍ, *Ibid.,* p. 102, pl. xxix, fig. 4.
— ᴄᴀꜱᴘɪᴛᴏꜱᴜᴍ, *Lonsdale,* Sil. Syst., p. 690, pl. xvi, fig. 10, 1839. (Not Goldfuss.)
— ᴅɪᴀɴᴛʜᴜꜱ (pars)? *Lonsdale,* Ibid., pl. xvi, fig. 12e (*Cæt. excl.*) (Not Goldfuss.)
— ᴄᴀꜱᴘɪᴛᴏꜱᴜᴍ, *Eichwald,* Sil. Syst. in Esthland, p. 203, 1840.
— — *D'Orbigny,* Prodr. de Paléont., vol. i, p. 47, 1850.
Sᴛʀᴇᴘʜᴏᴅᴇꜱ ᴄʀᴀɪɢᴇɴꜱɪꜱ, *M'Coy,* Ann. and Mag. of Nat. Hist., 2d ser., vol. vi, p. 275, 1850.
— — *M'Coy,* Brit. Palæoz. Foss., p. 30, pl. i c, fig. 10, 1851.
Cʏᴀᴛʜᴏᴘʜʏʟʟᴜᴍ ᴀʀᴛɪᴄᴜʟᴀᴛᴜᴍ, *Milne Edwards* and *Jules Haime,* Pol. Foss. des Terr. Palæoz. (Arch. du Mus., vol. v), p. 377, 1851.

Corallum composite, fasciculate; corallites closely set, subcylindrical, tall, presenting numerous prominent accretion ridges, and covered with a thin epitheca, through which the costæ are apparent. *Calices* circular, shallow. Septa about sixty in number, thin, equally developed, and somewhat closely set. Gemmation often very distinctly intra-calicular; the young corallites remain cylindrical, and rise side by side without uniting. The tabulæ are small, and irregularly placed. The lateral vesicles are almost as high as broad, and rather irregular in size. The septa are well developed.

Wenlock, Dudley. Craig Head near Girwan, Ayrshire (M'Coy); Gothland and Russia.

Specimens in the Collections of the Geological Society, of Mr. Fletcher, Mr. J. Gray, &c.

This species resembles *C. cæspitosum*[1] and *C. æquiseptatum,*[2] by the mode of aggregation of its corallites; but differs from them by the great development of its accretion ridges. In that respect it much resembles *C. Loveni,*[3] in which the costæ are thicker, and the septa more unequally developed.

We are inclined to think that the *Cladocora sulcata* of Mr. Lonsdale[4] is only a variety of the species in which the accretion ridges are less prominent. It was found at Benthall Edge, and, according to Professor M'Coy, at Ferriter's Cove and Dingle, in the county of Kerry.

The fossils described by Professor M'Coy under the name of *Cyathophyllum cæspitosum*[5] appear also to belong to this species, and were found at Ardaun, Cong, in the county of Galway; Chair of Kildare, Kildare, in the county of Kildare; Doonquin, Dingle, in the county of Kerry; Portrane, Malahide, in the county of Dublin.

[1] See tab. li, fig. 2.
[2] See tab. lii, fig. 2.
[3] See tab. lxvi, fig. 2.
[4] Silur. Syst., p. 692, pl. xvi bis, fig. 9.
[5] Silur. Foss. of Ireland, p. 61, 1846.

5. CYATHOPHYLLUM TRUNCATUM. Tab. LXVI, figs. 5, 5*a*, 5*b*, 5*c*.

> FUNGITÆ OCTO MAJORES, &c., *Magnus Bromel*, Acta Liter. Suec., vol. ii, p. 465, 1728.
> PORPITARUM PLURIUM, &c., *Ibid.*, pp. 466 and 467, figs. 1—5, 1728.
> MADREPORA COMPOSITA, &c., *Fougt*, Amæn. Acad., vol. i, p. 93, tab. iv, fig. 10, 1749.
> FUNGITES, *Thomas Pennant*, Phil. Trans., vol. xlix, 2d part, pp. 514 and 516, tab. xv, figs. 6 and 12, 1757.
> MADREPORA TRUNCATA, *Linné*, Syst. Nat., edit. 10, p. 795, 1758.
> — — *Linné*, Fauna Suec., p. 536, 1761.
> — — *Linné*, Syst. Nat., edit. 12, p. 1277, 1767.
> STROMBODES TRUNCATUS, *Schweigger*, Handb. der Naturg., p. 418, 1820.
> MADREPORITES TRUNCATUS, *Wahlenberg*, Nov. Act. Soc. Upsal., vol. viii, p. 97, 1821.
> STROMBODES TRUNCATUS, *Eichwald*, Zool. Spec., vol. i, p. 188, 1829.
> FLOSCULARIA COROLLIGERA, *Ibid.*, p. 188, pl. xi, fig. 4.
> STROMBASTRÆA TRUNCATA, *De Blainville*, Dict. des Sc. Nat., vol. lx, p. 342, 1830.—Man., p. 376.
> CARYOPHYLLIA EXPLANATA, *Hisinger*, Anteckningar, vol. v, p. 129, tab. viii, fig. 9, 1831.
> — — *Hisinger*, Leth. Suec., p. 101, tab. xxviii, fig. 13, 1837.
> CYATHOPHYLLUM DIANTHUS, *Lonsdale*, Silur. Syst., p. 690, pl. xvi, figs. 12, 12*d* (*Cæt. excl.*), 1839. (Not Goldfuss.)
> — — *var.* PROLIFERA, *Eichwald*, Silur. Syst. in Esthland, p. 200, 1840.
> — TRUNCATUM, *Bronn*, Index Pal., vol. i, p. 370, 1848.
> — SUBDIANTHUS, *D'Orbigny*, Prodr. de Pal., vol. i, p. 47, 1850.
> STREPHODES VERMICULOIDES, *M'Coy*, Ann. and Mag. of Nat. Hist., 2d ser., vol. vi, p. 275, 1850.
> — — *M'Coy*, Brit. Palæoz. Foss., p. 31, pl. i B, fig. 22, 1851.
> CYATHOPHYLLUM TRUNCATUM, *Milne Edwards* and *Jules Haime*, Pol. Foss. des Terr. Palæoz. (Arch. du Mus., vol. v), p. 379, 1851.

Corallum composite, forming a tall turbinate mass, the basis of which is occupied by the parent corallite, and the upper surface is convex, and occupied by calices that vary very much in size, and are generally free at their edges, but sometimes meet, and become partially soldered together. Gemmation exclusively calicinal. Corallites regularly turbinate, not very tall, and narrow at their basis. *Walls* covered with a very thin epitheca, and presenting numerous strongly marked accretion ridges. *Calices* circular, or deformed by the pressure of the neighbouring corallites, and terminated by thin, slightly everted edges. Central fossula large, and rather deep; outer part of the calice almost flat. *Septa* (50 or 60) very closely set and thick, towards the circumference, but rather thin towards the centre, rather unequal in length alternately; the largest reaching to the centre. The internal structure of the corallites very dense. The *tabulæ* are small, and not very distinct exteriorly from the regular dissepiments that divide obliquely the interseptal spaces. Height of the corallites 1 or $1\frac{1}{2}$ inch; diameter somewhat less.

It is found at Much Wenlock, and Benthall Edge, Ledbury, Haven near Aymestry, and Wenlock Edge (Murchison); Portrane and Malahide, Dublin (M'Coy). Sweden and Russia.

Specimens are in the Collections of the Bristol Museum, of the Museum of Practical Geology, of the Geological Society of London, of Mr. Bowerbank, M. Bouchard-Chantereaux, and M. de Verneuil.

This coral is remarkable by its mode of gemmation, exclusively calicinal, and by the everted form of its calices.

It most resembles *C. regium*[1] and *C. helianthoides,*[2] but its septa are thicker and less numerous than in those large corals.

6. CYATHOPHYLLUM FLEXUOSUM. Tab. LXVII, figs. 2, 2a.

MADREPORA COMPOSITA, &c., *Fougt*, Amæn. Acad., vol. i, p. 96, tab. iv, figs. 5 and 13, 1749.
— FLEXUOSA, *Linné*, Syst. Nat., edit. 12, p. 1278, 1767.
CARYOPHYLLIA FLEXUOSA, *Lonsdale*, in Murchison, Sil. Syst., p. 689, pl. xvi, fig. 7, 1839.
(Not Lamarck.)
DIPHYPHYLLUM FLEXUOSUM, *D'Orbigny*, Prodr. de Paléont., vol. i, p. 38, 1850.
CYATHOPHYLLUM FLEXUOSUM, *Milne Edwards* and *Jules Haime*, Pol. Foss. des Terr. Palæoz. (Arch. du Mus., vol. v), p. 386, 1851.

Corallum dendroid; gemmation calicular. Corallites cylindrical, tall; epitheca feeble; costal striæ not numerous (about 20). Diameter of the large individuals about 1½ line or 2 lines.

Wenlock Shale, Malvern. Ferriter's Cove and Dingle, Kerry (M'Coy). Gothland.

Specimens are in the Collection of the Geological Society of London.

This species much resembles *C. parricida,*[3] in which, however, the corallites are turbinate and the septa less developed.

7. CYATHOPHYLLUM TROCHIFORME.

STREPHODES TROCHIFORMIS, *M'Coy*, Ann. and Mag. of Nat. Hist., 2d ser., vol. vi, p. 275, 1850.
— — *M'Coy*, Brit. Palæoz. Foss., p. 31, pl. i B, fig. 21, 1851.

" *Corallum* simple, slightly curved, widely turbinate; average length one inch three lines, and width at mouth one inch one line, with irregular swellings of growth; outer wall very thin, marked with equal lamellar sulci (6 in 3 lines at one and a quarter inch in diameter, or 83 all round); terminal cup very deep, conical, margin rounded, sides gradually sloping, lined by the thin alternately "longer and shorter uneven-edged lamellæ, the longest of which unite, and are irregularly blended at the centre, connected throughout by numerous curved transverse vesicular plates : *horizontal section* shows the same characters as the terminal cup, the alternate lamellæ extending about half way to the

[1] See tab. xxxii, figs. 1, 2, 3, 4.   [2] See tab. xli, fig. 1.   [3] See tab. xxxvii, fig. 1.

centre ; *vertical section,* apex filled with solid matter, centre with irregular vermicular lines (the sections of the complicated edges of the radiating lamellæ), from thence to the walls made up of small thick rounded vesicular plates, the obscure rows having a slight downward curve."

" Not uncommon in the Wenlock limestone of Dudley, Staffordshire." M'Coy, op. cit.

8. Cyathophyllum? vortex.

Clisiophyllum vortex, *M'Coy,* Ann. and Mag. of Nat. Hist., 2d ser., vol. vi, p. 277, 1850.
　—　　　—　*M'Coy,* Brit. Palæoz. Foss., p. 33, pl. i ʙ, fig. 18, 1851.

" *Corallum* simple, conic, slightly curved, enlarging at the rate of one inch eight lines in three inches from the apex; outer wall thin, faintly marked with subequal longitudinal lamellar striæ (5 or 6 in the space of 3 lines at a diameter of one inch), and small concentric wrinkles; at a diameter of one inch the horizontal section shows an outer area of about 60 thick, equal, radiating lamellæ, barely reaching one fourth of the diameter towards the centre, connected by small irregular transverse vesicular plates; a few of the pairs have a very thin, short, marginal lamella between each of the larger pairs, and where this occurs the vesicular transverse plates become much more numerous; inner area rather more than half the diameter, forming a circular mass of confused vesicular tissue, crossed by a few arched radiating delicate lamellæ; *vertical section,* having the narrow outer area on each side (corresponding to the lamelliferous zone) of arched vesicular plates, forming large unequal horizontal or slightly inclined cells, 1 or 2 cells extending across the width of the area; wide inner area composed of small oval cells, arranged in much-curved transverse rows, the convexity of the curve upwards.

" Wenlock limestone, Wenlock, Shropshire."—M'Coy, op. cit.

Turbinolia fibrosa, Portlock,[1] is a cast found at Desertcreat, and appearing to belong to Cyathophyllum, but is not sufficiently characterised to be determined specifically.

The same remark is applicable to the fossil figured by Professor M'Coy under the name of *Petraia zigzag.*[2] It is a cast of the interior of the visceral chamber of a coral which, in all probability, belongs to the genus Cyathophyllum, and had very flexuous costæ. It is found in the Silurian formation at Ardaun, Cong, Galway.

Several other casts that are not determinable specifically, and which have been referred to the non-characterised genus Turbinolopsis or Petraia, appear also to belong to this group. For example :

Turbinolopsis elongata, Phillips,[3] a specimen of which belongs to the Collections of the Geological Society, and found at Leach Heath, Bromsgrove Lickey. We have figured

[1] Report on the Geol. of Londonderry, p. 329, pl. xx, fig. 7.
[2] Sil. Foss. of Ireland, p. 60, pl. iv, fig. 17.
[3] Brit. Palæoz. Foss., p. 6, pl. ii, fig. 6 ʙ, 1841.

it in this Monograph (Tab. LXVI, figs. 6, 6*a*), in order to give an exact idea of the form of such casts.

Petraia elongata, M'Coy,[1] from Bala, Merionethshire.

Petraia subduplicata, M'Coy.[2]

Petraia serialis, M'Coy.[3]

Turbinolopsis rugosa, Phillips,[4] from Snowdon.

3. *Genus* OMPHYMA (p. lxviii).

1. OMPHYMA TURBINATA. **Tab. LXIX, figs. 1, 1*a*.**

FUNGITES GOTHLANDICUS, &c., *Magnus Bromel*, Act. Liter. Suec., vol. ii, p. 461, 1728.

— MAJOR, &c., *Ibid.*, p. 462.

MADREPORA SIMPLEX, TURBINATA, &c., *Fougt*, Amæn. Acad., vol. i, p. 87, tab. iv, figs. 1, 2 (*Cæt. excl.*), 1749.

— TURBINATA, *Linné*, Fauna Suec., p. 536, 1761.

— — *Linné*, Syst. Nat., ed. 12, p. 1272, 1767.

— — *Esper*, Pflanz. (Petref.), tab. ii, figs. 1, 2 (*Cæt. excl.*).

TURBINATED MADREPORITE, *Parkinson*, Org. Remains, vol. ii, pl. iv, fig. 2, 1808.

TURBINOLIA TURBINATA (pars), *Lamarck*, Hist. des Anim. sans Vert., vol. ii, p. 231, 1816; 2d edit., p. 360.

— CYATHOIDES, *Lamarck*, Ibid., p. 231.

MADREPORITES TURBINATUS (pars), *Wahlenberg*, Nov. Act. Soc. Upsal., vol. viii, p. 96, 1821.

TURBINOLIA CYATHOIDES and TURBINATA, *Lamouroux*, Exp. Meth., p. 51, 1821.

— — *Deslongchamps*, Encyc. (Zooph.), p. 750, 1824.

— — *Defrance*, Dict. Sc. Nat., vol. lvi, p. 91, 1828.

CARYOPHYLLIA TURBINATA (pars), *Al. Brongniart*, Tabl. des Terrains, p. 431, 1829.

CANINIA LATA, *M'Coy*, Ann. and Mag. of Nat. Hist., 2d ser., vol. vi, p. 277, 1850.

— — *M'Coy*, Brit. Palæoz. Foss., p. 28, pl. i c, fig. 13, and perhaps also fig. 12, 1851.

OMPHYMA TURBINATA, *Milne Edwards* and *Jules Haime* (Introd., p. lxviii, 1850), Pol. Foss. des Terr. Palæoz. (Arch. du Mus., vol. v), p. 400, 1851.

Corallum simple, turbinate, straight, short, often twice as broad as high, subpedicillated, and bearing radiciform appendices in its lower half only. Epitheca thin, and presenting in general only slight accretion wrinkles. *Calice* subcircular; its edge slightly lamellated. Calicinal cavity large and deep; the uppermost *tabula* presenting an extensive smooth surface in the middle, and four well-characterised septal fossulæ, two of which are larger than the others. *Septa* (100 or 120) thin, not tall, and resembling simple folds, somewhat unequal in dimensions alternately, straight or slightly flexuous towards the centre of the calice. The lower *tabulæ* are horizontal, broad, and well developed. The lateral parts of

Brit. Palæoz. Foss., p. 40.   [2] Ibid., p. 40, pl. i B, fig. 6.   [3] Ibid., p. 41, pl. i B, fig. 25.
[4] Palæoz. Foss. p. 7, pl. ii, fig. 7 C.

the visceral chambers are occupied by large oblique vesicles. Height of the corallum 1½ or 2 inches; diameter of the calice often 3 inches or more; depth almost 1 inch.

Wenlock Edge, Benthall Edge, Dormington Wood, Wren's Nest near Dudley. Gothland.

Specimens are in the Collections of the Geological Society of London, of the Museum of Practical Geology, of the Bristol Museum, of Mr. Fletcher, Mr. John Gray, of M. Bouchard-Chantereaux, of MM. de Verneuil, D'Archiac, &c.

This species is easily distinguished from *O. subturbinata*[1] and *O. Murchisoni*,[2] by its short, cyathoidal form. It much resembles *O. grandis*,[3] in which, however, the calicular margin is more lamellated, the septal fossulæ less developed, and the interseptal spaces filled with vesicles that are apparent exteriorly as well as in the interior of the corallum.

The fossil of the Silurian formation of Ireland, which Professor M'Coy[4] refers to the *Cyathophyllum turbinatum*, Goldfuss, probably belongs to this species. It is found in various localities of the counties of Galway, Kildare, Mayo, and Dublin.

2. OMPHYMA SUBTURBINATA. Tab. LXVIII, figs. 1, 1*a*, 1*b*, 1*c*.

> MADREPORA SIMPLEX, TURBINATA, &c. (pars), *Fougt*, Amæn. Acad., vol. i, tab. iv, fig. 3 (*Cæt. excl.*), 1749.
> — TURBINATA (pars), *Esper*, Pflanz. (Petref.), tab. ii, fig. 4; tab. iii, fig. 5.
> TURBINATED MADREPORITE, *Parkinson*, Org. Remains, vol. ii, pl. iv, fig. 1, and perhaps also fig. 3, 1808.
> TURBINOLIA VERRUCOSA and ECHINATA, *Hisinger*, Anteckningar, vol. v, p. 128, pl. viii, figs. 5 and 6, 1831.
> — TURBINATA, *var.* VERRUCOSA and ECHINATA, *Hisinger*, Leth. Suec., p. 100, tab. xxviii, figs. 7 & 8, 1837.
> CYATHOPHYLLUM TURBINATUM, *Lonsdale*, in Murchison, Silur. Syst., p. 690, pl. xvi, figs. 11, 11*a*, 1839.
> — — *Eichwald*, Sil. Syst. in Esthland, p. 200, 1840.
> — SUBTURBINATUM, *D'Orbigny*, Prodr. de Paléont., vol. i, p. 47, 1850.
> CANINIA TURBINATA, *M'Coy*, Brit. Palæoz. Foss., p. 28, 1851.
> OMPHYMA SUBTURBINATA, *Milne Edwards* and *Jules Haime*, Pol. Foss. des Terr. Palæoz. (Arch. du Mus., vol. v), p. 401, 1851.

Corallum tall, straight, or slightly bent at its basis, cylindro-turbinate, with a slender peduncle, and presenting round its under half large radiciform appendices. Accretion ridges of the wall in general well developed; epitheca very thin. Calice circular, not very deep; its edge rather thick and not lamellated. Central smooth space of the upper tabula small: septal fossulæ smaller, and not as deep as in *O. turbinata*: septa (80) delicate, and somewhat unequal alternately. A vertical section shows that the tabulæ are very large,

---

[1] See tab. lxviii, fig. 1.

[2] See tab. lxvii, fig. 3.

[3] Milne Edwards and Jules Haime, Pol. Foss. des Terr. Palæoz., p. 403, 1851.

[4] Syn. of Sil. Foss. of Ireland, p. 61, 1846.

very numerous, strong, horizontal towards the centre, and directed upwards towards the circumference. The vesicles that occupy the exterior part of the visceral chamber are large, oblique, and very unequal in size. These corals become sometimes very large; some specimens are about 7 or 8 inches high.

Wenlock Edge, Walsall, Benthall Edge, Gleedon's Hill, Dudley. Lincoln Hill, Kinsham near Aymestry, Ledbury, Malvern, the Valley of Woolhope, and Prolimoor Well (Sir R. Murchison). Mathyrafal, south of Meifod, Montgomeryshire; Craig Head, Ayrshire (M'Coy).

It is also found in Gothland and in Russia.

Specimens are in the Collections of the Geological Society of London, of the Museum of Practical Geology, of the Bristol Museum, of the Parisian Museum, of Mr. Bowerbank, M. Bouchard-Chantereaux, M. de Verneuil, M. D'Archiac, and M. Michelin.

This species, which till of late was confounded with the preceding one, differs from it by its general form, the proportions of the smooth and radiated parts of the calice, the number of septa, and many other characters.

3. OMPHYMA MURCHISONI. Tab. LXVII, figs. 3, 3*a*, 3*b*.

> CYSTIPHYLLUM SILURIENSE (pars), *Lonsdale*, in Murchison, Silur. Syst., p. 691, pl. xvi bis, fig. 2 (*Cæt. excl.*), 1839.
>
> OMPHYMA MURCHISONI, *Milne Edwards* and *Jules Haime*, Pol. Foss. des Terr. Palæoz. (Arch. du Mus., vol. v), p. 402, 1851.

Corallum turbinate, tall, slightly bent near its basis, and bearing radiciform appendices very high up but in very small number. Accretion ridges of the wall in general strongly characterised. Calice circular, not very deep; the smooth central part of which is pretty well developed. Septal fossulæ very distinct, but not large. Septa (about 60 in number) not closely set, slightly flexuous, and intermingled with some large, very apparent vesicles. A vertical section shows that the tabulæ are large, thick, and generally more or less obliquely placed. The vesicles occupying the outer part of the visceral chamber are very large, unequal in size, and very oblique. Height about 2 inches.

The only specimens that we have seen were found at Wenlock, and belong to the Collection of Mr. Bowerbank.

This coral much resembles *O. subturbinata*[1] by its general form, but differs from it by the small development of its septa, and the existence of very apparent vesicles on the surface of the calice; circumstances that induced Mr. Lonsdale to place it in the genus Cystiphyllum.

---

[1] See tab. lxviii, fig. 1.

4. *Genus* GONIOPHYLLUM (p. lxix).

1. GONIOPHYLLUM FLETCHERI.   Tab. LXVIII, figs. 3, 3*a*.

> GONIOPHYLLUM FLETCHERI, *Milne Edwards* and *Jules Haime*, Pol. Foss. des Terr. Palæoz.
> (Arch. du Mus., vol. v), p. 405, 1851.

Corallum simple, tall, almost straight, pyramidal, and quadrangular.   Epitheca presenting strong accretion folds.   Calice almost square, rather deep, and appearing to contain about 50 septa.   The septal fossulæ are not distinct in the only specimen that we have seen.   Height about 1 inch.

Dudley.   Collection of Mr. Fletcher.

2. GONIOPHYLLUM PYRAMIDALE.

> FUNGITÆ TETRAGONI, GOTHLANDICI, &c., *Magnus Bromel*, Acta Liter. Suec., vol. ii, p. 446,
> figs. *a*, *b*, 1728.
> TURBINOLIA PYRAMIDALIS, *Hisinger*, Anteckningar, vol. v, p. 128, tab. vii, fig. 5, 1831.
> —   —   *Hisinger*, Leth. Suec., p. 101, tab. xxviii, fig. 12, 1837.
> CALCEOLA PYRAMIDALIS, *Girard*, Jahrbuch für Miner. und Geol., p. 232, figs. *a, b, c*, 1842.
> PETRAIA QUADRATA, *M'Coy*, Syn. of the Sil. Foss. of Irel., p. 61, pl. iv, fig. 18, 1846.
> GONIOPHYLLUM PYRAMIDALE, *Milne Edwards* and *Jules Haime* (Introd., p. lxix, 1850), Pol.
> Foss. des Terr. Palæoz. (Arch. du Mus., vol. v), p. 404,
> pl. ii, figs. 4, 4*a*, 1851.

We refer to this species an apparently ill-preserved specimen found by Professor M'Coy in the Silurian deposits of Ireland, and placed by that palæontologist in the non-characterised genus Petraia.

In well-preserved specimens, such as those that are found in Gothland, and have been figured by Hisinger, and described by us in a former work, the following characters are seen : Corallum simple, pyramidal, slightly curved, with a very narrow peduncle ; rather thick epitheca and accretion ridges that sometimes constitute at the angles irregular tubercles, and even thus assume the appearance of short radiciform appendices.   Calice almost square, not very deep.   Septal fossulæ pretty well characterised, and corresponding to the angles of the visceral chamber.   Septa (72) rather thick, but very slightly prominent, and extending almost to the centre of the calice, where they are slightly flexuous.   Height about 1 inch.

Ardaun and Kilbride, Cong, in the county of Galway (M'Coy).   Gothland.

G. Fletcheri[1] differs from this species by its narrow elongated form, and by its septa being less numerous, and by several other characters.

[1] See tab. lxviii, fig. 3.

5. *Genus* CHONOPHYLLUM (p. lxix).

VI

CHONOPHYLLUM PERFOLIATUM ? Tab. LXVIII, figs. 2, 2a. (See p. 235, and Tab. IV, fig. 5.)

It is not without some hesitation that we refer to this species, already described in the preceding chapter as being common in the Devonian formation, a coral found by M. D'Archiac in the Silurian rocks at Wenlock. The only apparent difference between this fossil and the Torquay specimen consists in the form of the calice, the border of which is not everted.

6. *Genus* PTYCHOPHYLLUM (p. lxix).

PTYCHOPHYLLUM PATELLATUM. Tab. LXVII. figs. 4, 4a.

> FUNGITES MEDIÆ, &c., *Magnus Bromel*, Acta Liter. Suec., vol. ii, p. 463, 1728.
> — PATELLATUS, *Schlotheim*, Petref., 1st part, p. 247, 1820.
> — — *Kruger*, Gesch. der Urwelt, vol. ii, p. 253, 1823.
> — — *Hisinger*, Leth. Suec., p. 99, pl. xxviii, fig. 3, 1837.
> STROMBODES PLICATUM, *Lonsdale*, in Murchison, Sil. Syst., p. 691, pl. xvi bis, fig. 4, 1839.
> — PLICATUS, *M'Coy*, Sil. Foss. of Ireland, p. 61, 1846.
> CYATHOPHYLLUM PATELLATUM, *Bronn*, Index Pal., vol. i, p. 369, 1848.
> CYATHAXONIA PLICATA, *D'Orbigny*, Prodr. de Pal., vol. i, p. 48, 1850.
> PTYCHOPHYLLUM PATELLATUM, *Milne Edwards* and *Jules Haime*, Pol. Foss. des Terr. Palæoz.
> (Arch. du Mus., vol. v), p. 407, 1851.

Corallum simple, pedicellated, straight or slightly curved, and short. Epitheca wrinkled; borders of the *calice* lamellated and very much everted, so as to give to the corallum the form of a mushroom, and to produce a circular elevation around the central fossula, which is very deep. Pseudo-columella small. *Septa* nearly 100 in number, somewhat unequally developed alternately, rather thick exteriorly, but very slender towards the centre of the calice, where the large ones become strongly twisted, and rise up a little to form the columella. Height about 1½ inch; diameter of the calice twice the height or even more.

Brand Lodge, Malvern, Malvern Mountains. Doonquin and Dingle, in the county of Kerry (M'Coy). It is also met with in Gothland.

Specimens are in the Bristol Museum, Bonn Museum, in the Collections of the Geological Society of London, and of M. de Verneuil.

This species much resembles *P. extensum*[1] in its general form, but its septa are more numerous and more equally developed. It differs from *P. Stokesi*,[2] by the lesser size of its pseudo-columella.

---

[1] Milne Edwards and Jules Haime, Pol. Foss. des Terr. Palæoz., p. 408, pl. viii, fig. 2.
[2] Ibid., p. 407; Stokes, Geol. Trans., 2d ser., vol. i, pl. xxix, fig. 1 (*dextra*), 1824.

7. *Genus* ACERVULARIA (p. lxx).

ACERVULARIA LUXURIANS. Tab. LXIX, figs. 2, 2*a*, 2*b*, 2*c*, 2*d*, 2*e*, 2*f*.

MADREPORA COMPOSITA, &c., *Fougt*, Amæn. Acad., vol. i, p. 93, tab. iv, fig. 8, 1749.
FUNGITES, *Thomas Pennant*, Phil. Trans., vol. xlix, 2d part, p. 515, tab. xv, fig. 11, 1757.
MADREPORA ANANAS (pars), *Linné*, Syst. Nat., edit. 12, p. 1275, 1767.
— TRUNCATA, *Parkinson*, Org. Rem., vol. ii, pl. v, fig. 2, 1808. (Not Linné.)
— ANANAS, *Parkinson*, Ibid., pl. v, fig. 1. (Not Linné.)
FLOSCULARIA LUXURIANS, *Eichwald*, Zool. Spec., vol. i, p. 188, tab. xi, fig. 5, 1829.
ASTREA ANANAS, *Hisinger*, Leth. Suec., p. 98, pl. xxviii, fig. 1, 1837.
CARYOPHYLLIA TRUNCATA, *Hisinger*, Ibid., p. 101, pl. xxviii, fig. 14.
CYATHOPHYLLUM DIANTHUS (pars), *Lonsdale*, in Murchison, Sil. Syst., p. 690, pl. xvi,
figs. 12*a*, 12*d*, 1839. (Not Goldfuss.)
ASTREA ANANAS, *Lonsdale*, Ibid., p. 688, pl. xvi, fig. 6. (Not Lamarck.)
LITHOSTROTION LONSDALEI, *D'Orbigny*, Prodr. de Paléont., vol. i, p. 48, 1850.
ACERVULARIA ANANAS, *M'Coy*, Brit. Palæoz. Foss., p. 35, 1851.
— LUXURIANS, *Milne Edwards* and *Jules Haime*, Pol. Foss. des Terr. Palæoz.
(Arch. du Mus., vol. v), p. 415, 1851.

Corallum composite, massive, convex; in general tall. Corallites sometimes free laterally and cylindrical, but in general united by their walls, and compressed so as to become prismatical. In some cases the individuals so united diminish in diameter as they grow up, and so become again free and cylindrical. Epitheca thick in the aggregate as well as in the free corallites, and forming on the surface of the massive corallum delicate, zigzag, slightly prominent, polygonal lines, that separate the individuals from each other. Gemmation principally calicinal. *Calices* vary much in size, and are rather deep in the centre. *Septa* nearly equally thick exteriorly, but unequally developed inside of the interior wall, the large ones only extending to the centre; in general 54 in large corallites, and about 30 in the small ones. The area comprised between the two walls terminated by a flat or slightly concave surface. Breadth of the large polygonal individuals about 6 lines; the small ones about 2 lines. Diameter of the true calice or central fossula about half the diameter of the corallites in large specimens, but much more proportionally in the small individuals, where the space comprised between the two mural investments is but little developed.

Dudley, Wenlock. Ledbury, Herefordshire (M'Coy). It is also met with in Gothland and Dalecarlia.

Specimens are in the Collections of the Bristol Museum, Parisian Museum, of the Museum of Practical Geology, of Mr. Fletcher, Mr. John Gray, Mr. Bowerbank, M. Bouchard-Chantereaux, and M. de Verneuil.

This coral may be easily distinguished from the other species of the same genus by the development of its inner walls, and its mode of gemmation, which is almost entirely calicinal.

8. *Genus* STROMBODES (p. lxx).

1. STROMBODES TYPUS.   Tab. LXXI. figs. 1, 1*a*, 1*b*.

ARACHNOPHYLLUM TYPUS, *M'Coy*, Ann. and Mag. of Nat. Hist., 2d ser., vol. vi, p. 278, 1850.
—   — *M'Coy*, Brit. Palæoz. Foss., p. 38, pl. i B, fig. 27, 1851.
STROMBODES LABECHII, *Milne Edwards* and *Jules Haime*, Pol. Foss. des Terr. Palæoz.
(Arch. du Mus., vol. v), p. 427, 1851.

Corallum composite, massive, subturbinate; its upper surface slightly convex; its basal common plate covered with a very thin epitheca, and presenting very prominent and irregular accretion ridges. The terminal surface of the corallites irregularly polygonal, and separated from each other by strong ridges that are more prominent at their angles than elsewhere. Near the centre of these polygonal spaces a slight circular elevation, corresponding to the upper edge of the inner wall, and circumscribing the true calice. Septal radii not distinct from the costæ, and thus extending to the outer edge of the corallite. The total number of the *septo-costal radii* amounts to about 100, but one third of them only extend to the calicinal fossula, where they become curved and somewhat prominent; most of them appear to bear small paliform lobules. The marginal or costal radii are still slenderer than the preceding ones, and adhere to them at their inner edge. Very slender, closely set, unequally distinct dissepiments unite all these radii, so as to constitute delicate very regular quadrangular reticulations. Vertical and horizontal sections show that the structure of the corallum is essentially vesicular. In a section corresponding to the direction of the axis of the corallites, the different layers of that vesicular tissue being of different degrees of density, constitute undulated parallel horizontal lines, the direction of which correspond to that of the surface of the corallum. A horizontal section shows that the inner or true walls are circular and well constituted, and that the reticulations become larger and more regular towards the circumference of the corallites. In many parts no remains of the costæ septal radii are distinct, and the space comprised between the inner wall and the lateral surface of the corallites is completely cellular. Large diagonal of the corallites at the upper surface of the compound mass about 8 lines; diameter of the true calice about 3 lines.

Wenlock Edge. Aymestry, Herefordshire (M'Coy).

Specimens are in the Collections of the Paris Museum, Bristol Museum, of the Museum of Practical Geology, and M. de Verneuil.

2. STROMBODES MURCHISONI.   Tab. LXX, figs. 1, 1*a*, 1*b*, 1*c*, 1*d*.

ACERVULARIA BALTICA (pars), *Lonsdale*, in Murchison, Sil. Syst., p. 689, pl. xvi, figs. 8*b*,
8*c*, 8*d*, 8*e* (*Cæt. excl.*), 1839. (Not Schweigger.)
ACTINOCYATHUS BALTICUS, *D'Orbigny*, Prodr. de Paléont, vol. i, p. 48, 1850.
STROMBODES MURCHISONI, *Milne Edwards* and *Jules Haime*, Pol. Foss. des Terr. Palæoz.
(Arch. du Mus., vol. v), p. 428, 1851.

Corallum massive, subturbinate; common basal plate covered with a very thick

39

epitheca, and presenting very irregular accretion wrinkles. Upper surface slightly convex, and presenting large polygonal prominent reticulations formed by the line of junction of the corallites. Circular calicular protuberance small, but pretty distinct. Pseudo-columella somewhat elevated. Septo-costal radii extremely slender and numerous (about 100), but very unequally developed; about 50 of them reach almost to the centre, and about half of these appear to bear a rudimentary paliform lobule. Diagonal about 1 inch; diameter of the calicinal ring about 3 lines. A vertical section shows that the tissue is almost entirely vesicular, and it is only in the parts corresponding to the axis of the visceral chamber that some appearance of vertical striæ corresponding to the septa are visible. The transversal undulating lines that are strongly marked in this section are produced by the intermittent mode of growth of the corallum, and correspond to the different surfaces, which become rather dense after each period of activity. They are almost horizontal in the parts corresponding to the visceral chambers, and suddenly rise up in the parts corresponding to the line of demarcation between the adjoining corallites.

Dudley, Wenlock; Egool and Bellaghaderreen, Mayo (M'Coy).

Specimens in the Collections of the Geological Society, of the Bristol Museum, of Mr. Bowerbank, and of the Parisian Museum.

This species is very closely allied to *S. typus*,[1] but differs from it by its internal structure being more completely vesicular, and by its septa being more closely set.

3. STROMBODES PHILLIPSI. Tab. LXX, figs. 2, 2*a*.

ACERVULARIA BALTICA, *Phillips*, Palæoz. Foss. of Cornw., Devon, &c., p. 13, pl. vii, fig. 18 E, 1841.
ACTINOCYATHUS PHILLIPSII, *D'Orbigny*, Prodr. de Paléont., vol. i, p. 108, 1850.
STROMBODES PHILLIPSI, *Milne Edwards* and *Jules Haime*, Pol. Foss. des Terr. Palæoz. (Arch. du Mus., vol. v), p. 429, 1851.

Corallum much resembling the preceding ones, but differing from them by the existence of well-characterised paliform lobes. The circular elevation corresponding to the limits of the true calice, is also much larger in proportion to the breadth of the corallite, and the septa are less numerous.

Wenlock. Collection of Professor Phillips.

4. STROMBODES DIFFLUENS. Tab. LXXI, figs. 2, 2*a*.

ACERVULARIA BALTICA (pars), *Lonsdale*, in Murchison, Silur. Syst., pl. xvi, figs. 8, 8*a*, 1839.
STROMBODES DIFFLUENS, *Milne Edwards* and *Jules Haime*, Pol. Foss. des Terr. Palæoz. (Arch. du Mus., vol. v), p. 431, 1851.

---

[1] See tab. lxxi, fig. 1.

Corallum massive, subgibbose; epitheca thin. Calicinal circles small, and circumscribing a shallow, but well marked central fossula. Costo-septal radii very delicate, closely set, very flexuous, unequally developed alternately, extending to a considerable distance, and confluent with those of the adjoining individuals. Some appearance of a pseudo-columella produced by the inner extremity of the large septa; some of which are slightly curved. 35 or 40 of these occupy the calicinal space, but exteriorly the number of the septo-costal ridges augments greatly, and varies much. Diameter of the calicinal circles $2\frac{1}{2}$ or 3 lines; depth of the calicinal fossula about half a line. A vertical section shows that the vesicular tissue is rather dense, the cells somewhat unequal in size, and presenting vertical septo-costal striæ, in general distinct, although small.

Much Wenlock.

Specimens are in the Collections of the Museum of Practical Geology, of the Parisian Museum, of Mr. Bowerbank, and M. de Verneuil.

This coral differs from all the other species of the genus Strombodes in not having any trace of exterior walls between the individuals.

9. *Genus* SYRINGOPHYLLUM (p. lxxii).

SYRINGOPHYLLUM ORGANUM. Tab. LXXI, figs. 3, 3*a*, 3*b*.

MADREPORA COMPOSITA, &c., *Fougt*, Amæn. Acad., vol. i, p. 96, tab. iv, fig. 6, and No. 1, 1749.
— ORGANUM, *Linné*, Syst. Nat., edit. 12, p. 1278, 1767.
SARCINULA ORGANON, *Schweigger*, Handb. der Naturg., p. 420, 1820.
— ORGANUM, *Goldfuss*, Petref. Germ., vol. i, p. 73, tab. xxiv, fig. 10, 1826. (Not Lamarck.)
— ORGANON, *Eichwald*, Zool. Spec., vol. i, p. 189, 1829.
— ORGANUM, *Holl*, Handb. der Petref., p. 401, 1830.
— — *De Blainville*, Dic. Sc. Nat., vol. lx, p. 314, 1830.—Man., p. 348.
— — *Morren*, Descr. Cor. Belg., p. 67, 1832.
— — *Hisinger*, Leth. Suec., p. 97, tab. xxviii, fig. 8, 1837.
— — *Eichwald*, Sil. Syst. in Esthland, p. 199, 1840.
ASTREOPORA ORGANUM, *D'Orbigny*, Prodr. de Paléont., vol. i, p. 50, 1850.
SARCINULA ORGANUM, *M'Coy*, Brit. Palæoz. Foss., p. 37, 1851.
SYRINGOPHYLLUM ORGANUM, *Milne Edwards* and *Jules Haime* (Introd., p. lxii, 1850), Pol. Foss. des Terr. Palæoz. (Arch. du Mus., vol. 5), p. 490, 1851.

Corallum massive, astreiform, rather tall; gemmation lateral; calices of unequal size, circular, rather prominent, and placed at a distance from each other about equal to their diameter; costæ thin, but slightly prominent, equally developed, separated by large furrows, straight or slightly flexuous, and extending to the bottom of the inter calicular space, where they join those of the surrounding corallites. Calicular fossula rather shallow; in all the specimens we have seen it was much clogged up with extraneous matter, but there was still some appearance of a styliform, slightly compressed columella, and of a crucial mode of arrangement of the principal septa. Septa (24 or 26,) well developed, rather thick, of unequal size alternately, and slightly exsert. Diameter of the calice about 1 line.

The British specimens examined by us were found in the upper Silurian rocks at Dudley, and in the inferior Silurian deposits at Coniston. Professor M'Coy mentions its existence at Coniston Water Head, Lancashire; Sunny Brow near Coniston; High Haume, Dalton in Furness, Lancashire; Long Steddale, Westmoreland; Applethwaite Common, Westmoreland. It is also found in Gothland, in Groningue, and in Russia.

Specimens are in the Collections of the Paris Museum, Bristol Museum, of the Geological Society of London, and M. de Verneuil.

Professor M'Coy thinks that the name of Sarcinula ought to be applied to this genus, because Lamarck considered the recent coral for which he established it as being identical with the Madrepora Organum of Linné; but we cannot adopt his opinion. For when Lamarck formed his genus Sarcinula he had evidently in view the above-mentioned recent coral, to which alone its characters are applicable, and the blunder he made consists only in the misapplication of the Linnean name. This specimen, which still exists in the public collection of the Parisian Museum, and has been figured in a recent work[1], must therefore receive a new specific name, but its generic name cannot be transferred to a fossil that differs essentially from it, and that Lamarck had never an opportunity of examining.

10. *Genus* LONSDALEIA (p. lxxii[2]).

LONSDALEIA WENLOCKENSIS.

STROMBODES WENLOCKENSIS, *M'Coy*, Ann. and Mag. of Nat. Hist., 2d ser., vol. vi, p. 274, 1850.
— — — *M'Coy*, Brit. Palæoz. Foss., p. 34, pl. i B, fig. 28, 1851.

" *Corallum* forming large, irregular masses of polygonal stems, the mouth of which vary usually from 8 to 10 lines in diameter; boundary walls strong, prominent, vertically sulcated on the inside; stars depressed round the margin of the walls, forming a large circular convexity nearer the centre, within which is a concavity from which rises the thick prominent compound axis; radiating lamellæ 24 in small specimens, 30 in large ones, strongest and most prominent in the circular convexity of the star, where an equal number of small alternate ones disappear; a vertical section shows the thick central axis composed of

---

[1] Milne Edwards, in Règne Anim. de Cuvier, Zooph., pl. 85, fig. 1.
[2] Under the name of Lithostrotium.

irregularly twisted plates; inner area a little narrower than the outer area, from which it is separated by a solid vertical wall, crossed by loose vesicular structure, curving upwards and outwards, one, or rarely two, vesicular plates reaching across the area on each side; vesicular plates of the outer area more curved, slightly smaller, the rows inclining slightly upwards and outwards, scarcely three cells in a row. A star 9 lines in diameter, has the prominent circular portion, 7 lines in diameter, and the prominent axis rather more than 1 line in diameter.

"Not uncommon in the Wenlock limestone near Wenlock, Shropshire." M'Coy, op. cit.

Family CYSTIPHYLLIDÆ, (p. lxxii.)

Genus CYSTIPHYLLUM (p. lxxii).

1. CYSTIPHYLLUM CYLINDRICUM. Tab. LXXII, figs. 2, 2*a*, 2*b*. 2*c*.

> FUNGITES GOTHLANDICUS, &c., *Magnus Bromel*, Act. Liter. Suec., vol. ii, p. 464, No. 18, 1728.
> CYSTIPHYLLUM CYLINDRICUM, *Lonsdale*, in Murchison, Silur. Syst. p. 691, pl. xvi bis, fig. 3, 1839.
> — — *Milne Edwards* and *Jules Haime*, Pol. Foss. des Terr. Palæoz. (Arch. du Mus., vol. v), p. 464, 1851.

Corallum tall, turbinate when young, but becoming cylindrical by growth, and presenting numerous irregular accretion ridges; epitheca strong, and often presenting radici-form tubercles or appendices, which sometimes accidentally unite several individuals together. *Calice* circular, rather shallow, and entirely covered with large irregular vesicles, but still showing some appearance of septal striæ. A vertical section shows that the vesicles are rather small, and very irregular. Height about 2 inches.

Benthall Edge: Ardaun and Cong, Kerry (M'Coy.)

Specimens are in the Collections of the Geological Society of London and of M. de Verneuil.

2. CYSTIPHYLLUM GRAYI. Tab. LXXII, figs. 3, 3*a*.

> CYSTIPHYLLUM GRAYI, *Milne Edwards* and *Jules Haime*, Pol. Foss. des Terr. Palæoz. (Arch. du Mus., vol. v), p. 465, 1851.

This species is very nearly allied to the preceding one, but the vesicles of the calice are much smaller and less irregular, and the lateral circular accretion ridges are much less developed. Height about 2 inches, diameter of the calice 1¼ inch.

Dudley. Collection of the Geological Society, of Mr. Fletcher, Mr. John Gray, &c.

3. CYSTIPHYLLUM SILURIENSE. Tab. LXXII, figs. 1, 1a.

> CYSTIPHYLLUM SILURIENSE (pars), *Lonsdale*, in Murchison, Sil. Syst., p. 691, pl. xv bis,
> 1839. (Not the fig. 2, which is an Omphyma.)
> CYATHOPHYLLUM VESICULOSUM, *Eichwald*, Sil. Syst. in Esthl., p. 201, 1840.
> CYSTIPHYLLUM SILURIENSE, *Milne Edwards* and *Jules Haime*, Pol. Foss. des Terr. Palæoz.
> (Arch. du Mus., vol. v), p. 465, 1851.

Corallum turbinate, short, and very broad; epitheca thick, and presenting some radici-form prolongations. *Calice* subcircular, rather deep, very broad, presenting some obscure indications of septal striæ, and occupied by large unequal vesicles, the structure of which is very distinct in a vertical section. Height about 1½ or 2 inches; diameter of the calice somewhat more.

Wenlock. Dudley (Lonsdale); Ardaun and Cong (M'Coy); Russia (Eichwald). Collection of the Geological Society of London.

The Silurian fossil described by Professor M'Coy, under the name of *Fistulipora decipiens*,[1] much resembles a Heliolites, in which all traces of the septal apparatus have disappeared, and the cœnenchyma does not present the vesicular structure which is characteristic of the genus Fistulipora, but is made up of small vertical tubes, divided like the visceral chambers by numerous horizontal tubulæ.

The same palæontologist has recently added to the list of the British Silurian Corals, two other species, the zoological characters of which are still so imperfectly known, that we cannot give any decided opinion respecting their natural affinities. One of these,

---

[1] M'Coy, Ann. and Mag. of Nat. Hist., 2d ser., vol. vi, p. 285, 1850; ibid., Brit. Palæoz. Foss., p. 11, pl. i c, fig. 1, 1851.

"*Corallum* forming hemispherical or sub-cylindrical masses, three or four inches in diameter, concentrically wrinkled at base; cell-tubes straight, sub-parallel, with moderately thick walls, leaving clearly definite, circular, smooth-edged cells in the transverse section, very regular in size and disposition; usually slightly less than half a line in diameter, and averaging rather less than their diameter in the shortest line between adjacent cells, in which line there are usually two, or more, rarely three, of the intermediate vesicular cellules; about eighteen of the intermediate or polygonal cellules in the space of two lines; diaphragms in the small tubes slightly more or less than their diameter apart, two interdiaphragmal spaces in the large tubes slightly exceeding the diameter.

"Wenlock Limestone, near Aymestry, Herefordshire." (M'Coy, op. cit.)

Protovirgularia dichotoma,[1] may probably belong to the group of Sertularina, the other, *Pyritonema fasciculus,*[2] is a cylindrical bundle of small vertical tubes.

[1] Ann. and Mag. of Nat. Hist., 2d ser., vol. vi, p. 272, 1850; Brit. Palæoz. Foss., p. 10, pl. i B, figs. 11 and 12, 1851.

"One specimen, about two and a half inches long, branches twice at an angle of about 30°, and shows all the pinnules extended at right angles to the capillary axis, with a gentle upward curvature, like the living *Virgularia* in the same state; another simple fragment about the same length has them half extended, being nearly straight, and oblique to the axis; a third fragment has them quite contracted, resembling a bit of narrow braid, exactly like the contracted state of the recent *Virgularia mirabilis.* This one shows very plainly the transverse cell-ridging. Width rather less than one line; four pinnæ in the space of two lines.

"In the slate at Rockerby, Dumfriesshire." (M'Coy, loc. cit.)

[2] Ann. and Mag. of Nat. Hist., 2d ser., vol. vi, p. 273, 1850; Brit. Palæoz. Foss. p. 10, pl. i B, fig. 13.

"I have proposed the above name for a singular fragment of a fossil from the dark limestone of Tre Gil, S. of Llandeilo. It is nearly straight, about two and a half inches long, four lines wide, and one and a half line thick, and marked longitudinally with coarse thread-like ridges, about the third of a line in diameter, occasionally cut by small sharp transverse wrinkles; the whole having some resemblance to an *Ichthyodorulite* (*onchus* or *ctenacanthus*). On first seeing the specimen, I doubted this reference, from observing that the ridges, instead of being merely superficial, thicker, and more numerous at one end, as they should be on this view, seemed equally thick at each end, and clearly not in one plane, but those at the surface of one part plunging into the mass and giving place to others emerging from it. Owing to the skill and kindness of Mr. Anthony, of Caius College, two sections for the microscope were prepared, which proved that the whole mass was really a bundle of thread-like rods of silica, corresponding exactly in diameter with the external ridges, the sections of which exactly correspond with the others in the interior; the siliceous fibres are solid, cylindrical, with slight occasional transverse rugosities; they are less than their own diameter apart, and the interstices shew no organisation under a magnifying power of 300 diameters, the limestone being of a finer texture, and lighter colour than that of the matrix, as if there had been originally a soft animal matter in the spaces between, which kept out the coarse calcareous mud, but the space occupied by which became filled with fine material by percolation on its decomposition." (M'Coy, loc. cit.)

TABLE

OF THE

BRITISH FOSSIL CORALS

DESCRIBED AND FIGURED IN THIS WORK.

CRAG.

Sphenotrochus intermedius, p. 2, tab. i, fig. 1.
Flabellum Woodi, p. 6, tab. i, fig. 2.
Cryptangia Woodi, p. 8, tab. i, fig. 4.
Balanophyllia calyculus, p. 9, tab. i, fig. 3.

LONDON CLAY.

Turbinolia sulcata, p. 13, tab. iii, fig. 3.
   Dixoni, p. 15, tab. iii, fig. 1.
   Bowerbanki, p. 16, tab. ii, fig. 3.
   Fredericana, p. 17, tab. iii, fig. 2.
   humilis, p. 18, tab. iii, fig. 4.
   minor, p. 19, tab. ii, fig. 5.
   firma, p. 20, tab. ii, fig. 4.
   Prestwichi, p. 20, tab. iii, fig. 5.
Leptocyathus elegans, p. 21, tab. iii, fig. 6.
Trochocyathus sinuosus, p. 22.
Paracyathus crassus, p. 23, tab. iv, fig. 1.
   caryophyllus, p. 24, tab. iv, fig. 2.
   brevis, p. 25, tab. iv, fig. 3.
Dasmia Sowerbyi, p. 25, tab. iv, fig. 4.
Oculina conferta, p. 27, tab. ii, fig. 2.
Diplohelia papillosa, p. 28, tab. ii, fig. 1.
Stylocoenia emarciata, p. 30, tab. v, fig. 1.
   monticularia, p. 32, tab. v. fig. 2.
Astrocoenia pulchella, p. 33, tab. v, fig. 3.

Stephanophyllia discoïdes, p. 34, tab. iv, fig. 3.
Balanophyllia desmophyllum, p. 35, tab. vi, fig. 1.
Dendrophyllia dendrophylloides, p. 36, tab. vi, fig. 2.
Stereopsammia humilis, p. 37, tab. v, fig. 4.
Litharæa Websteri, p. 38, tab. vii, fig. 1.
(Holaræa) Axopora parisiensis, p. 40, tab. vi, fig. 2.
Graphularia Wetherelli, p. 41, tab. vii, fig. 4.
Mopsea costata, p. 42, tab. vii, fig. 3.
Websteria crisioides, p. 42, tab. vii, fig. 5.

UPPER CHALK.

(Cyathina) Caryophyllia lævigata, p. 44, tab. ix, fig. 1.
Parasmilia centralis, p. 47, tab. viii, fig. 1.
   Mantelli, p. 49, tab. viii, fig. 2.
   cylindrica, p. 50, tab. viii, fig. 5.
   Fittoni, p. 50, tab. ix, fig. 2.
   serpentina, p. 51, tab. viii, fig. 3.
Cœlosmilia laxa, p. 52, tab. viii, fig. 4.

LOWER CHALK.

Synhelia Sharpeana, p. 53, tab. ix, fig. 3.
Stephanophyllia Bowerbanki, p. 54, tab. ix, fig. 4.

UPPER GREEN SAND.

Peplosmilia Austeni, p. 57, tab. x, fig. 1.
(Trochosmilia) Smilotrochus tuberosus, p. 58, tab. x, fig. 2.

Parastræa stricta, p. 59, tab. x, fig. 3.
Micrabacia coronula, p. 60, tab. x, fig. 4.

GAULT.

(Cyathina) Caryophyllia Bowerbanki, p. 61, tab. xi, fig. 1.
Cyclocyathus Fittoni, p. 63, tab. xi, fig. 3.
Trochocyathus conulus, p. 63, tab. xi, fig. 5.
        Harveyanus, p. 65, tab. xi, fig. 4.
        ? Konigi, p. 66.
        ? Warburtoni, p. 67.
Bathycyathus Sowerbyi, p. 67, tab. xi, fig. 2.
Trochosmilia sulcata, p. 68, tab. xi, fig. 6.

LOWER GREEN SAND.

Holocystis elegans, p. 70, tab. x, fig. 5.

PORTLAND STONE.

Isastræa oblonga, p. 73, tab. xii, fig. 1.

CORAL RAG.

Stylina tubulifera, p. 76, tab. xiv, fig. 3.
      Labechei, p. 79, tab. xv, fig. 1.
Montlivaultia dispar, p. 80, tab. xiv, fig. 2.
Thecosmilia annularis, p. 84, tab. xiii, fig. 1, and tab. xiv, fig. 1.
Rhabdophyllia Phillipsi, p. 87, tab. xv, fig. 3.
Calamophyllia Stokesi, p. 89, tab. xvi, fig. 1.
Cladophyllia Conybearei, p. 91, tab. xvi, fig, 2.
Goniocora socialis, p. 92, tab. xv, fig. 2.
Isastræa explanata, p. 94, tab. xviii, fig. 1.
      Greenoughi, p. 96, tab. xviii, fig. 2.
Thamnastræa arachnoides, p. 97, tab. xvii, fig. 1.
      concinna, p. 100, tab. xviii, fig. 3.
Comoseris irradians, p. 101, tab. xix, fig. 1.
Protoseris Waltoni, p. 103, tab. xx, fig. 1.

GREAT OOLITE.

Stylina conifera, p. 105, tab. xxi, fig. 2.
      solida, p. 105, tab. xxii, fig. 3, and p. 128.
      Ploti, p. 106, tab. xxiii, fig. 1.
Cyathophora Lucensis, p. 107, tab. xxx, fig. 5.
      Pratti, p. 108, tab. xxi, fig. 3.
Convexastræa Waltoni, p. 109, tab. xxiii, figs. 5 and 6.
Montlivaultia Smithi, p. 110, tab. xxi, fig. 1.

Montlivaultia Waterhousei, p. 111, tab. xxvii, fig. 7.
Calamophyllia radiata, p. 111, tab. xxii, fig. 1.
Cladophyllia Babeana, p. 113, tab. xxii, fig. 2.
Isastræa Conybearei, p. 113, tab. xxii, fig. 4.
      limitata, p. 114, tab. xxiii, fig. 2, and tab. xxiv, figs. 4 and 5.
      explanulata, p. 115, tab. xxiv, fig. 3.
      serialis, p. 116, tab. xxiv, fig. 2.
Clausastrea Pratti, p. 117, tab. xxii, fig. 5.
Thamnastræa Lyelli, p. 118, tab. xxi, fig. 4.
      mammosa, p. 119, tab. xxiii, fig. 3.
      scita, p. 119, tab. xxiii, fig. 4.
      Waltoni, p. 120, tab. xxix, fig. 4.
Anabacia complanata (orbulites), p. 120, tab. xxix, fig. 3, and p. 142.
Comoseris vermicularis, p. 122, tab. xxiv, fig. 1, and p. 143.
Microsolena regularis, p. 122, tab. xxv, fig. 6.
      excelsa, p. 124, tab. xxv, fig. 5.

INFERIOR OOLITE.

Discocyathus Eudesi, p. 125, tab. xxix, fig. 1.
Trochocyathus Magnevilleanus, p. 126, tab. xxvi, fig. 1.
Axosmilia Wrighti, p. 128, tab. xxvii, fig. 6.
Montlivaultia trochoides, p. 129, tab. xxvi, figs. 2, 3, and 10, and tab. xxvii, figs. 2 and 4.
      tenuilamellosa, p. 130, tab. xxvi, fig. 11.
      Stutchburyi, p. 131, tab. xxvii, figs. 3 and 5.
      Wrighti, p. 131, tab. xxvi, fig. 12.
      cupuliformis, p. 132, tab. xxvii, fig. 1.
      Labechei, p. 132, tab. xxvi, fig. 5.
      lens, p. 133, tab. xxvi, figs. 7 and 8.
      depressa, p. 134, tab. xxix, fig. 5.
Thecosmilia gregaria, p. 135, tab. xxviii, fig. 1.
Latimæandra Flemingi, p. 136, tab. xxvii, fig. 9.
      Davidsoni, p. 137, tab. xxvii, fig. 10.
Isastræa Richardsoni, p. 138, tab. xxix, fig. 1.
      tenuistriata, p. 138, tab. xxx, fig. 1.
      Lonsdalei, p. 139.
Thamnastræa Defranceana, p. 139, tab. xxix, figs. 3 and 4.
      Terquemi, p. 140, tab. xxx, fig. 2.
      mettensis, p. 141, tab. xxx, fig. 3.
      fungiformis, p. 141, tab. xxx, fig. 4.
      M'Coyi, p. 141, tab. xxix, fig. 2.

Anabacia hemispherica, p. 142, tab. xxv, fig. 2.
Zaphrentis? Waltoni, p. 143, tab. xxvii, fig. 8.

LIAS.

Thecocyathus Moorei, p. 144, tab. xxx, fig. 6.
Trochocyathus? primus, p. 145, tab. xxx, fig. 8.
Cyathophyllum? novum, p. 145, tab. xxx, fig. 7.

PERMIAN FORMATION.

Chaetetes? Mackrothi, p. 147.
   ? columnaris, p. 148.
   ? Buchana, p. 148.
Polycœlia Donatiana, p. 149.
   profunda, p. 149.

MOUNTAIN LIMESTONE.

Fistulipora minor, p. 151.
   major, p. 152.
Propora? cyclostoma, p. 152.
Favosites parasitica, p. 153, tab. xlv, fig. 2.
(Chaetetes) Favosites? tumida, p. 159, tab. xlv, fig. 3.
Michelinia favosa, p. 154, tab xliv, fig. 2.
   tenuisepta, p. 155, tab. xliv, fig. 1.
   megastoma, p. 156, tab. xliv, fig. 3.
   antiqua, p. 156.
Alveolites? septosa, p. 157, tab. xlv, fig. 5.
Alveolites? depressa, p. 158, tab. xlv, fig. 4.
Chaetetes radians, p. 158.
Beaumontia Egertoni, p. 160, tab. xlv, fig. 1.
   laxa, p. 161.
Syringopora ramulosa, p. 161, tab. xlvi, fig. 3.
   reticulata, p. 162, tab. xlvi, fig. 1.
   geniculata, p. 163, tab. xlvi, figs. 2 and 4.
   catenata, p. 164.
   laxa, p. 164.
Rhabdopora megastoma, p. 165.
Pyrgia Labechei, p. 166, tab. xlvi, fig. 5.
Cyathaxonia cornu, p. 166.
Zaphrentis cornucopiæ, p. 167.
   Phillipsi, p. 168, tab. xxxiv, fig. 2.
   Griffithi, p. 169, tab. xxxiv, fig. 3.
   Enniskilleni, p. 170, tab. xxxiv, fig. 1.
   Bowerbanki, p. 170, tab. xxxiv, fig. 4.
   patula, p. 171.
   cylindrica, p. 171, tab. xxxv, fig. 1.

Zaphrentis? subibicina, p. 172.
Amplexus coralloides, p. 173, tab. xxxvi, fig. 1.
   cornubovis, p. 174.
   nodulosus, p. 175.
   spinosus, p. 176.
   Henslowi, p. 176, tab. xxxiv, fig. 5.
Lophophyllum? eruca, p. 177.
Cyathophyllum Murchisoni, p. 178, tab. xxxiii, fig. 3.
   Wrighti, p. 179, tab. xxxiv, fig. 6.
   Stutchburyi, p. 179, tab. xxxi, figs. 1 and 2, and tab. xxxiii, fig. 4.
   regium, p. 180, tab. xxxii, figs. 1, 2, 3, and 4.
   parricida, p. 181, tab. xxxvi, fig. 1.
   ? pseudovermiculare, p. 182.
   dianthoides, p. 182.
   Archiaci, p. 183, tab. xxxiv, fig. 7.
Campophyllum Murchisoni, p. 184, tab. xxxvi, figs. 2 and 3.
Clisiophyllum turbinatum, p. 184, tab. xxxiii, figs. 1 and 2.
   coniseptum, p. 185, tab. xxxvi, fig. 5.
   Bowerbanki, p. 186, tab. xxxvi, fig. 4.
   Keyserlingi, p. 186.
   costatum, p. 187.
   bipartitum, p. 187.
Aulophyllum fungites, p. 188, tab. xxxvi, fig. 3.
   Bowerbanki, p. 189, tab. xxxviii, fig. 1.
Lithostrotium basaltiforme, p. 190, tab. xxxviii, fig. 3
   ensifer, p. 193, tab. xxxviii, fig. 2.
   aranea, p. 193, tab. xxxix, fig. 1.
   Portlocki, p. 194, tab. xlii, fig. 1.
   M'Coyanum, p. 195, tab. xlii, fig. 2.
   ? concinnum, p. 195.
   ? septosum, p. 196.
   decipiens, p. 196.
   junceum, p. 196, tab. xl, fig. 1.
   Martini, p. 197, tab. xl, fig. 2.
   irregulare, p. 198, tab. xli, fig. 1.
   affine, p. 200, tab. xxxix, fig. 2.
   Phillipsi, p. 201, tab. xxxix, fig. 3.
   ? derbiense, p. 201.
   major, p. 201.
   arachnoideum, p. 202.
   Flemingi, p. 203.
Phillipsastræa radiata, p. 203, tab. xxxvii, fig. 2.
   tuberosa, p. 204.
Petalaxis Portlocki, p. 204, tab. xxxviii, fig. 4.

Axophyllum radicatum, p. 205.
Lonsdaleia floriformis, p. 205, tab. xliii, figs. 1 and 2.
    papillata, p. 207.
        rugosa, p. 208, tab. xxxviii, fig. 5.
        duplicata, p. 209.
Mortieria vertebralis, p. 209.
Heterophyllia grandis, p. 210.
        ornata, p. 210.

DEVONIAN FORMATION.

Heliolites porosa, p. 212, tab. xlvii, fig. 1.
Battersbyia inequalis, p. 213, tab. xlvii, fig. 2.
Favosites Goldfussi, p. 214, tab. xlvii, fig. 3.
    reticulata, p. 215, tab. xlviii, fig. 1.
    cervicornis, p. 216, tab. xlviii, fig. 2.
    dubia, p. 216.
    fibrosa, p. 217, tab. xlviii, fig. 3.
Emmonsia hemispherica, p. 218, tab. xlviii, fig. 4.
Alveolites suborbicularis, p. 219, tab. xlix, fig. 1.
    Battersbyi, p. 220, tab. xlix, fig. 2.
    vermicularis, p. 220, tab. xlviii, fig. 5.
    compressa, p. 221, tab. xlix, fig. 3.
Metriophyllum Battersbyi, p. 222, tab. xlix, fig. 4.
Amplexus tortuosus, p. 222, tab. xlix, fig. 5.
Hallia Pengellyi, p. 223, tab. xlix, fig. 6.
Cyathophyllum ceratites, p. 224, tab. l, fig. 2.
        Roemeri, p. 224, tab. l, fig. 3.
        obtortum, p. 225, tab. xlix, fig. 7.
        damnoniense, p. 225, tab. l, fig. 1.
        celticum, p. 226.
        Bucklandi, p. 226.
        helianthoides, p. 227, tab. li, fig. 1.
        hexagonum, p. 228, tab. l, fig. 4.
        cæspitosum, p. 229, tab. li, fig. 2.
        boloniense, p. 230, tab. lii, fig. 1.
        Marmini, p. 231, tab. lii, fig. 4.
        Sedgwicki, p. 231, tab. lii, fig. 3.
        æquiseptatum, p. 232, tab. lii, fig. 1.
        ? gracile, p. 232.
Endophyllum Bowerbanki, p. 233, tab. liii, fig. 1.
        abditum, p. 233, tab. lii, fig. 6.
Campophyllum flexuosum? p. 234.
Pachyphyllum devoniense, p. 234, tab. lii, fig. 5.
Chonophyllum perfoliatum, p. 235, tab. l, fig. 5.
Heliophyllum Halli, p. 235, tab. li, fig. 3.
Acervularia Goldfussi, p. 236, tab. liii, fig. 3.
        coronata, p. 237, tab. liii, fig. 4.

Acervularia intercellulosa, p. 237, tab. liii, fig. 2.
        pentagona, p. 238, tab. liii, fig. 5.
        limitata, p. 238, tab. liv, fig. 1.
        Battersbyi, p. 239, tab. liv, fig. 2.
        Roemeri, p. 239, tab. liv, fig. 3.
Smithia Hennahi, p. 240, tab. liv, fig. 4.
        Pengellyi, p. 241, tab. lv, fig. 1,
        Bowerbanki, p. 241, tab. lv, fig. 2.
Spongophyllum Sedgwicki, p. 242, tab. lvi, fig. 2.
Syringophyllum cantabricum, p. 242, tab. lv, fig. 3.
Cystiphyllum vesiculosum, p. 243, tab. lvi, fig. 1.

SILURIAN FORMATION.

Palæocyclus porpita, p. 246, tab. lvii, fig. 1.
        præacutus, p. 247, tab. lvii, fig. 2.
        Fletcheri, p. 248, tab. lvii, fig. 3.
        rugosus, p. 248, tab. lvii, fig. 4.
Heliolites interstincta, p. 249, tab. lvii, fig. 9.
        Murchisoni, p. 250, tab. lvii, fig. 6.
        megastoma, p. 251, tab. lviii, fig. 2.
        Grayi, p. 252, tab. lviii, fig. 1.
        inordinata, p. 253, tab. lvii, fig. 7.
Plasmopora petaliformis, p. 253, tab. lix, fig. 1.
        scita, p. 254, tab. lix, fig. 2.
Propora tubulata, p. 255, tab. lix, fig. 3.
Favosites gothlandica, p. 256, tab. lx, fig. 1.
        aspera, p. 257, tab. lx, fig. 3.
        multipora, p. 258, tab. lx, fig. 4.
        Forbesi, p. 258, tab. lx, fig. 2.
        Hisingeri, p. 259, tab. lxi, fig. 1.
        cristata, p. 260, tab. lxi, figs. 3 and 4.
        fibrosa, p. 261, tab. lxi, fig. 5.
        crassa, p. 261.
Alveolites Labechei, p. 262, tab. lxi, fig. 6.
        Grayi, p. 262, tab. lxi, fig. 2.
        repens, p. 263, tab. lxii, fig. 1.
        seriatoporoides, p. 263, tab. lxii, fig. 2.
Monticulipora petropolitana, p. 264.
        papillata, p. 266, tab. lxii, fig. 4.
        Fletcheri, p. 267, tab. lxii, fig. 3.
        pulchella, p. 267, tab. lxii, fig. 5.
        ? Bowerbanki, p. 268, tab. lxiii, fig. 1.
        explanata, p. 268.
        lens, p. 269.
Labecheia conferta, p. 269, tab. lxii, fig. 6.
Halysites catenularia, p. 270, tab lxiv, fig. 1.
        escharoïdes, p. 272, tab. lxiv, fig. 2.

Syringopora bifurcata, p. 273, tab. lxiv, fig. 3.

        fascicularis, p. 274, tab. lxv, fig. 1.

        serpens, p. 275, tab. lxv, fig. 2.

Coenites juniperinus, p. 276, tab. lxv, fig. 4.

     intertextus, p. 276, tab. lxv, fig. 5.

     linearis, p. 277, tab. lxv, fig. 3.

     labrosus, p. 277, tab. lxv, fig. 6.

     ? strigatus, p. 278.

Thecia Swindernana, p. 278, tab. lxv, fig. 7.

     Grayana, p. 279, tab. lxv, fig. 8.

Cyathaxonia? siluriensis, p. 279.

Aulacophyllum mitratum, p. 280, tab. lxvi, fig. 1.

Cyathophyllum? Loveni, p. 280, tab. lxvi, fig. 2.

       angustum, p. 281, tab lxvi, fig. 4.

       pseudoceratites, p. 282, tab. lxvi, fig. 3.

       articulatum, p. 282, tab. lxvii, fig. 1.

       truncatum, p. 284, tab. lxvi, fig. 5.

Cyathophyllum flexuosum, p. 285, tab. lxvii, fig. 2.

      trochiforme, p. 285.

      vortex, p. 286.

Cyathophyllum (Turbinolopsis) elongatum, p. 286, tab. lxvi, fig. 6.

Omphyma turbinata, p. 287, tab. lxix, fig. 1.

      subturbinata, p. 288, tab. lxviii, fig. 1.

      Murchisoni, p. 289, tab. lxvii, fig. 3.

Goniophyllum Fletcheri, p. 290, tab. lxviii, fig. 3.

      pyramidale, p. 290.

Chonophyllum perfoliatum? p. 291, tab. lxviii, fig. 2.

Ptychophyllum patellatum, p. 291, tab. lxvii, fig. 4.

Acervularia luxurians, p. 292, tab. lxix, fig. 2.

Strombodes typus, p. 293, tab. lxxi, fig. 1.

      Murchisoni, p. 293, tab. lxx, fig. 1.

      Phillipsi, p. 294, tab. lxx, fig. 2.

      diffluens, p. 294, tab. lxxi, fig. 2.

Syringophyllum organum, p. 295, tab. lxxi, fig. 3.

Lonsdaleia wenlockensis, p. 296.

Cystiphyllum cylindricum, p. 297, tab. lxxii, fig. 2.

      Grayi, p. 297, tab. lxxii, fig. 3.

      siluriense, p. 298, tab. lxxii, fig. 1.

Fistulipora? decipiens, p. 298.

Protovirgularia dichotoma, p. 299.

Pyritonema fasciculus, p. 299.

GENERAL INDEX.[1]

A.

	PAGE
ACANTHASTRÆA . . .	xlii
ACANTHOCYATHUS . . .	xiii
ACERVULARIA . . .	lxx
Acervularia ananas, *M'Coy* . .	292
„ „ *Michelin* .	238
„ baltica (pars), *Lonsdale*	293, 294
„ „ *Phillips* .	294
Acervularia Battersbyi . .	239
„ coronata . .	237
„ Goldfussi . .	236
„ intercellulosa . .	237
„ limitata . .	238
„ luxurians . .	292
„ pentagona . .	238
„ Roemeri . .	239
ACROHELIA . . .	xx
Actinocyathus balticus, *D'Orbigny* .	293
„ crenularis, *D'Orbigny*	181
„ Hennahii, *D'Orbigny*	240
„ Phillipsii, *D'Orbigny*	294
AGARICIA . . .	xlix
Agaricia lobata (pars), *Morris* .	76, 100
„ Swinderniana, *Goldfuss* .	278
ALCYONIUM . . .	lxxvii
ALVEOLITES . . .	lx
Alveolites Battersbyi . .	220
Alveolites Buchiana, *King* .	148
„ celleporatus, *D'Orbigny* .	216
„ cervicornis, *Michelin* .	217
„ „ *Blainville* .	216
Alveolites compressa . .	221

	PAGE
Alveolites depressa . .	158
Alveolites dubia, *Blainville* .	216
„ escharoidea, *Blainville* .	219
„ escharoides, *Lamarck* .	219
„ fibrosa, *Lonsdale* .	217
„ fibrosus, *D'Orbigny* .	217
Alveolites Grayi . .	262
Alveolites hemispherica, *D'Orbigny* .	218
„ irregularis, *De Koninck* .	159
Alveolites Labechei . .	262
Alveolites Lonsdalei, *D'Orbigny* .	260
„ parisiensis, *Michelin* .	40
„ producti, *Geinitz* .	148
Alveolites repens . .	263
Alveolites reticulata, *Blainville* .	215
„ scabra, *Michelin* .	159
Alveolites septosa . .	157
„ ? seriatoporoides .	263
Alveolites spongites, *D'Orbigny* .	215
„ „ *Steininger* .	219
Alveolites suborbicularis .	219
Alveolites tuberosa, *D'Orbigny* .	219
„ tumida, *Michelin* .	159
Alveolites vermicularis . .	220
ALVEOPORA . . .	lvii
Alveopora microsolena, *M'Coy* .	122
AMPHIHELIA . . .	xxi
AMPLEXUS . . .	lxiii
Amplexus coralloides . .	173
„ cornubovis . .	174
„ Henslowi . .	176
„ nodulosus . .	175

[1] The names printed in *italics* are those adopted in this work, the others are those quoted as synonyms.

PAGE

Amplexus serpuloides, *De Koninck* . 175
„ Sowerbyi, *Phillips* . . 173
Amplexus spinosus . . . 176
„ *tortuosus* . . . 222
Amplexus tortuosus, *M'Coy* . . 177
„ Yandelli (pars), *Milne Edwards* and
Jules Haime . . 222
ANABACIA xlvii
Anabacia bajociana, *D'Orbigny* . 121
Anabacia complanata (orbulites) . 120, 142
„ *hemispherica* . . 142
(ANGIA) CYLICIA . . . xliii
ANISOPHYLLUM . . . lxvi
ANTHELIA lxxvii
Anthophyllum obconicum, *Goldfuss* . 80
ANTIPATHES lxxiii
APHRASTRÆA xliii
Aplocyathus Magnevillianus, *D'Orbigny* . 126
APLOSMILIA xxvi
Arachnophyllum Hennahii, *M'Coy* . 240
„ typus, *M'Coy* . 293
ARÆACIS xxiii
ASPIDISCUS xxxv
ASTRÆA xxxix
ASTRÆOPORA . . . liv
ASTRANGIA xliv
Astrea ananas, *Hisinger* . . 292
„ „ *Roemer* . . 229
„ annularis, *Conybeare* and *Phillips* . 97
„ arachnoides, *Defrance* . . 190
„ „ *Fleming* . . 97
„ aranea, *M'Coy* . . 193
„ basaltiformis, *Conybeare* and *W.*
Phillips . 190
„ „ *Roemer* . 236
„ cadomensis, *Michelin* . . 143
„ carbonaria, *M'Coy* . . 180
„ concinna, *Goldfuss* . . 100
„ corona, *Morren* . . 249
„ crenularis, *M'Coy* . . 181
„ cylindrica, *Defrance* . . 30
„ decorata, *Michelin* . . 30
„ Defranciana, *Michelin* . . 139
„ elegans, *Fitton* . . 70
„ „ *Goldfuss* . . 59
„ emarciata, *Lamarck* . . 30
„ emarcida, *Fischer* . . 206

PAGE

Astrea escharoides, *Goldfuss* . . 59
„ explanata, *Goldfuss* . . 94
„ explanulata, *M'Coy* . . 115
„ favosioides, *Phillips* . . 94
„ florida, *Defrance* . . 205
„ gracilis, *M'Coy* . . 120
„ helianthoides, *M'Coy* . . 94
„ „ *Steininger* . 227
„ Hennahii (pars), *Lonsdale* 240, 241
„ „ (pars), *Phillips* . 203
„ „ *Roemer* . 239
„ hexagona, *Portlock* . . 190
„ „ var. minor, *Portlock* . 193
„ „ *Steininger* . 228
„ hystrix, *Defrance* . . 32
„ inæqualis, *Phillips* . . 104
„ intercellulosa, *Phillips* . . 237
„ irregularis, *Portlock* . . 194
„ limitata, *Lamouroux* . . 114
„ mamillaris, *Fischer* . . 206
„ micraston, *Phillips* . . 100
„ parallela, *Roemer* . . 239
„ pentagona, *Fischer* . . 206
„ „ (pars), *Lonsdale* . 238
„ porosa, *Goldfuss* . . 212
„ „ *Hisinger* . . 249
„ Portlocki, *Bronn* . . 194
„ stylopora, *Goldfuss* . . 30
„ tenuistriata, *M'Coy* . . 138
„ tisburiensis, *Fitton* . . 73
„ tubulifera, *Phillips* . . 76
„ tubulosa, *Morris* . . 76
„ varians, *Roemer* . . 100
„ Websteri, *Bowerbank* . . 38
Astreopora antiqua, *M'Coy* . . 152
„ expatiata, *D'Orbigny* . . 278
„ grandis, *D'Orbigny* . . 255
„ Lonsdalei, *D'Orbigny* . 255
„ organum, *D'Orbigny* . . 295
„ petaliformis, *D'Orbigny* . 253
„ tubulata, *D'Orbigny* . . 255
Astrocerium constrictum, *Hall* . . 261
„ venustum, *Hall* . . 259
ASTROCŒNIA . . . xxx
Astrocænia pulchella . . 33
ASTROHELIA . . . xx
ASTROIDES xli

PAGE

ASTROBIA . . . xxxvii
AULACOPHYLLUM . . . lxvii
Aulacophyllum mitratum . . 280
AULOPHYLLUM . . . lxx
Aulophyllum Bowerbanki . . 189
,, fungites . . . 188
Aulophyllum prolapsum, Milne Edwards and
Jules Haime . 188
AULOPORA . . : . lxxvi
Aulopora campanulata, M'Coy . 164
,, conglomerata, Lonsdale . 275
,, gigas, M'Coy . . 164
,, Lonsdalei, D'Orbigny . 275
,, serpens, Blainville . 274
,, ,, Wahlenberg . 273
,, tubæformis, Lonsdale . 274
,, Voigtiana, King . 150
AXOHELIA . . . xxi
AXOPHYLLUM . . . lxxii
Axophyllum radicatum . . 205
AXOPORA . . . lix
AXOSMILIA . . . xxvi
Axosmilia Wrighti . . 128

B.

BALANOPHYLLIA . . . lii
Balanophyllia calyculus . . 9
,, desmophyllum . 35
BARYASTRÆA . . . xli
BARYPHYLLUM . . . lxvi
BARYSMILIA . . . xxvii
BATHYCYATHUS . . . xiii
Bathycyathus Sowerbyi . . 67
Battersbyia inæqualis . . 213
Beaumontia Egertoni . . 160
,, laxa . . 161
BEBRYCE . . . lxxx
BLASTOTROCHUS . . . xviii
BRACHYCYATHUS . . . xiii
BRIAREUM . . . lxxxi

C.

CALAMOPHYLLIA . . . xxxiii
Calamophyllia nodosa, D'Orbigny . 88

PAGE

Calamophyllia prima, D'Orbigny . . 113
Calamophyllia radiata . . . 111
,, Stokesi . . . 89
Calamophyllia undata, D'Orbigny . . 88
Calamopora alveolaris (pars), Goldfuss . 257
,, basaltica (pars), Goldfuss . 258
,, ,, Hisinger . 256
,, dentifera, Phillips . . 153
,, fibrosa, Eichwald . . 264
,, ,, (pars), Goldfuss . 264
,, ,, var. gracilis, Goldfuss . 263
,, ,, var. tuberoso-ramosa,
Goldfuss . . 217
,, ,, De Koninck . . 159
,, Gothlandica (pars), Goldfuss 214, 256
,, ,, Hisinger . 258
,, imbricata, Michelin . . 219
,, incrustans, Phillips . . 153
,, inflata, De Koninck . . 159
,, Mackrothii, Geinitz . . 147
,, megastoma, Phillips . . 156
,, minutissima, Castelnau . 259
,, parasitica, Phillips . . 153
,, polymorpha, var. divaricata, Goldfuss . . 216
,, ,, var. gracilis, Goldfuss 216
,, ,, Hisinger . . 260
,, ,, Roemer . . 216
,, spongites, Eichwald . . 262
,, ,, var. ramosa, Goldfuss 215
,, ,, var. tuberosa, Goldfuss 219
,, ,, Hisinger . . 260
,, squamosa, Michelin . . 219
,, suborbicularis, Michelin . 219
,, tenuisepta, Phillips . . 155
,, tumida, Phillips . . 159
Calceola pyramidalis, Girard . . 290
Calice . . . vi
Calophyllum Donatianum, King . . 149
,, spinosum . . 176
CAMPOPHYLLUM . . . lxviii
Campophyllum flexuosum . . 234
,, Murchisoni . . 184
Caninia cornubovis, Michelin . . 174
,, cornucopiæ, Michelin . . 167
,, gigantea, Michelin . . 171
,, lata, M'Coy . . 287

GENERAL INDEX. 309

	PAGE
Caninia patula, *Michelin* . . 171	
,, subibicina, *M'Coy* . . 172	
,, turbinata, *M'Coy* . . 288	
Caryophyllia affinis, *Fleming* . . 200	
,, annularis, *Fleming* . . 84	
,, centralis, *Fleming* . . 47	
,, cespitosa, *Conybeare* and *W.*	
Phillips . . . 91	
,, conulus, *Phillips* . . 63	
,, convexa, *Phillips* . . 143	
,, cylindrica, *Phillips* . . 84	
,, dubia, *Blainville* . . 229	
,, duplicata, *Fleming* . . 209	
,, explanata, *Hisinger* . . 284	
,, fasciculata, *Blainville* . . 158	
,, ,, *Fleming* . . 197	
,, flexuosa, *Lonsdale* . . 285	
,, juncea, *Fleming* . . 196	
,, quadrifida, *Howse* . . 150	
,, sexdecimalis, *De Koninck* . 196	
,, truncata, *Hisinger* . . 292	
,, turbinata (pars), *Brongniart* . 287	
Caryophyllœa, *Conybeare* and *W. Phillips* . 198	
Catenipora agglomerata, *Hall* . . 271	
,, approximata, *Eichwald* . . 270	
,, axillaris, *Lamarck* . . 275	
,, communicans, *Eichwald* . 270	
,, compressa, *Edwards* and *J. Haime* 271	
,, distans, *Eichwald* . . 270	
,, escharoides, *Blainville* . . 270	
,, ,, *Lamarck* . . 272	
,, exilis, *Eichwald* . . 272	
,, gracilis, *Edwards* and *J. Haime* . 271	
,, labyrinthica, *Goldfuss* . . 270	
,, Michelini, *Castelnau* . . 271	
,, or tubipora, *Taylor* . . 270	
,, reticulata, *Eichwald* . . 272	
,, tubulosa, *Lamouroux* . . 270	
CAVERNULARIA . . lxxxiv	
Cellastrea hystrix, *Blainville* . . 32	
CERATOTROCHUS . . xvii	
Ceriopora granulosa, *Goldfuss* . . 269	
,, inflata, *D'Orbigny* . . 159	
,, irregularis, *D'Orbigny* . 159	
,, oculata, *Goldfuss* . . 261	
,, tumida, *D'Orbigny* . . 159	
CESPITULARIA . . lxxviii	

	PAGE
CHAETETES lxi	
Chaetetes? Bowerbanki, *Milne Edwards* and	
Jules Haime . . 268	
(*Chaetetes ?*) *Monticulipora Buchana* . 148	
Chaetetes capillaris, *Keyserling* . . 158	
(*Chaetetes ?*) *Monticulipora columnaris* . 148	
,, cylindricus, *Fischer* . . 158	
,, dilatatus, *Fischer* . . 158	
,, excentricus, *Fischer* . . 158	
,, Fletcheri, *Milne Edwards* and *Jules*	
Haime . . . 267	
,, jubatus, *Fischer* . . 158	
,, Koninckii, *D'Orbigny* . . 159	
,, lycoperdon (pars), *Hall* . . 265	
,, ,, (pars), *Hall* . . 267	
(*Chaetetes ?*) *Monticulipora ? Mackrothi* . 147	
Chaetetes petropolitanus, *Lonsdale* . 264	
,, pulchellus, *M. Edwards* and *Jules*	
Haime . . 267	
Chaetetes radians . . . 158	
Chaetetes repens, *D'Orbigny* . . 263	
,, rugosus, *Hall* . . 265	
,, septosus, *Keyserling* . . 157	
,, subfibrosus, *D'Orbigny* . . 265	
,, tuberculatus, *M. Edwards* and *J.*	
Haime . . 266	
(*Chaetetes*) *Monticulipora tumida* . . 159	
CHONOPHYLLUM . . . lxix	
Chonophyllum perfoliatum . 235, 291	
CIRRHIPATHES . . . lxiii	
(CLADOCHONUS) SYRINGOPORA . . lxxvi	
Cladochonus brevicollis, *M'Coy* . . 164	
,, tenuicollis, *M'Coy* . 164	
CLADOCORA . . . xxxviii	
Cladocora cariosa, *Lonsdale* . . 8	
,, duplicata, *Geinitz* . . 209	
,, fasciculata, *Geinitz* . . 197	
,, Goldfussi, *Geinitz* . . 229	
,, irregularis, *Morris* . . 198	
,, sexdecimalis, *Morris* . . 196	
,, sulcata, *Lonsdale* . . 283	
Cladophyllia Babeauana . . 113	
,, *Conybearei* . . 91	
Cladopora multipora, *Hall* . . 263	
,, seriata, *Hall* . . 263	
Clausastrœa Pratti . . . 117	
CLAVULARIA lxxv	

PAGE

CLISIOPHYLLUM . . . lxx
Clisiophyllum bipartitum . . 187
    ,,   Bowerbanki . . 186
    ,,   coniseptum . . 185
    ,,   costatum . . 187
    ,,   Keyserlingi . . 186
Clisiophyllum Konincki, M. Edwards and J.
        Haime . . 184
    ,,   multiplex, M'Coy . 183
    ,,   prolapsum, M'Coy . . 188
Clisiophyllum turbinatum . . 184
CŒLORIA xxxvi
CŒLOSMILIA . . . xxv
Cœlosmilia laxa . . . 52
Cœnenchyma . . . vi
Cœnites labrosus . . . 277
    ,,   linearis . . . 277
    ,,   ? strigatus . . 278
CŒNOCYATHUS . . . xii
CŒNOPSAMNIA . . . liii
COLPOPHYLLIA . . . xxxii
Columella v
Columnaria floriformis, Blainville . 205
    ,,   laxa, M'Coy . . 161
    ,,   senilis, De Koninck . 154
    ,,   striata, Blainville . 190
    ,,   Troosti, Castelnau . 207
COMBOPHYLLUM . . . lxvii
Comoseris irradians . . . 101
    ,,   vermicularis . 122, 143
Compound madrepora, Young . 94
    ,,   madreporite, Parkinson . 250
CONSTELLARIA . . . lxi
Convexastræa Waltoni . . 109
Coralliolites columnaris, Schlotheim . 148
Corallite vi
Corallite, Knorr and Walch . . 272
CORALLIUM . . . lxxxii
Corallium Gothlandicum, Fougt . 256
Coralloidea columnaria, Parkinson . 73
CORNULARIA . . . lxxv
Coronets v
COSCINARÆA . . . lv
Costæ v
CRYPTABACIA . . . xlvii
CRYPTANGIA . . . xliv
Cryptangia Woodi . . . 8

PAGE

Cryptocœnia luciensis, D'Orbigny . 107
CRYPTOHELIA . . . xxi
(CTENOPHYLLIA) PECTINIA . xxviii
CYATHAXONIA . . . lxv
Cyathaxonia conisepta, D'Orbigny . 185
Cyathaxonia cornu . . . 166
Cyathaxonia costata, M'Coy . . 187
    ,,   mitrata, D'Orbigny . 166
    ,,   plicata, D'Orbigny . 291
Cyathaxonia ? siluriensis . . 279
Cyathaxonia spinosa, Michelin . 176
(CYATHINA) CARYOPHYLLIA . xii
(Cyathina) Caryophyllia Bowerbanki . 61
    ,,   ,,   Bredai . 46
    ,,   ,,   cylindrica . 45
    ,,   ,,   Debeyana . 46
    ,,   ,,   lævigata . 44
CYATHOHELIA . . . xx
Cyathophora elegans, Lonsdale . 70
Cyathophora lucensis . . . 107
    ,,   Pratti . . 108
CYATHOPHYLLUM . . . lxviii
Cyathophyllum æquiseptatum . 232
Cyathophyllum ananas, Goldfuss . 236
Cyathophyllum angustum . . 281
    ,,   Archiaci . . 183
    ,,   articulatum . . 282
Cyathophyllum astrea, Bronn . 206
    ,,   basaltiforme, Phillips . 190
    ,,   binum, M. Edwards and J. Haime 227
Cyathophyllum boloniense . . 230
    ,,   Bucklandi . . 226
    ,,   celticum . . 226
    ,,   ceratites . . 224
    ,,   ,, (pars), Goldfuss . 224
    ,,   ,, (pars), Goldfuss . 243
    ,,   ,, Michelin . 176
    ,,   cæspitosum . . 229
Cyathophyllum cœspitosum, M'Coy . 283
    ,,   ,, Lonsdale . 283
    ,,   ,, Michelin . 231
    ,,   coniseptum, Keyserling . 185
    ,,   crenulare, Phillips . 181
Cyathophyllum damnoniense . 225
    ,,   dianthoides . . 182
Cyathophyllum dianthus (pars), Goldfuss . 224
    ,,   ,, (pars), Lonsdale 283, 284, 292

	PAGE
Cyathophyllum (Turbinolopsis) elongatum .	286
Cyathophyllum expansum, D'Orbigny .	179
„ „ Fischer .	206
Cyathophyllum flexuosum .	285
Cyathophyllum flexuosum, M'Coy .	281
„ „ Hisinger .	280
„ floriforme, Phillips .	205
„ fungites, Geinitz .	188
„ „ De Koninck .	184
„ „ Portlock .	171
Cyathophyllum helianthoides .	227
Cyathophyllum Hennahii, Bronn .	240
Cyathophyllum hexagonum .	228
Cyathophyllum hexagonum, (pars), Goldfuss	229
„ „ Michelin .	230
Cyathophyllum ? Loveni .	280
„ Marmini .	231
Cyathophyllum mitratum, Geinitz .	280
„ „ (pars), De Koninck	166, 174
Cyathophyllum Murchisoni .	178
„ ? novum .	145
„ obtortum .	225
Cyathophyllum papillatum, Fischer	207
Cyathophyllum parricida .	181
Cyathophyllum patellatum, Bronn	291
„ pentagonum, Goldfuss	238
„ perfoliatum, Goldfuss	235
„ plicatum, Goldfuss .	235
„ „ (pars), De Koninck	174
„ profundum, Germer .	149
„ „ Michelin .	231
Cyathophyllum pseudoceratites .	282
„ ? pseudovermiculare .	182
Cyathophyllum recurvum, M. Edwards and J.	
Haime . . .	282
Cyathophyllum regium .	180
„ Roemeri .	224
Cyathophyllum secundum, Goldfuss	243
Cyathophyllum Sedgwicki .	231
„ Stutchburyi .	179
Cyathophyllum subdianthus, D'Orbigny	284
„ subturbinatum, D'Orbigny	288
Cyathophyllum trochiforme .	285
„ truncatum .	284
Cyathophyllum turbinatum, M'Coy	288
„ „ (pars), Goldfuss	224
„ „ Hall .	235

	PAGE
Cyathophyllum turbinatum, Lonsdale	288
„ „ Phillips	234
„ vermiculare, Hisinger	282
Cyathophyllum vesiculosum, Eichwald	298
„ „ Goldfuss	243
Cyathophyllum Wrighti .	179
Cyathopsis cornubovis, D'Orbigny	174
„ cornucopiæ, M'Coy .	167
„ ? eruca, M'Coy .	177
„ fungites, M'Coy .	171
CYATHOSERIS . . .	xlix
CYCLOCRINITES . . .	lxxiv
CYCLOCYATHUS . . .	xiv
Cyclocyathus Fittoni . .	63
CYCLOLITES	xlvi
Cyclolites elliptica, Conybeare and Phillips .	143
„ Eudesii, Michelin .	125
„ lenticulata, Lonsdale .	247
„ lævis, Blainville .	121
„ numismalis, Lamarck .	246
„ præacuta, Lonsdale .	217
„ præacutus, Eichwald .	247
„ truncata, Defrance .	125
CYCLOSERIS . . .	xlix
CYLICOSMILIA . . .	xxiv
CYPHASTRÆA	xxxix
CYSTIPHYLLUM . . .	lxxii
Cystiphyllum brevilamellatum, M'Coy .	281
Cystiphyllum cylindricum .	297
Cystiphyllum damnoniense, Lonsdale .	225
Cystiphyllum Grayi . .	297
Cystiphyllum secundum, D'Orbigny .	243
Cystiphyllum siluriense . .	298
Cystiphyllum siluriense (pars), Lonsdale	289, 298
Cystiphyllum vesiculosum . .	243

D.

DANAIA	lxi
DASMIA	xix
Dasmia Sowerbyi . .	25
DASYPHYLLIA . . .	xxxiv
Decacœnia Michelini, D'Orbigny .	76
DELTOCYATHUS . . .	xv
DENDRACIS . . .	xxiii
DENDROGYRA . . .	xxviii
DENDROPHYLLIA . . .	liii

	PAGE
Dendrophyllia dendrophylloides . .	36
,, plicata, *M'Coy* .	92
DENDROPORA . . .	lxiii
Dendropora megastoma, *M'Coy* .	165
DENDROSMILIA . .	xxvii
Dentipora glomerata, *M'Coy* .	76
Dermic sclerenchyma . .	v
DESMOPHYLLUM . .	xvii
DIASERIS . . .	xlix
DICHOCŒNIA . . .	xxx
Dictyophyllia antiqua, *M'Coy* .	156
Diphyphyllum cæspitosum, *D'Orbigny*	229
,, concinnum, *Lonsdale*	195
,, fasciculatum, *D'Orbigny*	197
,, flexuosum, *D'Orbigny*	285
,, gracile, *M'Coy* .	199
,, irregulare, *D'Orbigny*	199
,, latiseptatum, *M'Coy*	195
,, longiconicum, *D'Orbigny*	200
,, pauciradiale, *D'Orbigny*	199
,, sexdecimale, *D'Orbigny*	196
,, sociale, *D'Orbigny* .	200
DIPLOCTENIUM .	xxv
DIPLOHELIA . . .	xxi
Diplohelia papillosa . .	28
DIPLORIA . .	xxxvii
DISCOCYATHUS .	xiii
Discocyathus Eudesi . .	125
Discophyllum helianthoides, *D'Orbigny*	228
,, lenticulatum, *D'Orbigny*	247
,, præacutum, *D'Orbigny*	247
Discopora squamata, *Lonsdale* .	268
DISCOTROCHUS .	xvii
Dissepiments . . .	vi
DISTICHOPORA	lxxix

E.

ECHINOPORA . .	xlv
Emmonsia hemispherica . .	218
ENALLOHELIA .	xxi
ENDOHELIA . .	xxii
ENDOPACHYS . .	lii
Endophyllum abditum . .	233
Endophyllum Bowerbanki .	233
ENDOPSAMMIA .	lii
Epidermic sclerenchyma .	v

	PAGE
Epitheca . . .	vi
ERIDOPHYLLUM . .	lxxi
Erismatolithus, *Martin* . .	197
,, affinis, *Martin* .	200
,, catenatus, *Martin* .	162, 164
,, duplicatus, *Martin* .	209
,, floriformis, *Martin* .	205
,, radiatus, *Martin* .	203
Escharites spongites, *Schlotheim* .	219
Eunomia Babeana, *D'Orbigny* .	113
,, radiata, *Lamouroux* .	111
EUPSAMMIA . . .	li
EUSMILIA . . .	xxvi
EXPLANARIA . . .	liv
Explanaria flexuosa, *Fleming* .	97
,, interstincta, *Geinitz* .	212

F.

	PAGE
Favastrea helianthoidea, *Blainville* .	227
,, hexagona, *Blainville* .	228
,, intercellulosa, *D'Orbigny* .	237
,, manon, *Blainville* .	154
,, pentagona, *Blainville* .	238
,, regia, *D'Orbigny* .	180
,, senilis, *D'Orbigny* .	154
FAVOSITES . . .	lx
Favosites, *Pander* .	259
,, alcyon, *Defrance* .	259
,, alveolaris, *M'Coy* .	258
,, ,, *Hall* .	218
,, ,, *Lonsdale* .	257
,, alveolata, *Geinitz* .	154
Favosites aspera . .	257
Favosites capillaris, *Phillips* .	158
Favosites cervicornis . .	216
,, *crassa* . .	261
,, *cristata* . .	260
Favosites cronigera, *D'Orbigny* .	216
,, dentifera, *D'Orbigny* .	153
Favosites dubia . .	216
Favosites depressus, *Fleming* .	158
,, excentrica, *Fischer* .	158
Favosites fibrosa .	217, 261
Favosites fibrosa (pars), *Lonsdale* .	217
Favosites Forbesi . .	258

PAGE

Favosites Goldfussi . . . 214
,, *gothlandica* . . . 256
Favosites gothlandica, *Blainville* . . 259
,, ,, *Lonsdale* . . 258
,, ,, *Phillips* . . 214
,, gothlandicus, *Eichwald* . 256
,, ,, *Steininger* . 214
,, hemispherica, *Yandell* and *Shumard* 218
,, hemisphericus, *Kutorga* . . 264
Favosites Hisingeri . . . 259
Favosites incrustans, *D'Orbigny* . . 153
,, inflata, *M'Coy* . . 159
,, lycopodites, *Vanuxem* . . 264
,, megastoma, *M'Coy* . . 156
,, microporus, *Steininger* . . 217
Favosites multipora . . . 258
Favosites Niagarensis, *Hall* . . 256
,, oculata, *M'Coy* . . 261
,, Orbignyana, *Verneuil* and *J. Haime* 215
Favosites parasitica . . . 153
Favosites petropolitana, *M'Coy* . . 264
,, petropolitanus, *Pander* . . 264
,, polymorpha, *Lonsdale* . . 260
,, ,, *Phillips* . . 217
,, radiata, *Blainville* . . 111
Favosites reticulata . . . 215
Favosites reticulum, *Eichwald* . . 256
,, scabra, *De Koninck* . . 159
,, septosus, *Fleming* . . 157
,, spongites (pars), *Lonsdale* 262, 267, 268
,, ,, (pars), *Phillips* . 219
,, subbasaltica, *D'Orbigny* . 256
,, suborbicularis, *D'Orbigny* . 219
,, tenuisepta, *M'Coy* . . 155
,, tumida, *Portlock* . . 159
FISTULIPORA lix
Fistulipora decipiens, *M'Coy* . . 298
Fistulipora major . . . 152
,, *minor* . . . 151
FLABELLUM xviii
Flabellum Woodi . . . 6
Floscularia corolligera, *Eichwald* . . 284
,, luxurians, *Eichwald* . 292
Fossile Querfurtense, *Buttners* . . 246
FUNGIA xlvi
Fungia astreata, *Goldfuss* . . 55
,, clathrata, *Geinitz* . . 60

Fungia coronula, *Goldfuss* . . 60
,, levis, *Goldfuss* . . 121
,, orbulites, *Lamouroux* . . 121
,, semilunata, *S. Wood* . .
Fungitæ octo majores, *Bromel* . . 284
,, tetragoni, *Bromel* . . 290
Fungitarum capitula, *Bromel* . . 246
Fungite, *Knorr* and *Walch* . 270, 272
Fungites, *Pennant* 250, 259, 280, 282, 284, 292
,, *D. Ure* . . . 188
,, Gothlandicus, *Bromel* . 287, 297
,, major, *Bromel* . . 287
,, mediæ, *Bromel* . . 291
,, patellatus, *Schlotheim* . . 291

G.

Gemmastrea limbata, *M'Coy* . . 105
GENABACIA xlvii
Geoporites intermedia, *D'Orbigny* . . 251
,, interstincta, *D'Orbigny* . 249
,, Lonsdalei, *D'Orbigny* . 249
,, Phillipsii, *D'Orbigny* . 212
,, porosa, *D'Orbigny* . . 212
,, pyriformis, *D'Orbigny* . 249
GONIASTRÆA xlii
Goniocora socialis . . . 92
GONIOPHYLLUM . . . lxix
Goniophyllum Fletcheri . . 290
Goniophyllum pyramidale . . 290
GONIOPORA lvi
GORGONIA lxxix
GRAPHULARIA . . . lxxxiii
Graphularia Wetherelli . . . 41
GRAPTOLITHUS . . . lxxxi

H.

HADROPHYLLUM . . . lxvii
HALLIA lxvii
Hallia Pengellyi . . . 223
HALOMITRA lxviii
HALOSERIS 1
HALYSITES lxii
Halysites agglomerata, *D'Orbigny* . . 271
Halysites attenuata, *Fischer* . . 270

	PAGE
Halysites catenularia . . .	270
Halysites catenulata, *Keyserling* .	273
„ catenulatus (pars), *M'Coy*	271
„ dichotoma, *Fischer* .	270
Halysites escharoides . .	272
Halysites Jacowickyi, *Fischer* .	272
„ labyrinthica, *Bronn* .	270
„ macrostoma, *Fischer* .	270
„ microstoma, *Fischer* .	270
„ stenostoma, *Fischer* .	270
(Harmodites) Syringopora .	lxii
Harmodites anglica, *D'Orbigny* .	274
„ bifurcata, *D'Orbigny* .	273
„ catenatus (pars), *Geinitz*	273
„ filiformis, *D'Orbigny* .	274
„ geniculata, *D'Orbigny* .	163
„ irregularis, *D'Orbigny*	274
„ Lonsdalei, *D'Orbigny* .	275
„ radians, *Bronn* .	162
„ ramulosus, *Keyserling* .	161
„ strues, *D'Orbigny* .	162
Heliolites	lviii
Heliolites elegans, *J. Hall* .	255
Heliolites Grayi . .	252
„ *inordinata* . .	253
„ *interstincta* . .	249
Heliolites macrostylus, *J. Hall* .	251
Heliolites megastoma . .	251
„ *Murchisoni* . .	250
„ *porosa* . .	212
Heliolites pyriformis, *Hall* .	249
„ spinipora, *J. Hall* .	255
Heliolithe pyriforme, *Guettard* .	212
Heliophyllum . . .	lxix
Heliophyllum Halli . .	235
Heliopora	lviii
Heliopora interstincta, *Bronn* .	212
„ „ *Eichwald* .	249
„ pyriformis, *Blainville* .	212
Herpetolitha . . .	xlvii
Heterocœnia . . .	xxxi
Heterocyathus . . .	xv
Heterophyllia . . .	lxxiii
Heterophyllia grandis . .	210
„ *ornata* . .	210
Heteropsammia . . .	lii
Hippurites mitratus (pars), *Schlotheim*	280

	PAGE
(Holaræa)	lvi
(*Holaræa*) *Axopora parisiensis* .	40
Holocystis . . .	lxiv
Holocystis elegans . .	70
Honey comb, *Parkinson* .	154
Hydnophora . .	xxxviii
Hydnophora? cyclostoma, *Phillips* .	152
„ Frieslebenii, *Fischer* .	76

I.

	PAGE
Isastræa Conybearei . .	113
„ *explanata* . .	94
„ *explanulata* . .	115
„ *Greenoughi* . .	96
„ *limitata* . .	114
„ *Lonsdalei* . .	139
„ *oblonga* . .	73
„ *Richardsoni* . .	138
„ *serialis* . .	116
„ *tenuistriata* . .	138
Isis	lxxxi

J.

	PAGE
Jania antiqua, *M'Coy* . .	164
„ bacillaria, *M'Coy* .	164
„ crassa, *M'Coy* .	164
Juncei lapidei, *Ure* . .	196

K.

	PAGE
Koninckia	lx

L.

	PAGE
Labecheia conferta . .	269
Lapis calcarius, *Bromel* .	256
Lasmocyathus aranea, *D'Orbigny* .	193
Lasmophyllia radisensis, *D'Orbigny*	80
Latimæandra . .	xxxiv
Latomæandra Davidsoni .	137
„ *Flemingi* .	136
Leiopathes . . .	lxxiii
Leptastræa . . .	xl

PAGE

Leptocyathus . . . xiv
Leptocyathus elegans . . . 21
Leptopsammia . . . lii
Leptoria . . . xxxvii
Leptoseris 1
(Leptosmilia) Euphyllia . . xxvi
Lithactinia xlviii
Litharæa lv
Litharæa Heberti . . . 39
   „   Websteri . . . 38
(Lithodendron) Lithostrotium . . lxxi
Lithodendron affine, Keferstein . 200
   „   annulare, Keferstein . 84
   „   astreatum, M'Coy . . 143
   „   centrale, Keferstein . 47
   „   coarctatum, Portlock . 196
   „   cœspitosum, M'Coy . 197
   „   „   Morren . 283
   „   dichotomum, M'Coy . 91
   „   dispar, Goldfuss . . 80
   „   Edwardsii, M'Coy . . 87
   „   eunomia, Michelin . 111
   „   fasciculatum, Keyserling 201
   „   „   Phillips . 197
   „   „   Portlock . 198
   „   gracile, Morris . . 69
   „   irregulare, Phillips . 198
   „   junceum, Keferstein . 196
   „   longiconicum, Phillips . 200
   „   pauciradiale, M'Coy . 198
   „   sexdecimale, Phillips . 196
   „   sociale, Roemer . . 92
   „   „   Phillips . 200
   „   trichotomum, Morris . 84
(Lithostrotion) Lonsdaleia . . lxxii
Lithostrotion, Lhwyd . . . 190
Lithostrotion affine . . . 200
Lithostrotion ananas (pars), D'Orbigny . 236
   „   arachnoides, D'Orbigny 230
Lithostrotion arachnöideum . . 202
   „   aranea . . . 193
Lithostrotion astroides, Lonsdale . 206
Lithostrotion basaltiforme . . 190
   „   ? concinnum . . 195
   „   decipiens . . . 196
   „   ? derbiense . . 201
Lithostrotion emarciatum, Lonsdale . 207

PAGE

Lithostrotion ensifer . . . 193
   „   Flemingi . . . 203
Lithostrotion floriforme, Fleming . . 205
   „   „   Lonsdale . . 207
Lithostrotion Hennahii, D'Orbigny . 240
Lithostrotion irregulare . . . 198
   „   junceum . . . 196
Lithostrotion Lonsdalei, D'Orbigny . 292
Lithostrotion, M'Coyanum . . 195
   „   majus . . . 201
Lithostrotion mamillare, Lonsdale . 206
Lithostrotion Martini . . . 197
Lithostrotion microphyllum, Keyserling . 190
   „   oblongum, Fleming . 73
   „   pauciradiale, Milne Edwards
         and Jules Haime . . 199
   „   pentagonum, D'Orbigny . 238
Lithostrotion Phillipsi . . . 201
   „   Portlocki . . 194
Lithostrotion profundum, D'Orbigny . 231
Lithostrotion septosum . . . 196
Lithostrotion striatum, Fleming . 190
Lituaria lxxxiv
Lobophyllia trichotoma, M'Coy . 84
Lobopora lix
Lobopsammia . . . liii
Loculi v
Lonsdaleia crassiconus, M'Coy . 209
Lonsdaleia duplicata . . . 209
   „   floriformis . . 205
Lonsdaleia inordinata, D'Orbigny . 253
Lonsdaleia papillata . . . 207
   „   rugosa . . . 208
Lonsdaleia? stylastræiformis, M'Coy . 209
Lonsdaleia wenlockensis . . . 296
Lophohelia xx
Lophophyllum . . . lxvi
Lophophyllum? eruca . . . 177
Lophoseris xlix
Lophosmilia . . . xxv
Lucernaria . . . lxxxv

M.

Madracis xxii
Madrepora liv
Madrepora, Knorr and Walch . . 200

	PAGE
Madrepora, *Parkinson* . . .	198
„ ananas (pars), *Linné* .	292
„ „ *Parkinson* .	292
„ arachnoides, *Parkinson* .	97
„ catenularia, *Esper* .	272
„ centralis, *Mantell* .	47
„ composita, *Fougt* .	284
„ „ *Fougt* .	285
„ „ *Fougt* .	292
„ „ *Fougt* .	295
„ fascicularis, *Parkinson* .	256
„ flexuosa, *Linné* .	285
„ „ *Smith* .	113
„ interstincta, *Linné* .	249
„ orbicularis, *Fougt* .	246
„ organum, *Linné* .	295
„ pectinata, *Parkinson* .	200
„ poris, &c., *Fougt* .	259
„ porpita, *Linné* .	246
„ porpites, *Smith* .	120
„ simplex, *Fougt* .	287
„ „ *Fougt* .	288
„ „ var. δ, *Fougt* .	280
„ „ var. ε, *Fougt* .	282
„ subrotunda, *Fougt* .	258
„ turbinata (pars), *Esper* .	282
„ „ (pars), *Esper* .	288
„ „ *Linné* .	287
„ „ *Smith* .	110
„ truncata, *Esper* .	228
„ „ *Linné* .	284
„ „ *Parkinson* .	292
Madreporites articulatus, *Wahlenberg* .	282
„ cristatus, *Blumenbach* .	260
„ interstinctus, *Wahlenberg* .	249
„ porpita, *Wahlenberg* .	246
„ turbinatus (pars), *Wahlenberg* .	287
„ truncatus, *Wahlenberg* .	284
MÆANDRINA . . .	xxxvi
MANICINA . . .	xxxvi
Manon favosum, *Goldfuss* .	154
Meandrina vermicularis, *M'Coy* .	122
MELITÆA . . .	lxxxii
MENOPHYLLUM . . .	lxvi
MERULINA . . .	xlv
METRIOPHYLLUM . . .	lxix
Metriophyllum Battersbyi . .	222

	PAGE
Michelinea glomerata, *M'Coy* .	155
„ grandis, *M'Coy* .	156
MICHELINIA . . .	lx
Michelinia antiqua . .	156
Michelinia compressa, *Michelin* .	156
Michelinia favosa . .	154
„ *megastoma* . .	156
„ *tenuisepta* . .	155
MICRABACIA . . .	xlvii
Micrabacia coronula . .	60
MICROSOLENA . . .	lvi
Microsolena excelsa . .	124
„ *regularis* . .	122
MILLEPORA . . .	lviii
Millepora Burtiniana, *Morren* .	263
„ cervicornis, *Wahlenberg* .	263
„ ramosa, *Hisinger* .	263
„ repens, *Fougt* .	263
„ „ *Hisinger* .	263
„ „ (pars), *Lonsdale* .	263
„ subrotunda, *Fougt* .	249
Milleporites repens, *Wahlenberg* .	263
Monocarya centralis (pars), *Lonsdale* .	44
„ „ „	47
Montastrea boloniensis, *Blainville* .	230
Monticularia conferta, *Lonsdale* .	269
„ areolata, *Steininger* .	227
Monticulipora? Bowerbanki .	268
„ *explanata* .	268
„ *Fletcheri* .	267
„ *lens* .	269
„ *papillata* .	266
„ *petropolitana* .	264
„ *pulchella* .	267
MONTIPORA . . .	lvii
Montlivaltia caryophyllata, *Bronn* .	129
„ obconica, *M. Edwards* and *J.* Haime .	80
MONTLIVAULTIA . . .	xxv
Montlivaultia cupuliformis .	132
Montlivaultia decipiens, *M'Coy* .	132
Montlivaultia depressa . .	134
Montlivaultia dilatata, *M'Coy* .	80
„ dispar .	80
„ gregaria, *M'Coy* .	135
Montlivaultia Labechei . .	132
„ *lens* .	133

	PAGE
Montlivaultia Moreausiaca, *M'Coy* .	. 80
Montlivaultia Smithi . .	. 110
„ Stutchburyi .	. 131
„ tenuilamellosa .	. 130
„ trochoides .	. 129
„ Waterhousei .	. 111
„ Wrighti .	. 131
MOPSEA . . .	. lxxxi
Mopsea costata . .	. 42
MORTIERIA . . .	. lxxiv
Mortieria vertebralis .	. 209
MURICEA . . .	. lxxx

N.

Nebulipora explanata, *M'Coy* .	. 268
„ lens, *M'Coy* . .	. 269
„ papillata, *M'Coy* .	. 266
Nemaphyllum aranea, *M'Coy* .	. 193
„ decipiens, *M'Coy* .	. 196
„ minus, *M'Coy* .	. 190
„ septosum, *M'Coy* .	. 196
(NEMATOPHYLLUM) STYLAXIS .	. lxxi
Nematophyllum arachnoideum, *M'Coy*	. 202
„ clisioides, *M'Coy* .	. 194
„ minus, *M'Coy* .	. 190
NEPHTHYA . . .	lxxviii

O.

OCULINA . . .	. xix
Oculina conferta . .	. 27
Oculina dendrophylloides, *Lonsdale*	. 36
OMPHYMA . . .	. lxviii
Omphyma Murchisoni .	. 289
„ subturbinata .	. 288
„ turbinata .	. 287
OULANGIA . . .	. xlv.
OULASTRÆA . .	. xxxix
OULOPHYLLIA . .	. xxxiv

P.

PACHYGYRA . . .	. xxviii
PACHYPHYLLUM . .	. lxviii
Pachyphyllum devoniense .	. 234
PACHYSERIS . .	. 1

	PAGE
Palastræa carbonaria, *M'Coy* .	. 180
PALÆOCYCLUS . .	. xlvi
Palæocyclus Fletcheri .	. 248
„ porpita .	. 246
„ præacutus .	. 247
„ rugosus .	. 248
Palæopora expatiata, *M'Coy*	. 278
„ favosa, *M'Coy*	. 250
„ interstincta, *M'Coy*	. 249
„ megastoma, *M'Coy*	. 251
„ petaliformis, *M'Coy*	. 253
„ pyriformis, *M'Coy*	. 212
„ subtilis, *M'Coy*	. 253
„ subtubulata, *M'Coy*	. 250
„ tubulata, *M'Coy*	. 255
Palæosmilia Murchisoni, *M. Edwards* and *J.*	
Haime . .	. 178
Pali . . .	. v
PARACYATHUS . .	. xiv
Paracyathus brevis .	. 25
„ caryophyllus .	. 24
„ crassus .	. 23
PARALCYONIUM . .	lxxviii
PARASMILIA . .	. xxiv
Parasmilia centralis .	. 47
„ cylindrica .	. 50
„ Fittoni .	. 50
„ Mantelli .	. 49
„ serpentina .	. 51
PARASTRÆA . .	. xliii
Parastræa stricta .	. 59
PAVONARIA . .	. lxxxiii
PENNATULA . .	. lxxxii
PEPLOSMILIA . .	. xxv
Peplosmilia Austeni .	. 57
Petalaxis Portlocki .	. 204
Petraia æquisulcata, *M'Coy*	. 280
„ bina, *M'Coy* .	. 227
„ celtica, *Lonsdale* .	. 226
„ dentalis, *King* .	. 149
„ elongata, *M'Coy*	. 287
„ gigas, *M'Coy* .	. 226
„ profunda, *King* .	. 150
„ quadrata, *M'Coy*	. 290
„ serialis, *M'Coy* .	. 287
„ subduplicata, *M'Coy*	. 287
„ zigzag, *M'Coy* .	. 286

	PAGE
PHILLIPSASTRÆA . . .	lxx
Phillipsastræa cantabrica, *J. Haime* and *de Verneuil* . .	242
„ Hennahii (pars), *D'Orbigny* .	203
„ Hennahii (pars), *D'Orbigny* .	240
„ parallela, *D'Orbigny* .	239
Phillipsastræa radiata . . .	203
„ *tuberosa* . .	204
PHYCOGORGIA . . .	lxxx
PHYLLANGIA . . .	xliv
PHYLLASTRÆA . . .	l
PHYLLOCŒNIA . . .	xxx
PHYLLOGORGIA . . .	lxxx
PHYMASTRÆA . . .	xl
PLACOCYATHUS . .	xvi
PLACOPHYLLIA . . .	xxvii
PLACOSMILIA . . .	xxiv
PLACOTROCHUS . . .	xviii
PLASMOPORA	lix
Plasmopora petaliformis . .	253
„ *scita* . .	254
PLATYTROCHUS . . .	xvii
PLEROGYRA	xxix
PLESIASTRÆA	xi
PLEUROCORA . . .	xxxviii
POCILLOPORA . . .	lxii
Pocillopora approximata, *Eichwald* .	263
PODABACIA . . .	xlviii
Polycælia Donatiana . .	149
„ *profunda* . .	149
POLYPHYLLIA . . .	xlviii
PORARÆA	lvi
PORITES	lv
Porites cellulosa, *Fleming* . .	154
„ expatiata, *Lonsdale* .	278
„ hemisphericus, *Schlotheim* .	246
„ inordinata, *Lonsdale* .	253
„ interstincta, *Keyserling* .	249
„ megastoma, *M'Coy* .	251
„ petaliformis, *Lonsdale* .	253
„ pyriformis, *Lonsdale* .	212
„ „ (pars), *Lonsdale*	249, 251
„ Swindernana, *Bronn* .	278
„ tubulata, *Lonsdale* .	255
Porpital madreporite, *Parkinson* .	249
„ „ *Parkinson* .	260
Porpitarum plurium, &c., *Bromel* .	284

	PAGE
PRIMNOA	lxxx
PRIONASTRÆA . . .	xli
Prionastrea alimena, *D'Orbigny* . .	114
„ explanata, *M. Edwards* and *J. Haime* . .	94
„ limitata, *M. Edwards* and *J. Haime*	114
„ luciensis, *D'Orbigny* .	114
PROPORA	lix
Propora? cyclostoma . .	152
„ *tubulata* . .	255
Protoseris Waltoni . .	103
Protovirgularia dichotoma, *M'Coy* .	299
PSAMMOCORA . . .	lvii
PTEROGORGIA . . .	lxxx
PTYCHOPHYLLUM . . .	lxix
Ptychophyllum patellatum .	291
Pyrgia Labechei . .	166
Pyritonema fasciculus, *M'Coy* .	299

R.

RENILLA . . .	lxxxiv
Rhabdophyllia nodosa . .	88
„ *Phillipsi* . .	87
„ *undata* . .	88
RHABDOPORA . . .	lxiii
Rhabdopora megastoma . .	165
Rhinopora tuberculosa, *Hall* .	266
RHIPIDOGYRA . . .	xxviii
RHIZANGIA	xliv
RHIZOTROCHUS . . .	xviii
RHIZOXENIA	lxxv
RHODARÆA	lvi

S.

(SARCINULA) GALAXEA . .	xxxi
Sarcinula angularis, *Fleming* .	260
„ organon, *Schweigger* .	295
„ organum, *Goldfuss* .	295
„ Phillipsii, *M'Coy* .	203
„ placenta, *M'Coy* .	203
„ punctata, *Fleming* .	249
„ tuberosa, *M'Coy* .	204
SARCODICTYUM . . .	lxxvi
SARCOPHYTON . . .	lxxviii

	PAGE
Scapophyllia . . .	xxxv
Sclerenchyma	iv
Screw stone, *Plot* . .	198
Septa	v
Seriatopora . .	lxiii
Siderastræa . .	xli
Siderastrea agariciaformis, *M'Coy* .	97
„ cadomensis, *M'Coy* .	143
„ explanata, *Blainville* .	94
„ incrustata, *M'Coy* .	124
„ Lamourouxi, *M'Coy* .	118
„ meandrinoides, *M'Coy*	101
„ Websteri, *Lonsdale* .	38
Silicified coral, *Conybeare* and *Phillips*	73
Siphonodendron aggregatum, *M'Coy*	199
„ fasciculatum, *M'Coy*	197
„ pauciradiale, *M'Coy*	198
„ sexdecimale, *M'Coy*	196
Siphonophyllia cylindrica, *Scouler* .	171
Smithia Bowerbanki . .	241
„ Hennahi . .	240
„ Pengellyi . .	241
Solanderia . .	lxxxi
Solenastræa . .	xl
Sphenotrochus . .	xvi
Sphenotrochus intermedius .	2
„ Roemeri . .	5
Spongophyllum Sedgwicki .	242
Spurious columella . .	v
Stauria . .	lxiv
(Stenopora) Monticulipora .	lixi
Stenopora columnaris, *King* .	148
„ crassa, *Howse* .	147
„ fibrosa, *M'Coy* .	261
„ granulosa, *M'Coy* .	269
„ incrustans, *King* .	148
„ independens, *King* .	147
„ inflata, *M'Coy* .	159
„ Mackothii, *Geinitz* .	148
„ tumida, *M'Coy* .	159
Stephanocœnia . .	xxx
Stephanocœnia concinna, *D'Orbigny* .	100
Stephanophyllia . .	lii
Stephanophyllia astreata .	55
„ Bowerbanki .	54
„ discoides .	34
Stephanophyllia Nysti . .	35

	PAGE
Stereopsammia . .	liii
Stereopsammia humilis . .	37
Stone found in Wales, *Lhwyd* .	205
Strephodes gracilis, *M'Coy* .	232
„ graigensis, *M'Coy* .	283
„ helianthoides, *M'Coy* .	228
„ multilamellatum, *M'Coy*	178
„ pseudoceratites, *M'Coy* .	282
„ trochiformis, *M'Coy* .	285
„ vermicularis, *M'Coy* .	225
„ vermiculoides, *M'Coy* .	284
Streptelasma . .	lxviii
Streptelasma bina, *D'Orbigny* .	227
Strombastræa truncata, *Blainville* .	284
Strombodes . .	lxx
Strombodes conaxis, *M'Coy* .	206
Strombodes diffluens . .	294
Strombodes emarciatum, *M'Coy* .	207
„ floriforme, *M'Coy* .	206
„ helianthoides, *Phillips* .	235
„ Labechii, *Milne Edwards* and
Jules Haime .	293
Strombodes Murchisoni . .	293
„ Phillipsi . .	294
Strombodes plicatum, *Lonsdale* .	291
„ truncatus, *Schweigger* .	284
Strombodes typus . .	293
Strombodes vermicularis, *Lonsdale* .	225
„ Wenlockensis, *M'Coy* .	296
Stylaster . .	xxii
Stylastrea basaltiformis, *M'Coy* .	190
„ irregularis, *M'Coy* .	201
Stylaxis arachnoidea, *Milne Edwards* and
Jules Haime . .	202
„ Flemingii, *M'Coy* .	203
„ irregularis, *M'Coy* .	201
„ major, *M'Coy* . .	201
„ Portlocki, *Milne Edwards* and *Jules*
Haime . .	204
Stylina . .	xxix
Stylina Babeana, *D'Orbigny* .	105
„ compound, *Parkinson* .	205
Stylina conifera . .	105
„ Labechei . .	79
Stylina luciensis, *Milne Edwards* and *Jules*
Haime . .	107
Stylina Ploti . .	106

	PAGE
Stylina simple, *Parkinson* . . 166
Stylina solida . . . 105, 128
„ *tubulifera* . . . 76
Stylina tubulosa, *Michelin* . . 76
STYLOCŒNIA xxix
Stylocænia emarciata . . 30
„ *monticularia* . . 32
STYLOPHORA xxii
Stylopora monticularia, *Schweigger* . 32
„ solida, *M'Coy* . . 105
STYLOSMILIA . . . xxvii
SYMPODIUM lxxvi
Synapticulæ vi
SYNASTRÆA xlii
Synastræa concinna, *Milne Edwards* and *Jules*
Haime . . . 100
„ Defranciana, *Milne Edwards* and
Jules Haime . . 139
SYNHELIA . . . xx
Synhelia Sharpeana . . 53
SYRINGOPHYLLUM . . . lxxii
Syringophyllum cantabricum . . 242
„ *organum* . . 295
Syringopora bifurcata . . 273
„ *catenata* . . 164
Syringopora catenata, *M'Coy* . 162
„ cespitosa, *Lonsdale* . 275
Syringopora fascicularis . . 274
Syringopora filiformis, *Goldfuss* . 274
Syringopora geniculata . . 163
„ laxa, *Phillips* . 164
„ Lonsdaleana, *M'Coy* . 275
Syringopora ramulosa . . 161
„ *reticulata* . . 162
Syringopora reticulata, *Hisinger* . 273
Syringopora serpens . . 275

T.

| *Tabulæ* . . . . vi |
| TELESTHO . . . lxxvii |
| THAMNASTRÆA . . . xlii |
| *Thamnastræa arachnoides* . . 97 |
| „ *cadomensis* . . 143 |
| „ *concinna* . . 100 |
| „ *Defranceana* . . 139 |
| „ *fungiformis* . . 141 |

	PAGE
„ *Lyelli* . . . 118
„ *Mac-Coyi* . . 141
„ *mammosa* . . . 119
„ *mettensis* . . . 141
„ *scita* . . . 119
„ *Terquemi* . . 140
„ *Waltoni* . . . 120
Thamnopora madreporacea, *Steininger* . 216
„ milleporacea (pars), *Steininger* . 216
THECIA lxiii
Thecia Grayana . . 279
„ *Swindernana* . . 278
THECOCYATHUS . . . xiv
Thecocyathus Moorei . . 144
Thecophyllia arduennensis, *D'Orbigny* . 80
THECOSMILIA . . . xxvi
Thecosmilia annularis . . 84
Thecosmilia cylindrica, *M. Edwards* and *J.*
Haime . . . 84
Thecosmilia gregaria . . 135
Thecosmilia trilobata, *M. Edwards* and *J.*
Haime . . . 84
THECOSTEGITES . . . lxii
Trabiculæ . . . vi
TRACHYPHYLLIA . . . xxxv
Tremocoenia varians, *D'Orbigny* . 100
TRIDACOPHYLLIA . . . xxxv
TROCHOCYATHUS . . . xiv
Trochocyathus conulus . . 63
„ *cupula* . . 64
„ *Harveyanus* . . 65
„ *Konigi* . . 66
„ *Magnevilleanus* . 126
„ *perarmatus* . . 66
„ *? primus* . . 145
„ *sinuosus* . . 22
„ *Warburtoni* . . 67
TROCHOPHYLLUM . . lxvii
TROCHOSERIS . . . xlix
TROCHOSMILIA . . . xxiv
Trochosmilia crassa . . 69
„ *sulcata* . . 68
Trochosmilia Smilotrochus tuberosus . 58
TROPIDOCYATHUS . . . xv
TRYMOHELIA xix
Tryplasma articulata, *Lonsdale* . 281
Tuber corallinus, *Buttners* . 256

	PAGE
TUBIPORA · · ·	lxxvii
Tubipora, *Knorr* and *Walch* ·	161
,, catenata, *Fleming* ·	163
,, catenularia, *Linné* ·	270
,, catenulata, *Gmelin* ·	270
,, ,, *Parkinson* ·	272
,, fascicularis, *Linné* ·	274
,, flabellaris, *Fabricius* ·	270
,, musica, *Parkinson* ·	163
,, prismatica, *Lamarck* ·	256
,, radiata, *Woodward* ·	203
,, ramulosa, *Woodward* ·	163
,, serpens, *Linné* ·	275
,, strues, *Parkinson* ·	162
Tubiporites catenarius *Schlotheim* ·	270
,, catenularia, *Wahlenberg* ·	272
,, fascicularis, *Wahlenberg* ·	273
,, serpens, *Krüger* ·	275
Tubularia fossilis, *Bromel* ·	259, 270
Turbinated madreporite, *Parkinson*	287, 288
TURBINOLIA · · ·	xvi
Turbinolia Bowerbanki · ·	16
Turbinolia caryophyllus, *Lamarck* ·	24
,, celtica, *Lamouroux* ·	226
,, centralis, *Roemer* ·	47
,, cernua, *Michelin* ·	69
,, compressa, *Morris* ·	58
,, conulus, *Michelin* ·	63
,, cupula, *Rouault* ·	64
,, cyathoides, *Lamarck* ·	287
,, didyma, *Morris* ·	104
,, dispar, *Phillips* ·	80
Turbinolia Dixoni · ·	15
Turbinolia dubia, *Defrance* ·	22
,, Donatiana, *King* ·	149
,, echinata, *Hisinger* ·	288
,, excavata, *Hagenow* ·	47
,, expansa, *M'Coy* ·	179
,, fibrosa, *Portlock* ·	286
Turbinolia firma · ·	20
,, Fredericana · ·	17
Turbinolia fungites, *Phillips* ·	179
,, ,, (pars), *Fleming*	184, 188
,, furcata, *Hisinger* ·	280
,, helianthoides, *Steininger*	227
Turbinolia humilis · ·	18
Turbinolia intermedia, *Goldfuss* ·	2

	PAGE
Turbinolia Magnevilliana, *Michelin* ·	126
,, Milletiana, *S. Wood* ·	2
Turbinolia minor · ·	19
Turbinolia mitrata, *Hisinger* ·	280
,, ,, *Portlock* ·	185
Turbinolia Nystana · ·	15
Turbinolia obliqua, *Hisinger* ·	280
,, perarmata, *Talavignes* ·	66
Turbinolia Prestwichi · ·	20
Turbinolia pyramidalis, *Hisinger* ·	290
,, sinuosa, *Brongniart* ·	22
Turbinolia sulcata · ·	13
Turbinolia sulcata, *Lonsdale* ·	15
,, turbinata, *Hisinger* ·	288
,, ,, (pars), *Lamarck*	22, 287
,, verrucosa, *Hisinger* ·	288
Turbinolite, 2d height, *Cuv.* and *Brongniart*	13
Turbinolopsis bina, *Lonsdale* ·	227
,, ,, *M'Coy* ·	183
,, celtica, *M'Coy* ·	183
,, ,, *Phillips* ·	226
,, elongata, *Phillips* ·	227, 286
,, ,, *Portlock* ·	282
,, pauciradialis, *M'Coy* ·	183
,, ,, *Phillips* ·	227
,, pluriradialis, *M'Coy* ·	183
,, ,, *Phillips* ·	227
,, rugosa, *Phillips* ·	227, 287
Turnip-shaped madrepora, *Young* ·	80

U.

UMBELLULARIA · ·	lxxxiii

V.

VERETILLUM · · ·	lxxxiii
Verticillopora dubia, *M'Coy* ·	160
VIRGULARIA · · ·	lxxxiii
Visceral chamber · ·	vi

W.

Walls · · ·	v
WEBSTERIA · · ·	lxxxiv
Websteria crisioides · ·	43

X.

	PAGE
Xenia . . .	lxxviii

Z.

	PAGE
Zaphrentis . . .	lxv
Zaphrentis Bowerbanki .	170

	PAGE
Zaphrentis cornucopiæ . .	167
,, cylindrica . .	171
,, Enniskilleni . .	170
,, Griffithi . .	169
,, patula . .	171
,, Phillipsi . .	168
,, subibicina . ,	172
,, Waltoni . .	143
Zoopilus . . .	xlviii

TAB. I.

CORALS FROM THE CRAG.

SPHENOTROCHUS INTERMEDIUS (p. 2).

Fig. 1.   An adult specimen ; natural size.

1 *a*. An adult specimen ; variety having a dilated basis ; natural size.

1 *b*. A magnified view of the specimen represented at fig. 1.

1 *c*. A vertical section of the same, corresponding to the short diameter of the calice ; magnified.

1 *d*. Calice of the same ; magnified.

1 *e*. A very young individual, magnified ; the natural size is indicated by the length of the line placed near this figure.

1 *f*, 1 *g*, 1 *h*, 1 *i*. A series of young individuals, at different periods of their growth ; magnified.

FLABELLUM WOODII (p. 6).

Fig. 2.   A side view of the corallum ; natural size.

2 *a*. A specimen magnified, and showing the mode of arrangement of the septa ; one half of the calice has been cut away down to the bottom of the fossula.

2 *b*. Calice entire ; natural size.

BALANOPHYLLIA CALYCULUS (p. 9).

Fig. 3.   Two individuals cemented together by their basis ; natural size.

3 *a*. A variety with a narrow basis ; natural size.

3 *b*. An individual remarkably tall, with its calicular extremity worn away ; natural size.

3 *c*. Horizontal section of the same, near the calice, magnified so as to show the mode of arrangement of the septa.

3 *d*. A fragment of the wall deprived of its epitheca, and much magnified.

CRYPTANGIA WOODII (p. 7).

Fig. 4.   A small aggregation of corallites imbedded in a mass of Cellepora ; natural size.

4 *a*. One of these corallites extracted from the mass of Cellepora, and showing its epitheca ; natural size.

4 *b*. One of the same separated from the extraneous mass, and having its epitheca dimpled by pressure of the surrounding Cellepora.

4 *c*. Calice, magnified.

4 *d*. One of the same corallites magnified, and having part of its calice cut away so as to show the denticulate edge of the septa ; the dimples of the epitheca are accidental and produced by the investing Cellepora.

TAB. II.

CORALS FROM THE LONDON CLAY.

DIPLHELIA PAPILLOSA (p. 28).

Fig. 1.   A branch of this compound corallum; natural size.

1 *a*. Terminal portion of one of the corallites, much magnified, and having half of its calice cut away so as to show the structure of the septa and the thickness of the walls.

1 *b*. Calice, magnified.

OCULINA CONFERTA (p. 27).

Fig. 2.   A small mass of this compound corallum; natural size.

2 *a*. Terminal portion of one of the corallites magnified, and cut down so as to show the structure of the columella, the pali, the septa, and the thickness of the walls.

2 *b*. Calice, magnified.

TURBINOLIA BOWERBANKII (p. 16).

Fig. 3.   Adult specimen; natural size.

3 *a*. The same magnified, so as to show the characters of the intercostal furrows.

3 *b*. Calice, magnified.

TURBINOLIA FIRMA (p. 20).

Fig. 4.   Adult specimen ; natural size.

4 *a*. The same, magnified.

4 *b*. Calice, magnified.

TURBINOLIA MINOR (p. 19).

Fig. 5.   Adult specimen ; natural size.

5 *a*. Magnified view of the same.

5 *b*. Calice, magnified.

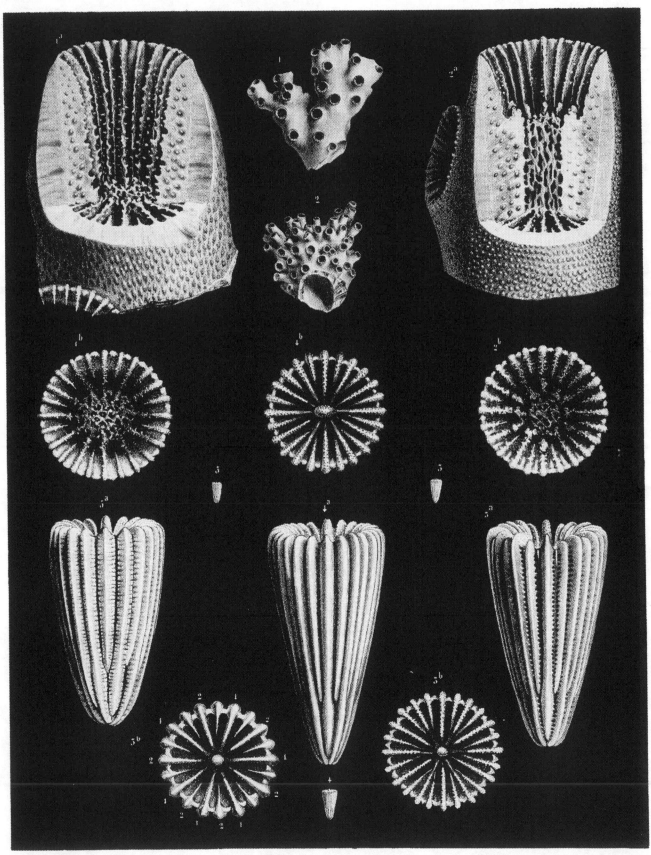

Day & Son, Lith.rs to The Queen.

TAB. III.

CORALS FROM THE LONDON CLAY.

TURBINOLIA DIXONII (p. 15).

Fig. 1. Adult specimen; natural size.

1 *a*. The same magnified, to show the characters furnished by the costæ, and the mural furrows.

1 *b*. Vertical section, showing the columella, the lateral surface of the septa, the wall, and the intercostal striæ.

1 *c*. Calice, magnified; the numbers surrounding the septa indicate the cycla to which each of these belong.

1 *d*. Fragment of the wall in which the lamellar costæ have been worn down, so as to show that the intercostal dimples are produced by transverse dissepiments, and are not pores, perforating the wall.

TURBINOLIA FREDERICIANA (p. 17).

Fig. 2. Adult corallum; natural size.

2 *a*. The same, magnified.

2 *b*. Calice, magnified.
In one of the systems the septa are numbered with reference to the cycla to which they belong.

TURBINOLIA SULCATA (p. 13).

Fig. 3. An adult corallum; natural size.

3 *a*. An individual showing a variety of forms.

3 *b*. The first of the preceding corals, magnified.

3 *c*. Calice, magnified.

TURBINOLIA HUMILIS (p. 18).

Fig. 4. An adult specimen; natural size.

4 *a*. The same magnified.

4 *b*. Calice, magnified.
By a mistake of the artist, the third cyclum of septa is here represented as being complete, whereas in reality these septa do not exist in two of the systems.

TURBINOLIA PRESTWICHII (p. 20).

Fig. 5. An adult specimen; natural size.

5 *a*. The same magnified.

5 *b*. Calice, magnified.

LEPTOCYATHUS ELEGANS (p. 21).

Fig. 6. Side view of the corallum; natural size.

6 *a*. The same magnified.

6 *b*. Under surface of the same, magnified.

6 *c*. Calice, magnified;—1, 1, 1, 1, 1, 1, Septa of the first order; 2, Septa of the second cyclum; 3, Septa of the third cyclum; 4, 5, Septa of the fourth and fifth orders constituting the fourth cyclum.

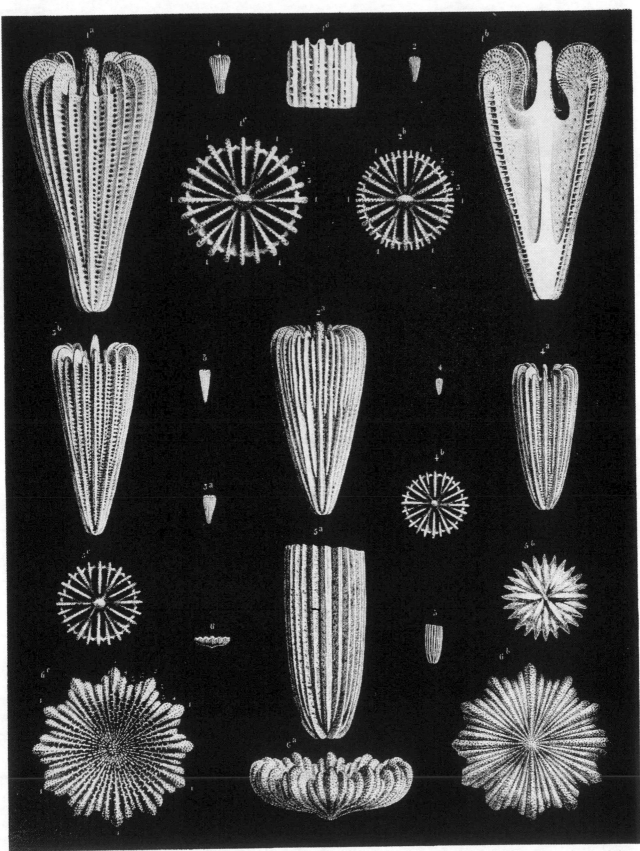

J. Delarue del

Day & Son, Lith.

TAB. IV.

CORALS FROM THE LONDON CLAY.

PARACYATHUS CRASSUS (p. 23).

Fig. 1. Two adult corals and a young one, cemented together by their basis; natural size.

1 a. A young specimen, with its basis spreading very much, viewed from above.

1 b. A young specimen, magnified to show the structure of the costæ.

1 c. Calice, magnified;—1, 1, 1, 1, 1, 1, Septa of the first cyclum.

PARACYATHUS CARYOPHYLLUS (p. 24).

Fig. 2. An adult specimen, complete; natural size.

2 a, 2 b, 2 c. Worn specimens; natural size.

2 d. Two young individuals cemented by their basis.

2 e. A remarkably short specimen.

2 f. A complete specimen, magnified so as to show the structure of the wall.

2 g. Calice, magnified.

PARACYATHUS BREVIS (p. 25).

Fig. 3. A specimen, the base of which is somewhat worn down; natural size.

3 a. The same, magnified.

3 b. Calice, magnified.

3 c. Interior cast of a corallum, which is taller than the preceding one, but appears to belong to the same species.

DASMIA SOWERBYI (p. 25).

Fig. 4. Side view of the corallum; natural size.

4 a. The same magnified, so as to show the structure of the costæ.

4 b. Calice, magnified; the central part is filled up with extraneous matter.

J. Delarue del.

Day & Son Lith^{rs} to The Queen

TAB. V.

CORALS FROM THE LONDON CLAY.

STYLOCŒNIA EMARCIATA (p. 30).

Fig. 1.   A small mass of this compound corallum ; natural size.

1 *a*. A portion of the calicular surface of the same, magnified.
This specimen is somewhat weather-worn.

STYLOCŒNIA MONTICULARIA (p. 32).

Fig. 2.   A small, somewhat gibbose mass of this compound corallum ; natural size.

2 *a*. A portion of the calicular surface, magnified so as to show the structure of the calices and of the marginal processes.

2 *b*. Transverse section of the compound corallum, slightly magnified, to show the cavity circumscribed by its under surface.

ASTROCŒNIA PULCHELLA (p. 33).

Fig. 3.   A small mass of this compound corallum, in which most of the corallites have been pressed together so that their calicular edges have become polygonal and completely united ; natural size.

3 *a*. Another group, in which most of the corallites have preserved their original circular form and free calicular margin.

3 *b*. A portion of the specimen fig. 3, magnified.

3 *c*. A portion of the specimen fig. 3 *a*, magnified.

STEREOPSAMMIA HUMILIS (p. 37).

Fig. 4.   A group of corallites ; natural size.

4 *a*. Terminal portion of some of these, magnified.

4 *b*. Calice, magnified.

T.5.

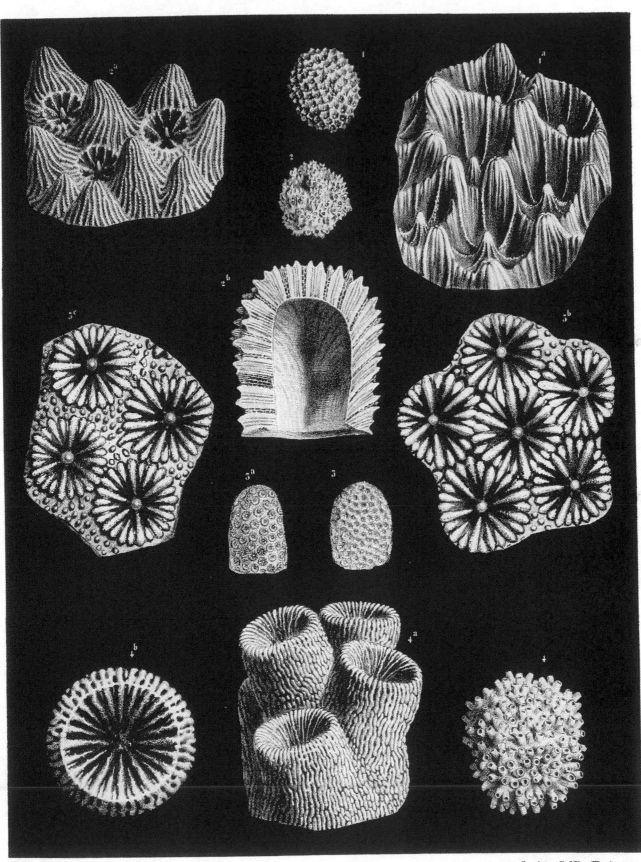

TAB. VI.

CORALS FROM THE LONDON CLAY.

BALANOPHYLLIA DESMOPHYLLUM (p. 35).

Fig. 1.   Side view of a complete specimen; natural size.

1 *a*. The same, magnified.

1 *b*. Calice, magnified.

1 *c*. Side view of the upper part of the same, with half of the calice cut away in order to show the structure of the septa and the depth of the fossula.

DENDROPHYLLIA DENDROPHYLLOÏDES (p. 36).

Fig. 2.   A large group; natural size.

2 *a*. Calice, magnified.

2 *b*. One of the branches, magnified to show the structure of the walls.

2 *c*. Two young individuals that have not yet produced young by gemmation, and are cemented together by their basis.

STEPHANOPHYLLIA DISCOIDES (p. 34).

Fig. 3.   Side view of a specimen, magnified; the natural size is indicated by the length of the line placed below.

3 *a*. Calicular surface, magnified.

3 *b*. Inferior surface, or mural disc, magnified.

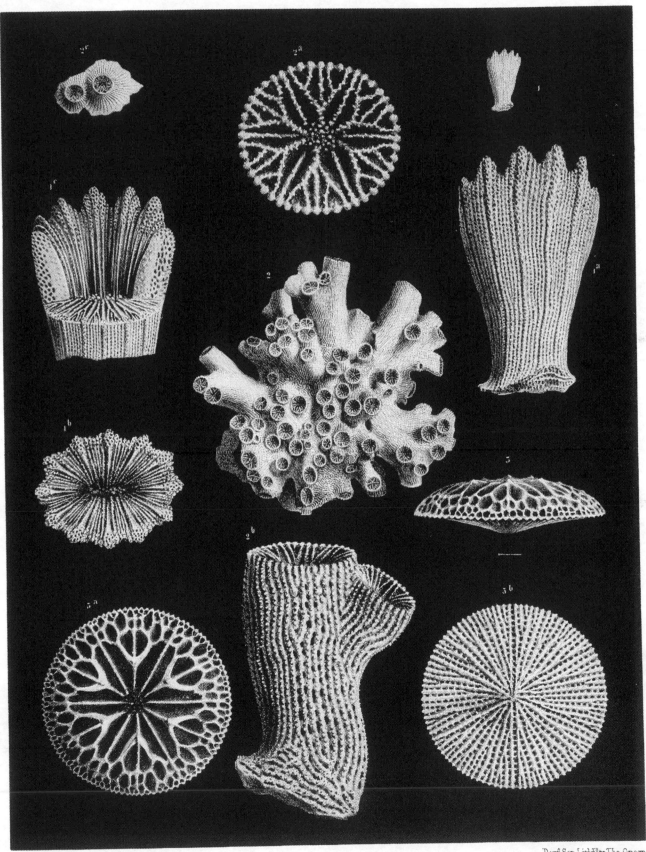

J. Delarue del.

Day & Son, Lith.rs to The Queen.

TAB. VII.

CORALS FROM THE LONDON CLAY.

LITHARÆA WEBSTERI (p. 38).

Fig. 1.   A mass of this compound corallum adhering to the surface of a pebble; natural size.

1 *a*. Calicular surface, magnified.

1 *b*. Vertical section of two corallites, magnified to show their internal structure.

1 *c*. A transverse section made at a considerable distance below the calice, and magnified so as to show the structure of the columella, the septa, and the walls.

HOLARÆA PARISIENSIS (p. 40).

Fig. 2.   A fragment of this cylindroid compound corallum magnified; the length of the line placed below indicates its real diameter.

2 *a*. Transverse section of the compound corallum, showing the vertical section of the corallites, magnified.

MOPSEA COSTATA (p. 42).

Fig. 3.   A large specimen; natural size.

3 *a*. A fragment, magnified.

GRAPHULARIA WETHERELLI (p. 41).

Fig. 4.   Fragments of the quadrangular portion of the sclerobasis; natural size.

4 *a*. A fragment of the cylindrical portion of the same.

4 *b*, 4 *c*. Fragments of both forms, magnified.

4 *d*, 4 *e*. Transverse sections of the same magnified, so as to show their radiate structure.

WEBSTERIA CRISIOÏDES (p. 43).

Fig. 5.   Fragment of a branch; natural size.

5 *a*. Portion of the same, magnified.

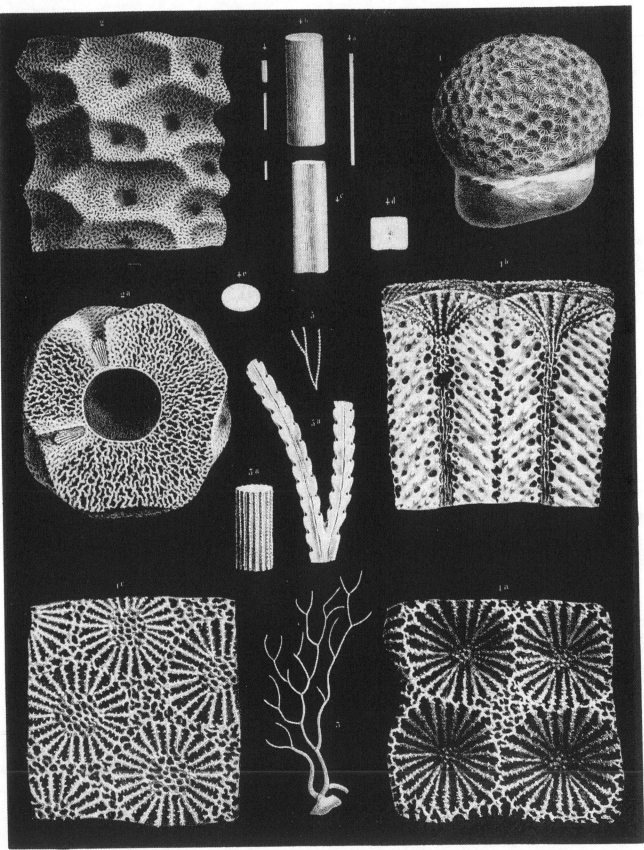

TAB. VIII.

CORALS FROM THE UPPER CHALK.

PARASMILIA CENTRALIS (p. 47).

Fig. 1.   A young specimen; natural size.

1 *a*. The same, magnified to show the structure of the wall.

1 *b*. Vertical section, magnified, so as to show the structure of the columella, the dissepiments, &c.

1 *c*. Calice, magnified; 1, septa of the first cyclum; 2 secondary septa; 3, septa of the third cyclum; 4, 5, septa of the fourth and fifth orders, constituting the fourth cyclum.

1 *d*, 1 *e*. Specimens remarkable by their great length.

PARASMILIA MANTELLI (p. 49).

Fig. 2.   Side view of the corallum; natural size.

2 *a*. The same, magnified.

PARASMILIA SERPENTINA (p. 51).

Fig. 3.   Side view of the corallum; natural size.

3 *a*. Calice, magnified.

3 *b*. Portion of the wall, magnified.

CŒLOSMILIA LAXA (p. 52).

Fig. 4.   Side view of the corallum; natural size.

4 *a*. A specimen, the growth of which has been intermittent; natural size.

4 *b*. The specimen No. 4, magnified, to show the structure of the wall.

4 *c*. Calice, magnified.

PARASMILIA CYLINDRICA (p. 50).

Fig. 5.   Side view of the corallum; natural size.

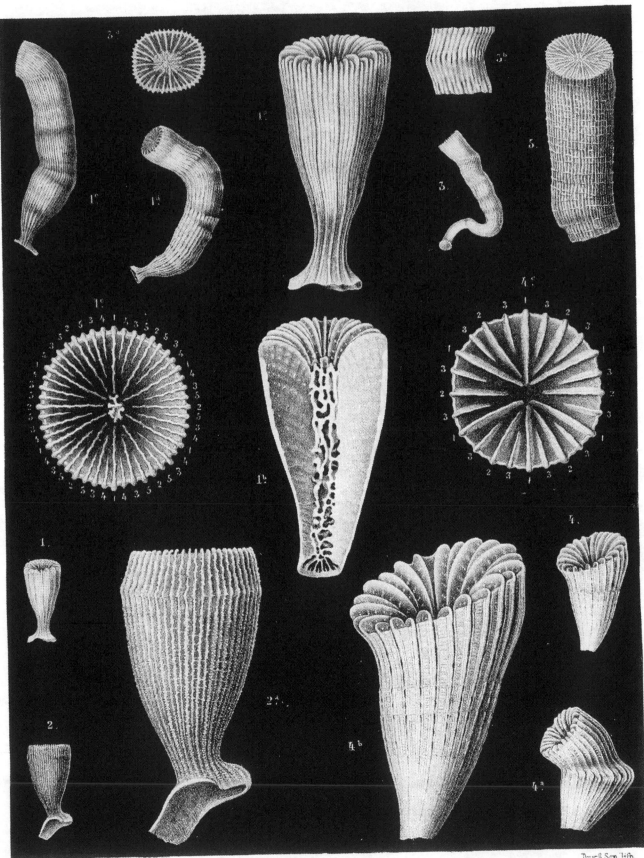

TAB. IX.

CORALS FROM THE UPPER CHALK.

CYATHINA LÆVIGATA (p. 44).

Fig. 1, 1 *a*, 1 *b*. Specimens of different forms; natural size.

1 *c*. Calice of the specimen fig. 1, magnified. It is to be remarked that in this specimen there are only nine pali; these organs not existing in the half systems, where the septa of the fourth cyclum are not developed.

1 *d*. Calice of the specimen fig. 1 *b*, magnified, and showing the twelve pali and the complete fourth cyclum of septa.

PARASMILIA FITTONI (p. 50).

Fig. 2. Side view of the corallum; natural size.

2 *a*. Specimen of a different form, magnified to show the structure of the wall.

2 *b*. Calice, magnified.

CORALS FROM THE LOWER CHALK.

SYNHELIA SHARPEANA (p. 53).

Fig. 3. A branch of this compound corallum; natural size.

3 *a*. Portion of the same magnified.

STEPHANOPHYLLIA BOWERBANKII (p. 54).

Fig. 4. Calicular surface; natural size.

4 *a*. Side view of the same, magnified.

4 *b*. Calice, magnified.

4 *c*. Mural disc, magnified.

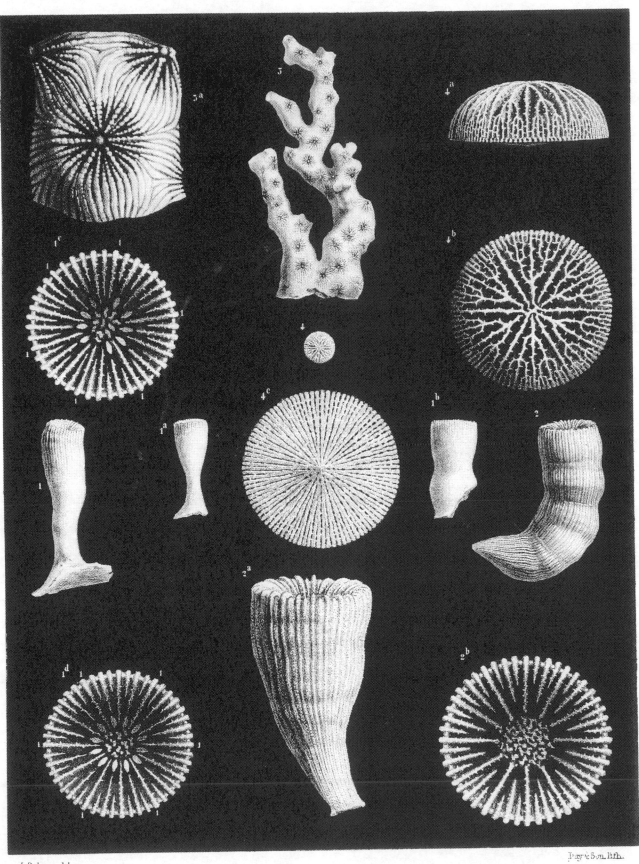

Day & Son, lith.

TAB. X.

CORALS FROM THE UPPER GREENSAND.

PEPLOSMILIA AUSTENI (p. 57).

Fig. 1.   Restored figure of the corallum; natural size.

1 a. Calice; natural size.

1 b. A broken specimen, showing part of the epitheca, the columella, and the structure of the septa.

TROCHOSMILIA TUBEROSA (p. 58).

Fig. 2.   Side view of the corallum; natural size.

2 a. Calice, magnified.

PARASTREA STRICTA (p. 59).

Fig. 3.   A mass of this compound corallum; natural size.

3 a. Portion of the calicular surface, magnified.

MICRABACIA CORONULA (p. 60).

Fig. 4.   Calicular surface; natural size.

4 a. Side view, magnified

4 b. Calice, magnified.

4 c. Mural disc, magnified.

CORALS FROM THE LOWER GREENSAND.

HOLOCYSTIS ELEGANS (p. 70).

Fig. 5.   A globose mass of this compound corallum; natural size.

5 a. Portion of the calicular surface, magnified.

5 b. Vertical section, magnified, of the visceral chambers in which the septa have been partly cut away, in order to show the tabular arrangement of the dissepiments.

TAB. XI.

CORALS FROM THE GAULT.

CYATHINA BOWERBANKII (p. 61).

Fig. 1.   A weather-worn specimen; natural size.

1 a. A specimen showing the wall, but not the calicular margin.

1 b. Horizontal section made near the calice, and magnified, to show the position of the pali, &c.

BATHYCYATHUS SOWERBYI (p. 67).

Fig. 2.   Side view of a specimen, magnified, so as to show the structure of the wall.
The line placed on the side shows the natural size of the corallum.

2 a. Calice magnified; the upper half is represented in its natural state, but in the under half the septa have been cut down; the centre is clogged up with extraneous matter.

CYCLOCYATHUS FITTONI (p. 63).

Fig. 3.   Side view of the corallum, magnified.
The line placed below shows the natural size of the specimen.

3 a. Calicular surface, magnified.

3 b. Under surface, or mural disc, magnified.

TROCHOCYATHUS HARVEYANUS (p. 65).

Fig. 4.   Side view of the corallum; natural size.

4 a. Calicular surface, magnified.

4 b. Under surface, magnified.

TROCHOCYATHUS CONULUS (p. 63).

Fig. 5.   Side view of the corallum; natural size.

5 a. Calice, magnified.

TROCHOSMILIA SULCATA (p. 68).

Fig. 6.   Side view of a specimen, the upper part of which is broken on one side; natural size.

6 a. Calice, magnified.

6 b. A restored specimen, magnified.

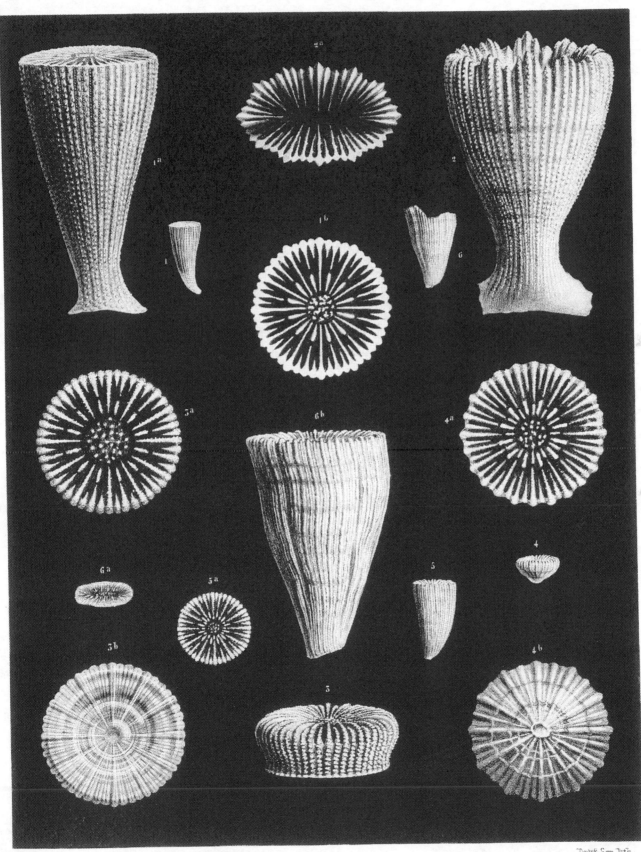

TAB. XII.

CORALS FROM THE PORTLAND STONE.

ISASTREA OBLONGA (p. 73).

Fig. 1.   A fragment silicified; natural size.

1 *a.* Calices (natural size), showing the different modifications produced by fossilization and wear; in some of the corallites the interseptal loculi are filled up with a calcareous deposit.

1 *b.* A horizontal section, natural size, showing the well-preserved corallites.

1 *c.* A portion of the same section magnified.

1 *d,* 1 *e.* Horizontal section of a portion of the corallum that is partly silicified and partly filled up with calcareous matter.   The gray parts are those in which the silex has replaced the tissue of the walls and the septa; the white radii correspond to the loculi of the visceral cavity, occupied by the calcareous deposit.   In fig. 1 *e* this deposit having been effected in an intermittent manner has produced the appearance of large pali.

1 *f.* A vertical section of a portion of one corallite, showing the thick simple walls, the dissepiments, the granulated surface of the septa, their denticulate inner edge, &c.

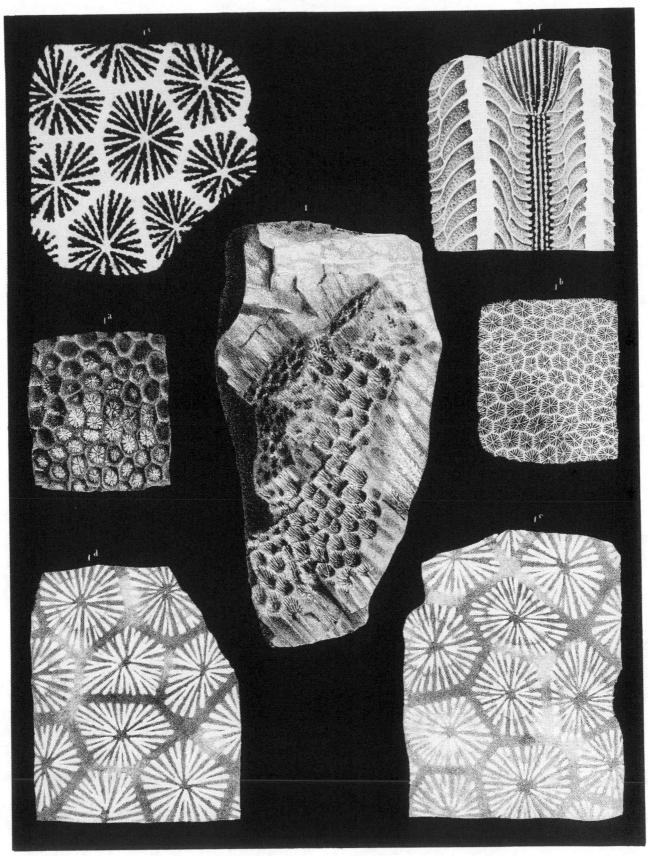

TAB. XIII.

CORALS FROM THE CORAL RAG.

THECOSMILIA ANNULARIS (p. 84).

Fig. 1. A group partly embedded in the stone; natural size.

1 *a*. A very young specimen; natural size.

1 *b*. A calice becoming deformed; from a young group.

1 *c*. Fissiparous multiplication of the calice in a young group, where that process goes on with great activity.

1 *d*. Calice of an adult corallite; natural size.

T.13.

TAB. XIV.

CORALS FROM THE CORAL RAG.

THECOSMILIA ANNULARIS (p. 84).

Fig. 1.   A young specimen; natural size.

1 *a*. A variety of the same species; natural size.

1 *b*. A younger specimen composed only of three individuals.

1 *c*. Calices of the same; natural size.

1 *d*. Calice of an adult corallite dividing by fissiparity; natural size.

MONTLIVALTIA DISPAR (p. 80).

Fig. 2.   An adult specimen which has lost its epitheca; natural size.

2 *a*. Horizontal section of the same, at a short distance from the calice, showing the septa and the dissepiments; natural size.

STYLINA TUBULIFERA (p. 76).

Fig. 3.   A small group; natural size.

3 *a*. A portion of the same seen from above, and showing the calices, the costal lines, &c.; natural size.

3 *b*. A portion of the same seen laterally, and showing the mural expansions.

3 *c*. Calicular surface of the same, magnified.

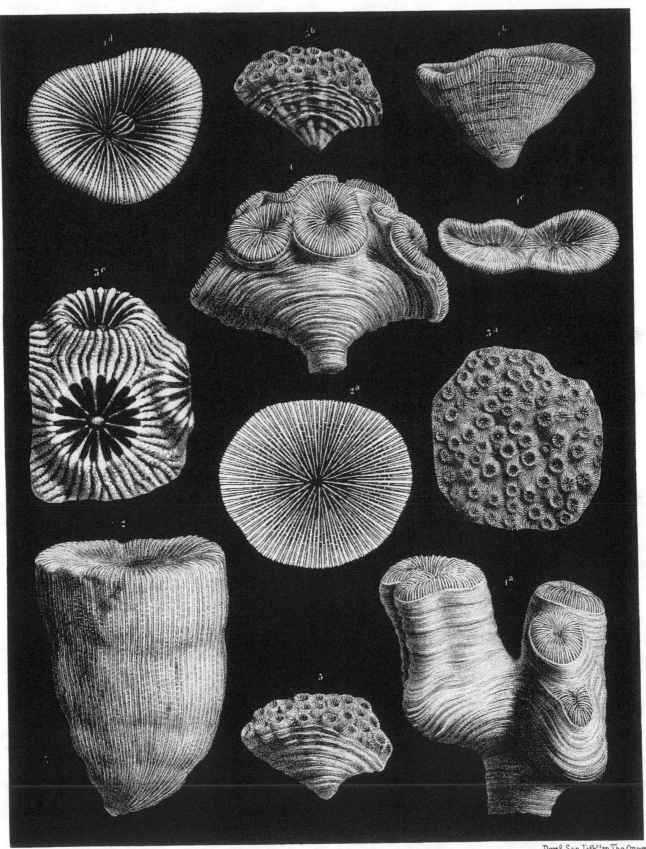

TAB. XV.

CORALS FROM THE CORAL RAG.

STYLINA DELABECHII (p. 79).

    Fig. 1.   A small group; the common basal plate has in part lost its epitheca.   Natural size.

        1 a. A few corallites magnified, showing the exotheca and some fragments of the epitheca.

        1 b. Calicular surface of the same corallum; natural size.

        1 c. Calices magnified.

        1 d. Variety with small calices; natural size.

GONIOCORA SOCIALIS (p. 92).

    Fig. 2.   A branch embedded in stone; natural size.

        2 a. A portion of the same magnified, showing the costæ and the mode of junction of the young individuals on the parent stem.

        2 b. Calice magnified.

RHABDOPHYLLIA PHILLIPSI (p. 87).

    Fig. 3.   A portion of this compound coral, embedded in stone; natural size.

        3 a. A fragment of the same magnified, and showing the structure of the costæ.

        3 b. A horizontal section of one of the branches at a short distance from the calice.

        3 c. A horizontal section of a corallite that is ready to divide by fissiparity.

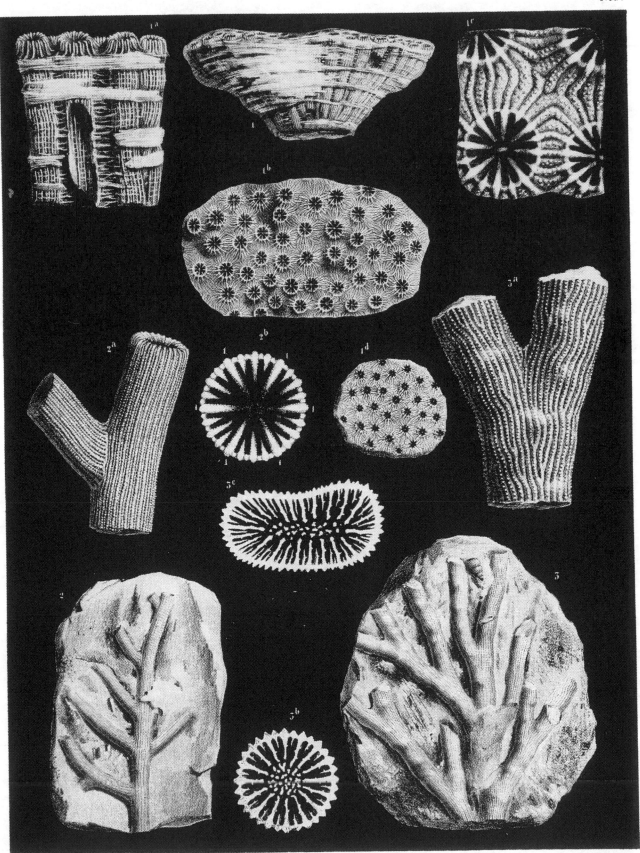

TAB. XVI.

CORALS FROM THE CORAL RAG.

CALAMOPHYLLIA STOKESI (p. 89).

Fig. 1.   Lateral view of the corallum ; natural size.

1 *a*. A fragment magnified, showing the structure of the walls.

1 *b*. Calice magnified, showing the denticulated edge of the septa.

1 *c*. A horizontal section of a specimen, in which the intermural spaces are filled up with extraneous matter.

1 *d*. A horizontal section of one of these corallites, magnified to show the endothecal dissepiments.

CLADOPHYLLIA CONYBEARII (p. 91).

Fig. 2.   Some small branches embedded in stone ; natural size.

2 *a*. A fragment, magnified.

2 *b*. Calice of an adult corallite, magnified.

2 *c*. Calice of a young corallite, magnified.

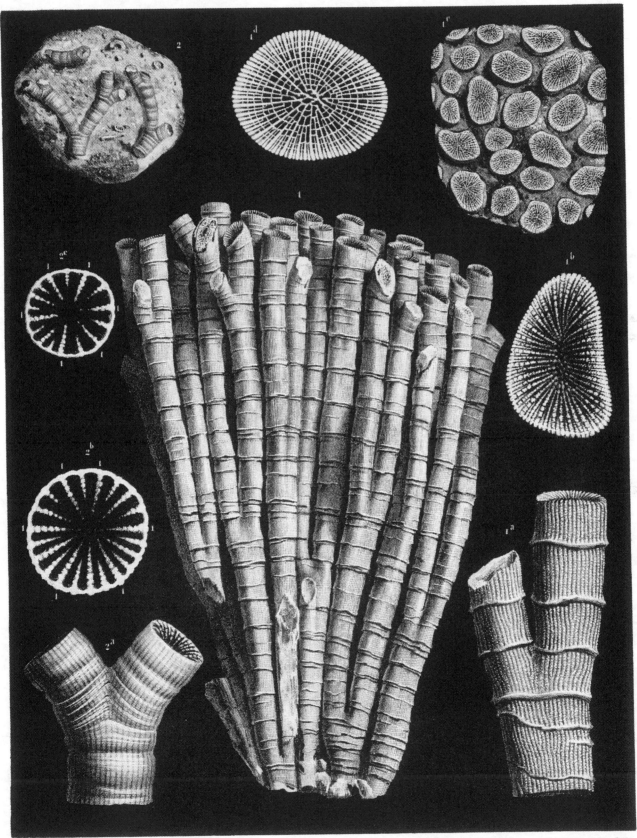

TAB. XVII.

CORALS FROM THE CORAL RAG.

THAMNASTREA ARACHNOIDES (p. 97).

Fig. 1. Side view of an adult specimen ; natural size.

1 *a*. A young specimen ; natural size.

1 *b*. Calicular surface of a young specimen, slightly magnified.

1 *c*. Side view of a young specimen, the growth of which has been intermittent.

1 *d*. Calices magnified.

1 *e*. Variety having the calices disposed in almost parallel series, and placed at a greater distance than usual ; natural size.

1 *f*. Some of the calices of this last variety, magnified.

1 *g*. A specimen much weather-worn, in which the terminal portion of the calices has disappeared and the walls have become prominent ; natural size.

1 *h*. Some well-preserved costæ, magnified.

1 *i*. Some weather-worn costæ, magnified.

1 *j*. A specimen the calicular surface of which is slightly worn down ; natural size.

1 *k*. A specimen much modified by the process of fossilization ; natural size.

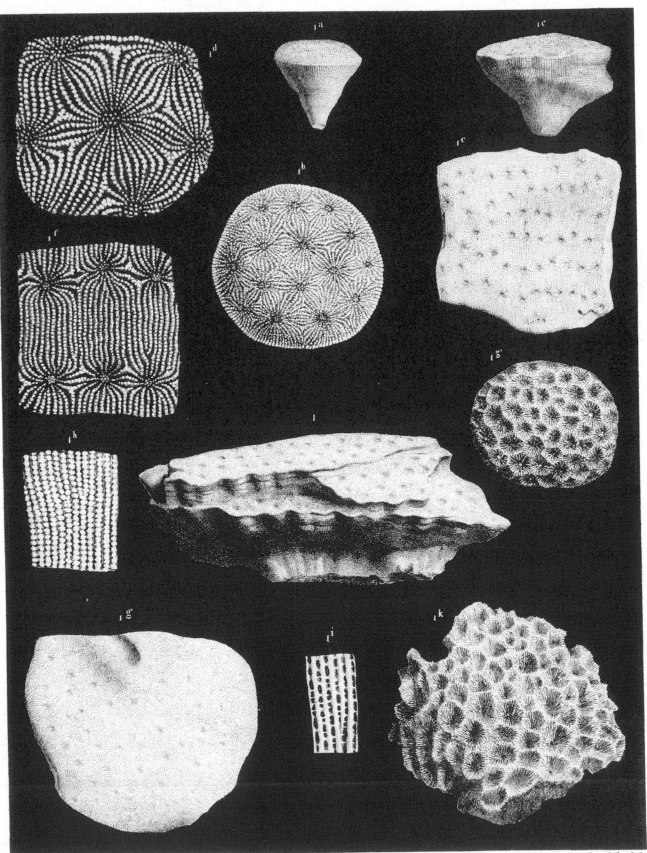

TAB. XVIII.

CORALS FROM THE CORAL RAG.

ISASTREA EXPLANATA (p. 94).

Fig. 1. Side view of a specimen, in which the epitheca has been partly removed in order to show the mode of arrangement of the costæ; natural size.

1 *a.* Calicular surface of the same; natural size.

1 *b.* Calice, magnified.

1 *c.* Horizontal section, magnified.

1 *d.* Specimen, in which the calicular fossula has been filled up by a calcareous deposit that assumes the appearance of a tuberculose columella.

ISASTREA GREENHOUGHI (p. 96).

Fig. 2. Portion of the calicular surface somewhat worn; natural size.

THAMNASTREA CONCINNA (p. 100).

Fig. 3. Side view of a large specimen; natural size.

3 *a.* A portion of the common basal plate, showing the wrinkled epitheca; natural size.

3 *b.* Calices in a good state of preservation, magnified.

3 *c.* Calices much weather-worn, magnified.

TAB. XIX.

CORALS FROM THE CORAL RAG.

COMOSERIS IRRADIANS (p. 101).

Fig. 1. Calicular surface of a specimen, in which the intercalicular ridges are not very numerous; natural size.

1 *a*. A specimen, in which the intercalicular ridges are very rare, and the aspect of the compound corallum more resembles that of *Astrea*; natural size.

1 *b*. A specimen, in which these ridges are very numerous, and give to the coral the aspect of a *Meandrina*.

1 *c*. A portion of the specimen, No. 1, magnified, to show the structure of the corallites.

1 *d*. Under surface, showing the epitheca covering the common basal plate; natural size.

TAB. XX.

CORALS FROM THE CORAL RAG.

PROTOSERIS WALTONI (p. 103).

Fig. 1.   Side view, showing the under surface or common basal plate of the compound
corallum ; natural size.

1 *a.* The same placed obliquely, so as to show the calicular surface.

1 *b.* Calice, magnified.

1 *c.* Costæ, magnified.

From a specimen belonging to M. Walton's collection.

TAB. XXI.

CORALS FROM THE GREAT OOLITE.

MONTLIVALTIA SMITHI (p. 110).

    Fig. 1.   Side view of an adult specimen; natural size.

      1 *a.* Another specimen much shorter.

      1 *b.* Calice; natural size.

STYLINA CONIFERA (p. 105).

    Fig. 2.   A fragment; natural size.

      2 *a.* A few corallites magnified.

CYATHOPHORA PRATTI (p. 108).

    Fig. 3.   A small mass; natural size.

      3 *a.* A few corallites magnified.

THAMNASTREA LYELLII (p. 118).

    Fig. 4.   Fragment, on part of which the calices are in their normal state; whereas towards the left side they are modified by the action of water and other agents.

      4 *a,* 4 *b.* Calices, situated in different parts of the same compound mass, magnified.

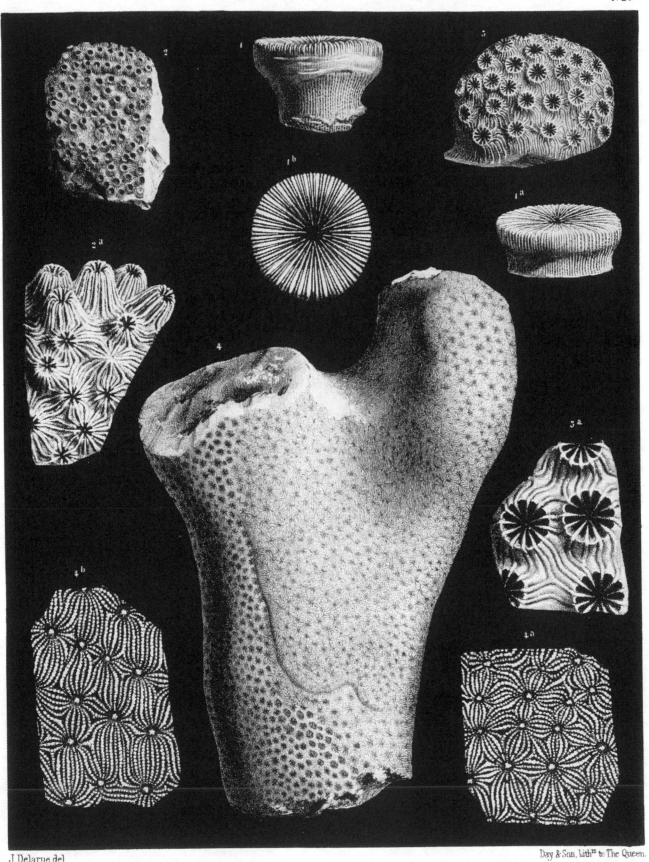

TAB. XXII.

CORALS FROM THE GREAT OOLITE.

CALAMOPHYLLIA RADIATA (p. 111).

Fig. 1. Fragment of the fasciculate mass of corallites; natural size.

1 a. A few of these corallites magnified.

1 b. A specimen, in which the spaces between the corallites have been filled up with calcareous matter, and the corallites themselves have afterwards been completely destroyed, so as to produce a fistulous mass.

1 c. Calices, magnified.

CLADOPHYLLIA BABEANA (p. 113).

Fig. 2. A group of small corallites; natural size.

2 a. Fragment, magnified; showing a calice and different horizontal sections of the corallites.

2 b. Calicular surface of a specimen, in which the spaces between the corallites have been filled up with extraneous matter.

STYLINA SOLIDA (p. 105).

Fig. 3. A natural cast.

3 a. Part of the same magnified.

3 b. Theoretic figure of the calices, magnified. (The artist, in attempting to restore this Coral, has given too much elevation to the calicular margins.)

ISASTREA CONYBEARII (p. 113).

Fig. 4. Fragment, showing the calices; natural size.

CLAUSASTREA PRATTI (p. 117).

Fig. 5. A fragment, much weather worn; natural size.

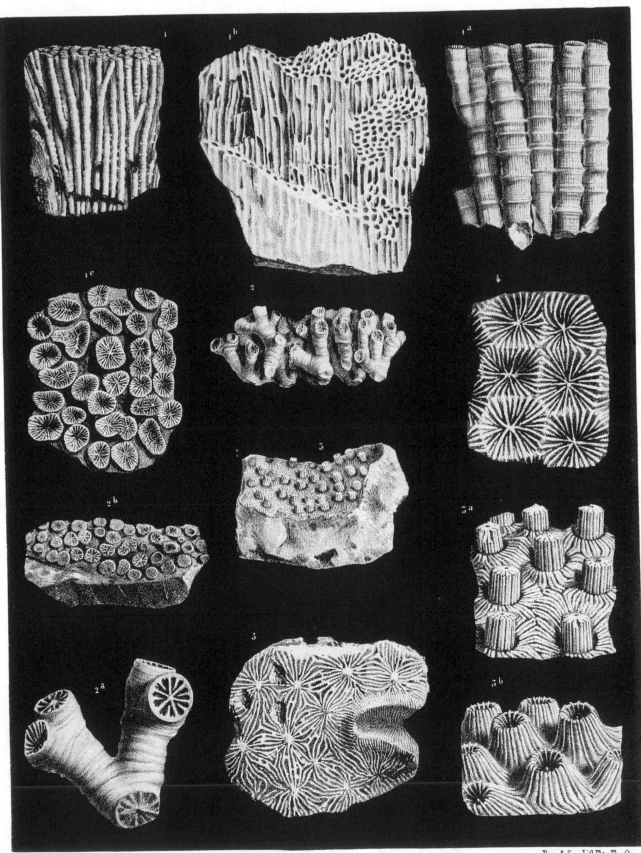

Day & Son, Lith.rs to The Queen.

TAB. XXIII.

CORALS FROM THE GREAT OOLITE.

STYLINA PLOTI (p. 106).

    Fig. 1.   A specimen, the surface of which is much worn; natural size.

ISASTREA LIMITATA (p. 114).

    Fig. 2.   A specimen, showing the calicular surface; natural size.
    2 *a*. Calices, magnified.

THAMNASTREA MAMMOSA (p. 119).

    Fig. 3.   A gibbose specimen; natural size.
    3 *a*. Calices, magnified.

THAMNASTREA SCITA (p. 119).

    Fig. 4.   A fragment, showing the calicular surface and the superposed layers that compose the mass of the corallum.
    4 *a*. Calices, magnified.

CONVEXASTREA WALTONI (p. 109).

    Fig. 5.   A specimen of a massive form; natural size.
    5 *a*. Calices, magnified.
    6.   A dendroid variety of the same specimen; natural size.

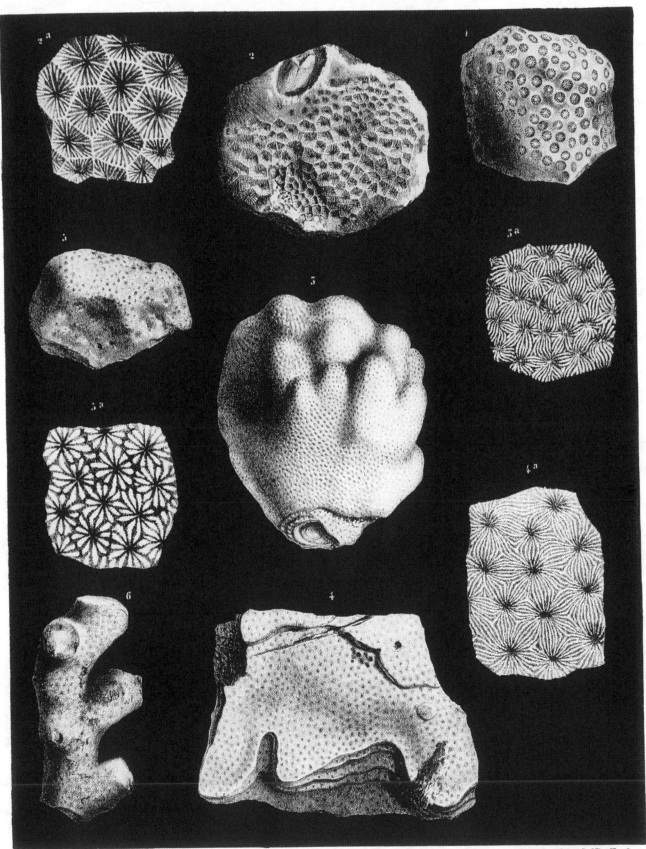

J.Delarue del.

Day & Son, Lith⁹ to The Queen.

TAB. XXIV.

CORALS FROM THE GREAT OOLITE.

COMOSERIS VERMICULARIS (p. 122).

Fig. 1. Calicular surface of an adult specimen ; natural size.
1 a. A part of the same magnified.

ISASTREA SERIALIS (p. 116).

Fig. 2. Calicular surface of a fragment ; natural size.
2 a. Calices, magnified.

ISASTREA EXPLANULATA (p. 115).

Fig. 3. A specimen, showing the calicular surface in different degrees of preservation ; natural size.
3 a. Calices, magnified.

ISASTREA LIMITATA (p. 114).

Fig. 4. A fragment, showing the calices in different states of preservation ; natural size.
4 a. Calices, magnified.
5. Horizontal section of a specimen, in which the visceral chambers had been filled up with white calcareous matter.

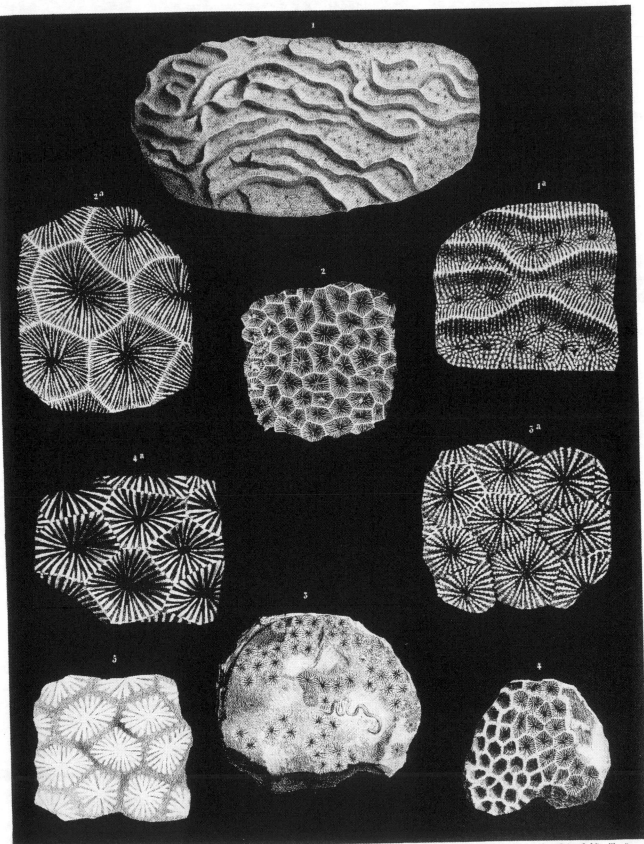

TAB. XXV.

CORALS FROM THE GREAT AND THE INFERIOR OOLITE.

DISCOCYATHUS EUDESI (p. 124).

From the Inferior Oolite.

Fig. 1.　Side view; natural size.

1 a.　Inferior surface of the same, showing the mural disc.

1 b.　Calice, magnified.

ANABACIA HEMISPHERICA (p. 142).

From the Inferior Oolite.

Fig. 2.　Side view; natural size.

2 a.　Calicular surface of the same.

ANABACIA ORBULITES (p. 120).

From the Great Oolite and the Inferior Oolite.

Fig. 3.　Side view; natural size.

3 a.　A smaller specimen, showing the calicular surface.

3 b.　The same, magnified.

3 c.　Inferior surface of a small specimen; natural size.

3 d.　Inferior surface, magnified.

3 e.　A large individual.

3 f.　Inferior surface of a worn specimen.

THAMNASTREA WALTONI (p. 120).

From the Great Oolite.

Fig. 4.　A small fragment; natural size.

4 a.　Calices, magnified.

MICROSOLENA EXCELSA (p. 124).

From the Great Oolite.

Fig. 5.　Side view of a group of these Corals; natural size.

5 a.　Calices, magnified.

MICROSOLENA REGULARIS (p. 122).

From the Great Oolite.

Fig. 6.　A specimen much weather worn; natural size.

6 a.　Calices, magnified.

6 b.　A part of the exterior surface of the compound corallum magnified, to show the regular granular structure of the costæ.

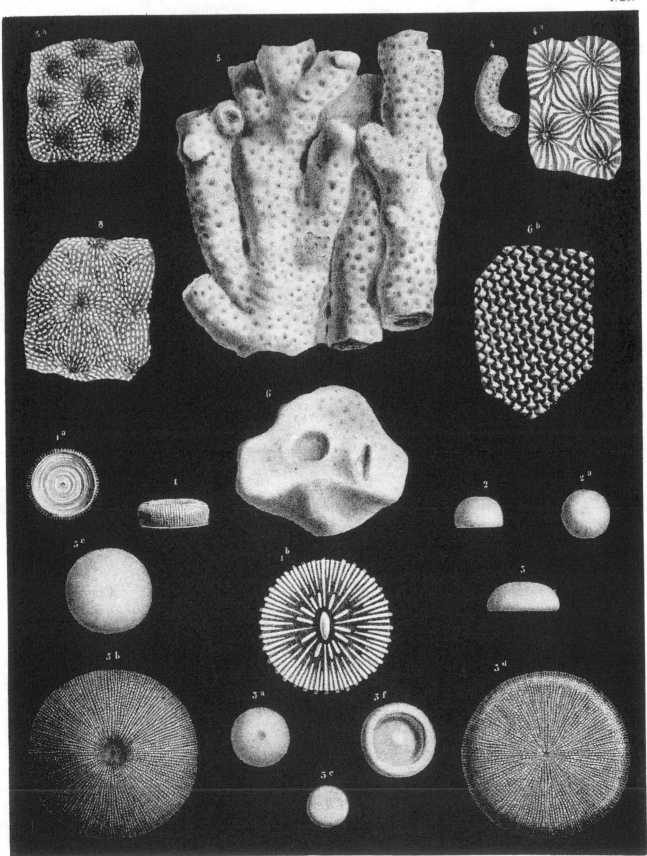

Day & Son, Lith[rs] to The Queen.

TAB. XXVI.

CORALS FROM THE INFERIOR OOLITE.

TROCHOCYATHUS MAGNEVILLIANUS (p. 126).

Fig. 1.   Side view; natural size.

1a.   Basal surface of the same.

1b.   Calice restored and magnified.

MONTLIVALTIA TROCHOIDES (p. 129).

Fig. 2.   Side view of a regularly-developed specimen; natural size.

2a.   Calice, natural size; the edge of the septa is worn so as to appear entire.

3.   An elongate variety of the same species; natural size.

3a.   Calice, natural size; the septa are well preserved, and show the denticulations of their upper edge.

4.   A short and broad variety of the same species.

10.   A variety with an irregular form.

MONTLIVALTIA DELABECHII (p. 132).

Fig. 5.   Side view; natural size.

5a.   Inferior surface of the same.

5b.   Calicular surface of the same.

6.   A young individual, belonging, probably, to the same species.

MONTLIVALTIA LENS (p. 133).

Fig. 7.   Side view; natural size.

7a.   Calicular surface of the same.

7b.   Inferior surface of the same.

7c.   Calicular surface magnified.

8.   A variety with thinner septa.

MONTLIVALTIA CUPULIFORMIS (p. 132).

Fig. 9. Young individuals, belonging, probably, to this species.

MONTLIVALTIA TENUILAMELLOSA (p. 130).

Fig. 11. Side view; natural size.

11a. Calice of the same.

MONTLIVALTIA WRIGHTI (p. 131).

Fig. 12. A specimen much weather-worn; natural size.

12a. Calice of the same.

J. Delarue del.

Day & Son, Lith^{rs} to The Queen.

TAB. XXVII.

CORALS FROM THE GREAT AND INFERIOR OOLITE.

MONTLIVALTIA CUPULIFORMIS (p. 132).

From the Inferior Oolite.

Fig. 1. Side view of an adult specimen; natural size.
1a. Calice; natural size.

MONTLIVALTIA TROCHOIDES (p. 129).

From the Inferior Oolite.

Fig. 2. Side view of a specimen, the form of which is somewhat irregular; natural size.
2a. Calice of the same.
4. A young individual; natural size.

MONTLIVALTIA STUTCHBURYI (p. 131).

From the Inferior Oolite.

Fig. 3. A specimen, fractured at its basis; natural size.
3a. A horizontal section, at a short distance from the calice.

AXOSMILIA WRIGHTI (p. 128).

From the Inferior Oolite.

Fig. 6. Side view; natural size.

MONTLIVALTIA WATERHOUSEI (p. 111).

From the Great Oolite.

Fig. 7. Side view; natural size.
7a. Calice of the same.

ZAPHRENTIS? WALTONI (p. 143).

Presumed to belong to the Inferior Oolite.

Fig. 8. Side view; natural size.
8a. Calice, partly clogged up with extraneous matter.

LATOMEANDRA FLEMINGI (p. 136).

From the Inferior Oolite.

Fig. 9. A fragment, showing the calice; natural size.
9a. Calice magnified.

LATOMEANDRA DAVIDSONI (p. 137).

From the Inferior Oolite.

Fig. 10. Calicular surface; natural size.
10a. Calices magnified.

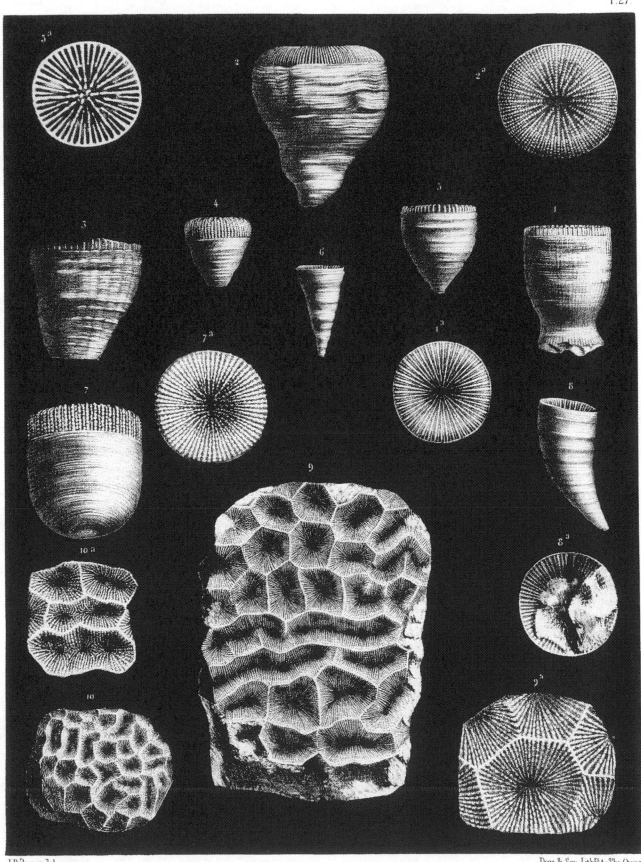

J. Delarue del.

Day & Son, Lith.rs to The Queen.

TAB. XXVIII.

CORALS FROM THE INFERIOR OOLITE.

THECOSMILIA GREGARIA (p. 135).

Fig. 1.   Oblique view of a large specimen, showing the mode of division of the mass; the epitheca, and the costal striæ; natural size.

1 a.   Upper surface of the same, showing the mode of arrangement of the corallites and the structure of the calices; natural size.

The fine specimen here figured belongs to the collection of Dr. Wright, of Cheltenham.

TAB. XXIX.

CORALS FROM THE INFERIOR OOLITE.

Isastrea Richardsoni (p. 138).

Fig. 1. A fragment showing the calices; natural size.

1 a. Calices, magnified.

Thamnastrea M‘Coyi (p. 141).

Fig. 2. A specimen much weather-worn; natural size.

2 a. Calices, magnified.

Thamnastrea Defranciana (p. 139).

Fig. 3. A large crateriform specimen; natural size.

3 a. Some well-preserved calices, magnified.

3 b. Some weather-worn calices, magnified.

4. A fragment much more weather-worn, natural size.

4 a. Calices of the same, magnified.

4 b. Some of the same modified corallites, slightly magnified.

Montlivaltia depressa (p. 134).

Fig. 5. A specimen much weather-worn; natural size.

5 a. Calice of the same.

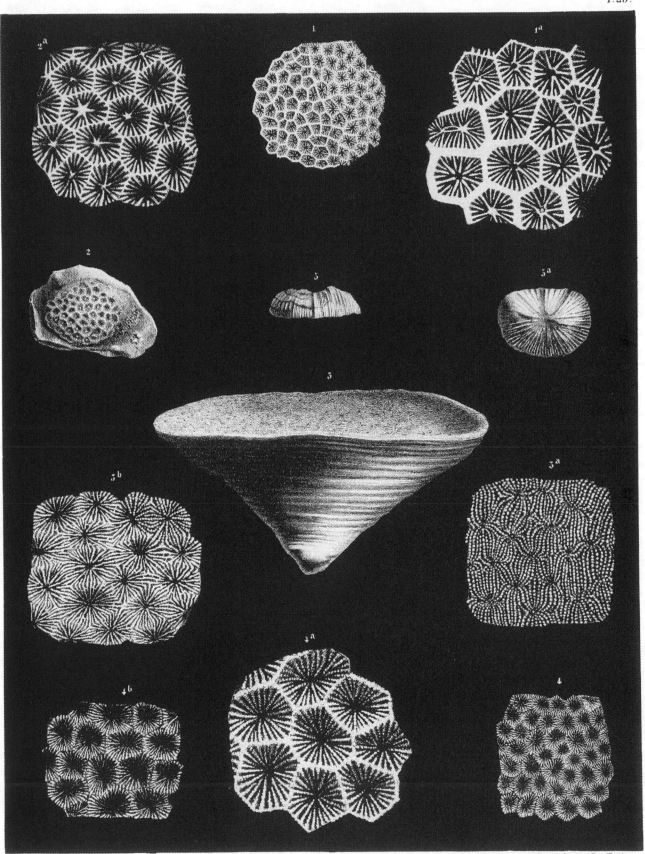

J.Delarue del.

Day & Son, Lith.rs to The Queen.

TAB. XXX.

CORALS FROM THE INFERIOR OOLITE.

ISASTREA TENUISTRIATA (p. 138).

Fig. 1. A fragment somewhat weather-worn, and showing the calices; natural size.
1 *a*. Calice magnified.

THAMNASTREA TERQUIEMI (p. 140).

Fig. 2. A fragment showing the calicular surface; natural size.
2 *a*. Calices magnified.
2 *b*. Inferior surface of a portion of the same coral, showing the common basal plate; natural size.

THAMNASTREA METTENSIS (p. 141).

Fig. 3. A fragment showing the calicular surface; natural size.
3 *a*. Calices magnified.

THAMNASTREA FUNGIFORMIS (p. 141).

Fig. 4. A young specimen; natural size.
4 *a*. Calices magnified.

CYATHOPHORA LUCIENSIS (p. 107).

Fig. 5. A small specimen; natural size,
5 *a*. Calices magnified.

CORALS FROM THE LIAS.

THECOCYATHUS MOORII (p. 144).

Fig. 6. Side view; natural size.
6 *a*. Calice magnified.

CYATHOPHYLLUM? NOVUM (p. 145).

Fig. 7. Front view of the coral; natural size.

TROCHOCYATHUS PRIMUS (p. 145).

Fig. 8. A specimen somewhat weather-worn; natural size.

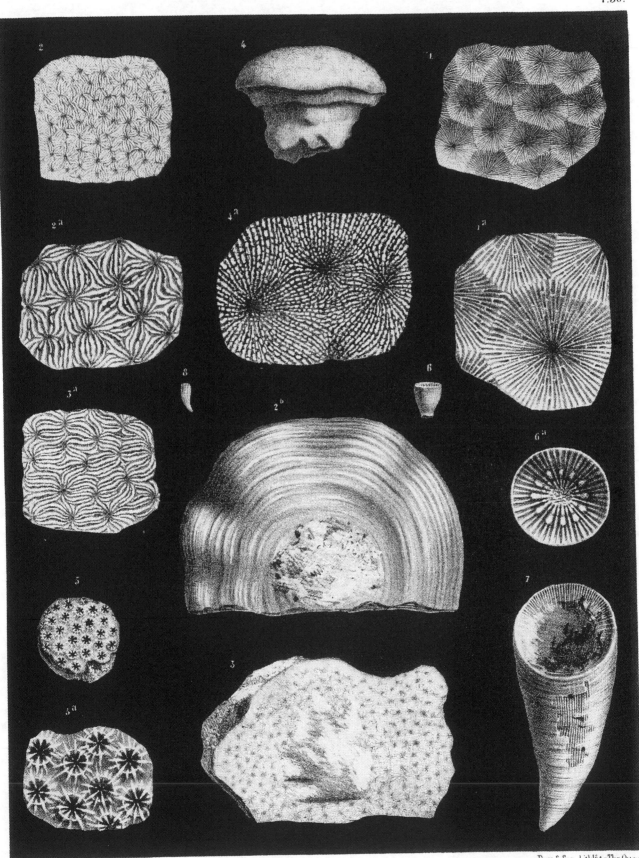

J Delarue del

Day & Son, Lith^{rs} to The Queen

TAB. XXXI.

CORALS FROM THE MOUNTAIN LIMESTONE.

CYATHOPHYLLUM STUTCHBURYI (p. 179).

Fig. 1.   Side view of a short specimen; natural size.

   1 *a.*   Calice of a short specimen; natural size.

   2.   Side view of a tall specimen; natural size.

   2 *a.*   Another tall specimen, rather worn, and showing accretion ridges; natural size.

TAB. XXXII.

CORALS FROM THE MOUNTAIN LIMESTONE.

CYATHOPHYLLUM REGIUM (p. 180.)

1. A massive specimen, showing unequal calices; the small ones on the upper front of the figure hide a large one; natural size.

1 a. Small portion of a vertical section; natural size.

2. Another specimen with thicker walls; natural size.

3. Another specimen, in which the mural ridges are very thin and sharp; natural size.

4. Two corallites issuing from the same parent, and showing exterior walls free laterally; natural size.

4 a. A calice of the same; natural size.

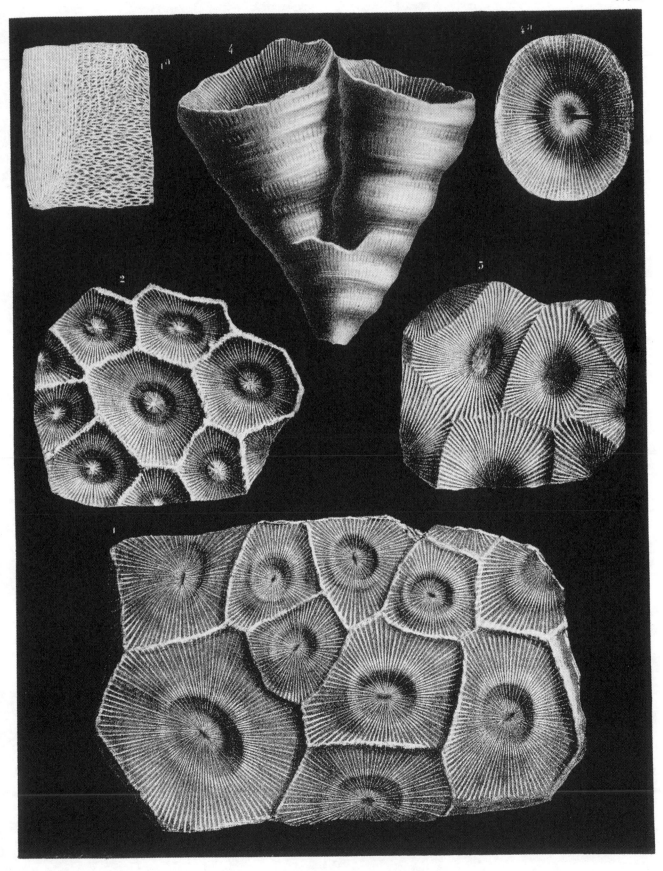

TAB. XXXIII.

CORALS FROM THE MOUNTAIN LIMESTONE.

CLISIOPHYLLUM TURBINATUM (p. 184).

Fig. 1.　A short and stout specimen ; natural size.

1 *a*.　Calice of the same ; natural size.

2.　A long specimen which has lost its epitheca ; natural size.

CYATHOPHYLLUM MURCHISONI (p. 178).

3.　Side view of a long specimen, somewhat weather-worn ; natural size.

3 *a*.　Horizontal section of the same ; natural size.

3 *b*.　Vertical section ; natural size.

CYATHOPHYLLUM STUTCHBURYI (p. 179).

4.　Part of a vertical section, magnified.

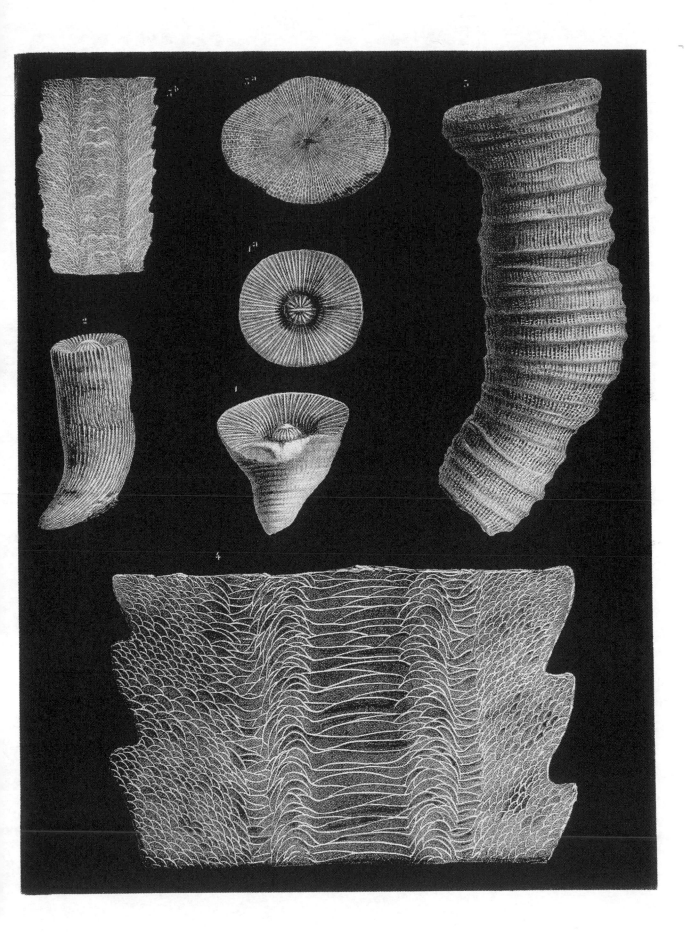

TAB. XXXIV.

CORALS FROM THE MOUNTAIN LIMESTONE.

ZAPHRENTIS ENNISKILLENI (p. 170).

Fig. 1.   A specimen, having half of its calice cut away, so as to show the septa and the septal fossule; natural size.

ZAPHRENTIS PHILLIPSI (p. 168).

2.   Side view of an adult specimen; natural size.

2 a.   A young specimen; natural size.

2 b.   Calice of an adult specimen; magnified.

ZAPHRENTIS GRIFFITHI (p. 169).

3.   Side view; natural size.

3 a.   Calice; natural size.

ZAPHRENTIS BOWERBANKI (p. 170).

4.   Side view; natural size.

4 a.   Calice, magnified.

AMPLEXUS HENSLOWI (p. 176).

5.   Side view; natural size.

5 a.   Vertical section; natural size.

CYATHOPHYLLUM WRIGHTI (p. 179).

6.   Side view; natural size.

6 a.   Calice; natural size.

CYATHOPHYLLUM ARCHIACI (p. 183).

7.   A specimen having half of the calice cut away; natural size.

TAB. XXXV.

CORALS FROM THE MOUNTAIN LIMESTONE.

ZAPHRENTIS CYLINDRICA (p. 171).

Fig. 1.   Side view of a straight specimen ; natural size.

     1 *a*. A fragment broken away, so as to show form and relative position of the septal fossulæ.

     1 *b*. Vertical section ; natural size.

TAB. XXXVI.

CORALS FROM THE MOUNTAIN LIMESTONE.

AMPLEXUS CORALLOIDES (p. 173).

Fig. 1. Side view of a part of a large specimen; natural size.

1 a. Vertical section; natural size.

1 b. Lower face of a tabula.

1 c. Side view of a part of a slender specimen; natural size.

1 d. Lower face of a tabula of the same.

1 e. Lower part of a young specimen.

CAMPOPHYLLUM MURCHISONI (p. 184).

Fig. 2. Side view; natural size.

2 a. Vertical section; natural size.

3. Horizontal section of a larger specimen; natural size.

T.36.

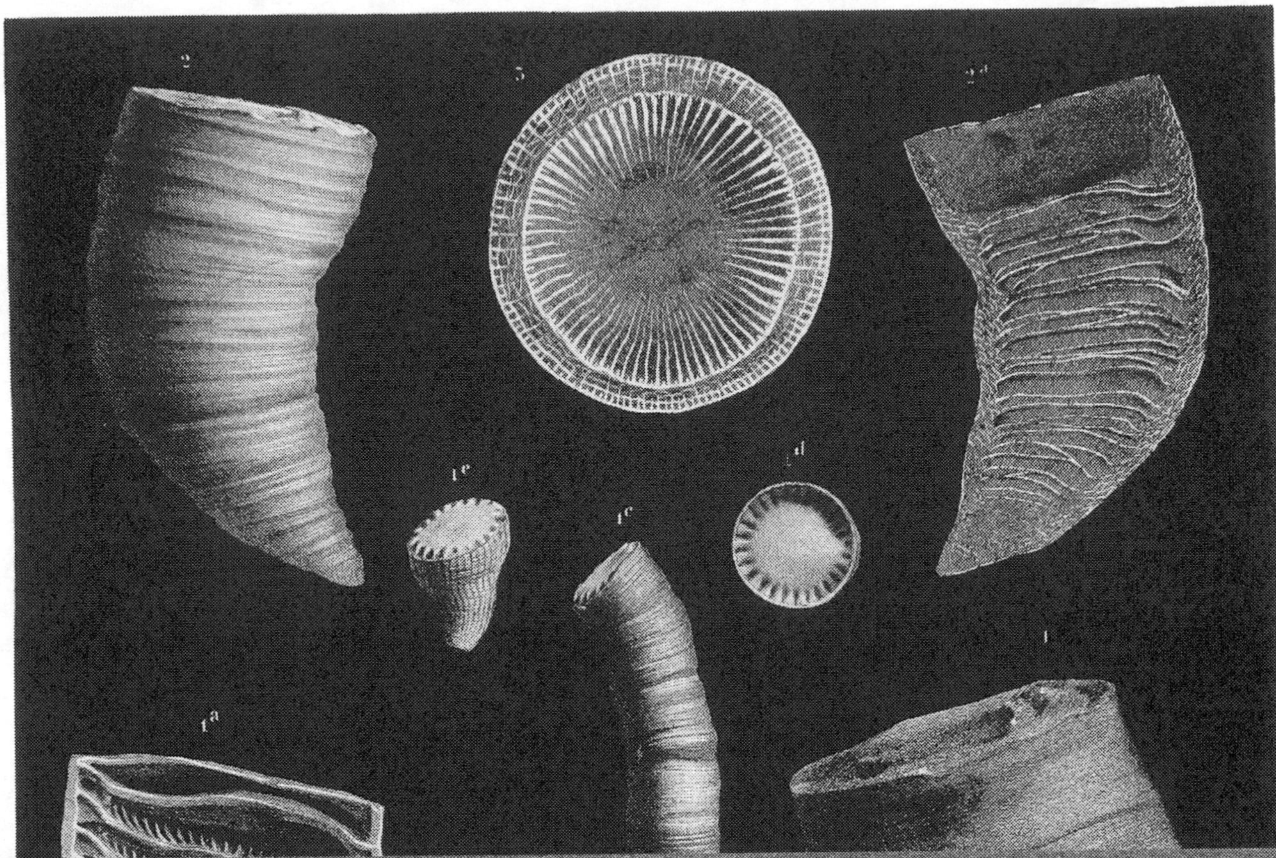

TAB. XXXVII.

CORALS FROM THE MOUNTAIN LIMESTONE.

CYATHOPHYLLUM PARRICIDA (p. 181).

Fig. 1.   A gemmiferous Corallite, bearing up four young ones; natural size.

1 *a*.   A separated Corallite; natural size.

1 *b*.   Its calice, magnified.

PHILLIPSASTRÆA RADIATA (p. 203).

Fig. 2.   A few calices; natural size.

2 *a*.   Part of a vertical section, showing the lateral surface of a septum and five septal edges, magnified.

AULOPHYLLUM FUNGITES (p. 188).

Fig. 3.   Side view; natural size.

CLISIOPHYLLUM BOWERBANKI (p. 186).

4.   Side view of a specimen, having the margins of its calice cut away; natural size.

4 *a*.   Its calice; natural size.

CLISIOPHYLLUM CONISEPTUM (p. 185).

5.   Side view of a specimen, partly broken at its two extremities; natural size.

5 *a*.   Upper part of a broken specimen; natural size.

TAB. XXXVIII.

CORALS FROM THE MOUNTAIN LIMESTONE.

AULOPHYLLUM BOWERBANKI (p. 189).

Fig. 1.   A weather-worn specimen; natural size.

LITHOSTROTION ENSIFER (p. 193).

Fig. 2.   Many calices; natural size.

2 a.   Calices, magnified.

LITHOSTROTION BASALTIFORME (p. 190).

3.   A few Corallites separated from a broken mass; natural size.

3 a.   Horizontal section; natural size.

3 b.   Part of a horizontal section, magnified.

PETALAXIS PORTLOCKI (p. 204).

Fig. 4.   A separated Corallite; natural size.

4 a.   Its calice, magnified.

LONSDALIA RUGOSA, (p. 208).

5.   Upper part of two Corallites, bearing young individuals; natural size.

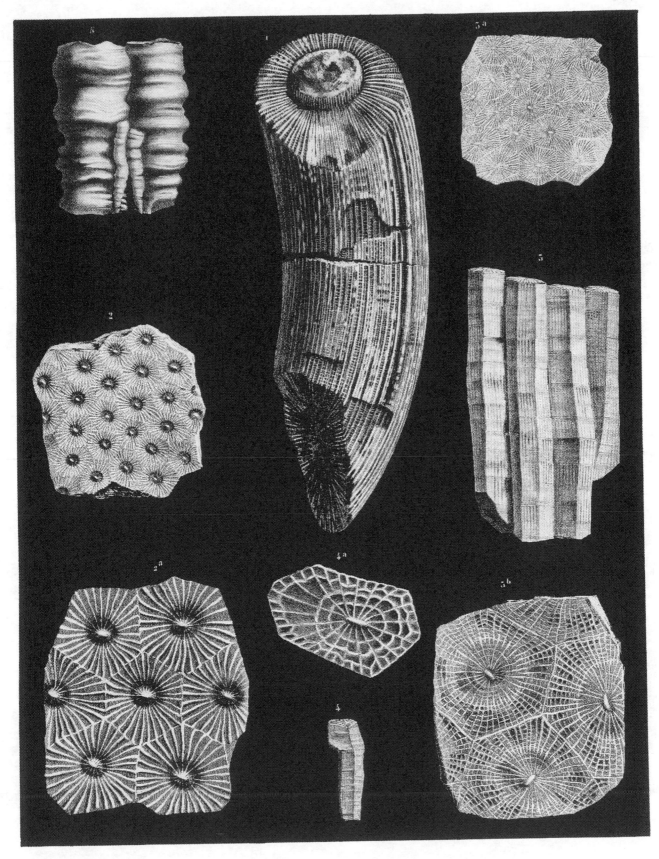

TAB. XXXIX.

CORALS FROM THE MOUNTAIN LIMESTONE.

LITHOSTROTION ARANEA (p. 193).

Fig. 1.   Horizontal section; natural size.

1 a. Part of a horizontal section, magnified.

LITHOSTROTION AFFINE (p. 200).

Fig. 2.   Side view of part of a group; natural size.

2 a. Calice magnified.

2 b. Oblique section, magnified.

LITHOSTROTION PHILLIPSI (p. 201).

Fig. 3.   Side view of part of a specimen; natural size.

3 a. Calice, magnified.

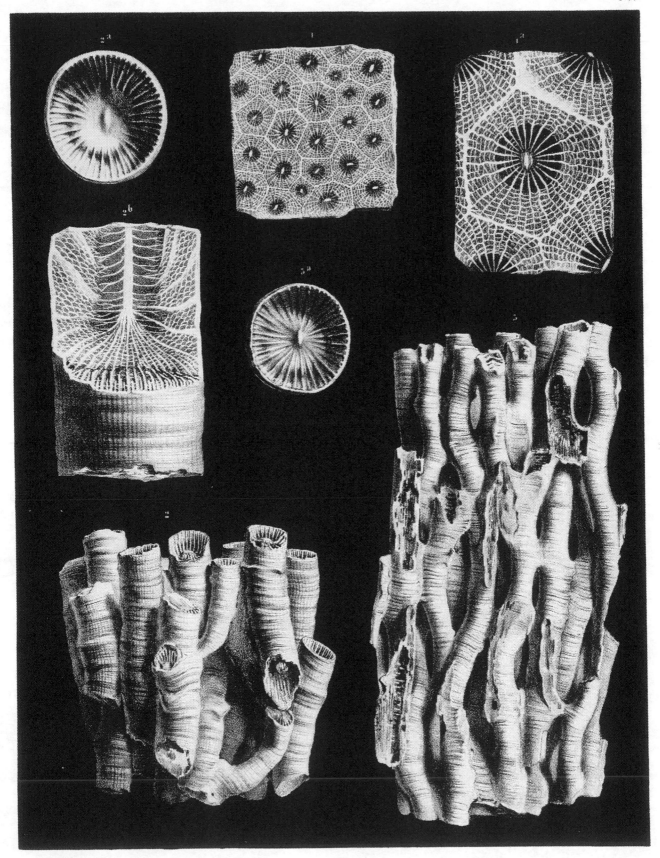

TAB. XL.

CORALS FROM THE MOUNTAIN LIMESTONE.

LITHOSTROTION JUNCEUM (p. 196).

Fig. 1.   Side view of part of a tuft; natural size.

1 *a*. Calice, magnified.

1 *b*. Vertical section of three Corallites, magnified.   In one of these the columella has disappeared.

LITHOSTROTION MARTINI (p. 197).

Fig. 2.   Side view of a tuft; natural size.

2 *a*. Calice, magnified.

2 *b*. A transversal section of a specimen, in which the intermural spaces are filled up with extraneous matter; natural size.

2 *c*. A vertical section of the same specimen; natural size.

2 *d*. A horizontal section of a calice, magnified.

2 *e*. Part of a vertical section of a Corallite, magnified.

2 *f*. A similar Corallite, but destitute of its columella.

2 *g*. Vertical section not passing through the central axis, magnified.

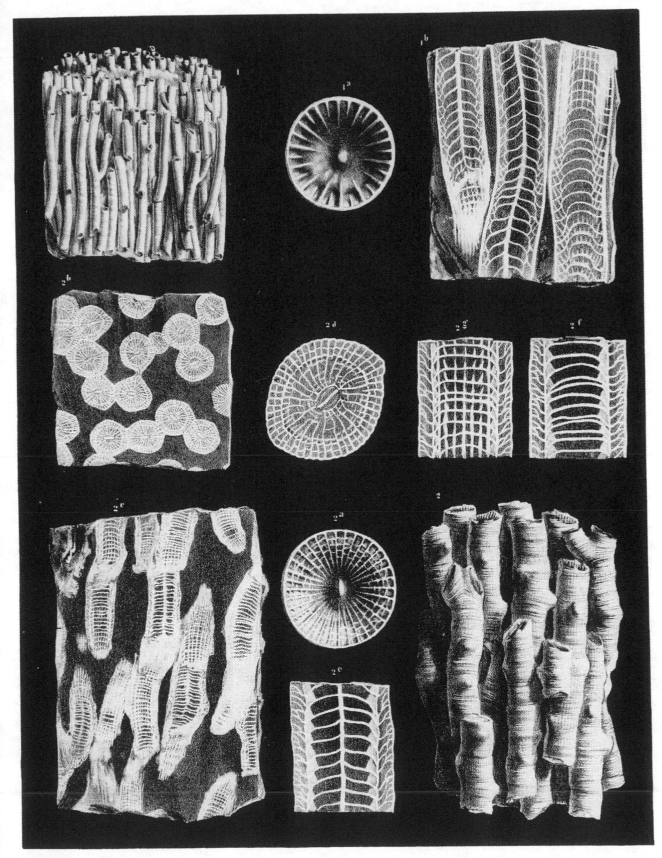

TAB. XLI.

CORALS FROM THE MOUNTAIN LIMESTONE.

Lithostrotion irregulare (p. 198).

Fig. 1. A specimen, partly broken; natural size.

1 *a*. Inferior part of a tuft; natural size.

1 *b*. Upper surface of a specimen, in which the intermural spaces are filled up with extraneous matter; natural size.

1 *c*. Calice, magnified.

1 *d*. Vertical section of a Corallite, magnified.

1 *e*. Vertical section of a Corallite not passing through the central axis, magnified.

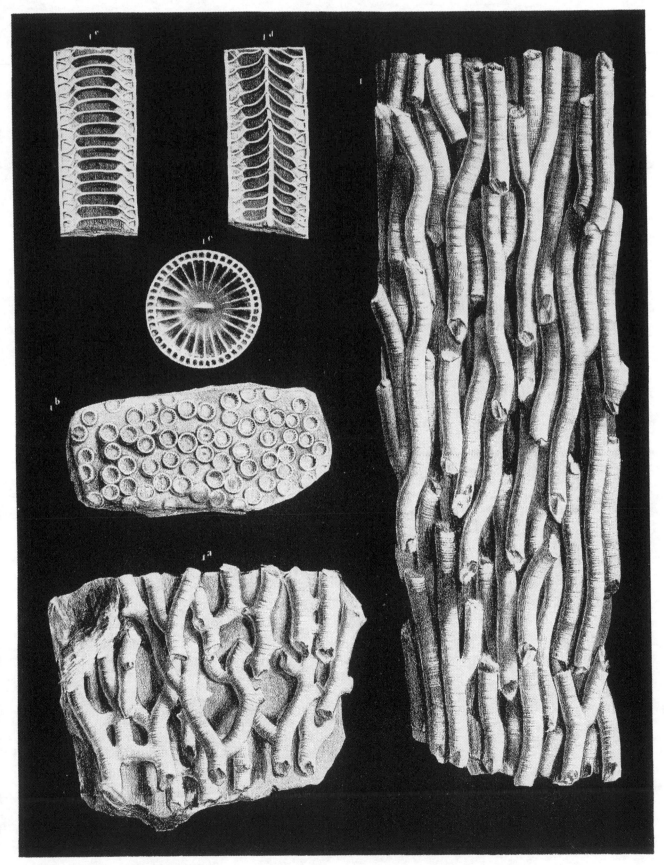

TAB. XLII.

CORALS FROM THE MOUNTAIN LIMESTONE.

LITHOSTROTION PORTLOCKI (p. 194).

Fig. 1. A specimen quite massive; natural size.

1 *a*. A massive specimen offering few corallites, free laterally; natural size.

1 *b*. Calice, magnified.

1 *c*. Some Corallites separated from a massive specimen; natural size.

1 *d*. One of these Corallites, magnified.

1 *e*. A vertical section, magnified.

1 *f*. Some calices having lost the columella.

1 *g*. A horizontal section, magnified.

LITHOSTROTION M'COYANUM (p. 195).

Fig. 2. A gibbose mass; natural size.

2 *a*. Under surface of a weather-worn specimen.

2 *b*. A few calices, magnified.

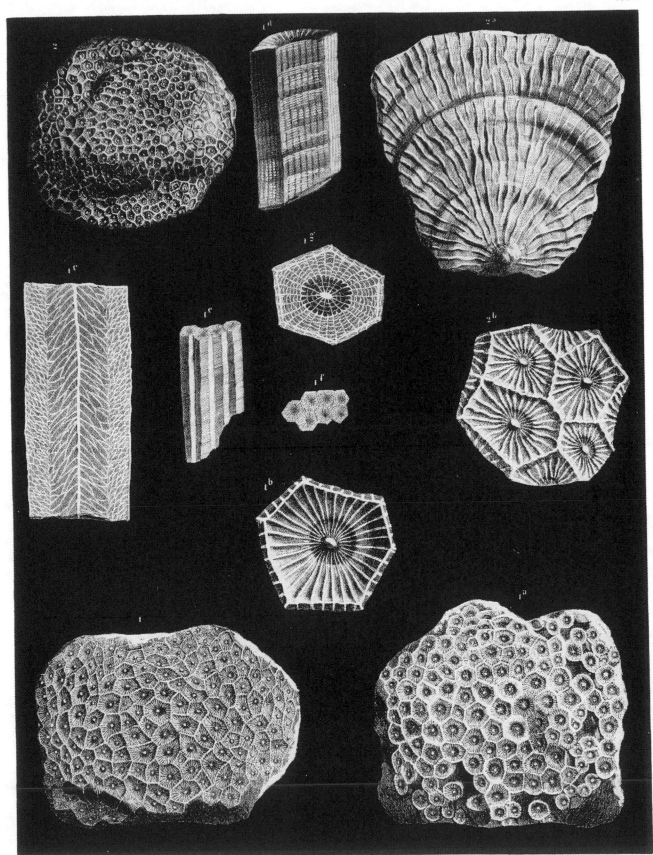

TAB. XLIII.

Lonsdalia floriformis (p. 205).

Fig. 1.   Part of a massive specimen; natural size.

1 *a*. Two Corallites issuing from a common parent; natural size.

1 *b*. Side view of a young group.

1 *c*. Side view of another young group; natural size.

1 *d*. Calice, magnified.

1 *e*. Part of a horizontal section, magnified.

2. *Varietas major.*   A few calices, somewhat broken; natural size.

TAB. XLIV.

CORALS FROM THE MOUNTAIN LIMESTONE.

MICHELINIA TENUISEPTA (p. 155).

Fig. 1.   Side view of a small mass; natural size.

1a.   Upper view of the same specimen.

1b.   Vertical section of a specimen embedded in extraneous matter; natural size.

MICHELINIA FAVOSA (p. 154).

Fig. 2.   A large broken specimen; natural size.

2a.   A few weather-worn calices, magnified.

2b.   Side view of some Corallites separated from the broken mass; natural size.

2c.   A few calices in a good state of preservation, magnified.

MICHELINIA MEGASTOMA (p. 156).

Fig. 3.   A subglobose specimen; natural size.

3a.   Some calices; natural size.

3b.   An oblique section; natural size.

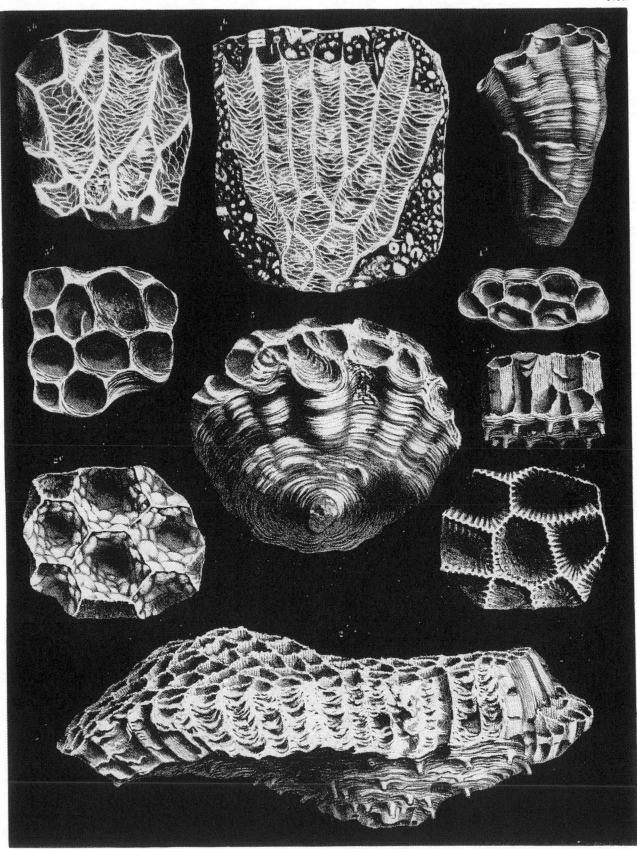

TAB. XLV.

CORALS FROM THE MOUNTAIN LIMESTONE.

BEAUMONTIA EGERTONI (p. 160).

    Fig. 1.   A broken mass ; natural size.

FAVOSITES PARASITICA (p. 153).

    Fig. 2.   A small globose specimen ; natural size.

CHÆTETES TUMIDUS (p. 159).

    Fig. 3.   A lobate specimen ; natural size.

        3*a*. A vertical section, somewhat magnified.

        3*b*. Calices, magnified.

ALVEOLITES DEPRESSA (p. 158).

    Fig. 4.   A gibbose mass ; natural size.

        4*a*. Some calices, magnified.

ALVEOLITES SEPTOSA (p. 157).

    Fig. 5.   A subglobose mass ; natural size.

        5*a*. A vertical section, magnified.

        5*b*. Some calices, magnified.

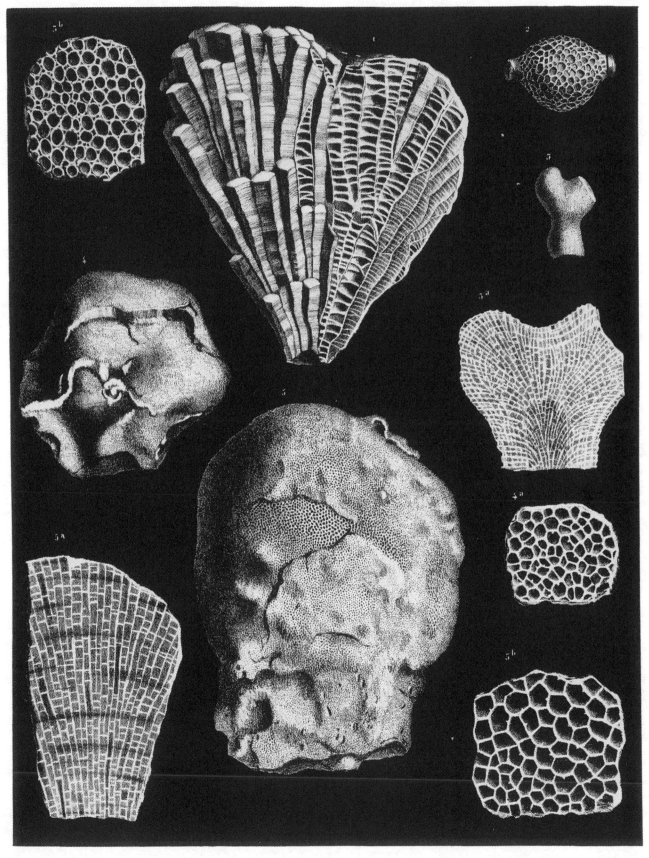

TAB. XLVI.

CORALS FROM THE MOUNTAIN LIMESTONE.

SYRINGOPORA RETICULATA (p. 162).

Fig. 1. A portion of a tuft; natural size.

 1a. A horizontal section of a specimen, in which the mural interspaces were filled up with extraneous matter.

SYRINGOPORA GENICULATA (p. 163).

Fig. 2. Side view of a broken tuft; natural size.

 2a. Upper view of the same; natural size.

SYRINGOPORA RAMULOSA (p. 161).

Fig. 3. Lateral view of the upper part of some Corallites embedded within extraneous matter; natural size.

 3a. A horizontal section, magnified.

 3b. A vertical section, magnified.

 3c. Lower part of some broken Corallites.

 4. A Corallum, supposed to be a small variety of *Syringopora geniculata*.

PYRGIA LABECHII (p. 166).

Fig. 5. Lateral view; natural size.

 5a. Lateral view, magnified.

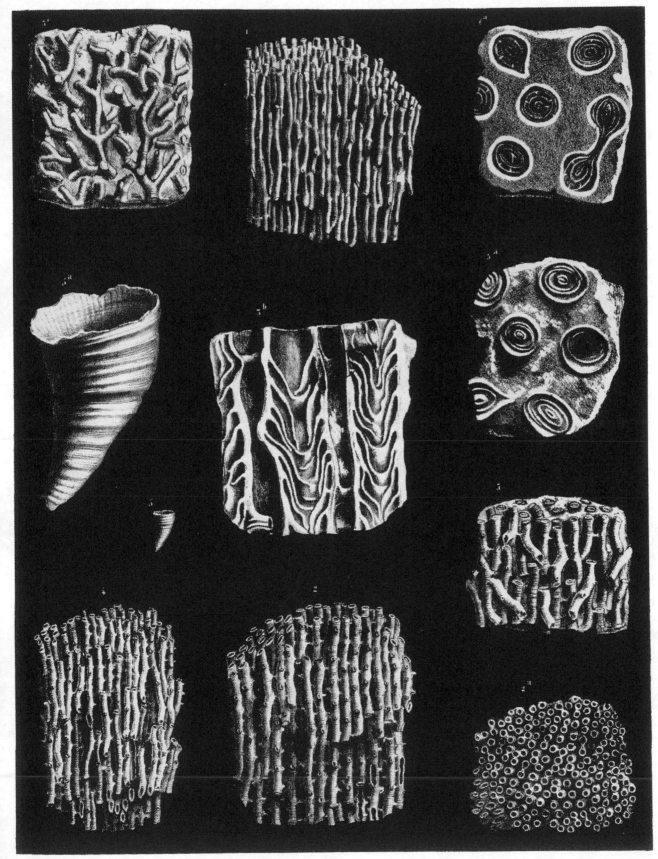

TAB. XLVII.

CORALS FROM THE DEVONIAN FORMATION.

HELIOLITES POROSA (p. 212).

Fig. 1.   A vertical section made through a globular specimen found at Torquay, and belonging to the collection of Mr. Bowerbank; natural size.

1*a*, 1*b*, 1*c*. Portions of different transverse sections, showing the variations in the size of the calices. Figs. 1*a* and 1*c* are from different parts of the same specimen; natural size.

1*d*. A magnified view of part of the specimen figured in 1*a*, showing the structure of the cænenchyma and the septa.

1*e*. A magnified view of a portion of the vertical section, fig. 1, showing the tabulæ of the corallites and the dissepiments of the columnal cænenchyma.

1*f*. A specimen of the same species, in which the substance of the coral has been destroyed and the visceral chambers of the corallites filled up with extraneous matter, constituting small prominent cylinders. Found at Torquay by Mr. Pengelly.

BATTERSBYIA INÆQUALIS (p. 213).

Fig. 2.   A transverse section polished; natural size. Specimen found at Teignmouth, and belonging to the collection of Dr. Battersby.

2*a*. A specimen showing the septa; natural size.

2*b*. The same, magnified.

FAVOSITES GOLDFUSSI (p. 214).

Fig. 3.   Vertical section polished; natural size. Specimen from Torquay, belonging to the collection of Mr. Bowerbank.

3*a*. Transverse section magnified, and showing some well-preserved calices with their septa.

3*b*. A fractured specimen, showing the prismatic form of the corallites; natural size. (Bowerbank collection.)

3*c*. The upper surface of the same specimen; natural size.

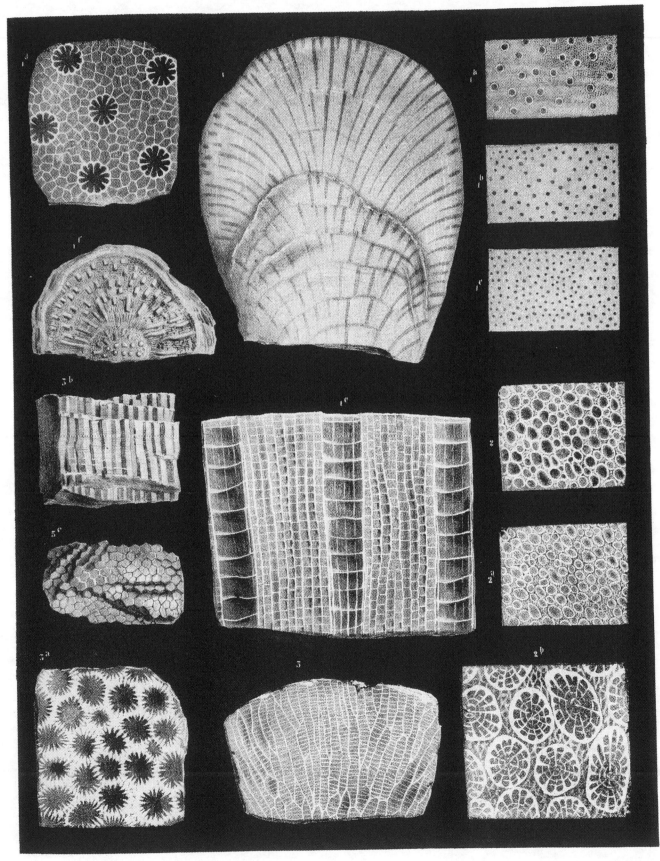

TAB. XLVIII.

CORALS FROM THE DEVONIAN FORMATION.

FAVOSITES RETICULATA (p. 215).

Fig. 1.   A specimen found at Torquay, and belonging to the collection of Dr. Battersby; natural size.

1*a*. A few calices of the same fossil, magnified.

1*b*. A vertical section of a specimen from Torquay, given to the Parisian Museum by Professor Milne Edwards.

FAVOSITES CERVICORNIS (p. 216).

Fig. 2.   A vertical section of a specimen from Torquay, belonging to the collection of the Geological Society; natural size.

FAVOSITES FIBROSA (p. 217).

Fig. 3.   Vertical section of a specimen from Torquay belonging to Dr. Battersby; natural size.

3*a*. A portion of the same, magnified, to show the septa and the visceral chambers of the corallites.

3*b*. A portion of the same, showing some calices that have been modified in their structure by the process of fossilization; magnified.

EMMONSIA HEMISPHERICA (p. 218).

Fig. 4.   Part of a vertical section of a specimen from Torquay, belonging to Dr. Battersby; natural size.

4*a*. Part of the same, magnified.

ALVEOLITES VERMICULARIS (p. 220).

Fig. 5.   Vertical section of a specimen from Torquay, belonging to the authors; natural size.

5*a*. Part of the same, magnified.

P.Lackerbauer ad nat in lap del.

Day & Son Lith.rs The Queen.

TAB. XLIX.

CORALS FROM THE DEVONIAN FORMATION.

ALVEOLITES SUBORBICULARIS (p. 219).

Fig. 1.   A polished vertical section of a specimen from Torquay belonging to Mr. Pengelly.

1a. An oblique section, magnified.

ALVEOLITES BATTERSBYI (p. 220).

Fig. 2.   A vertical section of a specimen from Torquay, belonging to Dr. Battersby; natural size.

2a. A part of the same, magnified, and showing the trabecular septa.

ALVEOLITES COMPRESSA (p. 221).

Fig. 3.   Part of a transverse section of a specimen from Torquay, belonging to Mr. Pengelly.

METRIOPHYLLUM BATTERSBYI (p. 222).

Fig. 4.   Transverse section of a specimen from Torquay, belonging to Dr. Battersby; double the natural size.

AMPLEXUS TORTUOSUS (p. 222).

Fig. 5a.   Vertical section of a specimen from Plymouth, belonging to Mr. Pengelly.

HALLIA PENGELLYI (p. 223).

Fig. 6.   Transverse section of a specimen from Torquay, belonging to Mr. Pengelly.

6a. A specimen from Petherwin, belonging to the collection of the Geological Society of London; natural size.

6b. A side view of the same.

CYATHOPHYLLUM OBTORTUM (p. 225).

Fig. 7.   Transverse section of a specimen from Torquay, belonging to Professor Phillips; double the natural size.

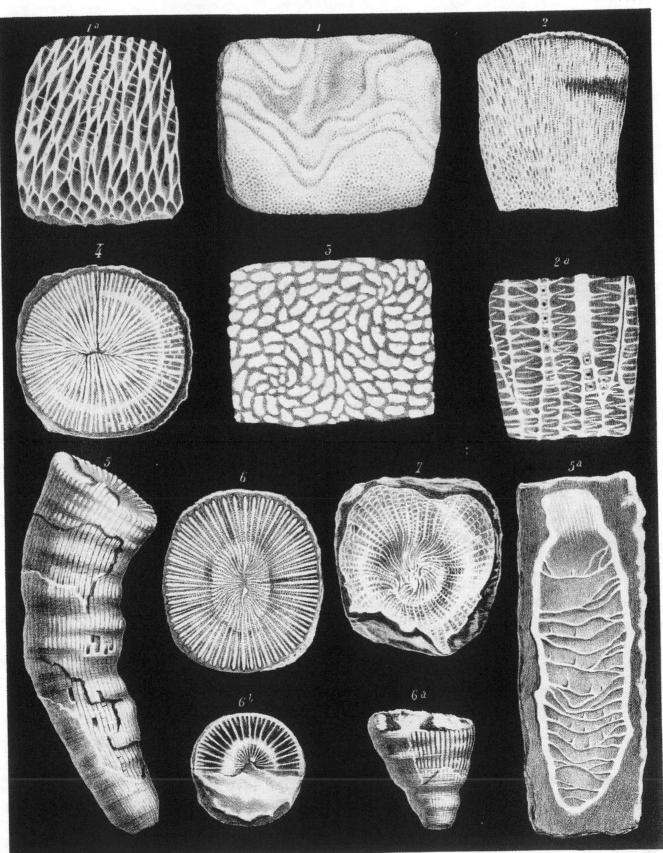

TAB. L.

CORALS FROM THE DEVONIAN FORMATION.

CYATHOPHYLLUM DAMNONIENSE (p. 225).

    Fig. 1.   An oblique section of a specimen from Torquay, belonging to Dr. Battersby; natural size.

CYATHOPHYLLUM CERATITES (p. 224).

    Fig. 2.   A young individual from Barton quarry, near Newton, belonging to Dr. Battersby; natural size.

CYATHOPHYLLUM ROEMERI (p. 224).

    Fig. 3.   Transverse section of a specimen from Torquay, belonging to Dr. Battersby; double the natural size.

CYATHOPHYLLUM HEXAGONUM (p. 228).

    Fig. 4.   Transverse section of a specimen from Torquay, belonging to Dr. Battersby; natural size.

      4a. One of the above calices magnified.

CHONOPHYLLUM PERFOLIATUM (p. 235).

    Fig. 5.   Vertical section of an individual imbedded in extraneous matter, from Torquay; natural size.   (Collection of Dr. Battersby.)

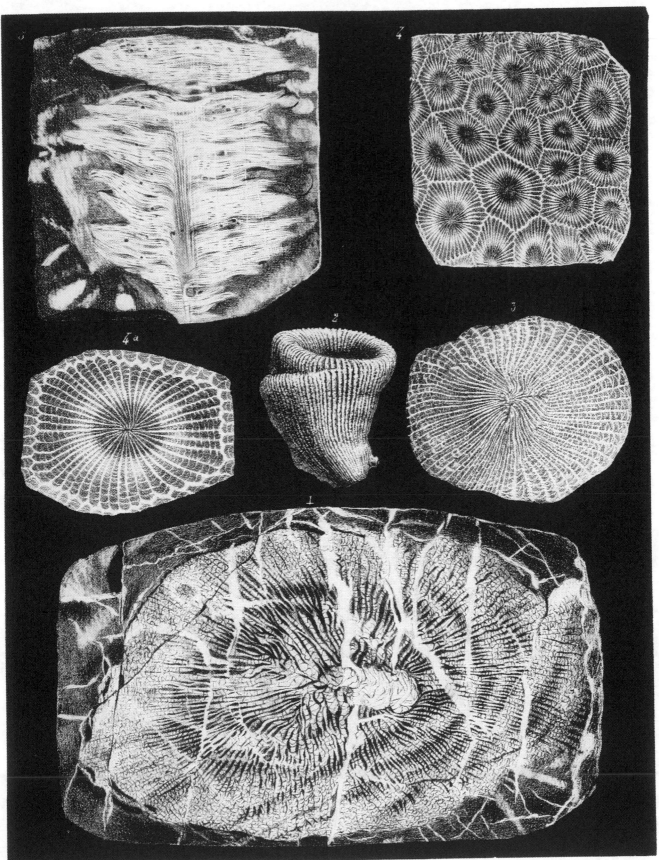

TAB. LI.

CORALS FROM THE DEVONIAN FORMATION.

CYATHOPHYLLUM HELIANTHOIDES (p. 227).

Fig. 1.   Transverse section of a simple corallum, from Torquay, belonging to Dr. Battersby; natural size.

1a. Transverse section of a compound corallum, found at Plymouth by Mr. Pengelly; natural size.

CYATHOPHYLLUM CÆSPITOSUM (p. 229.)

Fig. 2.   A fractured specimen, showing the structural characters; natural size.   From Torquay, Dr. Battersby's collection.

2a. A transverse section of some of the same corallites, somewhat magnified.

2b. A polished slab, showing a transverse section of a specimen from Teignmouth beach, belonging to the collection of Mr. Bowerbank.

HELIOPHYLLUM HALLI (p. 235).

Fig. 3.   Transverse section of a specimen from Torquay, belonging to Dr. Battersby; double the natural size.

P.Lackerbauer ad nat in lap del.

Day & Son Lith'to The Queen.

TAB. LII.

CORALS FROM THE DEVONIAN FORMATION.

CYATHOPHYLLUM BOLONIENSE (p. 230).

Fig. 1. Upper surface of a specimen modified by fossilization; found at Ogwell by Dr. Battersby.

1a. Part of a transverse section of a specimen from Torquay, magnified. (Collection of Dr. Battersby.)

CYATHOPHYLLUM ÆQUISEPTATUM (p. 232).

Fig. 2. A specimen showing the calices of a few corallites imbedded in extraneous matter; found at Ilfracombe, and belonging to the collection of the Geological Society of London; natural size.

CYATHOPHYLLUM SEDGWICKI (p. 231).

Fig. 3. A polished slab, showing a transverse section of a specimen found at Torquay, and belonging to Mr. Bowerbank's collection; natural size.

3a. A part of the same, magnified.

CYATHOPHYLLUM MARMINI (p. 231).

Fig. 4. A polished transverse section of a specimen from Torquay, belonging to Dr. Battersby; natural size.

4a. Part of the same, magnified.

PACHYPHYLLUM DEVONIENSE (p. 234).

Fig. 5. A polished transverse section of a specimen from Torquay, belonging to Dr. Battersby; natural size.

5a. Part of the same, magnified.

ENDOPHYLLUM ABDITUM (p. 233).

Fig. 6. Transverse section of a specimen from Teignmouth beach, belonging to Dr. Battersby; natural size.

34

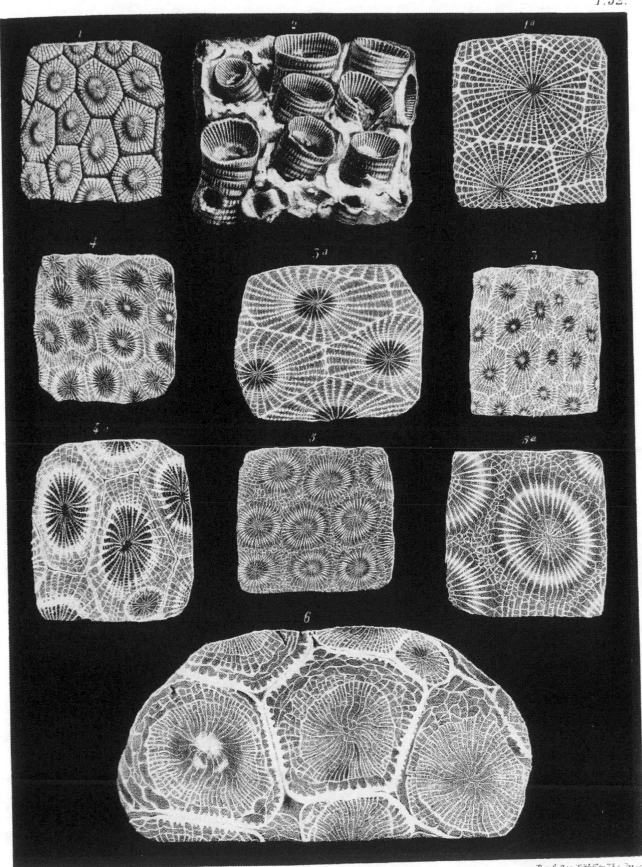

P. Lackerbauer ad nat in lap. del.

Day & Son, Lith.rs to The Queen

TAB. LIII.

CORALS FROM THE DEVONIAN FORMATION.

ENDOPHYLLUM BOWERBANKI (p. 233).

Fig. 1. A transverse section of a specimen found at Barton, and belonging to Mr. Bowerbank; natural size.

ACERVULARIA INTERCELLULOSA (p. 237).

Fig. 2. Upper surface of a specimen from Torquay, belonging to Mr. Pengelly natural size.

2a. Transverse section of another specimen from the same locality belonging to Dr. Battersby; magnified.

ACERVULARIA GOLDFUSSI (p. 236).

Fig. 3. Transverse section of a specimen from Torquay, belonging to Dr. Battersby; natural size.

3a. Part of the same slab, magnified.

ACERVULARIA CORONATA (p. 237).

Fig. 4. A specimen modified in its structure by the process of fossilization; from Torquay, and belonging to Mr. Pengelly; natural size.

4a. Transverse section of a specimen from Barton, belonging to Dr. Battersby; natural size.

4b. A portion of the same section, magnified.

ACERVULARIA PENTAGONA (p. 238).

Fig. 5. A specimen modified by the process of fossilization; from Torquay, and belonging to Mr. Pengelly; natural size.

5a. A specimen from Ogwell, belonging to Mr. Bowerbank's collection; natural size.

5b. A portion of a transverse section of another specimen from the same locality; (Mr. Bowerbank's collection); natural size.

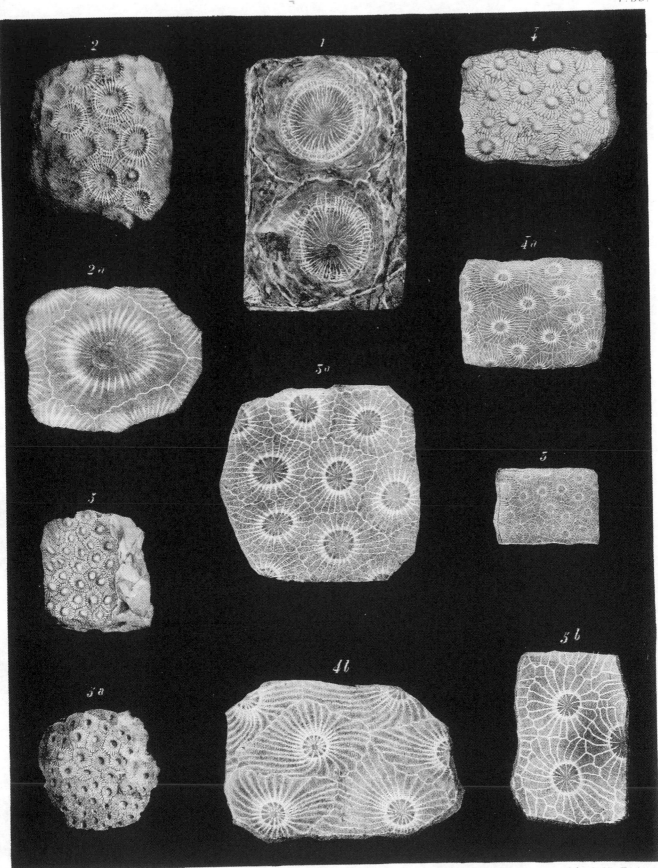

P.Lackerbauer ad nat. in lap. del.

Day & Son, Lithrs to The Queen.

TAB. LIV.

CORALS FROM THE DEVONIAN FORMATION.

ACERVULARIA LIMITATA (p. 238).

Fig. 1. Polished slab, showing a transverse section of a specimen from Newton, belonging to Mr. Pengelly; natural size.

1*a*. A part of the same, magnified.

ACERVULARIA BATTERSBYI (p. 239).

Fig. 2. Transverse section of a polished specimen from Torquay, belonging to Dr. Battersby; natural size.

2*a*. A part of the same slab, magnified.

ACERVULARIA RŒMERI (p. 239).

Fig. 3. Transverse section of a specimen from Torquay; natural size; Dr. Battersby's collection.

3*a*. Part of the same slab, magnified.

SMITHIA HENNAHI (p. 240).

Fig. 4. Transverse section of a specimen from Teignmouth beach; natural size; (Dr. Battersby's collection).

4*a*. Another specimen from Torquay.

4*b*. An oblique section of a specimen from Torquay; somewhat magnified.

4*c*. Transverse section of another specimen from Torquay; natural size; (Dr. Battersby's collection).

4*d*. Part of the same slab, magnified.

_____

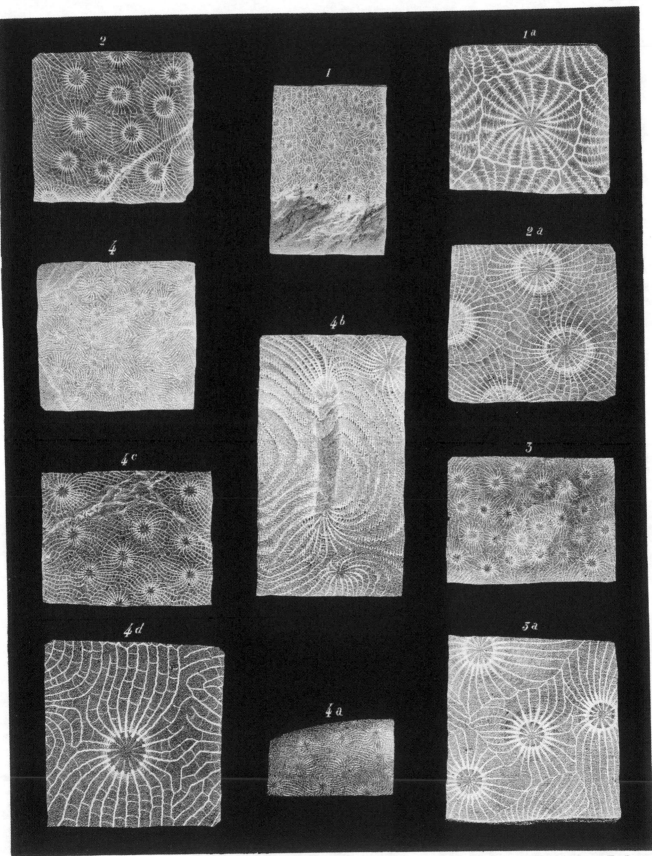

TAB. LV.

CORALS FROM THE DEVONIAN FORMATION.

SMITHIA PENGELLYI (p. 241).

Fig. 1. Transverse section of a specimen from Torquay, belonging to Mr. Bowerbank; natural size.

1a. A part of the same, magnified.

1b. Another specimen, from the same locality.

SMITHIA BOWERBANKI (p. 241).

Fig. 2. A polished specimen from Torquay, belonging to Dr. Battersby; natural size.

2a. Part of the same, magnified.

SYRINGOPHYLLUM CANTABRICUM (p. 242).

Fig. 3. Transverse section of a specimen from Teignmouth, belonging to the collection of the Geological Society of London; natural size.

3a. Part of the same, magnified.

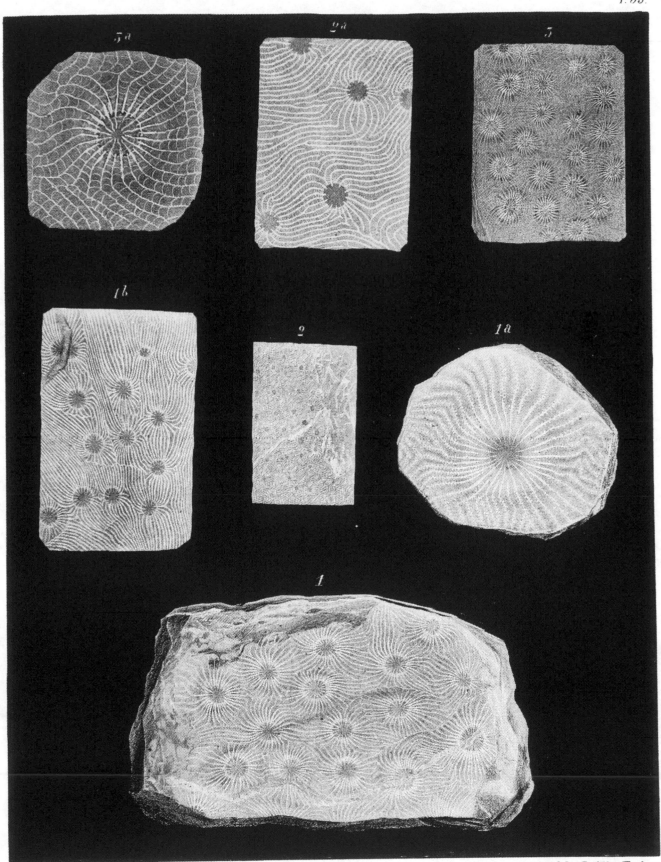

P.Lackerbauer ad nat in la p del.

Day & Son Lith.rs to The Queen

TAB. LVI.

CORALS FROM THE DEVONIAN FORMATION.

CYSTOPHYLLUM VESICULOSUM (p. 243).

Fig. 1. Side view of a coral from Torquay, belonging to Mr. Pengelly; natural size.

1*a*. A transverse section of another specimen from the same locality; natural size.

1*b*. Part of a vertical section of a specimen from Mudstone Bay, belonging to Dr. Battersby; natural size.

SPONGOPHYLLUM SEDGWICKI (p. 242).

Fig. 2. Transverse section of a specimen with thick walls, from Torquay; natural size.

2*a*. Part of the same, magnified.

2*b*. Specimen with thin walls; natural size.

2*c*. Part of the same, magnified.

2*d*. Another specimen from Torquay, with walls still thinner, and the vertical dissepiments stronger than in the preceding ones, magnified.

2*e*. A vertical section magnified.

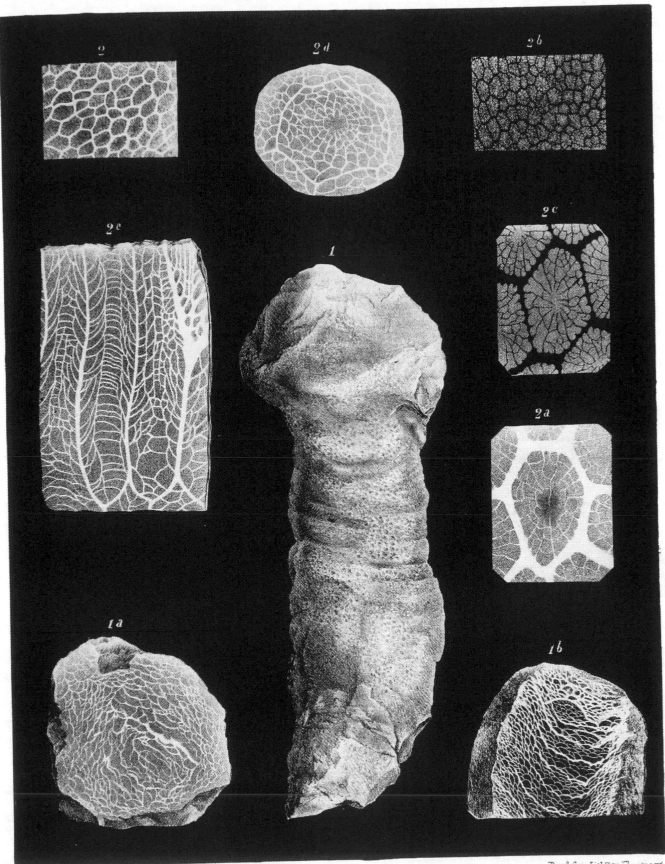

P.Lackerbauer ad nat in lap del[t]

Day & Son, Lith[rs] to The Queen.

TAB. LVII.

CORALS OF THE SILURIAN FORMATION.

PALÆOCYCLUS PORPITA (p. 246).

Fig. 1.   Upper surface of a specimen found at Dudley; Collection of Mr. Fletcher; natural size.

1a.   Lower surface of another specimen found at Dudley; Collection of Mr. Fletcher; natural size.

1b.   Side view of the specimen figured in 1; natural size.

1c.   Upper surface of the specimen figured in 1; magnified.

PALÆOCYCLUS PRÆACUTUS (p. 247).

Fig. 2.   Upper surface of a specimen found at Marloes Bay, and belonging to the Collection of the Geological Society of London; natural size.

2a.   Lower surface of the same; natural size.

2b.   Side view of the same; natural size.

2c.   A part of the upper surface, magnified.

PALÆOCYCLUS FLETCHERI (p. 248).

Fig. 3.   Side view of a young specimen; natural size.

3a.   Upper surface of the same; natural size.

3b.   Upper surface of an adult specimen; natural size.

3c.   Side view of a young elongate specimen.

3d.   Side view of an adult specimen; natural size.

3e.   A specimen bearing two young ones; natural size.

3f.   Part of the upper surface of 3b, magnified.

All these specimens were found at Dudley, and belong to the Collection of Mr. Fletcher.

PALÆOCYCLUS RUGOSUS (p. 248).

Fig. 4.   Side view of a small specimen; natural size.

4a.   Upper surface of another specimen; natural size.

4b, 4c.   Side views of two elongate ones; natural size.

4d.   Part of the upper surface of 4a, magnified.

These various specimens were found at Dudley, and belong to the Collection of Mr. Fletcher.

HELIOLITES INTERSTINCTA (p. 249).

Fig. 5.   A specimen found at Much Wenlock, belonging to the Collection of Mr. J. S. Bowerbank.

5a.   Portion of the surface; natural size.

5b.   Portion of the same surface, magnified.

5c.   A cast found at Applethwaite, belonging to the Collection of the Geological Society of London.

5d.   A part of the same surface, magnified.

HELIOLITES MURCHISONI (p. 250).

Fig. 6.   Under surface of a specimen found at Wenlock Edge, belonging to the Museum of Paris; natural size.

6a.   A part of the upper surface; natural size.

6b.   A part of the surface of a smaller specimen from Dudley, belonging to M. de Verneuil; natural size.

6c.   Vertical section of the specimen figured in 6; somewhat magnified.

HELIOLITES INORDINATA (p. 253).

Fig. 7.   A specimen from Robeston Walthen; in the Collection of the Geological Society of London; natural size.

7a.   Part of the same, magnified.

P.Lackerbauer, ad nat in lap. del.ᵗ

Day&Son, Lithʳˢ to the Queen

TAB. LVIII.

CORALS OF THE SILURIAN FORMATION.

HELIOLITES GRAYI (p. 252).

Fig. 1.   A portion of a large specimen found at Walsall, belonging to the Collection of Mr. John Gray ; natural size. (Restored by means of several fragments.)

1a.   Part of the surface, magnified.

HELIOLITES MEGASTOMA (p. 251).

Fig 2.   A portion of a specimen found at Wenlock Edge, in the Collection of the Museum of Paris ; natural size.

2a.   Part of its surface, magnified.

2b.   Vertical section of the same, magnified.

2c.   Surface of a cast found at Coniston, in the Collection of the Geological Society of London ; natural size.

2d.   Part of the same, magnified.

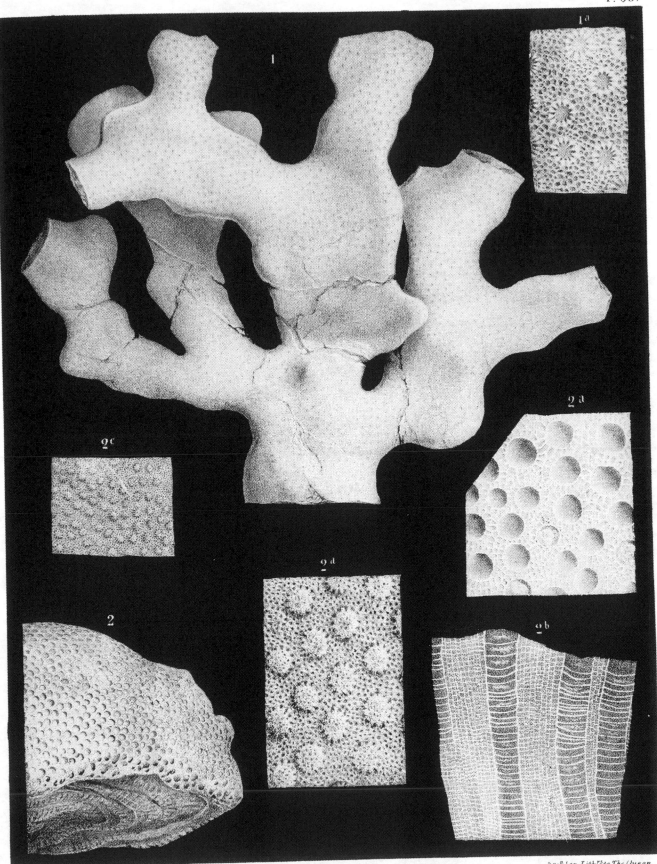

P Lackerbauer ad. nat in lap. del.ᵗ

Day &Son,Lithʳˢ to The Queen.

TAB. LIX.

CORALS OF THE SILURIAN FORMATION.

PLASMOPORA PETALIFORMIS (p. 253).

Fig. 1. Upper surface of a specimen found at Dudley, belonging to the Parisian Museum; natural size.

1a. Under surface of the same; natural size.

1b. A magnified portion of the surface of another specimen, found also at Dudley, and belonging to the Collection of Mr. John Gray, in which the calices are more distant, than in the preceding specimen.

1c. Vertical section of the specimen figured in 1 and 1a.

PLASMOPORA SCITA (p. 254).

Fig. 2. Upper surface of a specimen from Dudley, in the Collection of Mr. John Gray; natural size.

2a. Calices, magnified.

PROPORA TUBULATA (p. 255).

Fig. 3. Side view of a specimen from Wenlock, belonging to the Museum of Paris; natural size.

3a. Calices, magnified.

3b. Part of a vertical section, magnified.

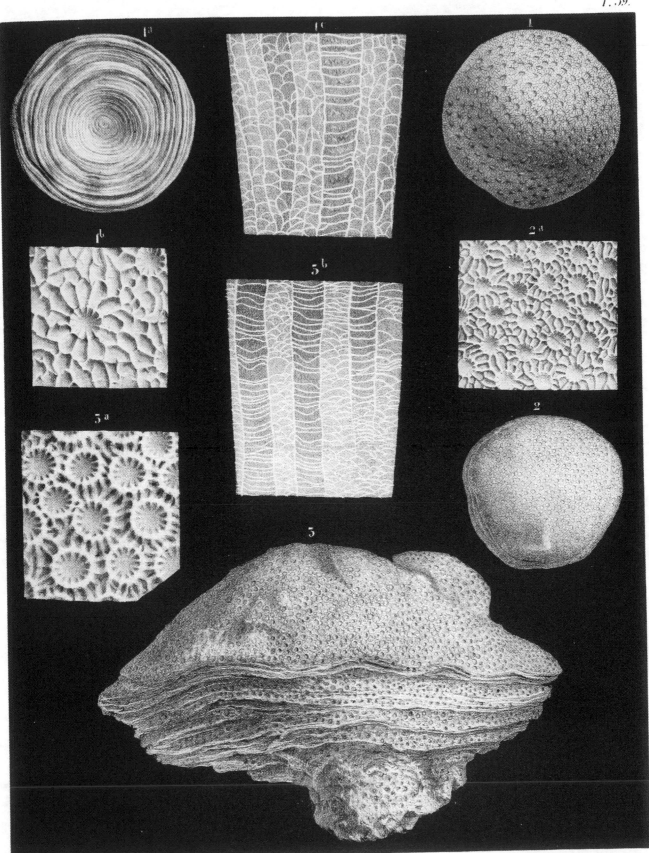

P Lackerbauer ad. nat. in lap. del.t

Day & Son, Lith.rs to The Queen.

TAB. LX.

CORALS FROM THE SILURIAN FORMATION.

FAVOSITES GOTHLANDICA (p. 256).

Fig. 1.   Side view of a broken specimen from Dinas court, belonging to the Collection of the Geological Society of London; natural size.

1a.   A cast from Cullimore's quarry, belonging to the same Collection; natural size.

FAVOSITES FORBESI (p. 258).

Fig. 2.   Upper surface of a young specimen found at Dudley, belonging to the Collection of Mr. T. W. Fletcher; natural size.

2a. and 2b.   Under surface and side view of other young specimens from the same locality and the same Collection; natural size.

2c.   Upper surface of a specimen found at Dudley, belonging to the Museum of Paris; natural size.

2d.   Portion of the surface of another specimen from Wren's Nest, in the Collection of the Geological Society; natural size.

2e.   Portion of the surface of another specimen from Dudley, belonging to the Museum of Paris; natural size.

2f.   Some corallites, magnified.   Specimen from Wenlock; Collection of the Geological Society.

2g.   Vertical section of the specimen figured in 2c; amplified.

FAVOSITES ASPERA (p. 257).

Fig. 3.   A broken specimen from Leinthall Earls, near Ludlow, belonging to the Collection of the Geological Society; natural size.

3a.   Some corallites; amplified.

FAVOSITES MULTIPORA (p. 258).

Fig. 4.   A cast found at Haverford West, belonging to the Collection of the Geological Society of London; natural size.

P. Hackerbauer ad. nat. in lap. del.t

Day & Son, Lith.rs to The Queen.

TAB. LXI.

CORALS OF THE SILURIAN FORMATION.

FAVOSITES HISINGERI (p. 259).

Fig. 1. Side view of a specimen found at Benthall Edge by M. Bouchard-Chantereaux; natural size.

1a. Calices, amplified.

1b. Vertical section, of amplified, a specimen from Wenlock Edge, belonging to the Museum of Paris.

ALVEOLITES GRAYI (p. 263).

Fig. 2. A portion of the surface of a specimen from Dudley, belonging to Mr. John Gray; natural size.

2a. Calices, magnified.

FAVOSITES CRISTATA (p. 260).

Fig. 3. A specimen from Dudley, belonging to the Collection of Mr. Bowerbank; natural size.

3a. A part of its surface, magnified.

FAVOSITES CRISTATA (*varietas major*, p. 260).

Fig. 4. A specimen from Wenlock, belonging to the Museum of Paris; natural size.

4a. A portion of its surface, magnified.

FAVOSITES FIBROSA (pp. 217, 261).

Fig. 5. A broken lobate specimen from Horderley, belonging to the Collection of the Geological Society of London; natural size.

5a. Corallites, amplified.

ALVEOLITES LABECHEI (p. 262).

Fig. 6. Side view of a specimen from Wenlock, in the Museum of Paris.

6a. Calices, magnified.

6b. A part of a vertical section, magnified.

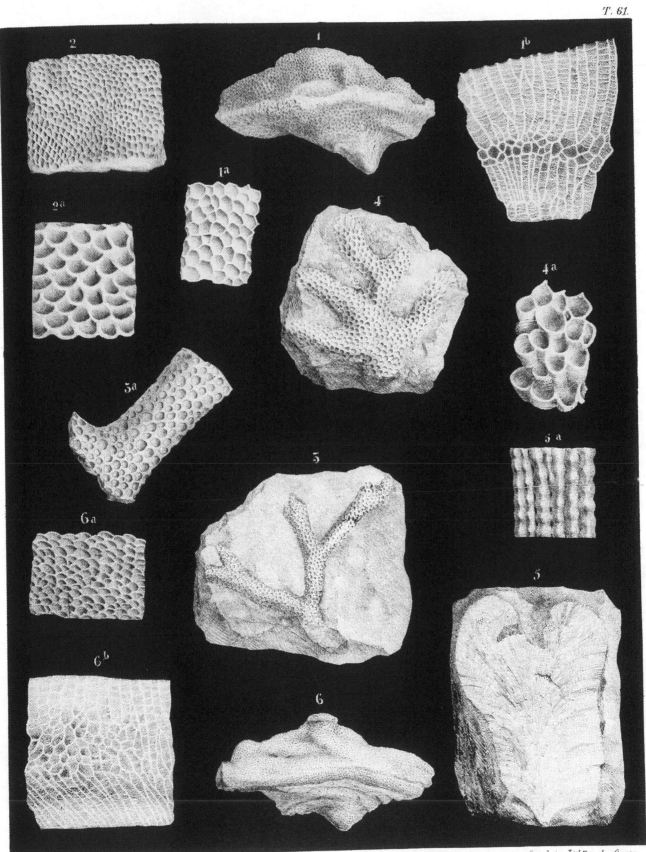

P.Lackerbauer ad nat. in lap del.ᵗ

Lxv & Son, Lithᵣˢ to the Queen.

TAB. LXII.

CORALS OF THE SILURIAN FORMATION.

ALVEOLITES REPENS (p. 263).

Fig. 1.   A specimen from Dudley, belonging to the Collection of Mr. Fletcher; natural size.

1a.   Branches, magnified.

ALVEOLITES? SERIATOPOROIDES (p. 263).

Fig. 2.   A specimen from Dudley, belonging to the Collection of Mr. Fletcher; natural size.

2a.   A branch, magnified.

MONTICULIPORA FLETCHERI (p. 267).

Fig. 3.   A small specimen found at Dudley, belonging to the Collection of the Geological Society of London; natural size.

3a.   Top of a branch, magnified.

MONTICULIPORA PAPILLATA (p. 266).

Fig. 4.   A young specimen found at Dudley by Mr. Fletcher; natural size.

4a.   A portion of its surface, magnified.

MONTICULIPORA PULCHELLA (p. 267).

Fig. 5.   A specimen from Dudley, belonging to the Collection of Mr. Fletcher; natural size.

5a and 5b.   Two branches, magnified.

LABECHEIA CONFERTA (p. 269).

Fig. 6.   Upper surface of a specimen found by us at Benthall Edge, and placed in the Museum of Paris.

6a.   The same surface, magnified.

6b.   Vertical section, magnified.

6c.   Under surface of the same specimen; natural size.

TAB. LXIII.

CORALS OF THE SILURIAN FORMATION.

MONTICULIPORA BOWERBANKI (p. 268).

Fig. 1.　A large specimen found at Dudley, belonging to the Collection of Mr. Fletcher.

1*a*.　A terminal portion, magnified.

1*b*.　A young specimen from the same locality and Collection.

1*c*.　A portion of its surface, magnified.

Day & Son, Lith^{rs} to the Queen.

P. Lackerbauer ad nat. in lap. del.

TAB. LXIV.

CORALS OF THE SILURIAN FORMATION.

HALYSITES CATENULARIA (p. 270).

Fig. 1. Side view of a vertical series of corallites; natural size. This specimen is from Dudley, and belongs to the Collection of the Geological Society of London.

1a. Upper surface of another specimen, from the same locality and Collection; natural size.

1b. Vertical section in a specimen from Dudley; natural size. (*Varietas major.*)

1c. Upper surface of a specimen found at Benthall Edge by M. Bouchard-Chantereaux; natural size. (*Var. agglomerata.*)

HALYSITES ESCHAROIDES (p. 272).

Fig. 2. Upper surface of a specimen found at Benthall Edge by M. Bouchard-Chantereaux; natural size.

2a. Side view of the same specimen.

SYRINGOPORA BIFURCATA (p. 273).

Fig. 3. Side view of a specimen found at Dudley, belonging to the Collection of Mr. Fletcher; natural size.

3a. Its upper surface; natural size.

3b. Some corallites, magnified.

T. 64

P.Lackerbauer ad. nat. in lap. del.t

Day & Son, Lith.rs to the Queen

TAB. LXV.

CORALS OF THE SILURIAN FORMATION.

SYRINGOPORA FASCICULARIS (p. 274).

Fig. 1. Young state of the corallum; natural size; Dudley, Collection of M. de Verneuil.

1a. Upper surface of a young mass, embedded in the extraneous matter; natural size; from the same locality and Collection.

1b. Upper surface of a specimen, found also at Dudley, belonging to the Collection of Mr. Fletcher; natural size.

1c. Under surface of a specimen from Dudley, belonging to M. de Verneuil; natural size.

SYRINGOPORA SERPENS (p. 275).

Fig. 2. Side view of a specimen from Dudley, belonging to the Collection of Mr. Fletcher; natural size.

2a. Upper surface of the same, somewhat magnified.

CŒNITES LINEARIS (p. 277).

Fig. 3. A part of a specimen found at Dudley, in the Collection of Mr. Fletcher; magnified.

CŒNITES JUNIPERINUS (p. 276).

Fig. 4. A specimen from Dudley, in the Collection of Mr. Fletcher; natural size.
4a. A portion of its surface, magnified.

CŒNITES INTERTEXTUS (p. 276).

Fig. 5. A specimen from Dudley, in the Collection of Mr. Fletcher; natural size.
5a. A part of its surface, magnified.

CŒNITES LABROSUS (p. 277).

Fig. 6. A specimen found at Dudley, belonging to the Collection of Mr. Fletcher; natural size.

6a. A part of its surface, magnified.

THECIA SWINDERNANA (p. 278).

Fig. 7. Upper surface of a specimen from Lincoln Hill, belonging to M. Bouchard-Chantereaux; natural size.

7a. Some calices, amplified.

THECIA GRAYANA (p. 279).

Fig. 8. A part of the upper surface of a specimen from Dudley, belonging to the Collection of Mr. John Gray; magnified.

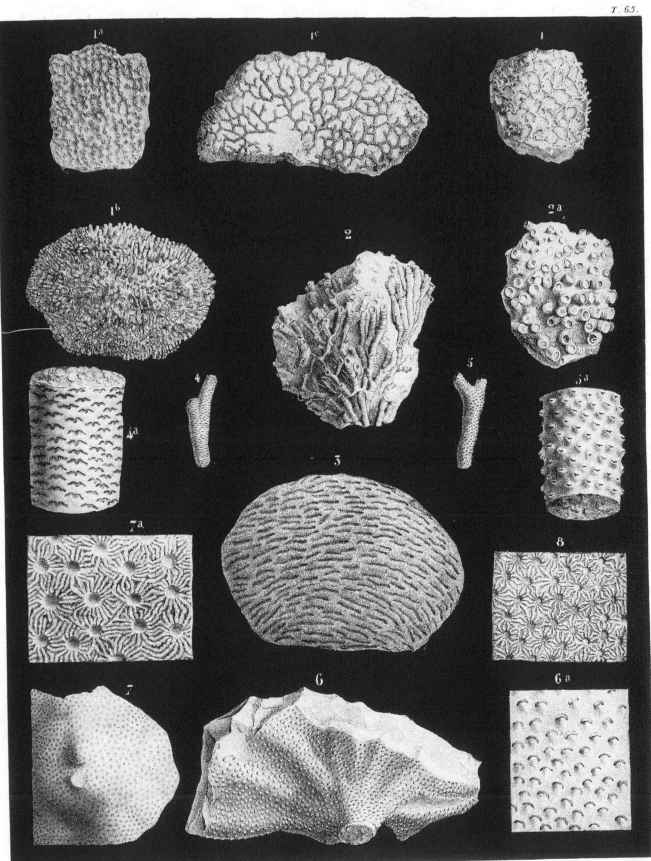

P. Lackerbauer ad nat in lap del?

Day & Son Lith?? to The Queen

TAB. LXVI.

CORALS OF THE SILURIAN FORMATION.

AULACOPHYLLUM MITRATUM (p. 280).

Fig. 1. A young specimen, found at Dudley, belonging to the Collection of Mr. John Gray; natural size.

1a. An adult specimen from the same locality, belonging to Mr. Fletcher's Collection; natural size.

1b. Calice of the specimen figured in 1, magnified.

CYATHOPHYLLUM LOVENI (p. 280).

Fig. 2. Side view of a specimen from Wren Nest, belonging to the Collection of M. Bouchard-Chantereaux; natural size.

2b. Calice, magnified.

CYATHOPHYLLUM PSEUDOCERATILES (p. 282).

Fig. 3. A young specimen found at Dudley, belonging to the Collection of Mr. Fletcher; natural size.

3a. Calice, magnified.

3b. An adult specimen from Wenlock, belonging to the Collection of M. Bouchard-Chantereaux.

CYATHOPHYLLUM AUGUSTUM (p. 281).

Fig. 4. A specimen from Attwood's Shaft, belonging to the Collection of the Geological Society of London; natural size.

4a. Vertical section, somewhat magnified.

CYATHOPHYLLUM TRUNCATUM (p. 284).

Fig. 5. A compound corallum found at Dudley, belonging to the Collection of Mr. Fletcher; natural size.

5a. Another specimen from the same locality, belonging to the Collection of the Geological Society of London; natural size.

5b. A calice, magnified.

5c. Another calice, magnified, and bearing some young corallites.

CYATHOPHYLLUM (TURBINOLOPSIS) ELONGATUM (p. 286).

Fig. 6. Side view of a cast found at Leach Heath, belonging to the Collection of the Geological Society; natural size.

6a. Under surface of the same; natural size.

TAB. LXVII.

CORALS OF THE SILURIAN FORMATION.

CYATHOPHYLLUM ARTICULATUM (p. 282).

Fig. 1. A large specimen found at Dudley, belonging to the Collection of Mr. Fletcher; natural size.

1a. A calice, magnified.

CYATHOPHYLLUM FLEXUOSUM (p. 285).

Fig. 2. A specimen from Much Wenlock, belonging to the Collection of Mr. Bowerbank; natural size.

2a. A corallite found at Malvern, in the Collection of the Geological Society of London, magnified.

OMPHYMA MURCHISONI (p. 289).

Fig. 3. Two corallites found at Wenlock, belonging to the Collection of Mr. Bowerbank; natural size.

3a. Calice, magnified.

3b. Vertical section, somewhat magnified.

PTYCHOPHYLLUM PARTELLATUM (p. 291).

Fig. 4. Side view of a specimen from Malvern, belonging to the Collection of the Geological Society of London; natural size.

4a. Upper surface of the same; natural size.

Day & Son, Lith.rs to the Queen.

P.Lackerbauer ad nat in lap. del.t

TAB. LXVIII.

CORALS OF THE SILURIAN FORMATION.

OMPHYMA SUBTURBINATA (p. 288).

Fig. 1. A large specimen found at Wenlock, belonging to the Collection of the Geological Society of London ; somewhat smaller than the natural size.

1a. A specimen found by us at Benthall Edge, and placed in the Museum of Paris ; natural size.

1b. Vertical section ; natural size.

1c. Calice ; natural size.

CHONOPHYLLUM PERFOLIATUM? (p. 235 and p. 291).

Fig. 2. Side view of a specimen from Wenlock, belonging to the Collection of the Viscount d'Archiac ; natural size.

2a. Its calice ; natural size.

GONIOPHYLLUM FLETCHERI (p. 290).

Fig. 3. Side view of a specimen found at Dudley, by Mr. Fletcher ; natural size.

3a. Its calice ; natural size.

P.Lackerbauer ad nat. in lap del.

Day & Son Lith.rs to the Queen

TAB. LXIX.

CORALS OF THE SILURIAN FORMATION.

OMPHYMA TURBINATA (p. 288).

Fig. 1.　Side view of a specimen from Benthall Edge, belonging to the Collection of M. Bouchard-Chantereaux; natural size.

1a.　Its calice; natural size.

ALCERVULARIA LUXURIANS (p. 292).

Fig. 2.　Under surface of a specimen found at Dudley.

2a.　Upper surface of a specimen from Dudley, belonging to the Collection of Mr. Fletcher; natural size.

2b.　Calice, magnified.

2c.　Upper surface of the specimen figured in 2; natural size.

2d.　Upper surface of a specimen from Dudley; natural size.

2e.　Calices of the same, magnified.

2f.　Some calices of a specimen from Dudley, belonging to the Collection of Mr. Bowerbank; natural size

P. Lackerbauer ad nat in lap del.

TAB. LXX.

CORALS OF THE SILURIAN FORMATION.

STROMBODES MURCHISONI (p. 293).

Fig. 1. Side view of a specimen found at Dudley, belonging to the Collection of the Museum at Paris ; natural size.

1a. Some calices ; natural size.

1b. A calice, magnified.

1c. Vertical section ; natural size.

1d. A portion of this vertical section, magnified.

STROMBODES PHILLIPSI (p. 294).

Fig. 2. Upper surface of a specimen found at Wenlock, belonging to the Collection of Professor Phillips ; natural size.

2a. A calice, magnified.

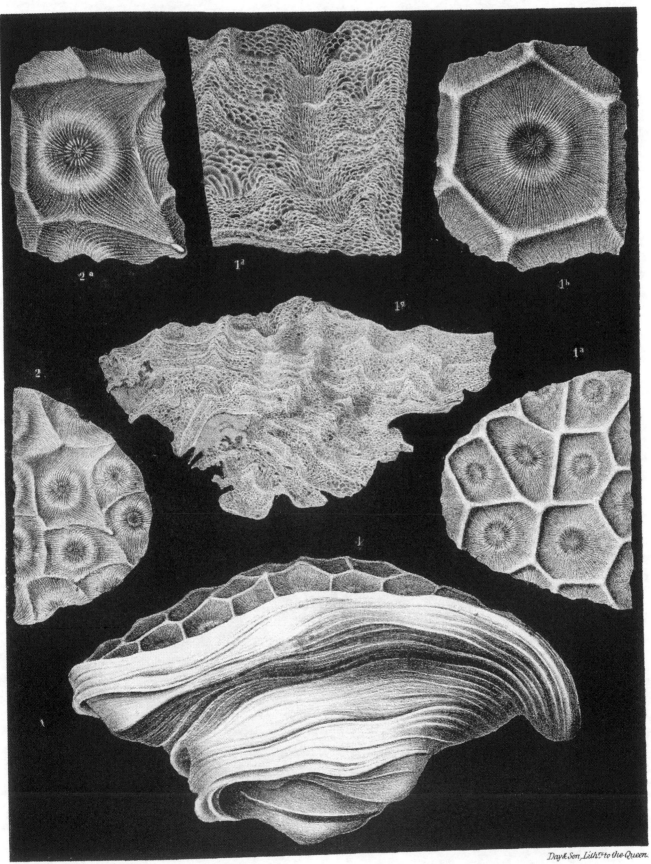

TAB. LXXI.

CORALS OF THE SILURIAN FORMATION.

STROMBODES TYPUS (p. 293).

Fig. 1. A part of the surface of a specimen from Wenlock, belonging to the Museum of Paris; natural size.

1a. A calice, magnified.

2b. A part of a vertical section, magnified.

STROMBODES DIFFLUENS (p. 294).

Fig. 2. Upper surface of a specimen found at Much Wenlock, belonging to the Parisian Museum; natural size.

2a. A calice, magnified.

SYRINGOPHYLLUM ORGANUM (p. 295).

Fig. 3. A large specimen from Dudley; natural size.

3a. Calice, magnified.

3b. A cast found at Coniston, belonging to the Collection of the Geological Society of London.

P. Lackerbauer ad. nat. in lap. del.ᵗ

Day & Son, Lithᵗ.ˢ to The Queen.

TAB. LXXII.

CORALS OF THE SILURIAN FORMATION.

CYSTIPHYLLUM SILURIENSE (p. 297).

Fig. 1. Side view of a specimen from Wenlock, belonging to the Geological Society of London; natural size.

1a. Vertical section; natural size.

CYSTIPHYLLUM CYLINDRICUM (p. 297).

Fig. 2. A subturbinate specimen from Dudley, belonging to the Geological Society of London; natural size.

2a. Vertical section, amplified.

2b. A broken specimen found at Dudley, by Mr. Fletcher; natural size.

2c. A young specimen from the same locality and Collection; natural size.

CYSTIPHYLLUM GRAYI (p. 298).

Fig. 3. A specimen from Dudley, belonging to the Collection of Mr. Fletcher; natural size.

3a. Its calice; natural size.

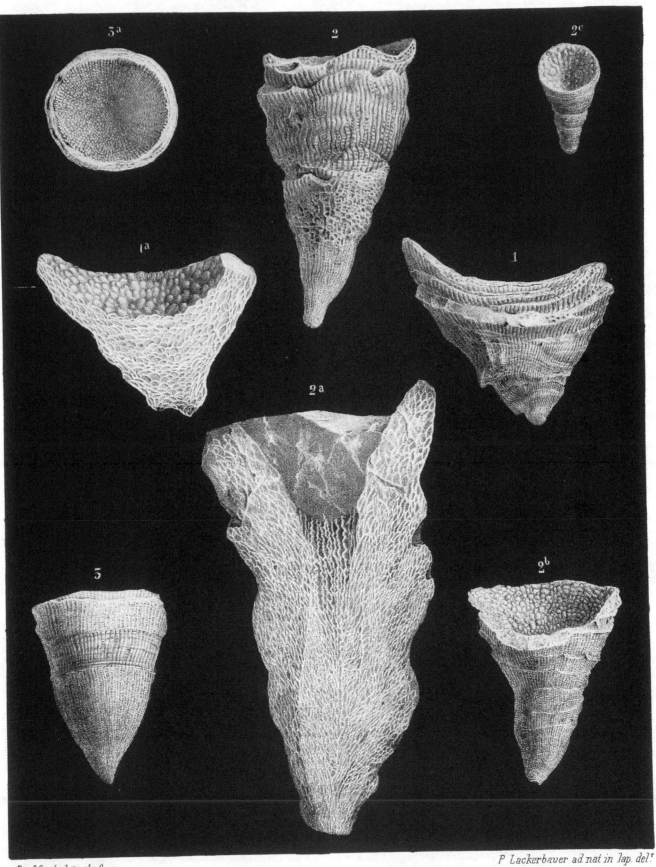

Day & Son Lith:s to the Queen

P. Lackerbauer ad nat in lap. del.t

Printed in the United States
By Bookmasters